CRC Series in The Biochemistry and Molecular Biology of the Cell Nucleus

Editor-in-Chief

Lubomir S. Hnilica, Ph.D.
Professor of Biochemistry and Pathology
Department of Biochemistry
Vanderbilt University School
of Medicine
Nashville, Tennessee

The Structure and Biological Function of Histones

Author

Lubomir S. Hnilica, Ph.D.

Chromosomal Nonhistone Proteins

Volume I: Biology
Volume II: Immunology
Volume III: Biochemistry
Volume IV: Structural Associations

Editor

Lubomir S. Hnilica, Ph.D.

Enzymes of Nucleic Acid Synthesis and Modification

Volume I: DNA Enzymes
Volume II: RNA Enzymes

Editor

Samson T. Jacob, Ph.D.
Professor
Department of Pharmacology
The Pennsylvania State University
The Milton S. Hershey Medical Center
Hershey, Pennsylvania

Chromosomal Nonhistone Proteins

Volume IV
Structural Associations

Editor

Lubomir S. Hnilica, Ph.D.
Professor of Biochemistry and Pathology
Department of Biochemistry
Vanderbilt University School of Medicine
Nashville, Tennessee

CRC Series in The Biochemistry and Molecular Biology of the Cell Nucleus

CRC Press, Inc.
Boca Raton, Florida

Library of Congress Cataloging in Publication Data

Main entry under title:

Chromosomal nonhistone proteins.

(CRC series in the biochemistry and molecular biology of the cell nucleus)
Includes bibliographies and indexes.
Contents: v.1. Biology. — v.2. Immunology. — v.3.
Biochemistry. — v.4. Structural Associations.
1. Nucleoproteins. 2. Chromosomes. I. Hnilica,
Lubomir S. II. Series. [DNLM: 1. Chromosomal
proteins, Non-Histone. QU 58 C5576]
QP625.N87C47 1983 574.87,322 82-17742
ISBN 0-8493-5511-7(v.1)
ISBN 0-8493-5512-5(v.2)
ISBN 0-8493-5513-3(v.3)
ISBN 0-8493-5514-1(v.4)

Direct all inquiries to CRC Press, Inc., 2000 Corporate Blvd., N.W., Boca Raton, Florida, 33431.

International Standard Book Number 0-8493-5511-7 (Volume I)
International Standard Book Number 0-8493-5512-5 (Volume II)
International Standard Book Number 0-8493-5513-3 (Volume III)
International Standard Book Number 0-8493-5514-1 (Volume IV)

Library of Congress Card Number 82-17742
Printed in the United States

INTRODUCTION

The fourth and last volume of the *Chromosomal Nonhistone Proteins* sequence attempts to confront these macromolecules with the major structural elements of the cell nucleus (with the exception of nucleosomes which will be treated separately in a later treatise). Proteins of the chromosomes, nucleoli, nuclear membrane, RNP particles, and nuclear matrix are described in detail. Unfortunately, the discussion of this rapidly expanding scientific discipline has taken the longest to reach the press. I deeply regret this unavoidable situation and I wish to thank those authors who submitted their contributions long before the closing of this volume for their patience.

THE EDITOR

Lubomir S. Hnilica is a professor of biochemistry and pathology at the Vanderbilt University School of Medicine, Nashville, Tennessee.

Dr. Hnilica received his degree in organic chemistry from Masaryk University, Brno, Czechoslovakia in 1952 and an advanced degree in biochemistry from the Czechoslovak Academy of Sciences. After postdoctoral training at the Czechoslovak National Cancer Institute in Bratislava, Dr. Hnilica continued his research as a WHO research fellow at the Chester Beatty Research Institute in London, England. After 2 years at the Baylor College of Medicine in Houston, Texas, Dr. Hnilica joined the staff of the University of Texas System Cancer Center, M.D. Anderson Hospital and Tumor Institute in Houston where he rose to the rank of Professor of Biochemistry and Chief of the Section of Biochemical Regulatory Mechanisms. In 1975 Dr. Hnilica joined the Vanderbilt University School of Medicine where he is Professor of Biochemistry and Pathology as well as director of the A. B. Hancock, Jr. Memorial Laboratory of the Vanderbilt University Cancer Center. His present research concerns proteins of the cell nucleus, their biochemistry and immunology, the interactions of nuclear proteins with DNA, and changes in gene expression during chemical carcinogenesis.

CONTRIBUTORS

Kenneth W. Adolph, Ph.D.
Assistant Professor
Department of Biochemistry
University of Minnesota
Medical School
Minneapolis, Minnesota

Ronald Berezney, Ph.D.
Assistant Professor
Division of Cell and
Molecular Biology
Department of Biological Sciences
State University of New York
Buffalo, New York

Harris Busch, Ph.D.
Distinguished Professor
and Chairman
Department of Pharmacology
Baylor College of Medicine
Houston, Texas

Viktor Holoubek, Ph.D.
Professor of Biochemistry
Department of Human Biological
Chemistry and Genetics
University of Texas Medical Branch
Galveston, Texas

Charles B. Kasper, Ph.D.
Associate Professor of Oncology
McArdle Laboratory for Cancer
Research
University of Wisconsin
Medical School
Madison, Wisconsin

Virginia P. Wray, Ph.D.
M.D. Anderson Hospital
and Tumor Institute
University of Texas Cancer Center
Houston, Texas

Wayne Wray, Ph.D.
Assistant Professor
Department of Cell Biology
Baylor College of Medicine
Houston, Texas

TABLE OF CONTENTS

CHROMOSOMAL NONHISTONE PROTEINS

TABLE OF CONTENTS

Volume IV

Chapter 1

NUCLEAR ENVELOPE PROTEINS: SELECTED BIOCHEMICAL ASPECTS*

Charles B. Kasper

TABLE OF CONTENTS

* The literature survey for this review article was completed in June 1981.

I. INTRODUCTION

Architecturally, the nuclear envelope is a double membrane system comprised of inner and outer leaflets. The outer leaflet is studded with ribosomes and is found to be continuous with the rough-surfaced endoplasmic reticulum, whereas the inner leaflet is devoid of polysomes and is usually found in close association with chromatin. Both the inner and outer membranes are joined wherever pore complexes are formed, and electron microscopic examination of these junctions in ultrathin sections reveals that the inner and outer membranes are continuous in these regions.[1] Hence, the two nuclear leaflets and the endoplasmic reticulum are physically connected, although unique biochemical functions appear to be associated with each. The general morphological features of highly purified nuclear envelope are presented in Figure 1. At low magnification (Figure 1A), the envelope appears as large, structurally well-preserved membranous sheets with numerous pore complexes. Upon closer examination the inner and outer leaflets are easily distinguished, and both tangential and transverse views of the pore complex are evident (Figure 1B). In certain instances, the central granule of the complex is also quite prominent (upper half of Figure 1B). Polysomes are absent since they are removed during the isolation procedure. Although the precise composition of the envelope varies, depending upon the method of preparation and the source, protein and phospholipid generally comprise 85 to 95% of the total membrane mass with carbohydrate, cholesterol and its esters, RNA, and in certain instances DNA, accounting for the remainder.[2-4]

The nuclear envelope has the distinction of being the membrane system responsible for the compartmentalization of the cell's genetic material and, as such, has as one of its important functions the regulation of nucleocytoplasmic transport — the nonrandom exchange of solutes between the cytoplasmic and nuclear compartments. In addition, much attention has recently been directed toward the role of the envelope in the initiation of replication, nuclear division, protein biosynthesis, metabolic activation of chemical carcinogens, receptor-mediated processes triggering intranuclear events, and chromosomal orientation. Also, the fact that the outer nuclear leaflet forms a morphological continuum with the endoplasmic reticulum has generated a great deal of interest in the functional relationship between these two distinct membrane systems. Over the past several years the nuclear envelope has been the subject of numerous reviews[1,2,5-12] in which particular emphasis was placed on envelope structure, composition, and function. The focus of this chapter will be on subject areas that have not been reviewed recently. Hence, with respect to nuclear envelope enzymology, which has been exhaustively evaluated[2,4,7,9,11] only specialized aspects will be presented, as they relate to the topics being developed.

II. LAMINS

Early studies on the fractionation of proteins from highly purified nuclear envelope revealed that three prominent polypeptides having molecular weights of 68,000, 74,000, and 78,000 accounted for approximately 25% of the total envelope protein, while a second group of proteins in the molecular weight range of 53,000 to 60,000 represented an additional 21%.[13] The electrophoretic pattern of envelope polypeptides, illustrating the size distribution, is presented in Figure 2. Gel filtration of SDS-solubilized nuclear envelope proteins on Sephadex® G-150 resulted in their separation into discrete molecular weight classes. Amino acid analysis of individual fractions indicated a twofold decrease in the ratio of acidic to basic residues upon progression from the high to low molecular weight proteins, and the content of hydrophobic amino acids ranged from 46 to 53 mol %.[13] A detailed electrophoretic analysis of chicken erythrocyte nuclear envelope demonstrated marked differences between the rat liver and erythrocyte nuclear envelopes in their protein compositions.[14-16] In addition to differences in polypeptide distribution, erythrocyte nuclear envelope was found not to contain any glycoproteins detectable by the periodate-Schiff stain.[14]

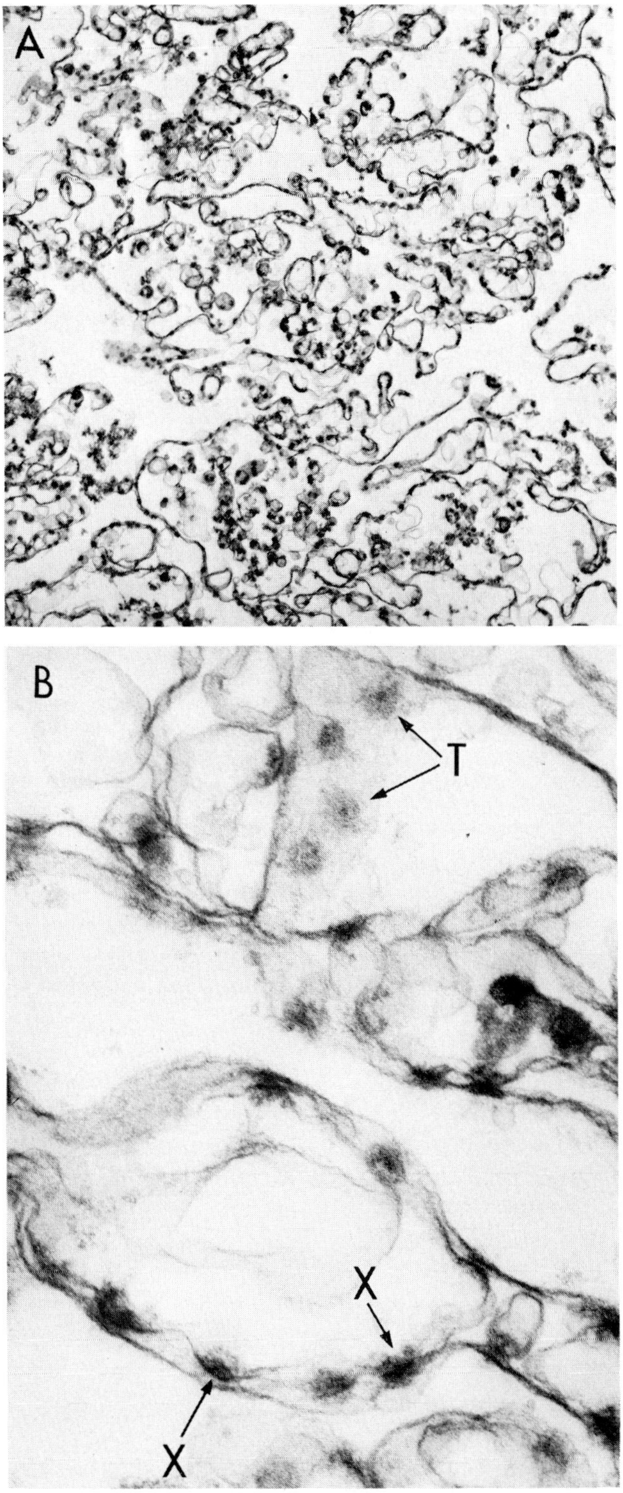

FIGURE 1. Electron micrographs of highly purified nuclear envelope (A, × 7,800; B, × 52,800). Transverse (X) and tangential (T) views of the pore complex are identified by arrows.

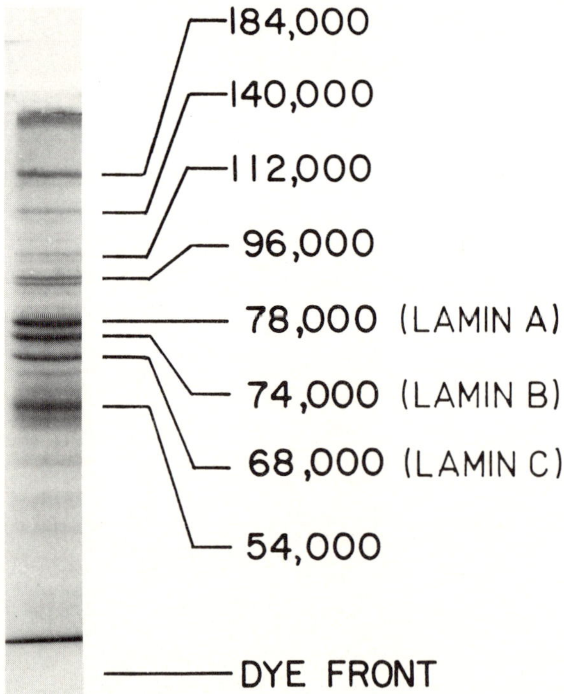

FIGURE 2. SDS-polyacrylamide gel electrophoretogram of nuclear envelope polypeptides.

To date it has not been possible to isolate a membrane fraction representative of either the outer or inner leaflet of the nuclear envelope. Selective extraction of the nuclear envelope, however, with buffers containing high salt concentrations and Triton® X-100 has yielded a unique morphological structure characterized by pore complexes associated with a 150-Å-thick lamina.[17,18] This nuclear pore-enriched preparation, which is 97% protein with only residual amounts of RNA and DNA and no phospholipid, has been designated the pore complex-lamina fraction. Kirschner et al.[19] have used high resolution scanning electron microscopy to further study the ultrastructural aspects of the lamina after detergent treatment of nuclei. Gel electrophoretic analysis of the lamina fraction reveals three major polypeptides having molecular weights of 70,000, 67,000, and 60,000;[20] according to recently suggested nomenclature, these proteins have been designated lamin A, lamin B, and lamin C, respectively.[21] The reported molecular weights for these three proteins from rat liver vary from 70,000[20,22] to 78,000[23,24] for lamin A, from 67,000[20,22] to 74,000[23,24] for lamin B, and from 58,000[20,22] to 68,000[23,24] for lamin C. In the case of chicken erythrocyte, values of 75,000, 71,000, and 61,000 to 68,000 have been given for lamins A, B, and C, respectively.[15,25] It should be emphasized that individual lamins have not been shown to be homogeneous proteins by physicochemical characterization. Hence, reference to a specific lamin refers to a protein representing a particular molecular weight class on SDS-polyacrylamide gel.

A. Topological Relationships and Sequence Homology

Studies designed to examine the molecular associations of the lamins in intact rat liver nuclear envelope indicate that a significant percentage of lamin B is cross-linked in the native membrane by disulfide bridges to form two electrophoretically distinct polymeric species (designated *a* and *b* in Figure 3), both of which appear to be trimers.[23] These trimers

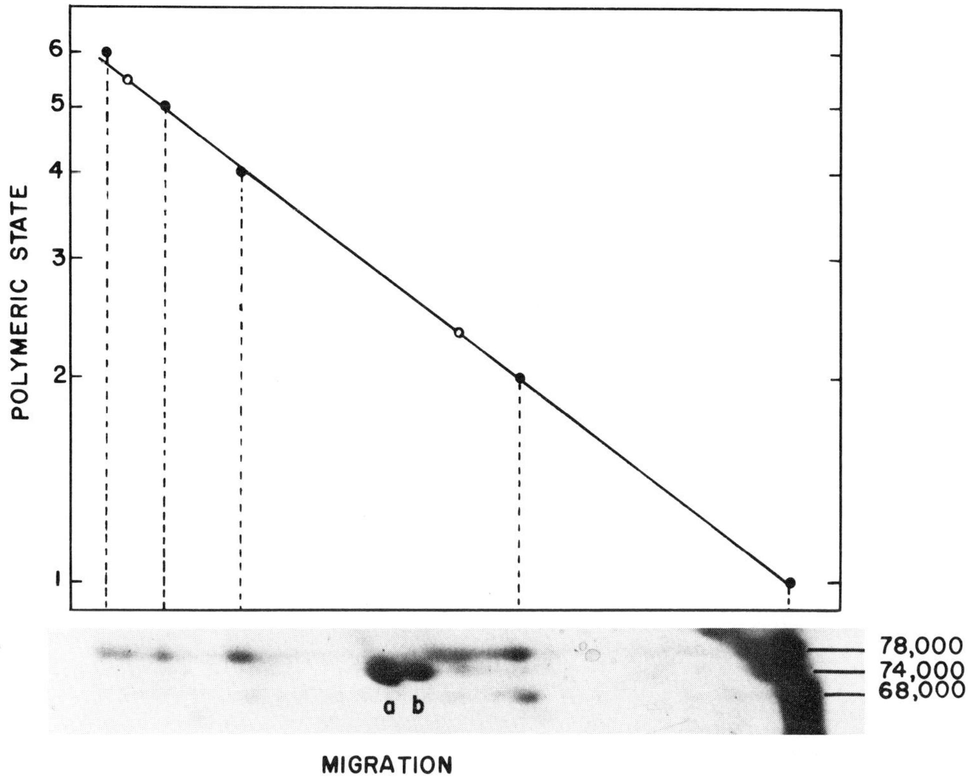

FIGURE 3. Polymeric states of lamin A (molecular weight = 78,000). The protein-stained gel segment below the graph represents a region of a two-dimensional polyacrylamide gel electrophoresis experiment. Nonreduced nuclear envelope proteins were electrophoresed in the first dimension (horizontal axis) prior to electrophoresis under reducing conditions in the second dimension (vertical axis). The proteins distributed along the horizontal axis represent monomers derived from disulfide-linked polymers. (From Lam, K. S. and Kasper, C. B., *J. Biol. Chem.*, 254, 11713, 1979. With permission.)

are homopolymers and probably differ in the extent of inter- or intramolecular, or both, disulfide bridging, thus allowing them to be separated on polyacrylamide gel on the basis of their compactness or symmetry. This interpretation gains support from the fact that *o*-phenanthroline-cupric ion complex-catalyzed oxidation of intrinsic sulfhydryl groups markedly increased the proportion of trimer *b* (Figure 3) relative to trimer *a* (data presented in Reference 23). Furthermore, the selective reduction of trimer *b* to trimer *a* and of trimer *a* to the monomeric 74,000 subunit could also be demonstrated.[23]

Lamin A (78,000, Figure 2) exists primarily as a monomer in the native envelope; however, a significant number of dimers, tetramers, pentamers, and higher-molecular-weight oligomers can be detected (Figure 3). Lamin C, on the other hand, is present almost exclusively in the monomeric state. In contrast to lamin B, the degree of cross-linking of lamins A and C is only minimally affected upon treatment with *o*-phenanthroline-cupric ion complex. It is also significant to note that proteins in the 49,000 to 60,000 mol wt region, which constitute nearly 25% of the total envelope protein,[2,13] are not cross-linked in the native membrane nor by the oxidative cross-linking procedure. This molecular weight class of proteins is also readily soluble in dilute detergent solution, unlike lamins A, B, and C. In fact, the extreme insolubility of the lamins in aqueous buffers and nonionic detergents suggests that they may form hydrophobic clusters in the membrane, which in the case of lamins A and B are stabilized by disulfide linkages to form homopolymers. Nuclear envelope prepared from nuclei that have been treated with iodoacetic acid yields a polymer distribution similar to that obtained with nonalkylated nuclei.[82] These results differ from those of Gerace and Blobel[21] which state that nuclear envelope prepared under nonoxidizing conditions yields "completely monomeric" lamins upon solubilization. The basis for this difference is unclear; however, the importance of the polymerization data is that they provide important clues to major protein-protein associations within the membrane. For example, the fact that only homopolymers of lamin B are detected strongly suggests that the 74,000 mol wt subunits are clustered in the envelope in such a way as to allow disulfide bridging between molecules.

Electrophoretic analysis of chemical and enzymic cleavage products of lamins A, B, and C has indicated a close sequence homology between lamins A and C. The results further indicate that peptides derived from lamin B are unique to this protein and do not correspond to cleavage products obtained from either lamin A or lamin C. The experiments presented in Figure 4 illustrate these points. Gels (a), (b), and (c) show the peptides generated by treatment of the nuclear envelope polypeptides with hydroxylamine, cyanogen bromide, and chymotrypsin, respectively. Since peptides originating from lamin B (74,000) migrate in close proximity to those derived from lamin A (78,000) when reducing conditions are used in both dimensions, gels were also run in the first dimension in the absence of reducing agent, under which conditions lamin B migrated as homotrimers (species *a* and *b* in left panel of Figure 4 (a) and (b)). This allowed peptides generated from lamins A and C to be compared critically. The major conclusions just presented dealing with the structural similarities and differences of the rat liver lamins are further supported by recent tryptic peptide mapping studies[16] as well as by unpublished experiments cited in Reference 21. Studies conducted with lamins from avian erythrocytes have produced similar results although these are not strictly analogous to those with the rat liver system.[15,16,25,26] An interesting difference is the near absence of lamin C in the avian envelope. Because of the significant differences in polypeptide compositions and fragmentation patterns between the avian and rat liver systems, caution should be exercised in discussing the two systems in interchangeable terms.

Immunological studies using Ouchterlony double-diffusion analysis show that each of the three purified lamins cross-reacts equally well with either antilamin A, B, or C.[20] In fact, each of the three antibody preparations produces a pattern of fused precipitin arcs for all lamins suggesting immunochemical identity of all three antigens. Since major structural differences are known to exist between lamin B and the other two lamins, incomplete fusion or spurring of the precipitation bands would have been expected. This would also have been expected when examining the immunological relatedness of lamins A and C, because they are definitely not identical molecules; but since this was not obtained the possibility must be considered whether immunization with a protein-SDS complex produces antibodies that recognize bound detergent rather than, or in addition to, unique antigenic determinants of the protein antigen. Presently, the immunochemical data support the peptide mapping studies that demonstrate extensive sequence homology between lamins A and C but are in contradiction with the fact that lamin B does not share structural similarities with lamins A and C.

B. Localization Within the Envelope

Studies devoted to determining the distribution of lamins A, B, and C within cells can be divided into two specific areas, those using immunochemical techniques[20,27,28] and those employing chemical labeling procedures.[22] Antibodies to individual lamins purified by preparative SDS-polyacrylamide gel electrophoresis were prepared in chickens[20,27,28] and indirect immunofluorescence microscopy was then used to detect the location of lamins in various cells. In the case of rat embryo fibroblasts,[28] mouse 3T3 cells,[27] and rat K-22 cells,[20] intense fluorescence was observed at the nuclear periphery, with no fluorescence of the nucleoplasm and little or no fluorescence of the cytoplasm. The nuclear envelope of rat hepatoma cells, mouse A9 cells, HeLa cells, and rat liver cells also strongly reacted with antibody.[28] Rat

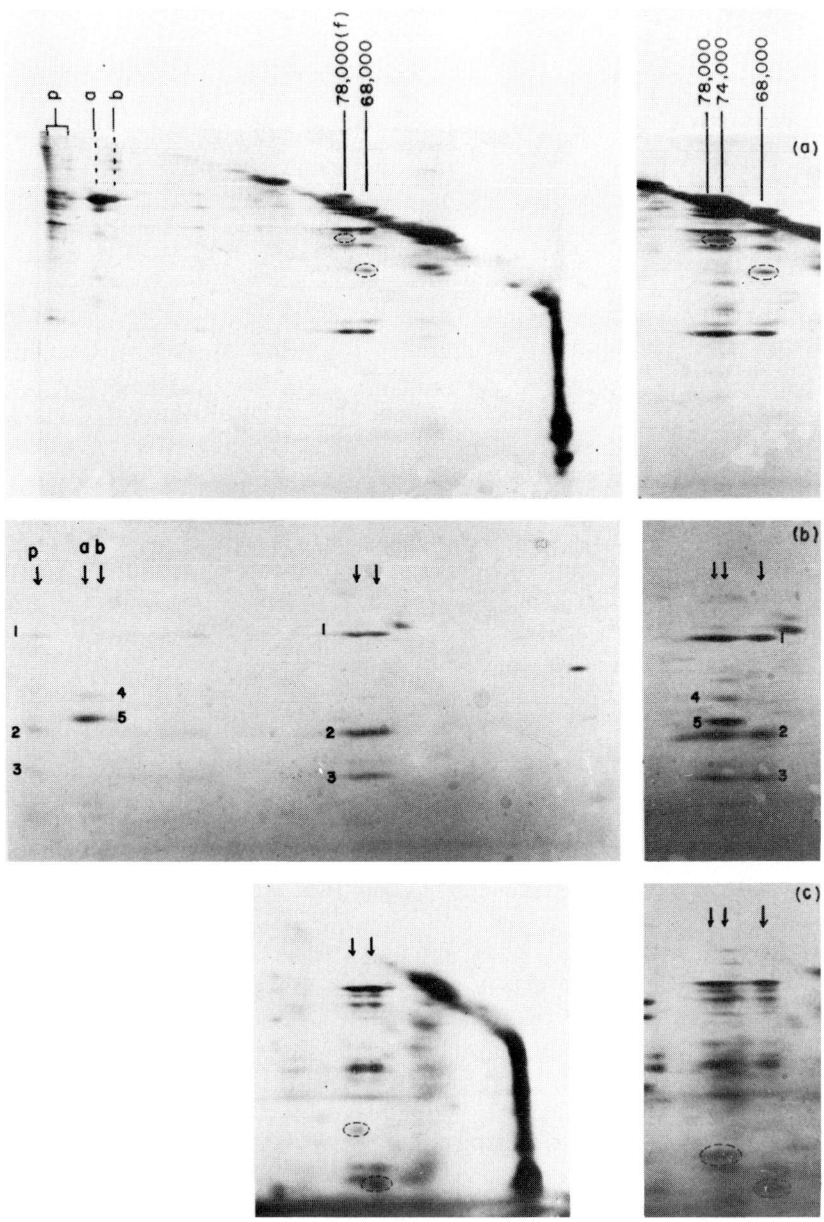

FIGURE 4. Analysis of chemical and enzymic digests of nuclear envelope proteins by two-dimensional gel electrophoresis. The following cleavage reagents were used: (1) hydroxylamine, (b) cyanogen bromide, and (c) chymotrypsin. Gels appearing in the *right vertical panel* contain protein samples that were electrophoresed under reducing conditions in the first dimension, while reducing agents were omitted from the first dimension of slab gels on the *left vertical panel*. (From Lam, K. S. and Kasper, C. B., *J. Biol. Chem.*, 254, 11713, 1979. With permission.)

liver nuclei treated with Triton® X-100 to remove the outer leaflet and membranous elements of the inner leaflet were also examined by high resolution indirect immunoperoxidase staining.[20] Such nuclei, having an exposed surface of lamina with associated pore complexes, reacted to yield a heavy deposit of the electron-dense peroxidase reaction product on the nuclear surface without any evidence of the reaction product inside the nucleus. Since detergent treatment of nuclei could conceivably reduce any penetration barrier that may exist for antibody entering the nucleus, the immunoperoxidase data have been taken as evidence that lamins A, B, and C are not structural elements of an internal nuclear protein matrix.[20] Furthermore, in Triton® X-100-treated nuclei, pore complexes extend prominently from the nuclear surface, making them easily distinguishable from the interconnecting lamina. Electron microscopic examination of these nuclei stained with antilamin A by the immunoperoxidase procedure revealed that reaction products did not concentrate in the region of the pore complex.[20] A similar conclusion is also drawn from the studies of Ely et al.[28] and Krohne et al.[27] Also, the marked similarity of the staining patterns for fixed cells and for the HeLa nucleoid,[28] which are basically equivalent to Triton® X-100-washed nuclei, strengthens the position that lamins A, B, and C are localized in the lamina underlying the inner nuclear membrane, although the precise distribution within the envelope has not been rigorously established.

This observation, coupled with the fact that all three antibodies stain the nuclear periphery evenly,[28] has prompted the conclusion that the lamins probably are not structural components of the pore complex.[20,28] Certain considerations, however, render this conclusion only tentative. For example, the tendency of lamins A and B to form disulfide-linked homopolymers could mask or reduce the availability of antigenic determinants on these two polypeptides so that enhanced binding of antibody to the pore complex would not be observed. In addition, because of the immunochemical cross-reactivity of these three proteins[20,28] the binding of individual antibodies to the fibrillar interconnecting network does not firmly establish the location of the three lamins. In fact, the presence of only one of these three proteins on the interconnecting lamina would produce the same result. Conceptually, there is no reason to exclude the possibility of the lamins being components of both the lamina as well as the pore complex, and it is intriguing to speculate that the properties of highly hydrophobic polypeptides such as lamins A, B, and C are well suited to generate and maintain a unique structure such as the pore complex. In this connection, it is relevant to note that analyses of pore complex-enriched fractions of *Xenopus* oocytes and rat liver led Krohne et al.[24] to suggest that a 74,000 mol wt protein, corresponding to lamin B, constitutes a major component of the pore complex.

A novel approach to the study of nuclear pore complex proteins has utilized solid state lactoperoxidase labeling of intact nuclei.[22] Conditions were chosen so as to selectively label only the cytoplasmic surface of nuclei, specifically, the outer nuclear leaflet and the exposed regions of the pore complex. The results indicated that lamins A and B, which were readily iodinated, are located in the nuclear pore complex along with at least ten other proteins ranging in molecular weight from 33,000 to 200,000.[22] Significantly, lamin C, the third major polypeptide of the nuclear envelope, was not labeled under these conditions; this peptide could be labeled only when nuclei were ruptured and the nucleoplasm removed. From this labeling behavior, Richardson and Maddy[22] concluded that lamin C is present only on the nucleoplasmic face of the envelope, presumably the fibrous lamina. It is interesting to find differences in reactivity of this magnitude between lamins A and C, considering that they share extensive sequence homology.

In summary, immunochemical data clearly show the association of one or more of the lamins with the fibrous lamina but fail to demonstrate an enrichment of any of the lamins in the pore complex. Furthermore, immunochemical studies indicate that all three lamins have a similar distribution and that they appear to be localized in the lamina underlying the inner nuclear membrane. In sharp contrast, iodination studies indicate that only lamin C is associated with the lamina and that lamins A and B are nuclear pore complex proteins. With each approach, a positive result is likely to provide useful organizational information, but a negative result remains inconclusive.

C. Changes in Distribution During the Cell Cycle

In most eukaryotic cells, the nuclear envelope is disaasembled and reconstructed during cell division. The breakdown is initiated during late prophase or prometaphase and is characterized at the macroscopic level by the dispersal of the nuclear envelope into small vesicles. These small membrane fragments, although consisting initially of a double membrane, eventually become indistinguishable from cytoplasmic vesicles derived from other intracellular membranes such as the endoplasmic reticulum and Golgi apparatus. The reassembly process begins during anaphase, at which time small vesicles are seen to condense on the chromosomal surface, ultimately generating a complete nuclear envelope. This reversible cyclic process is poorly understood at the molecular level, and until recently essentially all of our knowledge was derived from a microscopic viewing of the events (for reviews see References 1, 5, 6, 11, and 29). With the availability of antibodies against the three major polypeptides of the nuclear envelope, it has become possible to critically approach the crucial question of the fate of the envelope during cell division. Specifically, following the breakdown of the envelope, are the constituent polypeptides conserved and utilized for the reassembly of the envelope or are they degraded and synthesized *de novo*? With indirect immunofluorescence microscopy, dramatic changes in the distribution of lamins A, B, and C were noted during mitosis.[20,28] Coincident with the disassembly of the envelope at prophase was a shift from the characteristic interphase fluorescence pattern to one in which fluorescence was diffusely distributed throughout the cytoplasm. This pattern remains essentially unchanged until reformation of the envelope begins late in anaphase, at which time the antigens can be seen to migrate to the periphery of the daughter chromosomes where the new nuclear envelopes form. By the end of telophase, cytoplasmic fluorescence has disappeared and the antigens again appear to be localized only at the nuclear periphery.

In addition to these observations, immunoprecipitation studies performed on CHO cell populations grown in synchrony[21] reveal that upon dissolution of the envelope lamins A and C exist in a soluble form no longer associated with membranous elements. In contrast to this behavior, lamin B appears in mitotic cell homogenates still associated with membrane-like structures.[21] These studies present strong evidence that the nuclear envelope lamina undergoes reversible disassembly during mitosis and that the major polypeptides are conserved and used in the reconstruction of the new nuclear envelope. In this context, biochemical studies on the reformation of the nuclear envelope of CHO cells demonstrated that 60% or more of the nuclear envelope protein in early G_1 was present in the cell during the preceding G_2 phase of the cell cycle and was reutilized in the reconstruction of the nuclear envelope during late mitosis and early G_1.[30] A similar result was also found for nuclear envelope phospholipid. Furthermore, newly synthesized proteins of the daughter nuclear envelope were found to be synthesized during the early G_1 phase of the replicative cycle.[30]

Although the mechanism of the depolymerization process is unknown, the observation that lamin C is preferentially phosphorylated by an endogenous nuclear envelope protein kinase may have important biological implications.[31] The reaction is cyclic nucleotide-independent, and lamins A and B are only weakly phosphorylated. The fact that lamins A and C exhibit extensive sequence homology but yet only lamin C readily incorporates ^{32}P from [$\gamma - {}^{32}P$] ATP in vitro suggest either that structural features of lamin A make it a poorer substrate for the kinase or that lamin A is more highly phosphorylated in vivo and hence is only weakly phosphorylated in vitro. Gerace and Blobel[21] have reported that during mitosis the lamins are more extensively phosphorylated than their interphase counterparts, and they have suggested that the reversible phosphorylation of these major envelope proteins may play a role in the regulation of the dynamic behavior of the nuclear envelope during mitosis.

III. INDUCIBILITY AND BIOSYNTHESIS OF NUCLEAR ENVELOPE PROTEINS

Various compounds are known to increase the levels of specific microsomal enzymes active in drug metabolism (for reviews see References 32 to 34). Most notable of these compounds are phenobarbital and 3-methylcholanthrene, each representing a unique class of inducing agent. Comparative biochemical studies of the nuclear envelope and endoplasmic reticulum with these two agents produced intriguing results (Table 1). The basic observation is that phenobarbital serves as an effective inducer of specific microsomal enzymes while failing to induce the same biological activities associated with the nuclear envelope. For example, microsomal cytochrome P-450, NADPH-cytochrome *c* oxidoreductase, epoxide hydratase, aryl hydrocarbon hydroxylase, *N*-demethylase, and UDP-glucuronosyl transferase show a twofold or greater elevation above noninduced controls; however, the corresponding envelope activities show either no induction or an actual decrease in levels. It is important to note that four separate forms of cytochrome P-450 were identified in the nuclear envelope on the basis of their Type I difference spectra, and although phenobarbital did not increase the total level of envelope cytochrome P-450, a significant change in the relative proportions of these species did occur.[35] Similar observations regarding the failure of phenobarbital to induce various drug-metabolizing activities in intact nuclei[36-38] and nuclear envelope[39,40] have been made; however, others have obtained differing results.[41,42] The basis for this variance is not apparent at this time, although the purity of the nuclear envelope is of prime importance. In sharp contrast to phenobarbital, 3-methylcholanthrene produced a similar qualitative pattern of enzymic activities in both membrane systems. Interestingly, total benzo(a)pyrene oxidation was stimulated 60-fold in nuclear envelope but only 5-fold in microsomal membrane; however, when this activity is expressed as moles of substrate oxidized per mole of constitutive cytochrome P-450, both membranes have the same specific activity.[35] Cytochrome b_5 and NADPH-cytochrome *c* oxidoreductase levels remained unchanged by both inducing agents.

Since the question arises as to whether phenobarbital may in fact produce elevated levels of certain enzymes in the nuclear envelope which for some reason are not detected by activity measurements, epoxide hydratase and NADPH-cytochrome P-450 oxidoreductase levels were determined immunochemically to see if changes in enzymic specific activities correlated with the absolute amount of enzyme protein present (Table 2). Results show a close correspondence between the two procedures and indicate that, in these specific instances, activity measurements are a reliable indication of changes in protein levels.

In view of the differential response of the nuclear envelope to phenobarbital administration, it is worthwhile to examine what is known about the biosynthesis of envelope proteins. Several possibilities could explain this intriguing effect. First, the nuclear envelope and microsomal enzymes may be different molecular entities with similar activities. This is known to be the case for mitochondrial and microsomal cytochrome b_5.[43] Second, the microsomal and envelope enzymes may have different sites of synthesis. For example, the nuclear enzyme may be synthesized on nuclear polysomes, and the microsomal enzyme may be synthesized on the rough endoplasmic reticulum. Third, phenobarbital may affect the process by which the newly synthesized enzyme is inserted into the envelope matrix. Immunodiffusion studies clearly demonstrate that nuclear envelope and microsomal NADPH-cytochrome *c* oxidoreductase,[44] cytochrome P-448, and epoxide hydratase,[45] are indistinguishable when antibody to the microsomal enzymes is used. Furthermore, turnover studies have shown the microsomal and envelope oxidoreductases to have identical half-lives.[44] Hence, the evidence strongly suggests that the enzymes from the two membrane systems are similar if not identical. The possibility still exists, however, that the envelope enzymes may have slight structural differences since antigenic determinants different from the microsomal enzyme would not have been detected.

Recent studies have investigated the biosynthesis of epoxide hydratase and NADPH-cytochrome *c* oxidoreductase by polysomes isolated from the nuclear surface and by polysomes from the rough endoplasmic reticulum;[46] mRNAs coding for both enzymes fall in the category of low-abundance messages, representing approximately 0.016 and 0.024% of

Table 1
EFFECTS OF PHENOBARBITAL AND 3-METHYLCHOLANTHRENE ON THE INDUCIBILITY OF NUCLEAR ENVELOPE AND MICROSOMAL ENZYMES[a]

	Induced/control			
	Nuclear envelope		Microsomal membrane	
	PB	3-MC	PB	3-MC
NADPH-cytochrome *c* oxidoreductase	0.88	1.15	1.9	1.1
Cytochrome P-450	1.0	1.8	4.1	2.7
Epoxide hydratase				
Styrene oxide	0.70	N.D.[b]	2.5	N.D.
Benzo(a)pyrene-4,5-oxide	0.68	0.58	2.9	1.1
Arylhydrocarbon hydroxylase	0.95	60	2.3	5.4
N-Demethylase	1.2	N.D.	2.1	N.D.
UDP-Glucuronosyltransferase	1.1	2.2	1.8	2.2
NADH-Cytochrome *c* oxidoreductase	0.73	0.79	0.67	0.71
Cytochrome b_5	1.0	0.95	1.0	1.1

[a] Data are taken from References 2, 4, and 35.
[b] N.D. = not determined.

Table 2
IMMUNOCHEMICAL DETERMINATION OF EPOXIDE HYDRATASE AND NADPH-CYTOCHROME *c* OXIDOREDUCTASE IN MEMBRANES FROM PHENOBARBITAL-TREATED ANIMALS[a]

	% w/w	
	Epoxide hydratase	Cytochrome *c* oxidoreductase
Smooth Microsomal Membrane		
Control	2.13	0.71
Phenobarbital	6.42	3.2
Induced/control	3.0 (2.6)[b]	4.5
Rough Microsomal Membrane		
Control	1.96	0.58
Phenobarbital	5.34	0.64
Induced/control	2.7 (2.2)	1.1
Nuclear Envelope		
Control	0.79	0.21
Phenobarbital	0.96	0.18
Induced/control	1.2 (0.7)	0.85 (0.88)

[a] Data are obtained from References 44 and 83.
[b] The numbers in parentheses are the ratios (induced/control) for the enzymic specific activities.

translatable mRNA in nuclear and microsomal polysomes, respectively. Polysomes from the nuclear envelope also synthesized albumin, and in vitro translation studies indicated that albumin accounted for 1.5% of the total translation products. This compares with a value of 4.8% found for membrane-bound polysomes from the endoplasmic reticulum. RNA

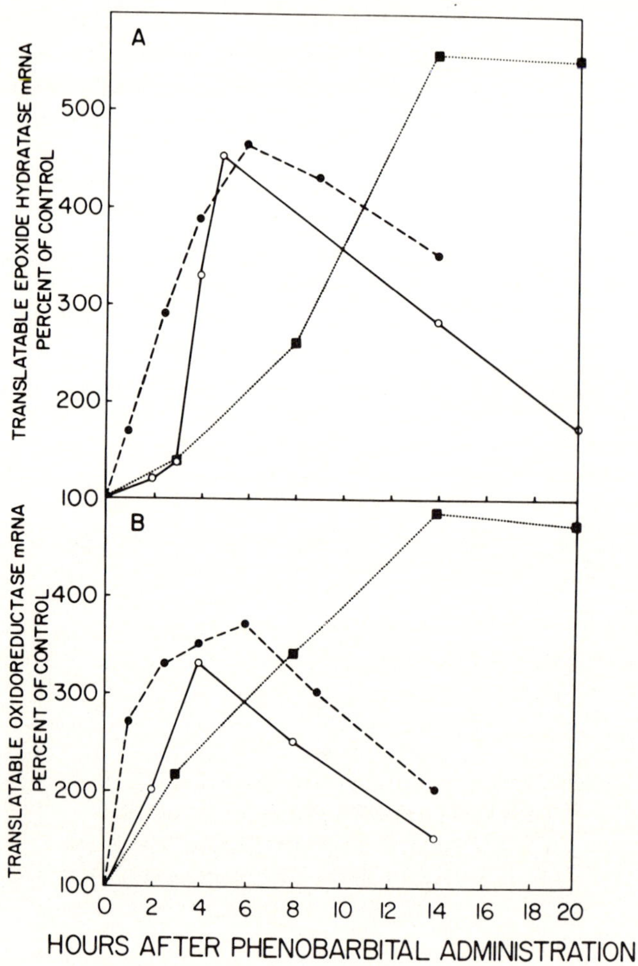

FIGURE 5. Quantitation of epoxide hydratase and NADPH-cytochrome *c* oxidoreductase mRNAs after phenobarbital administration. The percentage of control at each time point was determined for epoxide hydratase (A) and NADPH-cytochrome *c* oxidoreductase (B) translated from nuclear envelope bound (●), rough endoplasmic reticulum bound (○), and rapidly sedimenting endoplasmic reticulum bound (■) polysomal RNA. (From Gonzalez, F. J. and Kasper, C. B., *Biochemistry*, 20, 2292, 1981. With permission.)

isolated from free polysomes and polysomes loosely bound to the endoplasmic reticulum did not direct the synthesis of either epoxide hydratase or NADPH-cytochrome *c* oxidoreductase.[47,48] Hence, only those polysomes tightly bound to the nuclear and microsomal membranes produce these enzymes. Available evidence also indicates that the hydratase and oxidoreductase are synthesized without a cleavable signal sequence.[47,48] Interestingly, cytochrome b_5, an intrinsic nuclear and microscomal membrane protein, was found to be synthesized only from free and loosely membrane-bound polysomal RNA.[47-49]

Phenobarbital administration has been shown to dramatically increase the translatable levels of microsomal[46,48] and nuclear[46] NADPH-cytochrome *c* oxidoreductase mRNA. After a single dose of phenobarbital, elevated levels of these mRNAs first appeared in nuclear envelope-bound polysomes and then in polysomes bound to the rough endoplasmic reticulum (Figure 5). Maximum induction of the individual mRNAs was reached 4 to 6 hr after administration of the drug which corresponded to approximately three- and fivefold above control levels for the oxidoreductase and epoxide hydratase mRNAs, respectively. As these levels declined with half-lives in the range of 10 to 15 hr a sharp increase in the translatable levels of these mRNAs was noted in the rapidly sedimenting endoplasmic reticulum.[46] This is a fraction of the endoplasmic reticulum morphologically characterized by stacked parallel arrays of rough-surfaced membrane in close association with mitochondria.[50] The mRNAs remained elevated in this fraction at levels approximately fivefold above control values. The response to phenobarbital is markedly diminished by cordycepin and indicates that mRNA induction involves transcriptional and/or posttranscriptional events.[47,48] The mRNAs of proteins not induced by phenobarbital, such as cytochrome b_5 and serum albumin, are not increased. Hence, phenobarbital appears to trigger specific intranuclear events leading to increased production of those mRNAs coding for enzymes induced by the drug. Upon transport of the mRNAs to the cytoplasm and assembly into polysomes, the polysomes bind first to the nuclear envelope and are then sequentially translocated to adjacent rough endoplasmic reticulum and finally to that region of the reticulum operationally described as the rapidly sedimenting endoplasmic reticulum, where the induced mRNAs are sequestered and stabilized. It is interesting to note that at early times (1 to 2 hr) after phenobarbital administration, the induced level of oxidoreductase mRNA on the nuclear surface is significantly greater than the enrichment of epoxide hydratase mRNA, suggesting the asynchronous production, processing, or nucleocytoplasmic transport of the two mRNAs. This differential stimulation of two different mRNA species was not noted for the glucocorticoid induction of tyrosine aminotransferase and tryptophan oxygenase.[51]

Pulse-labeling studies with L−[^{3}H] leucine indicated that synthesis in vivo of nuclear envelope epoxide hydratase and oxidoreductase increased approximately twofold 6 to 8 hr after phenobarbital administration; however, this increase more closely paralleled the rate of increase of the respective mRNAs associated with the rough endoplasmic reticulum rather than the nuclear envelope. Hence, the transient association of the polysomes with the nuclear envelope may be of insufficient duration to allow these enzymes to be synthesized and inserted. This translocation phenomenon may partially explain the failure of phenobarbital-induced enzymes to accumulate in the envelope. Although the functional role of the rough endoplasmic reticulum-mitochondrial complex is far from firmly established, preliminary data suggest that it also may be the primary site of synthesis for apocytochrome P-450.[52,53] Thus, the enrichment of phenobarbital-induced mRNAs in this specialized region of the endoplasmic reticulum may be a generalized cellular response to phenobarbital administration.

IV. METABOLISM OF CHEMICAL CARCINOGENS

The conversion of polycyclic aromatic hydrocarbons to electrophilic reactants capable of covalently modifying DNA, RNA, and protein has been the focal point of many studies.[54-57] Until recently, the enzymes responsible for metabolic activation were thought to be exclusively associated with the endoplasmic reticulum; however, studies with highly purified nuclear envelope demonstrated that the enzymic machinery necessary for this activation step is also associated with the nuclear surface.[58,59] These and other reports[35,36,40,60-68] have focused attention on the importance of the nuclear envelope in initiating the metabolic sequences responsible for the modification of chromatin and, hence, for either mutagenesis or carcinogenesis.

The nuclear envelope contains two prominent electron transport chains, one utilizing NADPH and the other NADH.[2,4,8-11] The first includes NADPH-cytochrome *c* oxidoreductase and cytochrome P-450(s) and functions in the oxidation of a wide variety of compounds, while the second includes NADH-cytochrome b_5 oxidoreductase and cytochrome b_5 and is coupled to fatty acid desaturation via a cyanide-sensitive factor as well as the mixed-function oxidase system. A third flavoprotein, similar to, if not identical with, the microsomal amine oxidase described by Ziegler and co-workers[69,70] is also found in the nuclear envelope.[71,72] This enzyme utilizes either NADPH or NADH and catalyzes the *N*-oxidation of secondary

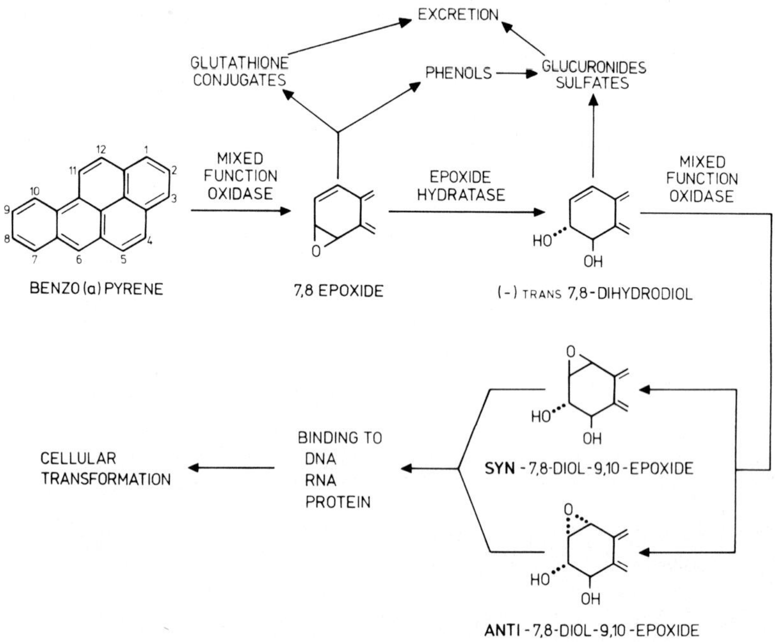

FIGURE 6. Scheme for the metabolic activation of benzo(a)pyrene.

and tertiary amines. In addition to the nuclear envelope localization of the mixed function oxidase system,[2,4,8-11,35,73] histochemical evidence for the intranuclear occurrence of cytochrome P-448, cytochrome P-450, and epoxide hydratase has also been presented.[74,75]

Although Matsuura et al.,[76] using ferritin immunoelectron microscopy, did not detect cytochrome P-450 inside nuclei from phenobarbital-treated animals, there exists an element of doubt as to whether the ferritin conjugate could penetrate the envelope.[75] The high resolution of this technique, however, makes it an excellent tool for studying this question and should be applied to nuclei whose envelope has either been removed or disrupted so as to allow suitable access of the probe to intranuclear contents. Earlier studies have estimated that 55 to 80% (uncorrected for recovery) of the total nuclear mixed-function oxidase activity is associated with the envelope,[59] and recent results[83] indicate that 80 to 100% of total nuclear epoxide hydrase, with both styrene oxide and benzo(a)pyrene used as substrate, is bound to the nuclear envelope.

Figure 6 illustrates the metabolic activation of benzo(a)pyrene that leads to the covalent modification of nucleic acids and proteins and subsequent transformation. This pathway is operative in both the nuclear envelope and endoplasmic reticulum and consists of four basic steps. The first involves the selective oxygenation of benzo(a)pyrene by cytochrome P-450 to form the 7,8-epoxide. The epoxide can either be acted upon by epoxide hydratase to form (−)-*trans*-7,8-dihydrodiol (second step), converted to the glutathione conjugate by glutathione S-epoxide transferase, or undergo a nonenzymic rearrangement to the 7-phenol, which can be further metabolized to the glucuronide and/or the sulfate. In the third step, the (−)-*trans*-7,8-dihydrodiol is converted by the mixed-function oxidase system to the highly reactive *syn* and *anti* 7,8-diol-9,10-epoxides. These diol-epoxides readily react with nucleophilic centers within the cell to produce transformation (fourth step). Although the chemical identity of the critical target(s) has yet to be established, nucleic acids, particularly DNA, have received the greatest attention since the attitude prevails that carcinogenesis involves genetic alterations. The modification of specific enzymes or regulatory factors as key events in carcinogenesis cannot be discounted, however.

Table 3
ENZYME CONTENT OF NUCLEAR ENVELOPE AND MICROSOMAL MEMBRANE FROM CONTROL, 3-MC-, AND PB-INDUCED ANIMALS

Enzyme source	Total benzo(a)pyrene oxidation (pmol/mg/min)	Cytochrome P-450 (nmol/mg)	λ_{max} (nm)	Specific benzo(a)pyrene oxidation (pmol/nmol P-450/min)
Nuclear envelope				
Control	10.3 ± 5.9[a]	0.18 ± 0.05	450.8 ± 0.4	57
PB	10.0 ± 4.3	0.20 ± 0.04	451.2 ± 0.3	50
3-MC	605 ± 82	0.34 ± 0.06	448.2 ± 0.3	1813
Microsomes				
Control	582	0.61 ± 0.03	450.3 ± 0.2	954
PB	1352	2.51 ± 0.09	450.0 ± 0.2	539
3-MC	3126	1.62 ± 0.06	448.4 ± 0.2	1929

[a] Mean ± standard deviation of values from at least four preparations of nuclear envelope or mean from two preparations of microsomes.

From Fahl, W. E., Jefcoate, C. R., and Kasper, C. B., *J. Biol. Chem.*, 253, 3106, 1978. With permission.

The distribution of benzo(a)pyrene metabolites produced upon incubation of benzo(a)pyrene with highly purified nuclear envelope and microsomal membrane has been analyzed, and the results indicate that both membrane systems produce similar metabolite patterns in the case of controls, phenobarbital-treated, and 3-methylcholanthrene-treated animals.[35] Since the metabolite pattern is primarily dependent upon the specific cytochrome P-450s present in the membranes,[77,78] it is interesting to note that spectrophotometric titrations with ligands known to produce a Type I spectral response (e.g., SKF 525A, lidocaine, testosterone, and camphor) when bound to cytochrome P450 revealed at least three separate forms of cytochrome P-450 in the nuclear envelope.[35] Although total cytochrome P-450 is not elevated in the nuclear envelope of phenobarbital-treated animals (Table 1 and 3), the drug did change the relative proportions of the individual species. In addition to showing the effects of two unique inducers on the levels of nuclear envelope and microsomal cytochrome P-450, Table 3 also reports the specific activities of benzo(a)pyrene oxidation for each membrane. As mentioned earlier, an important feature is the fact that benzo(a)pyrene oxidation per nanomole of cytochrome P-450 is the same for nuclear envelope and microsomal membrane from 3-methylcholanthrene-treated animals even though the microsomal membrane contains a fivefold higher level of the cytochrome.

Studies comparing benzo(a)pyrene metabolism by intact nuclei and microsomes demonstrated that the spectrum of metabolites generated in each case was quite similar.[40,66] Further studies dealing with the metabolism of chemical carcinogens by isolated nuclei and their subsequent binding to DNA, RNA, and proteins include such compounds as 2-acetylaminofluorene,[79] 3-methylcholanthrene,[80] and aflatoxin B_1.[38,81]

V. CONCLUDING REMARKS

Over the past few years, significant progress in the area of nuclear envelope structure and function has been made. These impressive achievements reflect an integrated multidisciplinary approach to many of the complex questions dealing with the role of the nuclear envelope in cell behavior. The topics covered in this review, such as membrane biogenesis, metabolic activation of chemical carcinogens, reformation of the nucle ır envelope during mitosis, and protein-protein interactions are representative of this rapid development. For example, the realization that certain structural elements of the nuclear envelope are reversibly depolymerized during the cell cycle sets the stage for the critical analysis of the biochemical events involved in the breakdown and reassembly process. Furthermore, the fact that enzymes associated with the nuclear envelope can metabolize potentially carcinogenic substances to ultimate reactive forms establishes the importance of the envelope in the transformation process. In addition, advances in our knowledge of the biosynthesis and degradation of intrinsic membrane proteins also provide a valuable framework within which membrane dynamics can be studied. Since all membrane proteins are not synthesized on tightly membrane-bound polysomes, experimental approaches to explore how proteins synthesized on free polysomes (e.g., cytochrome b_5) are transported to and inserted into the membrane matrix are needed.

Exciting areas of research continue to focus on the functional and structural organization of the nuclear envelope at the molecular level. This includes studies on the pore complex directed toward understanding its level of participation in nucleocytoplasmic exchange including its regulation of the transport process. Central to these goals is a detailed knowledge of the proteins comprising the pore complex (structural and catalytic) and how they interact to form this highly specific structure. Also of prime interest are questions dealing with the role of nuclear polysomes in the synthesis and maintenance of the nuclear envelope and the nature of receptor-mediated responses initiated at the nuclear surface. Based on the momentum developing in these and related areas, rapid advancement should be forthcoming.

REFERENCES

1. **Franke, W. W. and Scheer, U.,** Structures and functions of the nuclear envelope, in *The Cell Nucleus,* Vol. 1, Busch, H., Ed., Academic Press, New York, 1974, 220.
2. **Kasper, C. B.,** Chemical and biochemical properties of the nuclear envelope, in *The Cell Nucleus,* Vol. 1, Busch, H., Ed., Academic Press, New York, 1974, 349.
3. **Kasper, C. B.,** Isolation and properties of the nuclear envelope, *Methods Enzymol.,* 31, 279, 1974.
4. **Kasper, C. B.,** Chemical and enzymic composition of the nuclear envelope, in *Cell Biology,* Vol. 1, Altman, P. C. and Katz, D. D., Eds., Federation of American Societies for Experimental Biology, Bethesda, Md., 1976, 395.
5. **Feldherr, C. M.,** Structure and function of the nuclear envelope, in *Advances in Cell and Molecular Biology,* Vol. 2, DuPraw, E. J., Ed., Academic Press, New York, 1972, 273.
6. **Kay, R. R. and Johnston, I. R.,** The nuclear envelope: current problems of structure and function, *Sub. Cell. Biochem.,* 2, 127, 1973.
7. **Franke, W. W.,** Structure, biochemistry, and functions of the nuclear envelope, *Int. Rev. Cytol.,* Suppl. 4, 72, 1974.
8. **Franke, W. W.,** Nuclear envelopes, *Philos. Trans. R. Soc. London Ser. B,* 268, 67, 1974.
9. **Zbarsky, R. B.,** An enzyme profile of the nuclear envelope, *Int. Rev. Cytol.,* 54, 295, 1978.
10. **Harris, J. R.,** The biochemistry and ultrastructure of the nuclear envelope, *Biochim. Biophys. Acta,* 515, 55, 1978.
11. **Wunderlich, F., Berezney, R., and Kleinig, H.,** The nuclear envelope: an interdisciplinary analysis of its morphology, composition, and functions, in *Biological Membranes,* Vol. 3, Chapman, D. and Wallach, D., Eds., Academic Press, New York, 1976, 282.
12. **Maul, G. R.,** The nuclear and cytoplasmic pore complex: structure, dynamics, distribution, and evolution, *Int. Rev. Cytol.,* Suppl. 6, 75, 1977.
13. **Bornens, M. and Kasper, C. B.,** Fractionation and partial characterization of proteins of the bileaflet nuclear membrane from rat liver, *J. Biol. Chem.,* 248, 571, 1973.
14. **Jackson, R. C.,** Polypeptides of the nuclear envelope, *Biochemistry,* 15, 5641, 1976.
15. **Cochran, D. L., Yeoman, L. C., Egle, P. M., and Shelton, K. R.,** Comparison of polypeptides of the erythrocyte nuclear envelope, *J. Supramol. Struct.,* 10, 405, 1979.

16. **Shelton, K. R., Higgins, L. L., Cochran, D. L., Ruffolo, J. J., Jr., and Egle, P. M.,** Nuclear lamins of erythrocyte and liver, *J. Biol. Chem.*, 255, 10978, 1980.
17. **Aaronson, R. P. and Blobel, G.,** Isolation of nuclear pore complexes in association with a lamina, *Proc. Natl. Acad. Sci., U.S.A.*, 72, 1007, 1975.
18. **Dwyer, N. and Blobel, G.,** A modified procedure for the isolation of a pore complex-lamina fraction from rat liver nuclei, *J. Cell Biol.*, 70, 581, 1976.
19. **Kirschner, R. H., Rusli, M., and Martin, T. E.,** Characterization of the nuclear envelope, pore complexes, and dense lamina of mouse liver nuclei by high resolution scanning electron microscopy, *J. Cell Biol.*, 72, 118, 1977.
20. **Gerace, L., Blum, A., and Blobel, G.,** Immunocytochemical localization of the major polypeptides of the nuclear pore complex-lamina fraction, *J. Cell Biol.*, 79, 546, 1978.
21. **Gerace, L. and Blobel, G.,** The nuclear envelope lamina is reversibly depolymerized during mitosis, *Cell*, 19, 277, 1980.
22. **Richardson, J. C. W. and Maddy, A. H.,** The polypeptides of rat liver nuclear envelope. I. Examination of nuclear pore complex polypeptides by solid-state lactoperoxidase labelling, *J. Cell Sci.*, 43, 253, 1980.
23. **Lam, K. S. and Kasper, C. B.,** Electrophoretic analysis of three major nuclear envelope polypeptides. Topological relationship and sequence homology, *J. Biol. Chem.*, 254, 11713, 1979.
24. **Krohne, G., Franke, W. W., and Scheer, U.,** The major polypeptide of the nuclear pore complex, *Exp. Cell Res.*, 116, 85, 1978.
25. **Shelton, K. R., Guthrie, V. H., and Cochran, D. L.,** On the variation of the major nuclear envelope (lamina) polypeptides, *Biochem. Biophys. Res. Commun.*, 93, 867, 1980.
26. **Shelton, K. R. and Cochran, D. L.,** In vitro oxidation of intrinsic sulfhydryl groups yielding polymers of the two predominant polypeptides in the nuclear envelope fraction, *Biochemistry*, 17, 1212, 1978.
27. **Krohne, G., Franke, W. W., Ely, S., D'Arcy, A., and Jost, E.,** Localization of nuclear envelope-associated protein by indirect immunofluorescence microscopy using antibodies against a major polypeptide from rat liver fractions enriched in nuclear envelope-associated material, *Cytobiologie*, 18, 22, 1978.
28. **Ely, S., D'Arcy, A., and Jost, E.,** Interaction of antibodies against nuclear envelope-associated proteins from rat liver nuclei with rodent and human cells, *Exp. Cell Res.*, 116, 325, 1978.
29. **Stevens, B. J. and Andre, J.,** The nuclear envelope, in *Handbook of Molecular Cytology*, Lima-DeFaria, A., Ed., Interscience, New York, 1969, 837.
30. **Conner, G. E., Noonan, N. E., and Noonan, K. D.,** Nuclear envelope of Chinese hamster ovary cells: re-formation of the nuclear envelope following mitosis, *Biochemistry*, 19, 277, 1980.
31. **Lam, K. S. and Kasper, C. B.,** Selective phosphorylation of a nuclear envelope polypeptide by an endogenous protein kinase, *Biochemistry*, 18, 307, 1979.
32. **Conney, A. H.,** Pharmacological implications of microsomal enzyme induction, *Pharmacol. Rev.*, 19, 317, 1967.
33. **Gelehrter, T. D.,** Enzyme induction in mammals — an overview, in *The Induction of Drug Metabolism*, Estabrook, R. W. and Lindenlaub, E., Eds., F. K. Schattauer, Stuttgart, 1979, 7.
34. **Bresnick, E.,** Induction of the enzyme of detoxication, in *Enzymatic Basis of Detoxication*, Vol. 1, Jakoby, W. B., Ed., Academic Press, New York, 1980, 69.
35. **Fahl, W. E., Jefcoate, C. R., and Kasper, C. B.,** Characteristics of benzo(a)pyrene metabolism and cytochrome P-450 heterogeneity in rat liver nuclear envelope and comparison to microsomal membrane, *J. Biol. Chem.*, 253, 3106, 1978.
36. **Rogan, E. G., Mailander, P., and Cavalieri, E.,** Metabolic activation of aromatic hydrocarbons in purified rat liver nuclei: Induction of enzyme activities and binding to DNA with and without monooxygenase-catalyzed formation of active oxygen, *Proc. Natl. Acad. Sci. U.S.A.*, 73, 457, 1976.
37. **Viviani, A., Lutz, W. K., Schlatter, C.,** The course of the induction of aryl hydrocarbon hydroxylase in rat liver nuclei and microsomes by phenobarbital, 3-methylcholanthrene, 2,3,7,8-tetrachlorodibenzo-p-dioxin, dieldrin and other inducers, *Biochem. Pharm.*, 27, 2103, 1978.
38. **Guengerich, F. P.,** Similarity of nuclear and microsomal cytochromes P-450 in the *in vitro* activation of aflatoxin B_1, *Biochem. Pharm.*, 28, 2883, 1979.
39. **Alexander, K. and Frayssinet, C.,** Aryl and aniline hydroxylases in rat nuclear membrane after pretreatment with pregnenolone 16α-carbonitrile, phenobarbital, and methylcholanthrene, *Experientia*, 31, 778, 1975.
40. **Pezzuto, J. M., Yang, C. S., Yang, S. K., McCourt, D. W., and Gelboin, H. V.,** Metabolism of benzo[a]pyrene and (−) trans-7,8-dihydroxy-7,8-dihydrobenzo[a]pyrene by rat liver nuclei and microsomes, *Cancer Res.*, 38, 1241, 1978.
41. **Mukhtar, H., Elmamlouk, T. H., and Bend, J. R.,** Epoxide hydrase and mixed-function oxidase activities of rat liver nuclear membranes, *Arch. Biochem. Biophys.*, 192, 10, 1979.
42. **Mukhtar, H., Elmanlouk, T. H., Philpot, R. M., and Bend, J. R.,** Rat hepatic nuclear cytochrome P-450 and epoxide hydrase in membranes prepared by two methods: Similarities with the microsomal enzymes, *Mol. Pharmacol.*, 15, 192, 1972.

43. **Ito, A.,** Cytochrome b_5-like hemoprotein of outer mitochondrial membrane; OM cytochrome b. I. Purification of OM cytochrome b from rat liver mitochondria and comparison of its molecular properties with those of cytochrome b_5, *J. Biochem.*, 87, 63, 1980.
44. **Zimmerman, J. J. and Kasper, C. B.,** Immunological and biochemical characterization of nuclear envelope reduced nicotinamide adenine dinucleotide phosphate-cytochrome *c* oxidoreductase, *Arch. Biochem. Biophys.*, 190, 726, 1978.
45. **Thomas, P. E., Korzeniowski, E., Bresnick, E., Bornstein, W. A., Kasper, C. B., Fahl, W. E., Jefcoate, C. R., and Levin, W.,** Hepatic cytochrome P-448 and epoxide hydratase: enzymes of nuclear origin are immunochemically identical with those of microsomal origin, *Arch. Biochem. Biophys.*, 192, 22, 1979.
46. **Gonzalez, F. J. and Kasper, C. B.,** Sequential translocation of two phenobarbital-induced polysomal messenger ribonucleic acids from the nuclear envelope to the endoplasmic reticulum, *Biochemistry*, 20, 2292, 1981.
47. **Gonzalez, F. J. and Kasper, C. B.,** *In vitro* translation of epoxide hydratase messenger RNA, *Biochem. Biophys. Res. Commun.*, 93, 1254, 1980.
48. **Gonzalez, F. J. and Kasper, C. B.,** Phenobarbital induction of NADPH-cytochrome *c* (P-450) oxidoreductase messenger ribonucleic acid, *Biochemistry*, 19, 1790, 1980.
49. **Rachubinski, R. A., Verma, D. P. S., and Bergeron, J. J. M.,** Synthesis of rat liver microsomal cytochrome b_5 by free ribosomes, *J. Cell Biol.*, 84, 705, 1980.
50. **Shore, G. C. and Tata, J. R.,** Two fractions of rough endoplasmic reticulum from rat liver. I. Recovery of rapidly sedimenting endoplasmic reticulum in association with mitochondria, *J. Cell Biol.*, 72, 714, 1977.
51. **Roewekamp, W. G., Hofer, E., and Sekeris, C. E.,** Translation of mRNA from rat-liver polysomes into tyrosine aminotransferase and tryptophan oxygenase in a protein-synthesizing system from wheat germ, *Eur. J. Biochem.*, 70, 259, 1976.
52. **Meier, P. J., Spycher, M. A., and Meyer, U. A.,** Isolation of a subfraction of rough endoplasmic reticulum associated with mitochondria, *Exp. Cell Res.*, 111, 479, 1978.
53. **Pickett, C. B., Rosenstein, N. R., and Jeter, R. L.,** The physical association between rat liver mitochondria and rough endoplasmic reticulum. II. Possible role of RER-MT complexes play in the biosynthesis of cytochrome P-450, *Exp. Cell Res.*, 132, 225, 1981.
54. **Miller, E. C. and Miller, J. A.,** Biochemical mechanisms of chemical carcinogenesis, in *The Molecular Biology of Cancer*, Busch, H., Ed., Academic Press, New York, 1974, 377.
55. **Yang, S. K., Deutsch, J., and Gelboin, H. V.,** Benzo(a)pyrene metabolism: activation and detoxification, in *Polycyclic Hydrocarbons and Cancer*, Vol. 1, Gelboin, H. V. and Ts'o, P., Eds., Academic Press, New York, 1978, 205.
56. **Brookes, P., Newbold, R. F., and Osborne, M. R.,** Mechanism of the carcinogenicity of polycyclic hydrocarbons, in *Environmental Carcinogenesis*, Emmelot, P. and Kriek, E., Eds., Elsevier-North Holland, Amsterdam, 1979, 123.
57. **Gelboin, H. V.,** Benzo(a)pyrene metabolism, activation, and carcinogenesis: role and regulation of mixed-function oxidase and related enzymes, *Physiol. Rev.*, 60, 1107, 1980.
58. **Kasper, C. B.,** Biochemical distinctions between the nuclear and microsomal membranes from rat hepatocytes: The effect of phenobarbital administration, *J. Biol. Chem.*, 246, 577, 1971.
59. **Khandwala, A. S. and Kasper, C. B.,** Preferential induction of aryl hydroxylase activity in rat liver nuclear envelope by 3-methylcholanthrene, *Biochem. Biophys. Res. Commun.*, 54, 1241, 1973.
60. **Rogan, E. G. and Cavalieri, E.,** 3-Methylcholanthrene-inducible binding of aromatic hydrocarbons to DNA in purified rat liver nuclei, *Biochem. Biophys. Res. Commun.*, 58, 1119, 1974.
61. **Vaught, J. and Bresnick, E.,** Binding of polycyclic hydrocarbons to nuclear components *in vitro*, *Biochem. Biophys. Res. Commun.*, 69, 587, 1976.
62. **Alexandrov, K., Brookes, P., King, H. W. S., Osborne, M. R., and Thompson, M. H.,** Comparison of the metabolism of benzo[a]pyrene and binding to DNA caused by rat liver nuclei and microsomes, *Chem. Biol. Interact.*, 12, 269, 1976.
63. **Pezzuto, J. M., Lea, M. A., and Yang, C. S.,** Binding of metabolically activated benzo[a]pyrene to nuclear macromolecules, *Cancer Res.*, 36, 3647, 1976.
64. **Jernström, B., Vadi, H., and Orrenius, S.,** Formation in isolated rat liver microsomes and nuclei of benzo[a]pyrene metabolites that bind to DNA, *Cancer Res.*, 36, 4107, 1976.
65. **Pezzuto, J. M., Lea, M. A., and Yang, C. S.,** The role of microsomes and nuclear envelope in the metabolic activation of benzo[a]pyrene leading to binding with nuclear macromolecules, *Cancer Res.*, 37, 3427, 1977.
66. **Bresnick, E., Vaught, J. B., Chuang, A. H. L., Stoming, T. A., Bockman, D., and Mukhtar, H.,** Nuclear aryl hydrocarbon hydroxylase and interaction of polycyclic hydrocarbons with nuclear components, *Arch. Biochem. Biophys.*, 181, 257, 1977.

67. **Bresnick, E., Stoming, T. A., Vaught, J. B., Thakker, D. R., and Jerina, D. M.,** Nuclear metabolism of benzo[a]pyrene and of (±)-trans-7,8-dihydroxy-7,8-dihydrobenzo[a]pyrene, *Arch. Biochem. Biophys.*, 183, 31, 1977.
68. **Alexandrov, K. and Thompson, M. H.,** Influence of inducers and inhibitors of mixed-function oxidases on benzo[a]pyrene binding to the DNA of rat liver nuclei, *Cancer Res.*, 37, 1443, 1977.
69. **Ziegler, D. M. and Mitchell, C. H.,** Microsomal oxidase. IV. Properties of a mixed-function amine oxidase isolated from pig liver microsomes, *Arch. Biochem. Biophys.*, 150, 116, 1972.
70. **Prough, R. A. and Ziegler, D. M.,** The relative participation of liver microsomal amine oxidase and cytochrome P-450 in N-demethylation reactions, *Arch. Biochem. Biophys.*, 180, 363, 1977.
71. **Patton, S. E., Rosen, G. M., Ruckman, E. J., Graham, D. G., Small, B., and Ziegler, D. M.,** Hamster hepatic nuclear mixed-function amine oxidase: location and specific activity, *Mol. Pharmacol.*, 18, 151, 1980.
72. **Sum, C. Y. and Kasper, C. B.,** Mixed function amine oxidase of the rat hepatocyte nuclear envelope: Demonstration and induction studies, *Biochem. Pharmacol.*, in press.
73. **Jarasch, E. D. and Franke, W. W.,** Is cytochrome oxidase a constituent of nuclear membranes?, *J. Biol. Chem.*, 249, 7245, 1974.
74. **Bresnick, E., Boraker, D., Hassuk, B., Levin, W., and Thomas, P. E.,** Intranuclear localization of hepatic cytochrome P-448 by an immunochemical method, *Mol. Pharmacol.*, 16, 324, 1979.
75. **Bresnick, E., Hassuk, B., Liberator, P., Levin, W., and Thomas, P.,** Nucleolar cytochrome P-450, *Mol. Pharmacol.*, 18, 550, 1980.
76. **Matsuura, S., Fujii-Kuriyana, Y., and Tashiro, Y.,** Immunoelectron microscope localization of cytochrome P-450 on microsomes and other membrane structures of rat hepatocytes, *J. Cell Biol.*, 78, 503, 1978.
77. **Coon, M. J. and Persson, A. V.,** Microsomal cytochrome P-450: a central catalyst in detoxication reactions, in *Enzymatic Basis of Detoxification*, Vol. 1, Jakoby, W. B., Ed., Academic Press, New York, 1980, 117.
78. **Wislocki, P. G., Miwa, G. T., and Lu, A. Y. H.,** Reactions catalyzed by the cytochrome P-450 systems, in *Enzymatic Basis of Detoxification*, Vol. 1, Jakoby, W. B., Ed., Academic Press, New York, 1980, 135.
79. **Kawajiri, K., Yonekawa, H., Hara, E., and Tagashira, Y.,** Activation of 2-acetylaminofluorene in the nuclei of rat liver, *Cancer Res.*, 39, 1089, 1979.
80. **Tierney, B., Bresnick, E., Sims, P., and Grover, P. L.,** Microsomal and nuclear metabolism of 3-methylcholanthrene, *Biochem. Pharmacol.*, 28, 2607, 1979.
81. **Vaught, J. B., Klohs, W., and Gurtoo, H. L.,** In vitro metabolism of aflatoxin B_1 by rat liver nuclei, *Life Sci.*, 21, 1497, 1977.
82. **Lam, K. S. and Kasper, C. B.,** unpublished results.
83. **Gonzalez, F. J. and Kasper, C. B.,** Differential inducibility of nuclear envelope epoxide hydrase by *trans*-stirene oxide and phenobarbital, *Mol. Pharmacol.*, 21, 511, 1982.

Chapter 2

NUCLEAR RIBONUCLEOPROTEINS CONTAINING HETEROGENEOUS RNA

V. Holoubek

TABLE OF CONTENTS

"The Truth Is Rarely Pure and Never Simple"
Oscar Wilde

PREFACE

The discovery that the genetic information transcribed as *heterogeneous RNA (hnRNA)* is subjected to a complicated processing before some of the information is released into the cytoplasm as *messenger RNA (mRNA)* resulted in increased interest in the *nuclear ribonucleoproteins containing hnRNA (hnRNP)*. Recently, some fine reviews have been published which summarize the research done predominately in individual laboratories.[1-6] The goal of this review is to integrate, as far as possible, the picture emerging from the individual laboratories.

The findings of the association of *small nuclear RNA (snRNA)* with hnRNP particles, of the existence of the major proteins of these particles in many charged isomers, and of the close interrelationship between the nuclear matrix and the studied hnRNP particles are not yet adequately complete to propose a final model of the organization of hnRNA-protein interactions in the cell nucleus. But it is already possible to propose at least some interpretation of the diverse facts which have accumulated in the laboratories around the world. The work done is so extensive that a complete description would require publication of an independent book. The selection of papers discussed here is, to some extent, necessarily arbitrary, and the author hopes that he will be forgiven by the authors of many excellent papers which are not mentioned.

The main thesis proposed in this review is the following: the hnRNA binds immediately after its synthesis with proteins which are either closely associated with the nuclear network or are the network itself. In association with proteins, the RNA is transported from DNA to the nuclear membrane. On its way the hnRNA is processed, which means that the nucleotide sequences which will later form mRNA are assembled together and appropriately modified.

There are two basic types of arrangements of hnRNA with proteins in this *ribonucleoprotein* (RNP) structure. Parts of the hnRNA are associated with a very heterogeneous population of proteins with a molecular weight predominantly over 50,000. These sequences of the RNP are partially resistant to treatment with high concentrations of salts but are sensitive to treatment with a low concentration of ribonuclease (RNase). These proteins were named *β-proteins* in Jacob's laboratory.[7] In other segments of the RNP the hnRNA is folded together with a few major specific proteins with molecular weights below 40,000. The hnRNA associated with these proteins is more RNase-resistant than the hnRNA associated with β-proteins, but the proteins are more easily dissociated from hnRNA by salt treatment than are the β-proteins. Because of the characteristic amino acid composition of these proteins they were named the glycine-rich proteins.[8] The terms β-proteins and glycine-rich proteins will be used throughout this review. The hnRNP particles described in the literature are the result of fragmentation of the nuclear RNP network during particle isolation (by endogenous nucleases or physical forces) into particles composed either of segments of hnRNA associated with the glycine-rich proteins, or of segments of hnRNA associated with β-proteins, or of various mixtures of both types of segments. The proportions of both types of particles in the mixture are dependent on the procedures used for the disintegration of the nuclear network and the purification of the hnRNP particles. The *particles associated with the glycine-rich proteins* are for the sake of simplicity referred to in this text as the *Samarina particles*. The *particles containing hnRNA associated with the β-proteins* are referred to in the text as *Jacob particles*. Specific proteins are also associated with snRNA which are bound to the hnRNP particles, probably by hydrogen bonds between the hnRNA and snRNA. Other specific protein(s) is(are) associated with the poly(A) sequence present at the 3′ end of a fraction of the hnRNA molecules. Some of the proteins in the hnRNP complexes (perhaps only loosely and temporarily associated with the hnRNP) are enzymes probably involved in processing of hnRNA and in control of the function of the particle proteins.

I. INTRODUCTION

Many reports describing RNP particles in the nuclei of various types of cells can be found in the literature before 1966. At the time, the function of these particles was not understood and they were often considered "nuclear ribosomes" or viral particles. The evaluation of these older articles is beyond the scope of this review and more of historical interest.

Between 1965 and 1968 Samarina and co-workers showed in a series of publications[9-12] that newly synthesized premessenger RNA is present in the nucleus in the form of RNP particles. Extraction of nuclei from rat liver or Ehrlich ascites carcinoma cells with isotonic salt solution at slightly alkaline pH (10 m*M* Tris-HCl buffer, pH 8.0, containing 1 m*M* $MgCl_2$ and 0.14 *M* NaCl) removed from the nuclei particles containing pulse-labeled RNA. These particles sedimented in 15 to 30% linear sucrose density gradients as a homogeneous peak with a sedimentation coefficient of approximately 30 to 40 S. The ratio of RNA to protein in these particles was determined from the position of the RNA-protein complex in CsCl density gradient (after a formaldehyde fixation of the protein to RNA) to be 1:4. The size of the particles observed under electron microscopy was estimated to be 18 × 18 × 8 nm.[11,12] Soon afterwards in the same laboratory it was shown that RNA associated with the RNP particles contain nucleotide sequences which will be transported into the cytoplasm and also nucleotide sequences which will be degraded in the nucleus.[13] Based on these findings, the authors concluded that it seemed probable that the particles are involved in mRNA transport as well as in the cleavage of giant RNA chains, and in the selection of RNA which will be transported into the cytoplasm.[12,13]

Samarina and co-workers[12] proposed the term informofers to name these particles. This name has not been generally accepted, perhaps because of the controversy about the size and the composition of these particles and about their uniformity or diversity. A very similar name, informosomes, was given by Spirin to the free cytoplasmic RNP particles,[14] which could lead to confusion. The designation of the particles as 30 S nuclear RNP particles is also not completely appropriate because of the reported variation in the sedimentation constant of these particles. Neither is the term monoparticles adequate, since today we know that two different types of monoparticles containing hnRNA, which sediment together in linear sucrose density gradient, can be extracted from the nuclei.[15] The work of Samarina and co-workers not only opened a new field of study of the nuclear RNPs, but also gave an impact to the studies of RNA-protein interactions involved in the processing and transport of nuclear RNA. Therefore, perhaps it would not be inappropriate to name these 30 to 40 S mono-particles *Samarina particles*. This term will be used in this review.

When the RNP particles were extracted from the nuclei by isotonic salt buffer pH 8.0 in the presence of rat liver cytosolic RNase inhibitor, a heterogeneous population of particles sedimenting from 30 to over 200 S (polyparticles) was obtained. Mild treatment of these polyparticles with pancreatic RNase resulted in their complete breakage into 30 to 40 S monoparticles.[12] The authors concluded that the 30 S particles are monomers of much bigger complexes which are organized like polysomes. The monomer units are released because endogenous nucleases active in the liver cells during the particle isolation break the hnRNA which holds the polyparticle together.

In the broad 30 to 200 S peak of the polyparticles isolated from rat liver nuclei by isotonic salt buffer pH 8.0 extraction in the presence of rat liver cytosolic RNase inhibitor, it was possible to observe discrete peaks sedimenting with sedimentation coefficients of 30, 45, 55 to 60, and 70 to 75 S. Electron micrographs indicated that the 30 S peak contains monomers; the 45 S peak, dimers; the 55 to 60 S peak, trimers; and the 70 to 75 S peak, tetra- and pentamers of the particles. The buoyant densities in the CsCl gradient of the polyparticles of different sizes and of the monoparticles were similar. The authors interpreted these results as confirmation of their proposal that the RNA-to-protein ratio in the monomers is the same as in polymers and therefore the polyparticles structure is composed only of identical monomers. The buoyant density values given by the authors in their paper for monoparticles and for polyparticles are not in full agreement with this conclusion. For example, the buoyant density for 30 S monoparticles was found to be 1.397, but for 60 S polyparticles it was 1.392,[12] which would indicate that polyparticles contain more proteins in ratio to RNA than monoparticles. That the polyparticles are composed only from uniform monomers is not probable in view of more recent findings which are discussed later in this review.

Isolation and sizing of RNA present in monomers and polymers of the particles has shown that with the increasing size of the polymers, the polyparticles contain large RNA molecules.[12] For example, RNA isolated from 30 S monoparticles sedimented at 8.5 S; from 60 S polyparticles, at 17 S; and from 120 to 130 S polyparticles, at 32 S. This is in agreement with the conclusion that polyparticle structures contain one large hnRNA molecule.

The data of Samarina and co-workers, which later became very controversial, concern the protein composition of the particles. After the digestion of the particle RNA by RNase, the protein moiety of the particle was precipitated with 5% TCA,[12] or in other experiments separated from the RNA by elution of the particles in 6 *M* urea from DEAE-column.[13] After the removal of RNA and, according to the authors, not more than 5% of the particle proteins, the protein moiety of the particle was analyzed by polyacrylamide gel electrophoresis (PAGE) at pH 4.5 in the absence of SDS. The protein moiety separated into three main and a series of minor bands.[12] Reduction of S-S bonds by mercaptoethanol converted all the bands into one. According to the authors, only this one protein is present in the particles; its molecular weight was estimated by SDS-PAGE to be 40,000 and the name ''informatin'' was proposed for this protein.[16]

The possibility that the nuclear RNA complexes are involved in the selective release of RNA from the cell nucleus, which was indicated by the work of Samarina and co-workers, stimulated interest in the study of nuclear RNP particles in many laboratories. But the results of these studies have not led to uniform conclusions about the composition and structure of nuclear RNP particles. To the contrary, it has been more and more difficult to reconcile the discrepancies in the published data, and only recently has it become possible to explain at least some of the differences in the reported characterization of the nuclear RNP particles. An important step in this development was the realization of the close association of hnRNA with the nuclear matrix (nuclear RNP network) *in situ*.

II. NUCLEAR RIBONUCLEOPROTEIN NETWORK AND NUCLEAR MATRIX

Newly synthesized hnRNA is present in the cell in the form of fibrilar RNP particles 5 to 10 nm in diameter, and of granules approximately 30 to 50 nm in diameter: the ''perichromatin fibrils'' and ''perichromatin granules'' identified as RNPs in Bernhard's laboratory by observation of cells *in situ*.[17,18] The system of RNP granules and fibers which are present in cells is referred to as the nuclear RNP network,[19-21] nuclear RNA protein fibers,[22] or perichromosomal RNP.[23]

In the nucleus the macromolecules are organized in a highly ordered structure. In addition to chromatin and nuclear envelope (or membrane), it is possible to observe, after specific treatment, a proteinaceous, sponge-like network called nuclear matrix[24] or nuclear skeleton.[25] This network extends from the chromatin to the nuclear envelope, where it may be associated with the nuclear lamina.[22] In addition to a possible function of this structure as a nuclear scaffold, it could also serve as ''tracks'' for the movements of newly synthesized RNA from chromatin to the pore complexes in nuclear membrane. Changes observed in nuclear matrix after the inhibition or modification of RNA synthesis[26] indicate that nuclear matrix may also be involved in hnRNA processing. Excellent reviews on the morphology and chemistry of the nuclear matrix have been published recently.[26,27] Therefore, this discussion of nuclear matrix will be limited only to consideration of the interactions of nuclear RNP network with the nuclear matrix, and of possible overlap between the proteins of nuclear RNP particles and proteins of the nuclear matrix. Such overlap could be either functional or resulting from procedures used for the isolation of nuclear RNP particles.

Studies by Busch et al.[19,28-30] on nuclear structures remaining after the removal of deoxyribonucleoproteins from the nuclei of liver and Walker tumor cells by extraction with 2 *M* NaCl have shown that newly synthesized RNA is associated with the nuclear network.

Studies reported from the laboratory of Pogo[21] have confirmed and expanded these observations. Pogo and co-workers investigated the product-precursor relationship between nuclear RNP particles, which they studied earlier,[31,32] and the fibrogranular material of the nuclear RNP network first observed by electron microscopy in rat liver nuclei by Georgiev and Chentsov[33] and later studied in a variety of tissues in many laboratories.[18,19,34] When the rat liver or Krebs ascites cell nuclei are disrupted by compression in French pressure cells (which will cause the disintegration of nuclei mainly by shearing forces), the majority of the rapidly labeled RNA is found in nucleoplasm in the form of hnRNP particles.[31,35] The amount of released particles depends on the amount of stress to which nuclei are subjected. When the nuclei were compressed or decompressed in the French pressure cell under low pressure (3 to 5 $\times$ 10^3 lb/in.^{2}) or were ruptured by expansion during sudden decompression in the nitrogen cavitation bomb, only a small amount of the nuclear hnRNP particles was released into the nucleoplasm[21] (Table 1). The RNA which was not released could be fractionated by centrifugation on 10 to 30% sucrose gradient into RNA attached to the nuclear envelope and RNA which pelleted with chromatin. These experiments indicate that shearing forces could be responsible for the release of nuclear hnRNP particles from the more complex nuclear structures and that these structures could be connected with chromatin and nuclear membrane.

The existence of nuclear structures containing hnRNA was confirmed experimentally. Such structures were isolated by Pogo and co-workers[21,25] from rat liver and Krebs ascites cell nuclei by high salt treatment (0.8 *M* NaCl), a procedure which has been used by Mirsky and Pollister[36] to disrupt nuclei for the isolation of nucleoproteins from a wide variety of animal cells. Chromatin unfolded by the hypertonic salt solution was removed by digestion with deoxyribonuclease (150 μg/mℓ at 37°C). The procedure used for the digestion of chromatin was a modification of a method described by Penman[37] for removal of chromatin from HeLa cell nuclei. The modification consisted of lowering the concentration of Mg^{2+} from 50 to 8 m*M* to prevent aggregation of nuclear structures. The nucleoli were released from the digested chromatin by homogenization with four strokes in a Dounce homogenizer. Structures containing the newly synthesized RNA were purified by centrifugation in 10 to 30% sucrose density gradient[21] or by centrifugation on 30 to 68% sucrose gradient[25] and collected on a cushion of 71.5% sucrose. Inhibition of endogenous nuclear proteases by the addition of phenylmethyl sulfonyl chloride or phenylmethyl sulfonyl fluoride improved the yields and the integrity of isolated structures containing newly synthesized RNA.[25] In the presence of these inhibitors nuclear structures associated with the RNA were recovered in a narrow band at the sucrose cushion after centrifugation through 30 to 68% sucrose gradient in a SW 41 Spinco rotor at 18,000 r/min for 18 hr. In the absence of the inhibitors the RNA was found associated with structures heterodispersed in the gradient even after centrifugation at 36,000 r/min for 18 hr.[25] It is important that in the presence of the inhibitors of proteases the majority of labeled RNA, after the 18-hr centrifugation at 36,000 r/min, was layered at the sucrose cushion when the animals were labeled with [^{3}H] orotic acid for 30 min and also when they were labeled for only 5 min[25] (Figure 1). The coincidence between the UV-absorbing material and rapidly labeled RNA, regardless of the length of the pulse, indicated that structures containing RNA *in statu nascendi* and RNA in processing are not separated

Table 1
THE DEPENDENCE OF THE RELEASE OF HnRNP PARTICLES FROM THE NUCLEI ON THE AMOUNT OF THE SHEARING FORCE USED

	Krebs ascites nuclei			Rat liver nuclei		
	French pressure cell		Cavitation bomb	French pressure cell		Cavitation bomb
	a	*b*	*c*	*a*	*b*	*c*
Nucleoplasm	20.6	6.6	4.7	4.8	2.0	1.5
Pellet	28.0	42.0	44.0	4.0	6.0	6.9

Note: The labeled and isolated Krebs and rat liver nuclei were disrupted in the French pressure cell under high or low pressure or were ruptured by expansion during decompression in the nitrogen cavitation bomb: a = 15 to 20 $\times 10^3$ lb/in.^{2}, b = 3 to 5 $\times 10^3$ lb/in.^{2}, c = 1.2 to 1.4 $\times 10^3$ lb/in.^{2}. All values are 1 $\times 10^4$ cpm. The suspensions of disrupted nuclei were centrifuged at 3000 $\times$ g for 10 min. Radioactivities were determined in nucleoplasm, pellet, and intact nuclei. The radioactivity in intact Krebs nuclei was 49 $\times 10^4$ cpm and in intact rat liver nuclei was 9.6 $\times 10^4$ cpm. The Krebs cells were labeled for 30 min with [^{3}H] uridine, and the liver cells were labeled for 30 min with [^{3}H] orotic acid.

From Faiferman, I, and Pogo, A. O., *Biochemistry,* 14, 3808, 1975. With permission.

by the fractionation. Since the UV-absorbing material and the radioactive counts were also in the same position when lower centrifugal force was used (18,000 r/min for 18 hr) the possibility is ruled out that the radioactive profile could have been caused by very dense structures sedimenting to this position, whereas the UV profile could have been caused by structures which reached equilibrium at this position. These experiments strongly favor the conclusion that hnRNA is immediately associated with the nuclear network and remains associated with it during RNA processing.

Electron micrographs have shown that structures isolated from nuclei by high-salt-DNase treatment consist of nuclear membranes (envelopes) and of fibrogranular material.[21] The sponge-like skeletal network of the fibrogranular material expanded through the entire nucleus and appeared to have close association with nucleoli, inner membrane of the nuclear envelope, and with the nuclear pore complexes.[25] After the removal of chromatin the residual nuclei (named by the authors "achromatinic" nuclei) were composed of 79.5% protein, 11.0% RNA, 8.5% phospholipid, and 1.0% DNA. They retained 65% of phospholipid, 31% of RNA, 19% of protein, and 0.48% of DNA originally present in the nuclei.[25] The low concentration of DNA in the preparation demonstrated that the used high-salt-DNase treatment separated nuclear membranes and the nuclear RNP network from chromatin. Larger contamination of the isolated RNP network by chromatin or by nucleoli was also excluded by protein analysis, which did not show any histones or ribosomal proteins.[21] The complex isolated in the absence of protease inhibitors contained less protein (62.8%).[21]

The yield of the RNP nuclear network obtained from rat liver nuclei in the laboratory of Pogo was similar to the yield obtained in Berezney's laboratory.[38,39] Rat liver nuclei, prepared by fractionation of liver homogenate by centrifugation in 2.2 *M* sucrose, were, after a mild DNase I treatment (5 μg DNase/10^8 nuclei/mℓ/10 min), extracted with 0.2 m*M* $MgCl_2$ (in 10 m*M* Tris, pH 7.4 buffer) followed by extraction with 1% solution of Triton® X-100. This removed some of the proteins from the nuclear RNP network and the remaining structure, designated by the authors as nuclear matrix, contained approximately 10% of nuclear protein, 30% of RNA, and 2% of the DNA originally present in the nuclei (16% of nuclear protein,

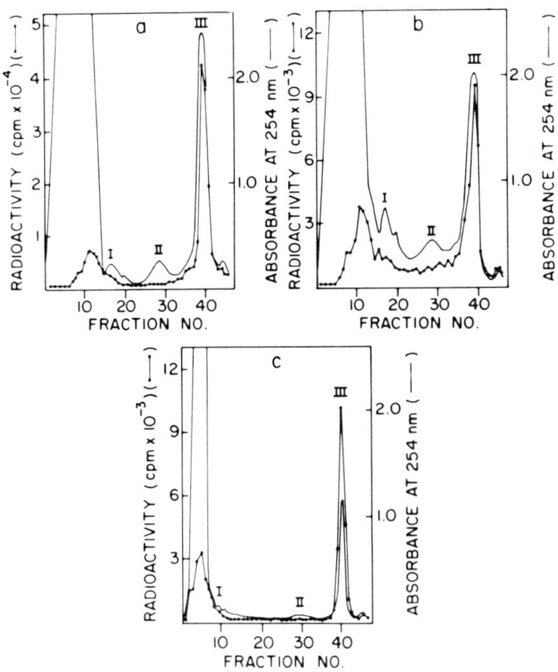

FIGURE 1. Isolation of nuclear structures containing hnRNA in the presence of the inhibitors of proteolytic enzymes. The rats were injected with [^{3}H] orotic acid and the nuclei isolated from rat liver. After treatment of the isolated nuclei with high salt and digestion with DNase the nuclear structures containing newly synthesized RNA were purified by centrifugation on 30 to 68% sucrose gradient. The cells were labeled either (a) for 30 min or (b) for 5 min, and nuclear digest centrifuged in a SW 41 Spinco rotor at 36,000 r/min for 18 hr; (c) the cells were labeled for 30 min and nuclear digest centrifuged in the same rotor at 18,000 r/min for 18 hr. The direction of sedimentation is from left to right. (From Miller, T. E., Huang, C.-Y., and Pogo, A. O., *J. Cell Biol.*, 76, 675, 1978. With permission.)

36% of RNA, and 2% of DNA before the treatment with Triton® X-100). The composition of the nuclear RNP network prepared by Faiferman and Pogo,[21] compared with the composition of the nuclear matrix prepared by Berezney and Coffey[38] shows the latter contains a much higher concentration of RNA (5.7:1 as compared with 4.7:1).

The nuclear RNP network is composed of a heterogeneous population of proteins with molecular weights predominantly over 40,000 (Figure 2).[21] There is a strong similarity in the electrophoretic distribution of proteins of the network and that of the β-proteins of the hnRNP particles. In the nuclear matrix isolated by the procedure of Berezney and Coffey[38] some of the heterogeneous high-molecular-weight proteins seen in the network are partly or completely removed, and three major polypeptides with an approximate molecular weight of 62,000, 66,000, and 69,000 predominate.[38] But this matrix is still associated with hnRNA, and also the glycine-rich proteins of the Samarina particles are still present among the matrix proteins. Peters and Comings,[40] by using two-dimensional gel electrophoresis (separation by nonequilibrium pH gradient in one dimension and by SDS-electrophoresis in the second dimension), could clearly identify the glycine-rich proteins of Samarina particles (Figure 3, polypeptides 34 and 36) in the preparation of matrix proteins. Therefore proteins of both types of the hnRNP particles, Samarina particles and Jacob particles, are at least partly present in the nuclear matrix.

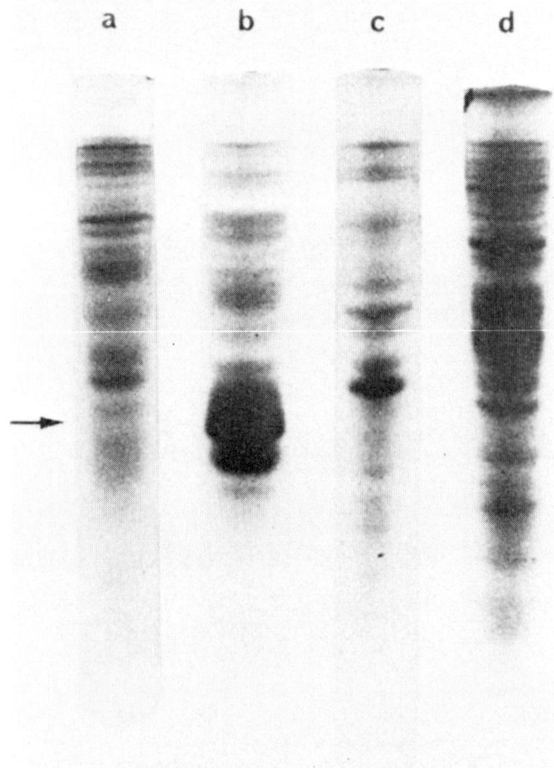

FIGURE 2. SDS-PAGE of rat liver nuclear proteins: (a) proteins of the RNP network, (b) proteins of the hnRNP particles, (c) proteins of the hnRNP particles treated with high salt, DNase and Triton® X-100, and (d) total soluble protein. The arrow indicates protein having 40,000 in molecular weight. (From Faiferman, I. and Pogo, A. O., *Biochemistry*, 14, 3808, 1975. With permission.)

The turnover of the matrix proteins is high. Hemminki,[41] who studied the incorporation of amino acids into nuclear proteins of lung, found that the incorporation of amino acids into matrix proteins exceeded threefold the incorporation into the next highest nonhistone protein fraction. In rat liver labeled for 1 hr with [^{3}H] leucine the specific activity of nuclear matrix proteins was about 3.5 times higher than the specific activity of total nuclear proteins.[26] By their high turnover, many of the matrix proteins are similar to β-proteins of the hnRNP particles. Also, many of the matrix proteins are phosphoproteins,[42] another characteristic which they have in common with the β-proteins of the mRNP particles.

The indication that there is an extensive overlap between proteins of the hnRNP particles and proteins of the nuclear matrix does not exclude the existence of a very simple protein skeleton in the cell nuclei which, together with the nuclear RNP network, forms a structure that could represent the functional nuclear matrix, i.e., a skeleton with muscles. Indeed, published experimental results support the notion that such a protein skeleton really exists.

One approach used to distinguish between the nuclear matrix proteins and the nuclear RNP network proteins was to isolate the simplest structure which still maintains the architectural framework of nuclear structures. Such a structure, composed of few fibrilar proteins only, is called matrixin by Comings and Okada[43] and matricin by Berezney.[39] An earlier observation by electron microscopy has shown that 3- to 5-nm fibers are components of various extrachromosomal structures as are nuclear envelope, nucleolar matrix, or interchromatinic structures.[18] According to Comings and Okada[43] these fibers are composed of

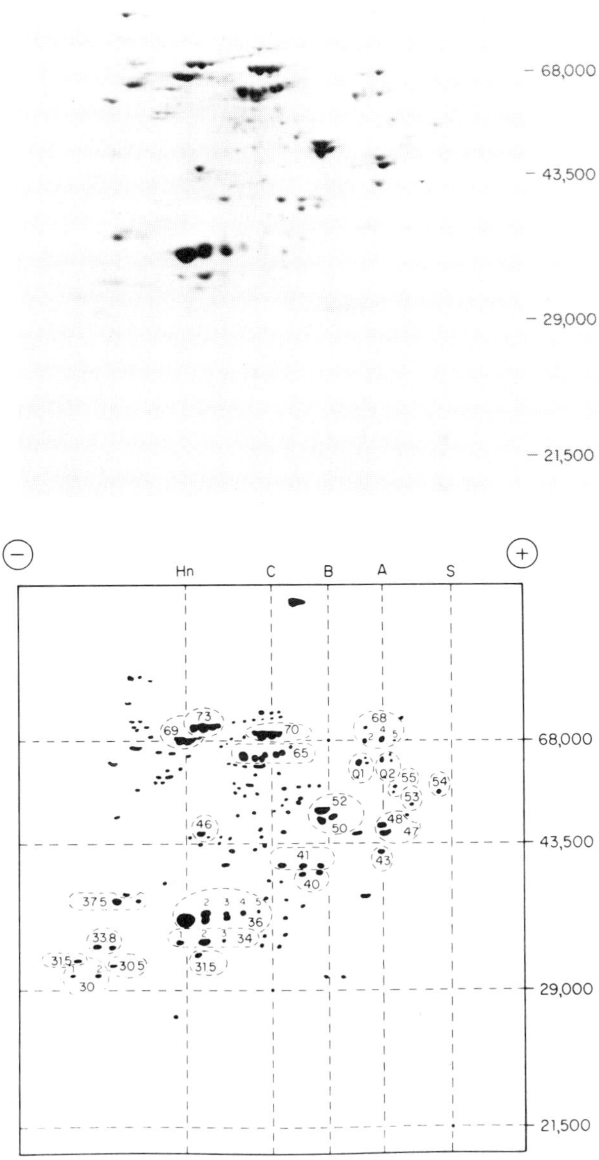

FIGURE 3. Two-dimensional gel electrophoresis of rat liver nuclear matrix proteins. The proteins were separated by nonequilibrium pH gradient electrophoresis in one direction and according to their molecular weight in the second direction: hn = position of the most basic of the charge isomers of the 36,000-dalton hnRNP particle protein, C = pH 7.4, B = pH 6.7, A = pH 6.1, and S = pH 5.5. (From Peters, K. E. and Comings, D. E., *J. Cell Biol.*, 86, 135, 1980. With permission.)

2-nm protein fibrils which can self-associate to produce 4- to 30-nm fibers. DNA, nuclear membrane, nuclear pores, nucleolus, and the nuclear RNP network are all attached to these fibers. Since this minimal structure represents the framework to which other structures are attached in the cell nucleus it can also be called the nuclear skeleton.

As already mentioned, analysis of the protein composition of the nuclear matrix fraction prepared from rat liver nuclei by Berezney and Coffey[38] has shown the presence of three main (primary) polypeptides (62,000, 66,000, and 69,000 daltons) which according to the authors represent the nuclear skeleton, to which are attached other nuclear structures. Similar major polypeptides have been reported in the nuclear matrix of HeLa cells,[44] avian erythroblasts,[45] and even of the macronuclei in *Tetrahymena*.[46] It can therefore be assumed that these polypeptides are characteristic for the matrix of eukaryotic cell nuclei. The rat liver nuclear matrix preparation of Berezney and Coffey[38] contained, in addition to the three primary polypeptides, still a larger amount (≃50%) of minor (secondary) polypeptides (molecular weights of 44,000 and higher) and even traces of the main polypeptides of the nuclear RNP monoparticles. To further purify the nuclear skeleton, Berezney[39] prepared crude nuclear matrix fraction from rat liver nuclei by DNase digestion and high-salt extraction followed by washing with Triton® X-100, omitting the treatment with RNase. The crude nuclear matrix prepared by this procedure was associated with the nuclear RNP network. To separate this network from the nuclear skeleton, Berezney[39] disrupted the crude nuclear matrix fraction by sonication and separated the RNP network from the nuclear skeleton by centrifugation. After the separation and additional digestion with DNase, the nuclear skeleton (matricin) was composed only of the three primary polypeptides of the crude matrix fraction. Two-dimensional gel electrophoresis of the proteins of the crude matrix fraction has shown that each of the three primary polypeptides is composed of multiple-charge isomers.[40] In a two-dimensional gel electrophoresis system the molecular weights of these three primary polypeptides of rat liver nuclear matrix were determined to be 65,000, 70,000, and 73,000.[40]

It is difficult to draw a border between the proteins of the nuclear RNP particles and the matrix proteins. It is even possible that such a border may be artificial and that nuclear matrix and nuclear RNP network represent one functional structure. In addition to the already-discussed results from the laboratory of Pogo,[21] Berezney,[39] and Peters and Comings,[40] the same conclusion is strongly suggested by the work of Herman and co-workers.[22] Treatment of HeLa cell nuclei by DNase and ammonium sulfate removed from the residual nuclei, in addition to chromatin, some of the proteins characteristic of Samarina particles. Simultaneous treatment with DNase and RNase followed by ammonium sulfate extraction removed from the residual nuclei all the proteins of Samarina particles. However, many of the proteins which are found in Jacob particles and are larger than 43,000 daltons were left associated with the residual nuclear structures despite the RNase treatment (Figure 4).[22]

Close relationship between the nuclear matrix and hnRNP particles is also evident from the firm attachment of hnRNA to nuclear matrix, from which it can be released only by procedures degrading the nuclear matrix.[47,48]

In a study of the attachment of hnRNA to nuclear residual structures, Herman et al.[22] found that after the disruption of the integrity of the hnRNP fibers by RNases (RNase A 10 μg/mℓ and RNase T_1 10 units/mℓ at 25° for 10 min), approximately 98% of the labeled hnRNA was removed from the nuclei by exposure to high ionic strength and DNase I, but that half of the poly(A) sequences and half of the native double-stranded nuclear RNA remained attached to the residual nuclear structures. The poly(A) sequences were attached to the structures by EDTA-sensitive linkage and could be eluted by 0.01 *M* EDTA in the form of 12 S RNP particles. The amount of double-stranded RNA remaining in the residual nucleus was not altered by washing with 0.01 *M* EDTA.

The conclusion that hnRNA is directly associated with the nuclear matrix is also supported by other experiments. Maundrell and co-workers[45] found in the residual protein matrix, prepared from avian erythroblasts by a modification of the procedure of Berezney and Coffey[38] (with the omission of the ribonuclease digestion), many of the hnRNA sequences including globin premessenger RNA. Two proteins of the hnRNP particles were present: the 73,000-dalton protein could be identical with the poly(A) binding protein of the hnRNP particles reported in the studies of Herman et al.,[22] the molecular weight of the second protein was 43,000. The protein composition of the nuclear matrix remained unchanged

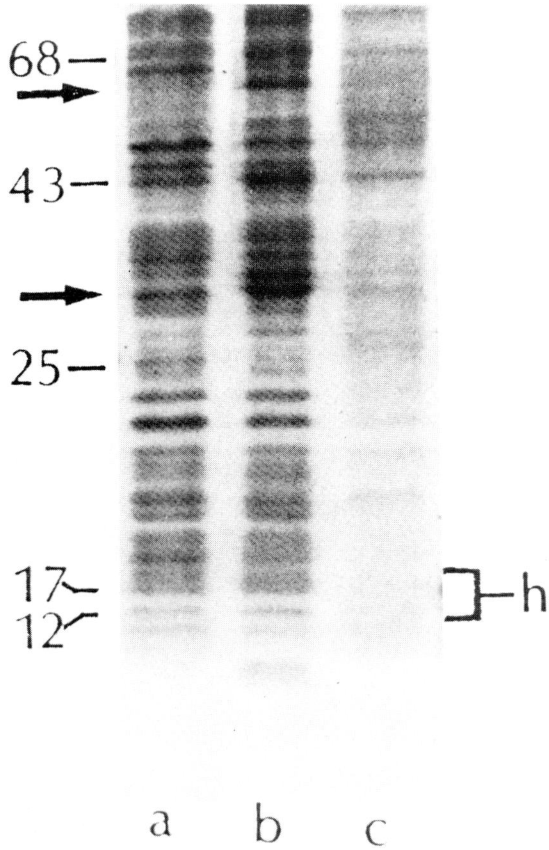

FIGURE 4. SDS-PAGE of proteins of HeLa cell nuclear structures. (a) Proteins of the nuclear lamina prepared by DNase digestion, detergent treatment and ammonium sulfate elution of nuclei, (b) proteins of hnRNP particles, and (c) proteins remaining in nuclei after the digestion with DNase and RNase followed by ammonium sulfate treatment. The arrows indicate the position of the 32,000-, 34,000-, and 63,000-dalton proteins of the hnRNP particles. The numbers indicate the positions of the molecular weight standards. The position where the histones migrate is marked (h). (From Herman, R., Weymouth, L., and Penman, S., *J. Cell Biol.*, 78, 663, 1978. With permission.)

after the RNA was removed by digestion with RNase.[45] The finding of a 43,000-dalton protein in the residual nuclear matrix and in 40 S hnRNP particles is very important if considered together with the results of the following experiments.

Van Eekelen and van Venrooij[48] used cross-linking by UV irradiation of hnRNA to the proteins of nuclear matrix to determine which proteins of the matrix were closely associated with hnRNA in vivo. They irradiated HeLa cells at 254 nm with a 8 $\times$ 10^4 ergs/mm^2 dose and prepared their nuclear matrix by DNase digestion and high-salt extraction. After such treatment, when special care was taken for the used DNase not to be contaminated with RNase, undegraded hnRNA could be quantitatively recovered from the nuclei in the form associated with the purified matrix. After separation from the matrix, the hnRNA had the same sedimentation profile in 10 to 41% isokinetic glycerol gradients as hnRNA isolated from whole nuclei. To distinguish between matrix proteins and other proteins bound to hnRNA in the nuclear RNP network, the purified matrix associated with undegraded hnRNA

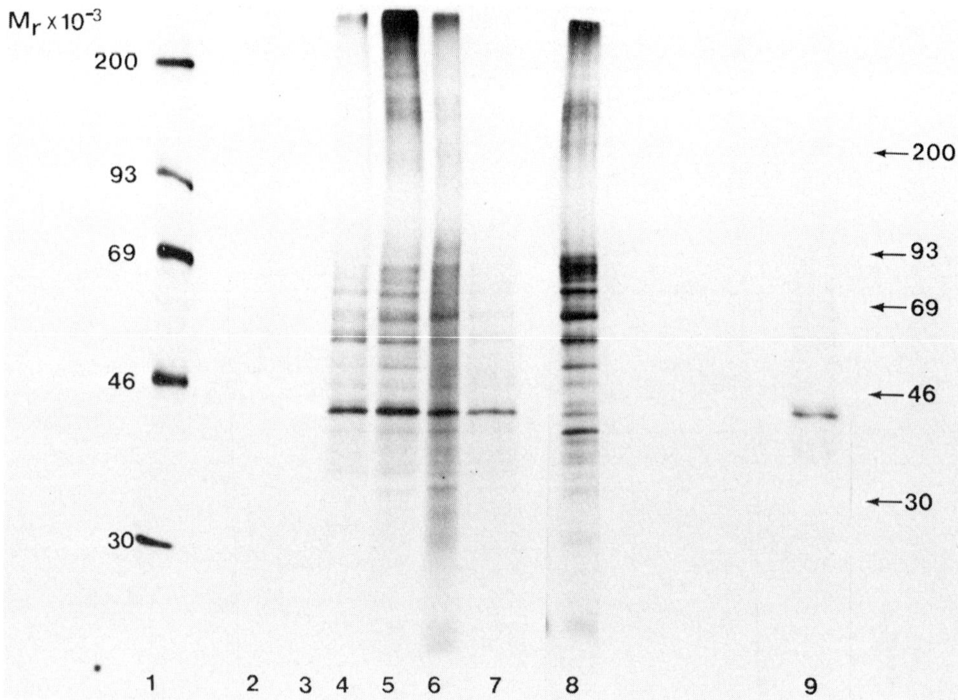

FIGURE 5. Analysis of proteins from HeLa cells labeled overnight with [^{35}S] methionine. The proteins were cross-linked to poly(A)$^+$ and poly(A)$^-$ hnRNA by UV irradiation. The figure shows fluorographs of 10% SDS-polyacrylamide gels: (1) molecular weight marker proteins (for gels 2 to 8); (2) proteins covalently associated with poly(A)$^-$ hnRNA from unirradiated cells; (3) proteins covalently associated with poly(A)$^+$ hnRNA from unirradiated cells; (4) proteins cross-linked to poly(A)$^+$ hnRNA from cells irradiated with UV light for 3 min; (5) proteins cross-linked to poly(A)$^+$ hnRNA from isolated matrices irradiated with UV light for 3 min; (6) as (5), but poly(A)$^-$ hnRNA; (7) as (4), but poly(A)$^-$ hnRNA; (8) total nuclear matrix proteins; (9) proteins covalently associated with poly(A)$^+$ hnRNA from irradiated cells labeled by their residual [^3H] RNA components. Arrows indicate the position of marker proteins for gel (9) only. (From van Eekelen, C. A. G., and van Venrooij, W. J., *J. Cell Biol.*, 88, 554, 1981. With permission.)

was extensively digested with RNase A (100 μg/mℓ for 20 min at 37°). Since the matrix itself is RNase resistant, only proteins associated with hnRNA and not with the matrix proper should be released after this treatment. Only 1 to 2% of the total matrix proteins were released. To determine which proteins were responsible for the attachment of the hnRNA to nuclear matrix, the matrix proteins and hnRNA were cross-linked by UV irradiation of cells in vivo, and also by in vitro irradiation of isolated matrix-hnRNA complexes. By dissociation of these complexes by boiling for 2 min in 1% SDS, and separation of the cross-linked RNPs from noncross-linked proteins by centrifugation in discontinuous sucrose-D_2O-H_2O gradient, two minor matrix proteins (41,500 and 43,000 daltons) were found predominantly cross-linked to hnRNA. A separation of poly(A)$^+$ and poly(A)$^-$ hnRNA by oligo(dT) cellulose chromatography has shown that both proteins were cross-linked to poly(A)$^+$ and to poly(A)$^-$ hnRNA by in vivo UV irradiation as well as by in vitro UV irradiation of isolated matrices (Figure 5). It is of interest that at the dose used, the major core proteins found in hnRNP particles were not covalently linked to the matrix RNA in cells irradiated in vivo.

In more recent experiments, only briefly mentioned in the discussion of their paper, the authors[48] found that at least two core proteins of the hnRNP particles were also cross-linked to hnRNA, although less efficiently than the 41,500- and 43,000-dalton proteins. As is also discussed by van Eekelen and van Venrooij,[48] the 41,500- and 43,000-dalton proteins could correspond to the C-proteins of Beyer et al.[49] The C-proteins are not dissociated from the RNA if the hnRNA particles are treated with 0.4 *M* NaCl and traces of the proteins remain attached to RNA even at 0.6 *M* NaCl concentration.[50] In the van Eekelen and van Venrooij experiments,[48] washing of nuclear matrix with 0.4 *M* $(NH_4)_2SO_4$ during its preparation did not release the 41,500- and 43,000-dalton proteins. It is probable that these proteins are part of the nuclear matrix since extensive RNase treatment of the matrix did not release these proteins. This fully agrees with results published earlier by Augenlicht et al.[51] These authors isolated a complex composed of a 40,000-dalton polypeptide and 26 ribonucleotides after RNase digestion of nuclear RNP polyparticles prepared from human colon carcinoma cell line.

Despite recent progress in the characterization of the nuclear matrix and the nuclear RNP particles, the question of which proteins should be identified with the nuclear matrix or with RNP particles cannot be answered definitively. A product-precursor relationship between the nuclear particles and the RNP network was proposed by Faiferman and Pogo.[21] A similar conclusion, that the nuclear RNPs are intimately attached to the matrix, was also reached by others.[22,40,45,48] Therefore, it is possible that the RNP particles and the nuclear matrix do not exist as defined morphological entities and should be considered only as operational definitions. A few years ago, one leading molecular biologist defined chromatin as anything which he will define as chromatin. Nuclear RNPs, even today, are defined by the decision of individual workers according to the procedure they will use for their isolation. The problem of which proteins represent the RNP proteins will not be solved before the function of individual proteins associated with the nuclear matrix-RNP network are known. Perhaps only the nuclear skeleton scaffold composed of the three polypeptides should be considered the matrix proper: a scaffold which maintains the architecture of the nucleus and on which the other proteins involved in RNA processing (or other nuclear functions) are assembled. In the case of RNA processing, this scaffold could also serve as a track on which processed RNA is transported to the nuclear membrane for release into the cytoplasm. That the core matrix is a structure by which the transcriptional events of chromatin, processing, intranuclear transport, and release of RNA into the cytoplasm via the nuclear pore complex are coordinated was proposed by Berezney.[26]

In conclusion it is possible to say that all studies of the nuclear matrix strongly indicate that nuclear hnRNP particles are subfractions of the hnRNP network which is tightly associated with all or part of the nuclear matrix, and that they originate by the fragmentation of the network in the process of the isolation of the particles. The nuclear hnRNP particles are then in the same relation to the nuclear network as are the microsomes to the endoplasmic reticulum. This proposal is also supported by findings that hnRNP particles and nuclear matrix contain the same species of snRNA (discussed in detail in Section XI).

It is possible that in some specialized types of cells such as newt oocytes, the nuclear RNP network is not associated with the nuclear skeleton,[52] but this could be an exceptional situation. In most cell types, the hnRNA is firmly associated with the matrix, and native hnRNA can be released from the nuclear matrix only by disruption of this structure.[47,48] Because RNase treatment will remove hnRNA from the nuclei without degradation of the nuclear skeleton[45] it is probable that the extraction of hnRNP particles from intact nuclei by isotonic salt at pH 8.0 will achieve better separation of proteins of nuclear skeleton from proteins of the hnRNP network than fractionation of nuclei disintegrated by physical forces.

As *post scriptum* to this section it should be mentioned that the nuclear RNP network is also closely associated with chromatin and therefore some cross-contamination between the proteins of the hnRNP particles and chromatin proteins should be expected.[40,53-57] Furthermore, proteins with mobilities identical to those of chromatin proteins were identified in hnRNP polyparticles.[58]

A breakage of hnRNP fibers during separation of chromatin and hnRNP particles with some of the fibers remaining associated with chromatin through RNA *in statu nascendi* is strongly indicated by the findings that newly synthesized hnRNA can be found immediately associated with chromatin to the same extent as steady-state hnRNA.

Augenlicht and Lipkin[55] separated chromatin from the RNPs by centrifugation in sucrose density gradient after the sonication of nuclei. The presence of all labeled RNA in the chromatin fraction after a 30-sec labeling, and the presence of 60 to 70% of labeled RNA in the chromatin fraction regardless of the length of labeling (if the cells were labeled for more than 2 min), indicates that during the separation of chromatin and hnRNP fraction the nascent RNP fibers could be broken (perhaps by nicking with endogenous ribonucleases), with parts of the RNP fibers remaining bound to chromatin by RNA *in statu nascendi*. The RNA isolated from hnRNP or from the chromatin fraction was smaller than RNA isolated from the whole nuclei. Electron microscopic observations of spread *Drosophila melanogaster*[59] chromatin have shown, after dissolution of the nuclear skeleton by detergent, that the nonnucleolar RNP fibrils remained at one end, bound to chromatin.

Similar results were obtained by Herman et al.[22] When HeLa cell nuclei were treated by DNase at low salt concentration (0.11 *M* NaCl) a subsequent washing with 0.4 *M* $(NH_4)_2SO_4$ released approximately 96% DNA from the nuclei. After digestion and elution of chromatin from the nuclei about the same amount of labeled RNA (from 60 to 73%) remained associated with the nuclear fraction regardless of the length of the pulse of [^{3}H] uridine which was used for RNA labeling (pulses of 15 to 210 sec were used). The same amount of labeled RNA was also found associated with the chromatin-free nuclear fraction after steady-state labeling for 3 hr. The association of hnRNA pulse-labeled for 15 sec with the same nuclear fraction indicates that this association must occur while the molecules are still nascent. The binding of hnRNP with nuclear structures remaining after the removal of chromatin probably reflects the actual state found in the cell in vivo. Nonspecific association through electrostatic interactions seems very unlikely in view of the very high ionic strength used to prepare the nuclei depleted of chromatin, and only a small amount of hnRNA becomes associated with the nuclear ultrastructures when purified hnRNA is added to the nuclei during the removal of chromatin.

The direct association of nuclear network with diffuse chromatin is also indicated by the appearance of such structures in electron micrographs of achromatinic nuclei.[25] After the removal of chromatin, the nuclear RNP network is visible as electron-dense material arranged in zigzag (flexons) configuration, with connection to the nuclear complexes clearly visible. In the cell nucleus the chromatin nucleosomes are coiled into a 20- to 30-nm-diameter chromatin cable[60] which is stabilized by interaction between H1 histones.[61] Spaces of 19 to 25 nm observed along the heavier electron-dense structure of the achromatinic nuclei could be spaces left after the removal of the chromatin cable. Close association of the RNP-nuclear matrix complex with chromatin was also confirmed in additional experiments in which the use of high-salt extraction was avoided during the removal of DNA from nuclei.[47]

When nuclei from differentiated or undifferentiated Friend erythroleukemia cells were lysed with Triton® X-100-saponin mixture (final concentration, 0.25% each), and DNA removed from the nuclei by digestion with 500 μg/mℓ of DNase I in low (0.1 *M*) salt concentration buffer, 95% of DNA was removed after 60 min or 120 min incubation at 10°. The time of incubation was dependent on the extent of condensation of chromatin and was longer for differentiated cells. Under these conditions of DNA digestion, histones (with the exception of H_1 and $H_1{}^o$ histones) and most of the nonhistone proteins were found associated with the nuclear matrix, together with the RNPs (Figure 6). The hnRNA extracted from these DNA-depleted nuclei had the same sedimentation velocity in a formamide-sucrose gradient as hnRNA extracted from the whole cells (the size of RNA was not given). The DNA-depleted nuclei from differentiated and undifferentiated cells preserved the topographical distribution of the condensed and diffused regions. Heparin, despite its strong anionic

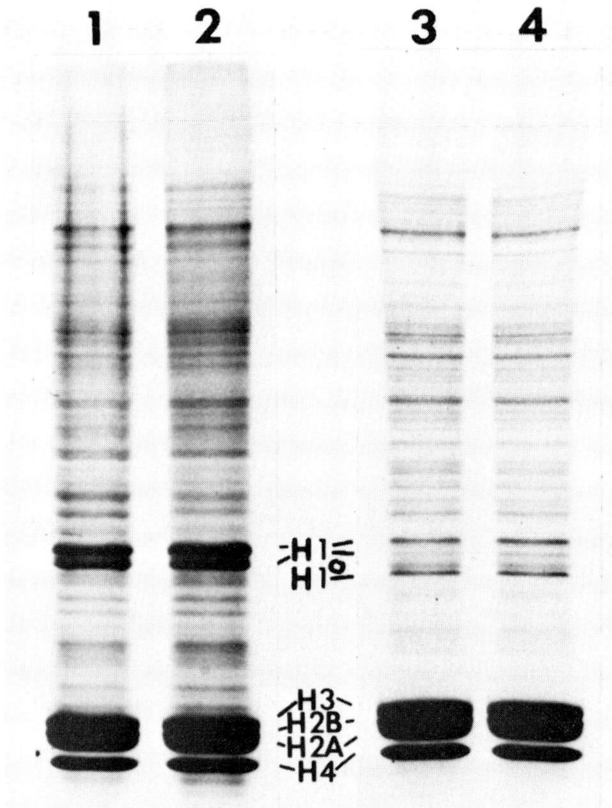

FIGURE 6. SDS-PAGE of proteins of DNA-depleted nuclei. The nuclei were isolated from Friend erythroleukemia cells. Lanes 1 and 2: proteins of nuclei from undifferentiated cells. Lanes 3 and 4: proteins of nuclei from differentiated cells. The nuclei analyzed in lanes 2 and 4 were exposed to hypotonic buffer to release histones aggregated onto the matrix. (From Long, B. H., Huang, C.-Y., and Pogo, A. O., *Cell*, 18, 1079, 1979. With permission)

character, was unable to compete with the matrix for core histones. These findings, together with the fact that H1 histones were not found associated with the matrix, speaks against the possibility that binding of histones to nuclear matrix is an artifact. Furthermore, experiments in which nonhistone proteins and histones were cross-linked in normal and DNA-depleted nuclei favored the conclusion that there are proteins in the matrix which interact directly with nucleosome proteins.[63] The authors concluded[47] that the thicker chromatin fibers are interwoven with the thinner matrix fibers, forming one complex. At high-salt concentrations the matrix fibrils, associated by nonionic bonds with the RNP complex, are pulled together with this complex from the unwinding chromatin. In the chromatin-nuclear matrix-RNP supercomplex the histones are bound to the matrix through ionic bonds or saline-labile elements. HnRNA is associated with the matrix fibrils by hydrophobic and hydrogen bonds. The same bonds are involved in the maintenance of the matrix structure. Breakage of these bonds will also produce a lysis of the matrix.[25,32] Especially tight attachment between chromatin and nuclear matrix was shown in a tissue with active DNA replication. Smith and Berezney[63] found that nuclear matrices in actively replicating regenerating liver, but not in normal liver, contain a significant amount of DNA polymerase α, an indication that functional replicational complexes are assembled on the matrix.

The association of newly synthesized RNA with chromatin has to be expected since chromatin is the site of RNA synthesis. Nascent RNA is quickly covered with proteins. Electron microscopic evidence using a variety of cytochemical techniques has shown that RNA synthesized on DNA loops on chromomeres of lampbrush chromosomes is associated with proteins. The nascent RNA molecules appear to be considerably shortened within the protein coats.[64] Electron micrographs of HeLa cell genome dispersed by detergent have RNP attached to chromatin fibrils with RNP granules appearing at the end of the nascent fibrils.[64] Nascent RNP must also be immediately associated with the nuclear matrix. This is indicated by the already discussed findings that newly synthesized hnRNA is associated with nuclear matrix to the same extent as steady-state RNA.

Because of the close spatial and functional interactions of nuclear structures which play a role in transcription, processing, and transportation of the genetic information, deciding which proteins should be considered nuclear RNP is difficult and will continue to be somewhat arbitrary.

III. THE REALITY OF NUCLEAR RIBONUCLEOPROTEINS

In 1966 Girard and Baltimore showed that RNPs are formed when RNA is added to nuclear cytosol.[65] Artificial RNA-protein complexes are also formed by mixing hnRNA with nuclear proteins.[66] It was therefore important to show that nuclear RNP complexes are real.

One of the criteria is the stability of RNP complexes. Artificial RNA-protein complexes are much less stable if treated with monocationic salts than native RNP complexes isolated from the cytoplasm[14] or from cell nuclei.[66] The association of hnRNA proteins in vivo has also been demonstrated in numerous electron microscopic studies.[18,67-71] Another strong argument against the formation of nuclear RNP complexes during the fractionation of the cell nuclei is the result of experiments in which labeled hnRNA was added to nuclei prior to the isolation of the hnRNP particles. The added RNA was not recovered in the form of RNPs.[72-74]

To rigorously exclude the possibility of macromolecular rearrangement or any nonspecific RNA-protein association during cell fractionation or during the preparation of cell specimens, Mayrand and Pederson[75] studied the photochemical cross-linking of nuclear RNA to the proteins in intact HeLa and Friend erythroleukemia cells. The formation of covalent bonds between nucleic acids and proteins was photo-induced only when the reacting amino acids and nucleotides were one bond-length apart. Therefore, the reaction will link molecules which are present in the cell in very tight complexes. Hence, the authors prepared hnRNP particles from nuclei of HeLa and mouse erythroleukemia cells which were irradiated with 254-nm UV light at doses of 10^1 to 10^5 ergs/mm^2. From nuclei disrupted by mild sonication, the hnRNP particles were isolated on sucrose gradients after the removal of nucleoli by low-speed centrifugation. The isolated polyparticles sedimented between 40 and 250 S in 15 to 30% linear sucrose gradient. The recovery of hnRNP proteins from nuclei in the form of polyparticles by the procedure used was not affected by cross-linking. The polyparticles isolated from nuclei after cross-linking of the RNA to proteins by UV radiation had the same resistance to digestion by pancreatic RNase and the same RNA:protein ratio (determined by Cs_2SO_4 banding) as native hnRNP particles. The native and cross-linked particles were also equally digested by endogenous nuclease associated with the hnRNP particles. After a 2-hr incubation at 37°, 22% of the particle RNA digested to acid solubility in both types of particles. These results rigorously established that hnRNP particles isolated by nuclear fractionation represent native structures.

IV. ISOLATION OF NUCLEAR RIBONUCLEOPROTEIN PARTICLES CONTAINING HETEROGENEOUS NUCLEAR RNA

All procedures used for isolation of the hnRNP particles start with purified nuclei. This minimizes the contamination of hnRNP particles by cytoplasmic proteins and ribosomes.

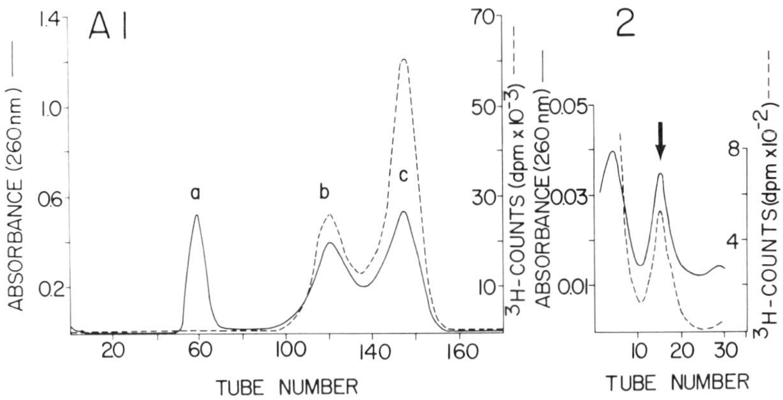

FIGURE 7. Purification of the hnRNP particles from rat liver nuclei. (A1) elution of the 0.1 *M* NaCl, pH 8.0, extract of nuclei from Bio-Gel® A-50 m; (2) sedimentation of the RNP particles eluted in the second (b) peak from Bio-Gel® A-50 m in 15 to 30% sucrose gradient. The animals were labeled for 30 min with [^{3}H] orotic acid. (From Patel, N. T. and Holoubek, V., *Biochem. Biophys. Acta,* 474, 524, 1977. With permission.)

Because, in contrast to cytoplasmic ribosomes, the nuclear hnRNP particles are resistant to EDTA treatment and removal of Mg^{2+},[73] ribosomes attached to nuclear membrane can be removed by washing nuclei with EDTA. Use of detergents for the removal of nuclei-attached ribosomes[21] is not recommended since it could activate nuclear ribonucleases causing extensive degradation of particle RNA.[76]

There are two procedures used for the isolation of hnRNP particles. One is an extraction procedure which leaves nuclei intact, the other releases hnRNP particles by nuclear disruption. The first method, based on the original Samarina procedure,[12] is widely used in many laboratories. To remove preribosomal particles, nuclei are first extracted with isotonic salt buffer at pH 7.0. Then the hnRNP particles are released by extraction with the same buffer at somewhat elevated pH. The slightly alkaline pH facilitates dissociation of RNPs from chromatin. Usually, a buffer of pH 8.0[12] is used, but extraction of nuclei with pH 9.0 buffer has also been reported.[77] The extraction of hnRNP particles from intact nuclei depends on nicking the hnRNP by nuclear RNases; from cells with low levels of nuclear RNase activity the particles can be prepared only by extraction at higher temperatures.[49,73,77-79]

RNP particles released from the nuclei are usually purified by centrifugation in a sucrose density gradient.[12] It is also possible to purify the hnRNP particles by gel filtration,[80] or both procedures can be combined[81] (Figure 7). Furthermore, the hnRNP particles can also be precipitated by addition of ammonium sulfate to 50% final concentration.[82] At this concentration, ammonium sulfate will not dissociate the glycine-rich proteins of Samarina particles, perhaps because the dehydration of the particles by SO_4^{--} ions increases the hydrophobic interactions between the particle macromolecules. This notion is supported by findings that the buoyant density of noncross-linked hnRNP particles can be determined by their centrifugation through a preformed Cs_2SO_4 gradient,[74] but that cross-linking is required when a CsCl gradient is used.

If rat liver cytoplasmic RNase inhibitor is added to the extraction buffers the RNP particles can be extracted from nuclei in the form of large (30 to 300 S) RNP complexes[12] (Figure 8). Ducamp and Jeanteur,[83] who were first to study hnRNP particles isolated from HeLa cells, observed that reduction of time required for preparation of the nuclei will increase the sedimentation coefficient of RNP particles extracted without the use of rat liver cytosol inhibitor. However, a complete inhibition of nuclear RNase is probably not desirable, since the isolation of nuclear hnRNP particles requires that they be released from RNA associated

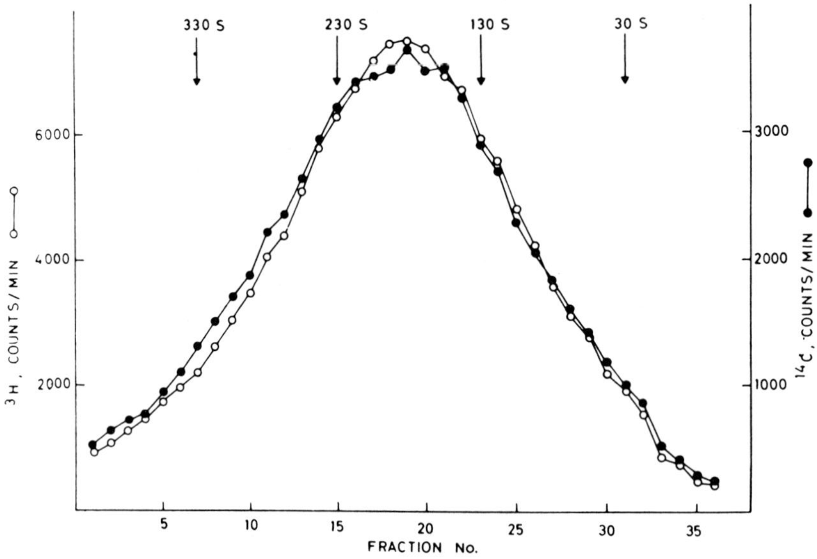

FIGURE 8. Sedimentation profile in 15 to 30% sucrose gradient of hnRNP polyparticles extracted from rat liver nuclei in the presence of rat liver cytosolic RNase inhibitor. The hnRNA was labeled in vivo with [^{3}H] orotic acid for 4 min (o——o) and with [^{14}C] orotic acid for 4 hr (●——●). (From Niessing, J. and Sekeris, C.E., *Biochem. Biophys. Acta,* 247, 391, 1971. With permission.)

with other nuclear structures such as RNA polymerase and DNA template on one side and the nuclear matrix on the other. It is conceivable that limited activity of endogenous ribonucleases is required to release hnRNP particles from these structures.

Release of RNP particles was also described after incubation of isolated nuclei in sucrose solutions in vitro[84-87] or by perfusion of immobilized nuclei with 0.25 *M* sucrose buffer pH 7.4.[88] Such release was energy and temperature dependent, and RNPs were released as polydispersed particles with a majority of them sedimenting with a sedimentation coefficient around 30 to 40 S. This probably represents nuclear cytoplasmic transportation of mRNA and not the release of nuclear hnRNP particles. Only a small fraction of the pulse-labeled nuclear RNA was released in the reported experiments, and the extent of the RNA release was dependent on the source of cytosol. For example, Schumm and Webb[84] reported the release of 6% of pulse-labeled RNA when nuclei from regenerating rat liver were incubated with homologous cytosol at 30° for 30 min, this amount decreased to 50% or even less, if the same nuclei were incubated with cytosol from control rat liver. Patel et al.[87] observed, after incubation in vitro of normal rat liver nuclei suspended in 0.25 *M* sucrose in the presence of homologous cytosol at 37°C for 30 min, a release of 4% of the pulse-labeled nuclear RNA. The RNA was released in the form of RNP particles with a mean sedimentation coefficient of 30 to 35 S[87] (Figure 9). In both experiments the buoyant density in CsCl of the released RNP particles was 1.47 g/cm^3, which is characteristic for cytoplasmic messenger RNP particles (informosomes).[89,90] Because of the low yield of the RNP particles released in the in vitro incubation system, no electrophoretic characterization of particle proteins was given in any of the discussed reports.

A slightly higher release of RNP particles from rat liver nuclei was obtained by Ishikawa and co-workers[91] who used a different system for the incubation of nuclei in vitro. In their experiments the nuclei were incubated with 0.88 *M* sucrose in the presence of rat liver cytoplasmic RNase inhibitor, but in the absence of rat liver cytosol. The pH of the incubation mixture varied from 6.0 to 8.0. The release was energy and temperature dependent. After

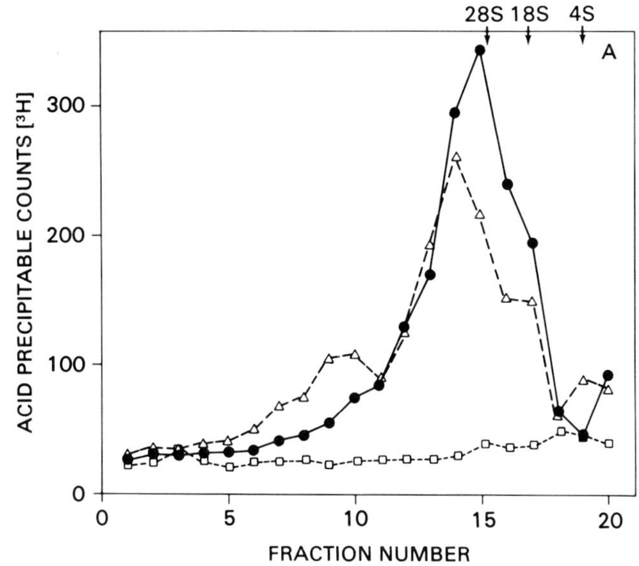

FIGURE 9. Size distribution of labeled ribonucleoprotein particles released in vitro from isolated rat liver nuclei prelabeled with [^{3}H] orotic acid. Particles released after 20 min (●——●) and 60 min (△ - - △)of incubation at 37°C and after 60 min of incubation at 0° (□ - - □). (From Patel, N. T., Yoshida, M., and Holoubek, V., *Cancer Res.*, 41, 743, 1981. With permission.)

a 90-min incubation of nuclei at 20°, approximately 15% of the total RNA labeled with a 40-min pulse of [^{14}C] orotic acid was released. The sedimentation coefficient of the released particles was 45 S, the buoyant density in CsCl approximately 1.40 g/cm^3 and the RNA-to-protein ratio (determined by chemical analyses), 1:5.6. The released 45 S RNP particles were similar in their total amino acid composition to the major proteins of Samarina particles (high content of glycine, glutamic and aspartic acid, or of their amides). Unfortunately, the electrophorogram obtained by analysis of the particle proteins on one-dimensional polyacrylamide gel in alanine-acetate buffer pH 4.1 and without β-mercaptoethanol, was difficult to evaluate. The bands, all close to the top of the gel, could represent aggregates of proteins.[92] Since the incubation system of Ishikawa and co-workers[91] utilized no spermidine or cytosol (under such incubation conditions Schumm and Webb[84] observed 72% lysis of nuclei), and no data were provided concerning the intactness of the nuclei during incubation, it is possible that this procedure involves disruption of the nuclei and not the extraction or release of RNP particles. Because of the relatively low yields of the particles, and the long time and high temperatures involved in particle release, this reviewer does not recommend this procedure for the isolation of nuclear hnRNP particles.

The other methods used for the isolation of hnRNP particles are based on their extraction from disrupted nuclei. The most commonly used procedure is sonication.[53,73,83,93-97] Its advantage is rapid release of hnRNP structures as compared to prolonged time or increased temperature required for isotonic salt extraction. The disadvantage is that nuclear disruption could lead to the breakage of different nuclear structures and to the release of nuclear enzymes, with resulting contamination and damage of isolated nuclear RNP particles. To prevent such side effects, the sonication should be sufficient to break 90 to 95% of the nuclei, but not to break nucleoli. After sonication the nucleoli are removed by centrifugation. Nucleoli and unbroken nuclei are pelleted, and the supernatant which contains chromatin, RNP particles, nuclear membrane fragments, and soluble macromolecules is further fractionated in a linear sucrose density gradient.

For example, Stévenin and Jacob[95] sonicated rat brain nuclei in the presence of rat liver cytosolic inhibitor in 10 m*M* Tris-HCl buffer pH 8.0, containing 0.1 *M* NaCl, 25 m*M* KCl, and 1.0 m*M* $MgCl_2$. The nuclear lysate was centrifuged for 5 min at 100,000 × g and the hnRNP particles were isolated from the supernatant by centrifugation for 3.5 hr at 23,000 r/min in a SW 25-2 Spinco rotor on a 10 to 25% linear sucrose density gradient. Principally the same procedure was used by Ducamp and Jeanteur[83] to prepare hnRNP particles from HeLa cells. As another example, Kish and Pederson[4] and Pederson[73] used different procedures for the isolation of nuclear hnRNP particles from HeLa cell nuclei and rat liver nuclei. The HeLa cell nuclei were disrupted by sonication in low-salt 10 m*M* Tris-HCl buffer pH 7.0 or 7.2, containing 10 m*M* NaCl and 1.5 m*M* $MgCl_2$, then the nucleoli were removed by centrifugation through 30% sucrose in the same buffer (4500 × g, 15 min) and the hnRNP particles were isolated by centrifugation on 15 to 30% linear sucrose density gradient in the same buffer. Centrifugation at 40,000 × g for 17 hr or at 286,000 × g for 2.5 hr was used. Isotonic 10 m*M* Tris-HCl pH 7.0 buffer containing 0.15 *M* NaCl and 1.5 *M* $MgCl_2$ was used to prepare the hnRNP particles from liver cell nuclei. With the exception of this buffer, all other procedures were the same as used for the isolation of hnRNP particles from HeLa cell nuclei.

The extraction of hnRNP particles from sonicated nuclei at low salt concentration is not suitable for all cell types. The hnRNP particles prepared from HeLa cell nuclei as described by Pederson could be quantitatively separated from chromatin by centrifugation through 15 to 30% linear density gradient. Augenlicht and Lipkin,[55] who also used low-salt concentration buffer (the same as Pederson) during sonication of HT-29 human adenocarcinoma cells and during the separation of hnRNP particles by centrifugation through sucrose gradient, obtained hnRNP particles contaminated with approximately 15% of chromatin proteins.

Preparation of nuclear hnRNP particles by extraction of sonicated nuclei with isotonic salt buffer solution yields particles only slightly contaminated with chromatin proteins, as judged by the absence of histones from hnRNP particles isolated in isotonic buffer from sonicated rat liver nuclei by Pederson.[73] Similarly, Gross et al.[98] could detect less than 1% of contaminating histones in the hnRNP particles prepared from rat liver nuclei by a procedure similar to that used by Pederson.[73] Rat brain nuclear hnRNP particles prepared by Gallinaro-Matringe et al.,[99] also by isotonic salt extraction of nuclei disrupted by sonication, were maximally contaminated with 1.4% of nonhistone chromatin proteins and 3% histones.

The lysis of nuclei in high-salt buffer (0.5 to 0.8 *M* NaCl) followed by DNase digestion to reduce the viscosity of nucleohistones (as was first proposed by Penman[37]) is not suitable for the preparation of hnRNP particles. Particles prepared by this procedure are heavily contaminated with histones and other nuclear proteins which are not a part of the hnRNP particles.[73] The extraction of hnRNP particles with buffer containing 0.3 *M* NaCl had already resulted in substantial contamination of hnRNP particles with histones.[73] In addition, higher salt concentrations will release from the hnRNP particles glycine-rich proteins, which are the major proteins of the Samarina particles. Pederson[73] reported a preferential loss of the major polypeptide of the particles. A loss of the same polypeptide was also observed by Faiferman and Pogo[21] after the isolation of hnRNP particles from Krebs ascites cell nuclei by high-salt buffer, 0.8 *M* NaCl-DNase procedure.

However, even low concentrations of salt can result in substantial losses of the major glycine-rich proteins of the particles. Gallinaro-Matringe et al.[99] have shown already that 0.25 *N* NaCl will release about 30 to 40% of proteins from rat brain hnRNP particles, the released proteins belonging predominantly to the glycine-rich class of the Samarina particles. Dissociation of the main proteins of the HeLa Samarina particles in 0.3 *M* NaCl and their nearly complete loss in 0.5 *M* NaCl was described by Beyer and co-workers[49] (Figure 10).

Disruption of the nuclei by detergent is even less satisfactory. The use of sodium deoxycholate[95] leads to the dissociation of particles. Extent of the dissociation correlates

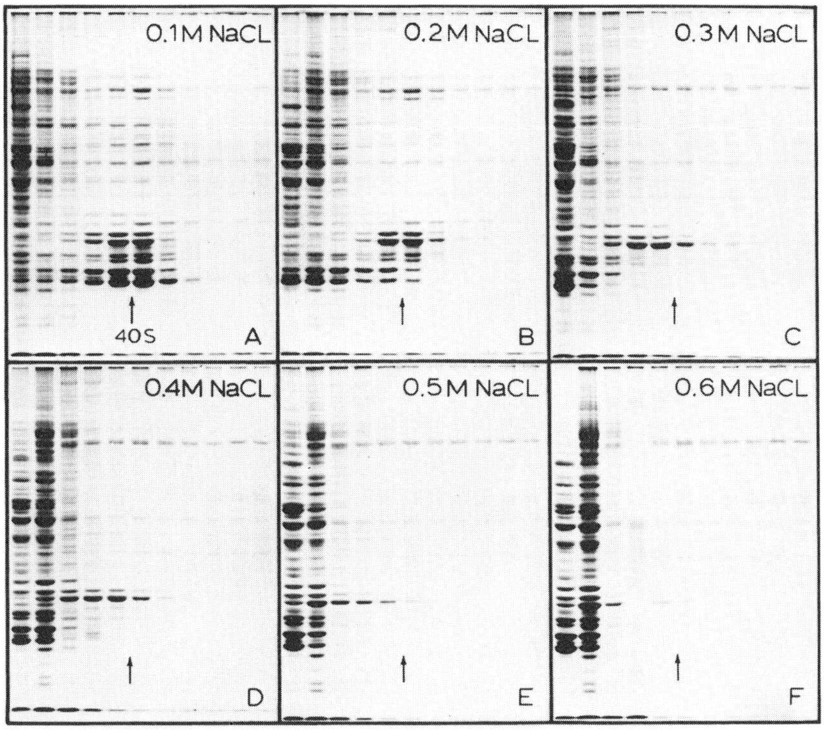

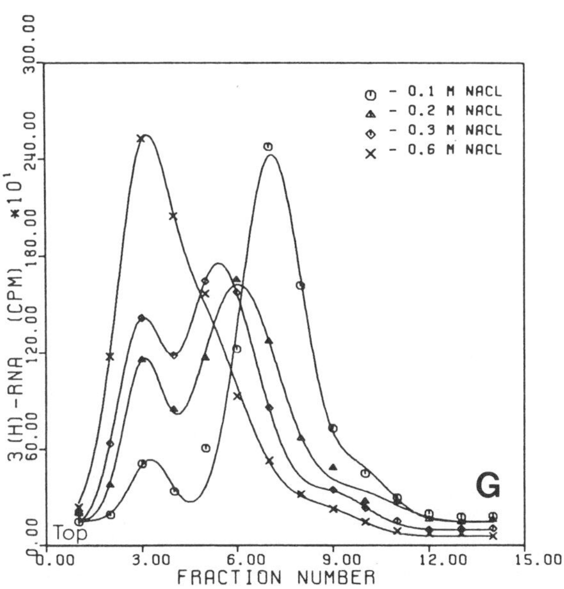

FIGURE 10. Salt dissociation of 40 S hnRNP particles. The particles were isolated from HeLa cell nuclei by extraction with 0.1 *M* NaCl at pH 8.0. The amount of NaCl in the pH 8.0 buffer was increased to the indicated concentration, and the particles were centrifuged in 15 to 30% sucrose gradient. Each gradient was fractionated into 14 equal fractions and the fractions were analyzed by SDS-PAGE (A to F). In a parallel experiment sucrose gradients of increasing ionic strength were fractionated into 14 fractions which were analyzed for [^{3}H]-uridine radioactivity (G). (From Beyer, A. L., Christensen, M. E., Walker, B.W ., and LeStourgeon, W. M., *Cell*, 11, 127, 1977. With permission.)

with the concentration of sodium deoxycholate used.[32,101,102] Faiferman and co-workers[32] suggested that the high sensitivity of particles to sodium deoxycholate treatment indicates that proteins are bound to the RNA mainly by hydrophobic bonds. Triton® X-100 will not cause the disintegration of nuclear hnRNP particles, but the use of this detergent is also not recommended. The catalytic activities of nuclear RNases (and of other enzymes) are enhanced in rat liver nuclei treated with Triton® X-100.[76]

The Samarina particles can also be released by digestion of nuclei with micrococcal nuclease, which cleaves DNA as well as RNA.[103,104] Webb-Walker and co-workers[105] found that after a 2-min digestion at 37° of HeLa cells with this nuclease, the 40S hnRNP particles were released from the nuclei together with nucleosomes. The composition of the 40 S hnRNP particles was the same as those obtained in earlier experiments by the extraction of intact nuclei.[49]

The major conclusions concerning isolation of the nuclear hnRNP particles can be summarized as follows. Since the nuclear hnRNP particles represent structural components of the nuclear RNP network, it is necessary for their isolation to release the network from its association with chromatin and nuclear membranes, and to disrupt the network itself. Shearing forces and nuclear RNases are involved in the formation and release of the hnRNP particles. Faiferman and Pogo[21] have shown that the yields of particles from disrupted nuclei are proportional to the shearing forces applied. Shearing forces are important for the release of hnRNP particles in all procedures which break the nuclei (e.g., sonication or French pressure cell). The extraction of hnRNP particles from intact nuclei depends on the action of nuclear RNases and on the slightly alkaline pH which is required for the release of the particles from nuclear structures. For the extraction of hnRNP particles disrupted by shearing forces, the alkaline pH is not required.[37,53] Since the two types of particles forming the RNP network (Samarina and Jacob particles) contain different sets of proteins and are sensitive to RNase (see later), exposure to endogenous nuclear RNases will preferentially remove Jacob particles from the resulting particle extract. At an appropriate level of digestion only Samarina particles will be obtained by extraction of nuclear hnRNP from intact nuclei or even from nuclei disrupted by sonication.[105] Since the endogenous ribonuclease(s) is(are) associated with proteins of the Jacob particles (see later discussion), only the Samarina particles will be left after repeated purification of the hnRNP particle extract. The Samarina particles do not contain endogenous RNase activity, and after their separation from proteins of the Jacob particles they are stable.[100] However, extensive digestion by RNase will also affect the Samarina particles, stripping them of RNA with the resulting formation of slightly larger particles composed predominantly of proteins.[106] The presence of such particles in the sucrose density gradient peak containing RNP particles can lead to a discrepancy between the RNA-to-protein ratio determined by chemical analysis of all the particles present in the peak and the same ratio calculated from the buoyant density of the RNP particles as determined from the position of labeled RNA in CsCl or Cs_2SO_4 gradient. Enrichment of the extracted hnRNP particles by Samarina particles can also result from an incomplete disintegration of nuclear structures by shearing forces. Under such conditions, more of the heterogeneous Jacob particles than of the Samarina particles will pellet, still being associated with nuclear membrane[21] or chromatin.[55]

V. DISCREPANCIES IN THE REPORTED PROTEIN COMPOSITION OF THE HNRNP PARTICLES

To analyze the protein moiety of rat liver nuclear hnRNP particles, Samarina et al.[12] used PAGE at pH 4.5 in the presence of urea and in the absence of β-mercaptoethanol. The particle proteins separated into three bands in this gel. In later experiments it was shown in the same laboratory that two of the three bands were the result of intramolecular and intermolecular S–S bond formation, and only one band was seen after treatment of the proteins with β-mercaptoethanol.[107] The proposed unique protein of the particle was called informatin.[16]

The notion of a very simple composition for the particle proteins deduced from these results was soon challenged. After analysis of proteins of rat liver hnRNP particles by SDS-PAGE, Faiferman et al.[32] concluded that the protein composition of nuclear hnRNP particles may be more complex than indicated by the simple electrophoretic pattern observed in the acid-urea gels. The results are difficult to compare since Faiferman et al.[32] used a different procedure for the isolation of nuclear hnRNP particles. Niessing and Sekeris,[108] using the same procedure for isolation of rat liver nuclear hnRNP particles as the Samarina group,[12] also demonstrated a more complex composition of the particle proteins by using two different electrophoretic procedures for their analysis: electrophoresis at pH 11.6 in the presence of 4 *M* urea and SDS-PAGE. In the presence of SDS the authors observed three major polypeptides with apparent molecular weights of 36,000, 39,000, and 42,000 and a larger number of heterogeneous peptides ranging in molecular weight from 45,000 to over 200,000 (Figure 11). A similar electrophoretic distribution of particle proteins, but with higher ratio of the high-molecular-weight proteins to the major lower-molecular-weight polypeptides, was obtained by Niessing and Sekeris[109] when they analyzed proteins of the 30 to 180 S polyparticles prepared from rat liver nuclei in the presence of cytoplasmic RNase inhibitor.

Analysis of the particle proteins by SDS-PAGE soon completely replaced the acid-urea gel electrophoresis used earlier,[12,32,91,110-113] especially after Gallinaro-Matringe and Jacob[114] demonstrated that phosphorylated and possibly other acidic particle proteins will not enter the gel at pH 4.5. But the controversy persisted. According to an earlier report by Krichevskaya and Georgiev,[16] the protein of the rat liver nuclear RNP particles migrated even in SDS-PAGE as one band with an approximate molecular weight of 40,000. The authors concluded that hnRNA is combined with this specific type of protein. In a recent publication from the same laboratory, the protein moiety of the 30 S rat liver nuclear hnRNP particle was resolved by SDS-PAGE into two informatin components of very close molecular weights, which comprised over 90% of the total particle proteins.[82] However, finding of only one or at most two components migrating closely together is not without exception even in the Samarina laboratory. Molnar and Samarina[115] reported preparation, by the standard procedure used in the laboratory, of 30 S hnRNP particles from rat liver nuclei containing proteins which separated on SDS-PAGE in a complex pattern of heterogeneous polypeptides ranging in molecular weight from approximately 35,000 to over 150,000.

Only two distinct polypeptides (34,000 and 38,000 daltons) were originally found by Martin and co-workers[116] in the 30 S hnRNP particles prepared by isotonic salt (pH 8.0) extraction. Likewise, only two proteins were identified by Schweiger and Kostka[117] in rat liver and hepatoma hnRNP particles prepared by isotonic salt extraction at pH 7.0 of sonicated nuclei. In these experiments, only 37,000- and 40,000-dalton polypeptides could be eluted with increasing concentrations of NaCl from a column, when 30 S hnRNP particles, disintegrated by spermidine, were chromatographed on polyadenylated Sepharose® gel.

In contrast, most other authors reported heterogeneous polypeptide composition of the hnRNP particles similar to the one originally found by Niessing and Sekeris.[108] For example, Albrecht and van Zyl,[118] who prepared rat liver nuclei particles by the isotonic salt extraction procedure of Samarina,[12] separated proteins of these particles by SDS-PAGE into three major bands with molecular weights between 37,000 and 44,500 and into many minor bands with molecular weights reaching 185,000. With the exception of an apparent decrease in the amount of a 125,000-dalton polypeptide, the same electrophoretic pattern was also seen when the authors analyzed hnRNP particles isolated from hepatoma.

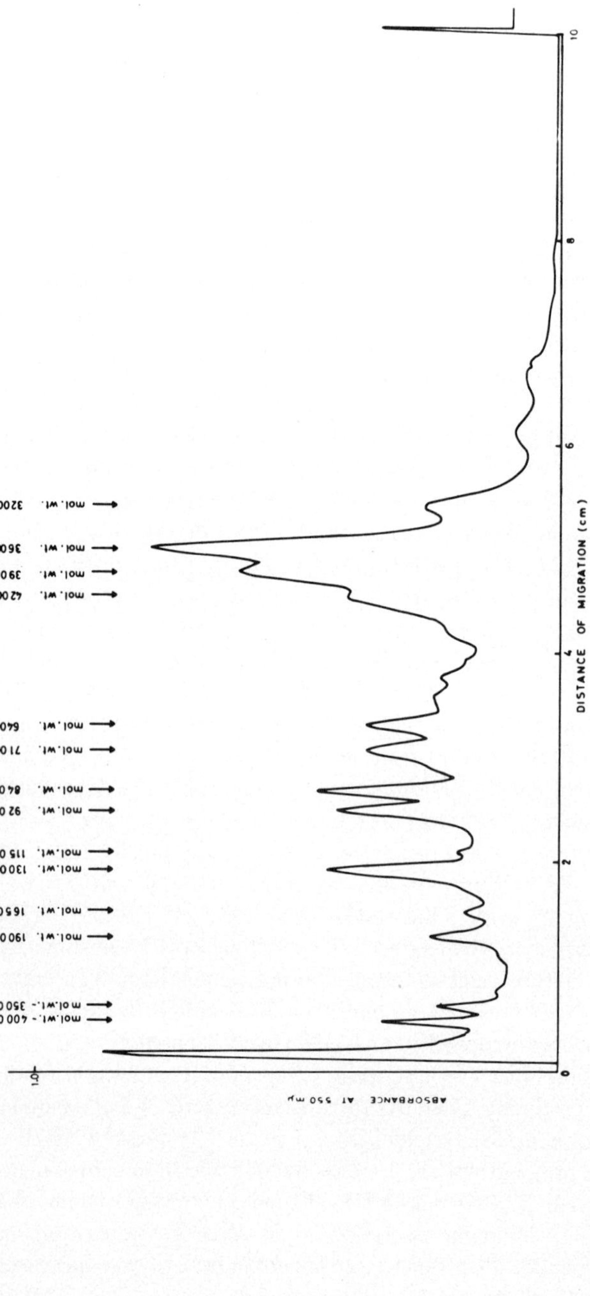

FIGURE 11. Densitometric tracing of a SDS-polyacrylamide gel after the electrophoresis of proteins of the rat liver 30 S hnRNP particles. (From Niessing, J. and Sekeris, C. E., *FEBS Lett.*, 18, 39, 1971. With permission.)

Ducamp and Jeanteur,[83] who first analyzed the protein composition of HeLa cell hnRNP particles, have shown that the particle proteins separated by SDS-PAGE into a complex pattern of proteins ranging in molecular weight from 35,000 to 150,000.

Pederson[73] analyzed RNPs isolated by isotonic salt (pH 7.0) extraction from intact liver cell nuclei and from HeLa cell nuclei disrupted by sonication. One-dimensional SDS-PAGE separated the protein moiety of the liver hnRNP particles into three main bands with molecular weights between 36,000 and 40,000 and a heterogeneous population of minor bands of molecular weights between 40,000 and 130,000. In contrast, the hnRNP particles from HeLa cells showed the presence of only one major band of approximate molecular weight of 40,000, in addition to some minor bands with molecular weights between 40,000 and 130,000. Later, in the same laboratory, the protein moiety of hnRNP polyparticles of Friend erythroleukemia cells was found similar to that of rat liver.[74]

Heterogeneous composition of hnRNP particle proteins was also reported by Jacob et al.[58,119] (Figure 12). The particles were isolated from rat liver or brain nuclei, sonicated in the presence of cytosolic RNase inhibitor, and extracted with an isotonic pH 8.0 buffer. The released polyparticles were further purified by centrifugation through a 10 to 25% linear sucrose density gradient. The particle proteins separated by SDS-PAGE into a cluster of nine to ten bands with molecular weights ranging from 28,000 to 38,000 (α-proteins) and heterogeneous group of bands of higher molecular weight (β-proteins). The results differed from the findings of Pederson[73] by the much larger proportion of the high-molecular-weight proteins in the particle and more complex pattern of the major low-molecular-weight bands. It was obvious from the electrophoretic patterns obtained in the Jacob's laboratory that of the three major proteins bands which migrated with approximate apparent molecular weights 33,000, 35,000, and 37,000, two were not homogeneous, the second and third being a doublet. In addition, a 33,600-dalton band could be identified in the hnRNP particles from brain, which was absent from liver. The proteins migrating with apparent molecular weights ranging from 28,000 to 38,000 were characteristic for the hnRNP particles. In contrast, a majority of the heterogeneous bands migrating with molecular weight over 40,000 had identical electrophoretic mobilities with proteins which were also found in the nucleosol or in chromatin fraction of the nuclei.[58] A very similar protein distribution of the rat liver hnRNP polyparticles on SDS-polyacrylamide gel after electrophoresis was also observed by Northemann et al.,[97] with the difference that the low-molecular-weight main proteins (α-proteins) separated only into four to five peaks. It is of interest that when in the same laboratory the hnRNP particles were prepared in the absence of the cytosolic RNase inhibitor, such monoparticles contained a much lower ratio of the heterogeneous high-molecular-weight proteins (β-proteins) to the low-molecular-weight α-proteins as the polyparticles.[97,100]

The amount of heterogeneous minor high-molecular-weight polypeptides found in the electrophorograms of hnRNP particles is different in monoparticles and in polyparticles. The hnRNP particles isolated from rat liver nuclei in Heinrich's laboratory separated on one-dimensional SDS-PAGE into three major bands with molecular weights between 34,000 and 42,000 and a series of high-molecular-weight minor bands reaching over 150,000. The ratio of protein present in the high-molecular-weight polypeptides to the amount of proteins migrating as the three lower-molecular-weight bands was low in the 38 S RNP particles prepared without RNase inhibitor[98,100] and high in polyparticles prepared in the presence of rat liver cytosolic RNase inhibitor.[97,120] This suggests that in the polyparticles stretches of RNA associated with the major proteins are interrupted by stretches of RNA associated with the minor heterogeneous proteins. Fragmentation of the polyparticles would result in the release of both types of particles.

Patel and Holoubek[81] have shown that the amount of minor high-molecular-weight polypeptides found in the protein moiety of the particle depends on the extent of particle puri-

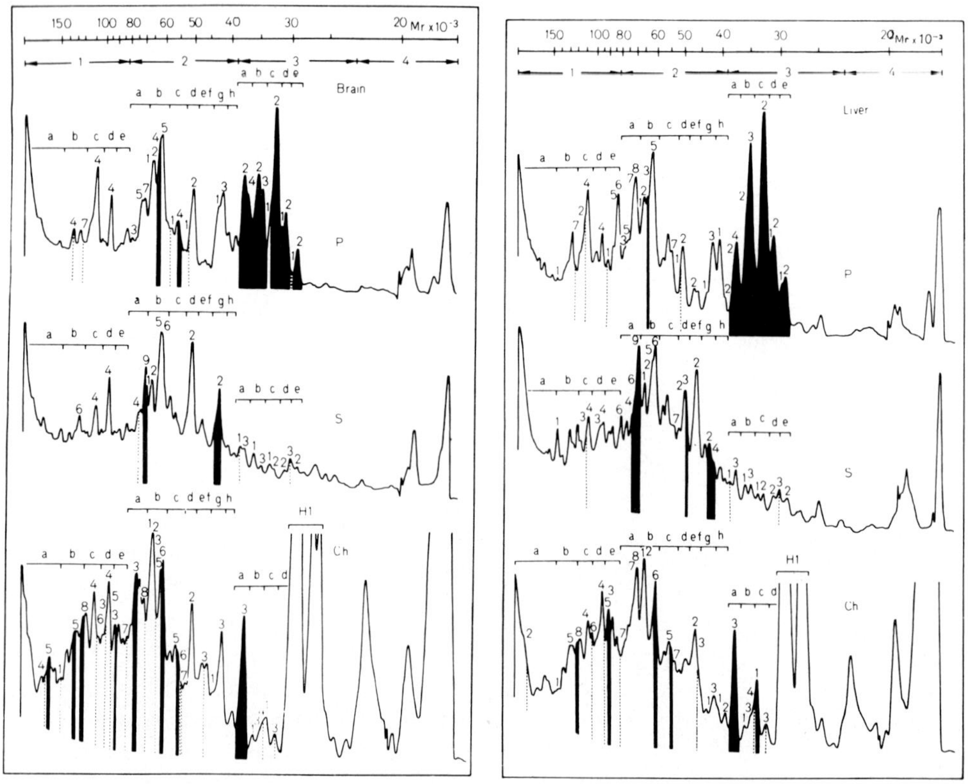

FIGURE 12. Comparison of the proteins from rat liver and brain nuclear fractions. Electrophoretic profiles of the proteins of (P) the hnRNP polyparticles, (S) nucleosol, and (Ch) total chromatin. The peaks of the proteins specific to the hnRNP particles, chromatin, or soluble proteins are either shaded (major proteins) or shown by dotted lines (faint bands). (From Stévenin, J., Gattoni, R., Gallinaro-Matringe, H., and Jacob, M., *Eur. J. Biochem.*, 84, 541, 1978.

fication. The authors isolated the hnRNP particles from rat liver nuclei by isotonic salt, pH 8.0, extraction according to Samarina et al.[12] SDS-PAGE separated the particle proteins into three major bands (38,000, 40,500, and 42,000 daltons) and a series of heterogeneous bands with molecular weights over 42,000. The amount of polypeptides with molecular weights over 42,000 was dependent on the extent of the particle purification. In particles purified by only one centrifugation in a linear 15 to 30% sucrose density gradient the major polypeptides represented about half of the total particle proteins. When the particles were first purified by gel filtration through Bio-Gel® A 50 m followed by centrifugation in the sucrose density gradient, the major polypeptides represented 80% of all the particle proteins (Figure 13). By repeated purification the particles were freed of the heterogeneous proteins of the polyparticles (with molecular weight over 42,000) without any extensive release of the major polypeptides from the Samarina particles.

VI. TWO TYPES OF HNRNP PARTICLES

The most obvious explanation for the differences in particle protein composition reported from different laboratories is the presence of more than one type of particle in the analyzed mixture. This was proposed in 1974 by Pederson, based on his observation of different

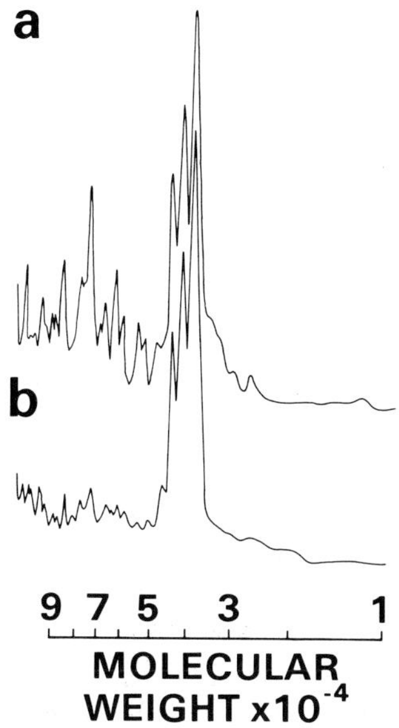

FIGURE 13. Densitometric tracings of SDS-electrophorograms of the proteins of the hnRNP particles from rat liver. The particles were extracted with 0.1 *M* NaCl at pH 8.0 and purified (a) by centrifugation through a 15 to 30% sucrose gradient and (b) by Bio-Gel® A-50 m filtration followed by centrifugation through 15 to 30% sucrose gradient.

levels of stability of HeLa hnRNP particles to exposure to high salt concentrations.[73,121] Recently, in the same laboratory, the existence of multiple classes of hnRNP particles having somewhat different structures was also suggested on the dose-response relationships of RNA-protein cross-linking in vivo.[75] But the existence of more than one type of hnRNP particles is not an adequate explanation for the reported differences in protein composition of the particles. It is still necessary to explain the variation in the proportions of the two or more types of particles in the particle mixture.

Some of the reported discrepancies in protein composition of the nuclear RNP particles can be explained if one considers the results obtained in the laboratory of Jacob.[7,15,95,99,106,123-125] The RNP particles studied in this laboratory were isolated from rat brain nuclei. The brain cells were used because of their low level of endogenous nuclear RNases. The standard method was based on the release of RNP polyparticles from the nuclei after their disruption by sonication in a 10 m*M* Tris-HCl, pH 8.0 buffer (containing 0.1 *M* NaCl, 25 m*M* KCl, and 1.0 m*M* $MgCl_2$) in the presence of a rat liver cytosolic RNase inhibitor. In this preparation the chromatin and nuclei were removed by centrifugation, and the RNP particles, present in the supernatant, were separated from the smaller molecules by centrifugation in a 10 to 25% (w/w) linear sucrose density gradient. Prepared under these conditions, the nuclear RNP polyparticles sedimented in a broad peak with a sedimentation coefficient from 30 to 200 S. In a series of experiments, Stévenin and co-workers studied the effect of RNase digestion on the size, density, and protein composition of the RNP particles.[7,15,95,106] When the RNPs were released by the standard procedure, but in the absence of the rat liver cytosolic RNase inhibitor, about 80% of the labeled, newly synthesized RNA sedimented in the 30 to 50 S region of the gradient, the rest of the labeled RNA sedimented as RNPs, heterogeneous in size but larger than 50 S.[106] These results confirmed findings by Samarina and co-workers[12] and Pederson[73] that monoparticles are released from nuclear polyparticle structures after nicking the RNA by endogenous nucleases. Treatment of the nuclear extracts with a low concentration of RNase (0.1 to 0.5 μg/mℓ, 15°, 20 min), before centrifugation in a linear sucrose density gradient increased the amount of particles sedimenting in the 35 to 45 S region,[15] and from 20 to 40% of the ^{32}P radioactivity was rendered acid-soluble. Upon treatment with a medium concentration of RNase (2 μg/mℓ, 15°, 20 min), 80 to 90% of the RNA was hydrolyzed, and only a small amount of radioactivity sedimented in the 35 to 45 S region or in the higher than 60 S region of the sucrose density gradient. After hydrolysis with 20 μg RNase, 90% of the ^{32}P radioactivity was no longer precipitable with trichloroacetic acid, and no peak could be observed in the 35 to 45 S region.

Determination of the buoyant density in CsCl of the particles recovered from a sucrose density gradient has shown that the RNP particles sedimenting in the 30 to 60 S region of the gradient lost approximately 20% of their RNA after treatment with low concentration of RNase.[15] An unexpected finding was that after RNase treatment the particles which were labeled with ^{32}P showed higher buoyant density (1.380 g/cm^3) than the particles which were labeled with ^3H amino acids (buoyant density = 1.374 g/cm^3). The buoyant density of particles which were not treated with RNase was the same regardless of the type of labeling (1.392 g/cm^3). These differences in the buoyant density were accentuated if the nuclear extracts were treated with 2 μg of RNase. The ^{32}P-labeled 35 to 50 S particles had a buoyant density of 1.373 g/cm^3, as compared with only 1.357 g/cm^3 of the 35 to 50 S particles labeled with ^3H amino acids. This represents an approximate loss of 40 to 50% RNA. These results indicated that the particles sedimenting at 35 to 50 S were composed of two types of moieties, one type containing metabolically more active proteins and more sensitive to digestion with RNase. There was no difference in the buoyant densities of ^3H- and ^{32}P-labeled RNP particles in complexes larger than 90 S. The observed buoyant density of 1.354 g/cm^3 indicated that the larger than 90 S complexes have the same sensitivity to RNase as the particles containing metabolically more active proteins and sedimenting in the 35 to 50 S fraction of the sucrose gradient (Figure 14). Some of the RNA which sedimented with the larger than 60 S fraction of the gradient was protected from digestion by RNase and not removed after digestion with 20 μg/mℓ RNase at 15° for 20 min.[15]

Protein analysis of the particles by one-dimensional SDS-PAGE supported the concept of different types of RNP particles being present in the nuclear extract. The particles which sedimented in the 35 to 45 S region of the sucrose density gradient were rich in 28,000- to 38,000-dalton proteins.[15] To this group belong specific glycine-rich proteins present in the particle in many predominantly basic charge isomers.[8] Using pulse labeling with [^{3}H]tryptophan, Augenlicht and Lipkin[55] have shown that the majority of these proteins is metabolically relatively stable, whereas the heterogeneous fraction of the high-molecular-weight proteins turns over rapidly. The high metabolic stability of the 35,000- to 38,000-dalton proteins (they have slower turnover rates than H1 histone) was also confirmed by Martin et al.[122] The differential turnover of the lower and higher molecular weight particle proteins nicely complements the data of Stévenin et al.,[15] who showed the differential resistance to RNase of RNP particles labeled with [^{32}P] and with [^{3}H] amino acids. The heterogeneous complexes which sedimented with a coefficient higher than 60 S were predominantly composed of high-molecular weight-proteins and sensitive to RNase. Extensive RNase hydrolysis led to a decrease in the amount of the 35 to 45 S monoparticles with the protein moiety of the remaining monoparticles composed almost exclusively of the 28,000- to 38,000-dalton proteins[15] (Figure 15).

How does one interpret the results obtained after further RNase digestion of the 30 to 50 S monoparticles? Stévenin and co-workers[106] first proposed that two types of proteins were present in the 30 to 50 S monoparticles, the 28,000- to 38,000-dalton polypeptides (α-proteins), which form particles relatively resistant to hydrolysis by RNase, and the higher-molecular-weight, very heterogeneous proteins readily released upon RNase hydrolysis (β-proteins). The authors considered two possibilities for the arrangement of these proteins in the polyparticles. The polyparticles could be composed of two classes of monoparticles

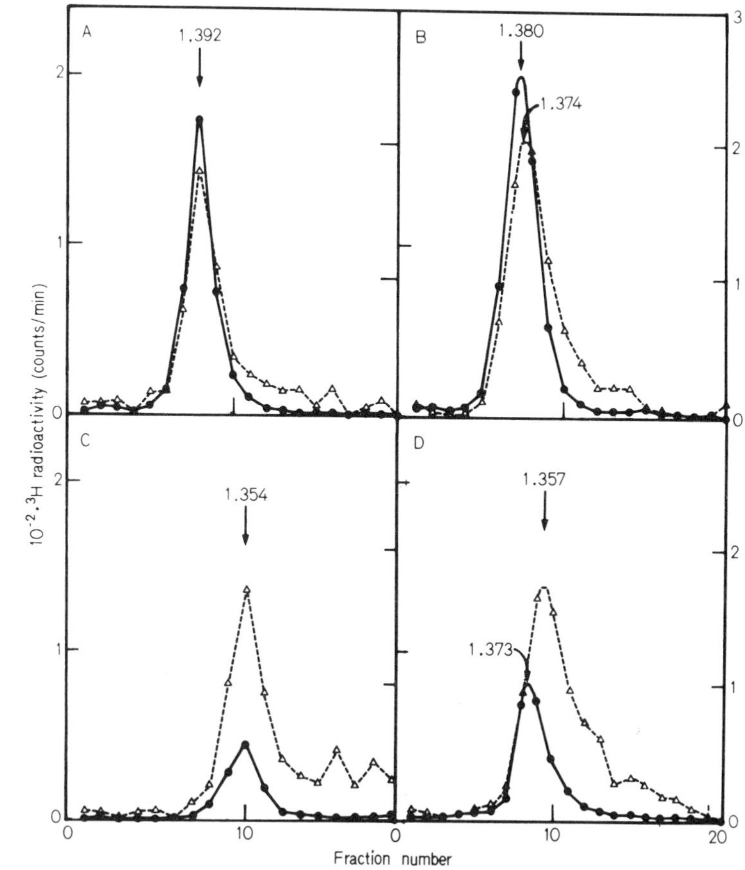

FIGURE 14. Buoyant densities of rat brain hnRNP polyparticles in CsCl density gradient. The particles were double-labeled in vivo for 4 hr with [^{3}H] amino acids and [^{32}P] orthophosphate: (A) 60 to 80 S control particles; (B) 50 to 60 S particles obtained after the treatment of the polyparticles with 0.1 μg/mℓ of RNase; (C) 90 to 120 S particles obtained after the treatment of the polyparticles with 2 μg/mℓ of RNase; (D) 35 to 50 S particles obtained after the treatment of the polyparticles with 2 μg/mℓ of RNase; (△ - - △) = ^{3}H; (●——●) = ^{32}P. (From Stévenin, J., Gallinaro-Matringe, H., Gattoni, R., and Jacob, M., *Eur. J. Biochem.*, 74, 589, 1977. With permission.)

(designated by the authors as Mα and Mβ monoparticles). As both the α and β proteins were found sedimenting in a range between 30 and 50 S,[123] the authors assumed that both types of particles were heterogeneous in size. It was also possible that there was only one type of monoparticle sedimenting between 30 and 50 S which was composed of a resistant core (composed of α- proteins) surrounded by β-proteins. To choose between these two possibilities, Stévenin and co-workers[106] hydrolyzed the 30 to 50 S monoparticles by RNase under conditions which would release only the more sensitive β-proteins (or destroy only the particles composed of β-proteins). It could be expected that if the same particles were composed of the RNase-resistant core of α-proteins associated to a different extent with β-proteins present in an arrangement which is sensitive to treatment with RNase, homogeneous and smaller particles composed only of α-proteins should be found after incubation with low concentration of RNase. Or, if two types of particles (Mα and Mβ), both heterogeneous in size, coexisted, particles with the same sedimentation coefficient as before but containing only α-proteins should be recovered after RNase hydrolysis. In the experiment, the broader peak of monoparticles was separated into three fractions containing approximately 30, 40,

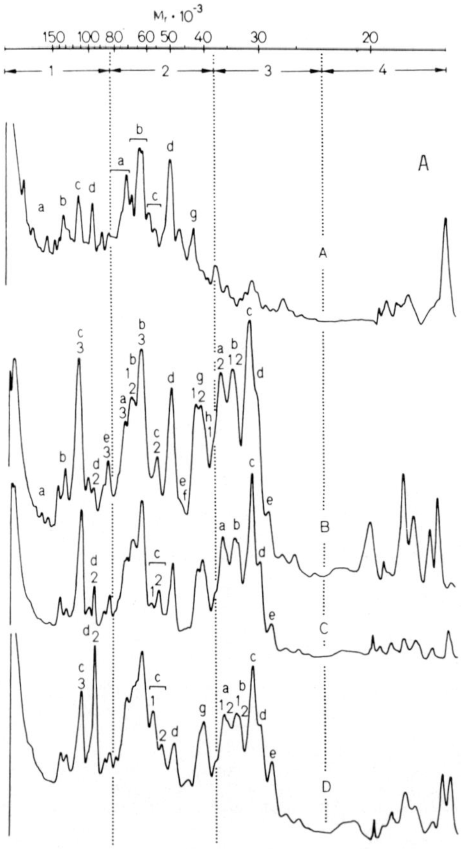

A

FIGURE 15. Densitometric tracings of SDS-polyacrylamide gel electrophorograms of proteins of rat brain hnRNP particles and soluble nuclear proteins: (a) control particles, (b) particles treated with 0.5 μg/mℓ of RNase, and (c) particles treated with 2 μg/mℓ of RNase. (A) = soluble nuclear proteins, (B) = 35 to 45 S hnRNP particles, (C) = 60 to 80 S hnRNP particles, and (D) = 130 to 190 S hnRNP particles. (From Stévenin, J., Gallinaro-Matringe, H., Gattoni, R., and Jacob, M., *Eur. J. Biochem.*, 74, 589, 1977. With permission.)

and 50 S monoparticles, and each fraction was digested with 0.25 μg/mℓ RNase for 20 min at 15°. About 70 to 80% of particles in each group were destroyed by such treatment, with the remaining particles in all three fractions sedimenting with a 40 to 45 S sedimentation coefficient in the sucrose gradient. In other words, there was a substantial increase in the sedimentation coefficient of the 30 S particles and a small increase in the sedimentation of the 40 S particles. Repeated centrifugation of the 50 S particles in the sucrose density gradient, without treatment with RNase, decreased the sedimentation coefficient of the 50 S particles to between 40 to 45 S. These 40 to 45 S particles were very stable, and only minimal changes in their sedimentation coefficient were observed as the result of RNase treatment. The shifting in the sedimentation coefficient occurred gradually and was related to the extent of RNA hydrolysis. The 30 S monoparticles shifted to 36 S after the digestion of 60% of particle RNA and to 41 S after the digestion of 85% of particle RNA. When the

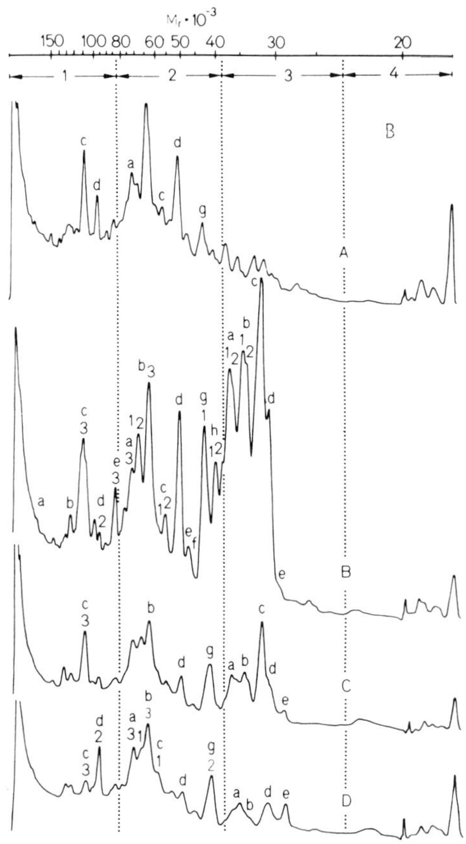

FIGURE 15B

protein moieties of the 30, 40, and 50 S fractions of the polyparticles were analyzed by SDS-PAGE all contained the proteins of the α and β class, but the proportion of both types of proteins was dependent on the sedimentation coefficient of the particles. The 50 S particles contained a higher proportion of the β-proteins. Hydrolysis by RNase released from all three groups of particles the β proteins (with some proteins which could belong to the α-class). The resulting 40 to 45 S particles contained almost exclusively α-proteins. Based on these results, Stévenin and co-workers concluded[106] that the 40 to 45 particles composed of α-proteins do not preexist in vivo and are likely to be a product of rearrangements in vitro in response to the treatment with RNase. The gradual increase in the sedimentation coefficient of the particles observed after increasing removal of RNA from the particles by ribonuclease is related to the progressive availability of new sites for protein-protein interaction after the removal of RNA.

The warning of Stévinin and co-workers[106] that the Samarina particles could be the result of the rearrangement of nuclear RNPs during the particle isolation deserves very serious consideration. Particles with a sedimentation coefficient varying between 30 and 45 S and isolated from various organs are often considered as the major or even the unique constituent of the heterogeneous complexes of polyparticles. Since these particles are often isolated from cells with a high endogenous RNase activity (e.g., liver), or by extraction at a higher temperature it is possible that they could be formed or at least rearranged as a result of the action of endogenous RNases.

The existence of RNA-free Samarina particles is a matter of controversy. In 1972 Lukanidin and co-workers[79] published their findings that Samarina particles (called "informofers" in

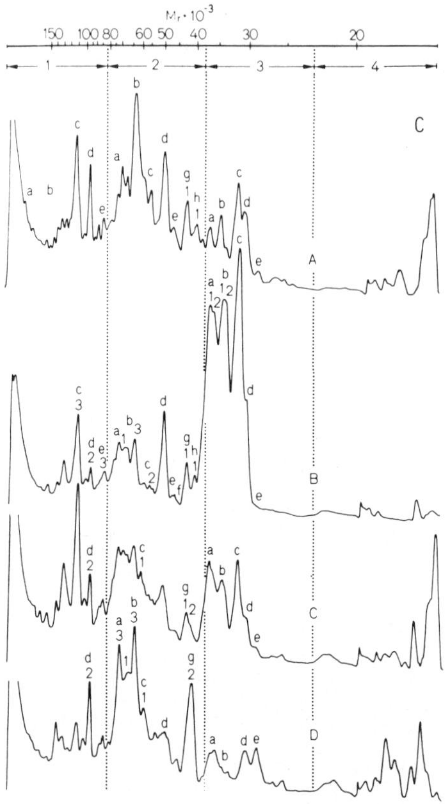

FIGURE 15C

the paper) can survive the removal of RNA by treatment with 2 *M* NaCl. In the experiments, the particles were labeled in vitro with ^{125}I for detection of the particle proteins. If RNA and proteins of the Samarina particles were dissociated in 2 *M* NaCl and sedimented in 15 to 30% linear sucrose density gradient a majority of the particle proteins (60 to 70%) sedimented as a homogeneous narrow peak with a sedimentation coefficient of 30 S, like the original Samarina particles. No protein particles sedimenting at 40 to 45 S were observed. By increasing the temperature to 35° in the presence of Triton® X-100, the authors were also able to isolate 30 S particles from FL human cells grown in vitro. Likewise, it was possible to remove RNA from these particles by 2 *M* NaCl and preserve the 30 S protein particles. In electron micrographs the particles appeared homogeneous, regularly shaped, and 20 nm in diameter. After mixing the protein particles with the particle RNA in 2 *M* NaCl and reduction of the salt concentration by dialysis, the Samarina particles were reconstituted. The Samarina particles could also be reconstituted from FL human cells 30 S protein particles and virus-specific RNA synthesized in adenovirus-infected cells. From the fact that RNA could be removed from the Samarina particles without destroying their protein structure and that the 30 S protein particles could be reassociated with pre-messenger RNA, the authors concluded that RNA is probably localized on the surface of Samarina particles. It should be mentioned that the authors labeled the particle RNA in vivo by a 45-min pulse of [^{14}C]-orotic acid, conditions under which hnRNA but not snRNA will be labeled. Therefore, their claim that all RNA was removed from the particle proteins by treatment with 2 *M* NaCl is valid only for hnRNA.

The existence of RNA-free Samarina particles as prearranged building blocks around which newly synthesized RNA could wrap would contradict the possibility that Samarina particles are artifacts produced during the extraction of the nuclear RNPs. However, the existence of RNA-free Samarina particles could not be confirmed by other authors. Stévenin and co-workers[95,124,125] observed a gradual dissociation of hnRNP proteins upon salt treatment without remaining 30 S structures. Neither could such structures be detected in other laboratories.[49,73,126] To explain this discrepancy Kulguskin and co-workers[82] tested the effect of particle concentration on the dissociation of particles in protein and RNA in the presence of 2 *M* NaCl. Samarina particles prepared from rat liver, when incubated with 2 *M* NaCl at a protein concentration of 20 mg/mℓ dissociated into RNA, 30 S protein particles, and some protein which sedimented with a lower sedimentation coefficient. At a lower particle protein concentration the amount of 30 S protein particles increased in ratio to the free protein with increasing protein concentration. The authors concluded that there is a concentration dependency of transition between the specific 30 S protein particles and the dissociated proteins. They explained the preservation of 30 S protein particles at lower concentration of particle proteins observed in the earlier experiments[79] by the stabilization effect of iodination on the proteins of 30 S particles. Simultaneously with Kulguskin et al.,[127] the stabilization effect of protein iodination on the α-proteins of Samarina particles was also described by Martin et al.[128] who found that the 30 S hnRNP particles containing iodinated proteins will lose RNA if treated with high salt concentration but that the protein moiety will remain intact and sediment as 30 S particles, even at a salt concentration as high as 2 *M* NaCl. Noniodinated hnRNP particles did not contain a salt-stable core. Since 3.3 to 4.6 mol % of tyrosine residues were found when the amino acid composition of the protein moiety of Samarina particles was determined,[15,49,116,129] and formation of dityrosyl bridges during iodination conditions is probable,[130] Martin et al.[128] also concluded that RNA binding sites must be exposed on the surface of the protein core, but the authors kept open the question whether the surface location of RNA binding sites reflects existence of preformed hnRNA binding protein complexes in vivo or is the result of secondary folding of hnRNA-protein complexes in Samarina particles under experimental conditions. Another stabilization effect observed by Kulguskin et al.[82,127] was the aging of the Samarina particles. During storage at 0°C for several days, the 30 S protein particles resistant to dissociation accumulated. The authors found only RNP particles reconstituted from RNA and 30 S protein particles similar to native Samarina particles. The particles reconstituted from RNA and the slowly sedimenting "free" proteins were rather stable in CsCl density gradient without fixation with formaldehyde, and the RNA of these particles was resistant to pancreatic RNase. In contrast, RNA in native Samarina particles and in particles reconstituted from RNA and 30 S protein particles was sensitive to pancreatic RNase, and without fixation with formaldehyde the particles dissociated in CsCl density gradient. The authors assumed that in the first case, RNA binds to the individual protein molecules and then folds irregularly in such a way that a significant part of the RNA is protected by localization inside the particles. In the second case, the RNA interacts with the surface of preformed protein particles. It should be mentioned that in discussion of these experiments the possible role of snRNA in the particle structure was not considered.

The conclusion of the authors that 30 S protein particles are assembled in vivo first and only after the RNA is wrapped on their surface, as proposed originally by Samarina,[12,79] is rather attractive in view of another report from the same laboratory.[131] In these experiments, the reversible dissociation and reassociation of the Samarina particle proteins in course of the reconstitution procedure was excluded by treatment of the Samarina particles with bifunctional reagents dimethylsuberimidate or dimethyl-3,3′-dithiobispropionimidate to cross-link the particle proteins together. It was shown that RNA can be easily removed from the Samarina particles without previous dissociation of the proteins as well as returned to the cross-linked 30 S protein particle with reconstitution of the original particle.

Explanation of the observed discrepancies in the dissociation of Samarina particles in

NaCl solutions proposed by Kulguskin and co-workers[82,127] is plausible. Difference in the concentration of proteins and stabilization of the protein particles by formation of a limited number of cross-links between protein molecules by the iodination process could explain the variation of results reported in the literature. The fact that proteins which correspond to the proteins of Samarina particles are the most easily dissociated from nuclear RNPs by increasing salt concentration[49,50,73,126] is also not contradictory to the presence of preformed protein particles in the RNPs. However, it is too early to definitely close this controversy.

In conclusion, are Samarina particles present in the cell in vivo, or are they the product of techniques used for the preparation of nuclear RNP particles? In the opinion of the reviewer, the particles are real subunits of the nuclear RNP network which is involved in RNA processing and transport. It is obvious from the results of Gallinaro-Matringe et al.[119] that there are two types of proteins associated with hnRNA in the cell. The hnRNA-β-protein complex is more susceptible to RNase digestion than the hnRNA associated with the α-proteins which are characteristic for the Samarina particles. The levels of endogenous nucleases in the nuclei of various types of cells are rather different, e.g., activity in the nuclei of liver cells is such that it will destroy most of the RNA associated with β proteins. Particles composed of α-protein and RNA — the Samarina particles — are left. In some other types of cells the same level of nuclear RNase activity could be reached by extraction of the nuclear RNPs at elevated temperature. Only when the RNase activity destroys the RNA completely are the α-proteins liberated. Whether the 40 to 45 S protein particles observed by Stévenin et al.[106] are unspecific aggregates or dimers of the 30 S protein particles of Kulguskin and co-workers[82] is not known. Some specificity of the interaction of particle proteins is indicated. According to Stévenin et al.[106] the formed 40 to 45 S protein particles were the most stable aggregates. After the treatment of 30, 40, or 50 S particles with a concentration of RNase which removes 80 to 90% of the particle RNA there was only a small change in the sedimentation coefficient of the 40 S particles, an increase to 40 to 45 S in the sedimentation coefficient of the 30 S particles, but a decrease after resedimentation in sucrose gradient of the sedimentation coefficient of the 50 S particles by 5 to 10 S.

The conditions under which Stévenin and co-workers[106] observed the formation of the stable 40 to 45 S protein particles are not likely to be encountered during procedures used for isolation of the nuclear hnRNP monoparticles, even in cells with high levels of endogenous RNase activity. As we will see later, more problematic could be the isolation of polyparticles, even when cytosolic RNase inhibitor is present during the disintegration of nuclei. A shift in the sedimentation coefficient of 30 S monoparticles to 36 S was observed only after the hydrolysis of 60% of the particle RNA, which also resulted in the destruction of some 60 to 70% of the 30 S particles. A shift in the sedimentation coefficient to 41 S was observed only after 85% of the particle RNA was hydrolyzed and over 90% of the particles disintegrated.[106] The high level of RNase activity required for the formation of the 40 to 50 S protein-particles is obvious from the finding that incubation of the rat brain nuclear RNP particles with 0.1 μg/mℓ of pancreatic RNase at 15° for 20 min solubilized only about 20% of the hnRNA particle.[15]

Since many laboratories are studying Samarina particles isolated from rat liver, which is an organ with a high level of endogenous RNases, studies of the effects of RNases on nuclear RNP particles from rat liver are of special importance for the evaluation whether Samarina particles are indeed present in liver cells. Seifert et al.[132] studied the effects of RNase on polyparticles prepared by isotonic salt extraction at pH 8.0 from rat liver nuclei disrupted by sonication. Chromatin and nucleoli were separated from RNP particles by centrifugation at 12,000 × g for 10 min, and RNP particles were purified from the supernatant by centrifugation in 15 to 30% linear sucrose density gradient. When the nuclei were sonicated in the presence of the rat liver cytoplasmic RNase inhibitor the RNP sedimented in the linear sucrose density gradient as a broad peak with a maximum of approximately 75 S. In the absence of the inhibitor, the particles sedimented as 38 S monoparticles. Experiments in which ^{14}C-labeled RNase inhibitor was used during the extraction of RNP particles from the cell nuclei have shown that most of the inhibitor is separated from the particles during their purification. Only 1% of the label present in the inhibitor fractions sedimented in the particle peak in linear sucrose density gradient.[98] No degradation of pulse-labeled RNA (30 min pulse with [^3H] orotic acid) into acid-soluble nucleotides was detected in the presence of the inhibitor during the sonication of nuclei and separation of the ribonucleoprotein fraction from chromatin by centrifugation.[98] But the protection provided by the RNase inhibitor was only limited. As has been first shown by Niessing and Sekeris,[133] an endogenous RNase activity is associated with polyparticles. Perhaps as the result of removal of the cytosolic RNase inhibitor from the particles by centrifugation in a linear 15 to 30% sucrose density gradient, incubation of polyparticles at 37° for 15 min resulted in a nearly complete degradation of hnRNA, despite the original presence of cytosolic RNase inhibitor during sonication of the nuclei.[132] It is of interest that the activity of this endogenous ribonuclease(s) associated with the polyparticles did not result in the conversion of polyparticles into monoparticles.[132] This endogenous RNase activity also did not degrade snRNA associated with the polyparticles.[132] Whether these last two observations are related, cannot be decided at the present time. The presence of endogenous nuclease activity has been likewise demonstrated in hnRNP polyparticles from HeLa and mouse erythroleukemia cells.[75] When hnRNP polyparticles prepared from these cells were incubated at 37° for 2 hr in low-salt pH 7.2 Tris-HCl buffer, 22% of the particles-associated RNA was digested to acid solubility. The kinetics of this digestion was not detectably altered when the concentration of hnRNP particles was reduced tenfold, suggesting that the responsible nuclease(s) is(are) particle bound.

Only traces of such endogenous activity were found associated with monoparticles. Monoparticles isolated from rat liver nuclei by centrifugation on 15 to 30% linear sucrose density gradient of nuclear supernatant prepared in the absence of RNase inhibitor by centrifugation of sonicated nuclei were completely free of RNase activity which would degrade poly(A) polymers or yeast RNA.[100] After incubation of the purified monoparticles at 37° for 1 hr, only 16% of the trichloroacetic acid-precipitable radioactivity of the particle hnRNA was lost.[98]

It is perhaps important that monoparticles larger than 30 S, and also that the discrepancy in the RNA-to-protein ratios determined from buoyant density of the particles or by their direct chemical analyses, were found mostly by authors who isolated the particles in a linear sucrose density gradient of nuclear extracts prepared from sonicated nuclei. Such a procedure will release both types of particles (the Jacob particles containing β-proteins and the Samarina particles containing α-proteins) from the nuclei. These particles will be released either in the form of polyparticles composed of both types of particles (if extracted in the presence of rat liver cytosolic RNase inhibitor) or as a mixture of both types of particles after the fragmentation of the polyparticles in the absence of the RNase inhibitor. During the centrifugation in a linear sucrose density gradient, the endogenous nucleases present in the particles containing the β-proteins could act not only on RNA of these particles, but also on the RNA of the Samarina particles which co-sediment with the particles containing the β-proteins in the gradient. In most reported experiments the RNA was pulse labeled, therefore, after the fixation of RNA to the proteins the determined buoyant density represents the density of hnRNP particles still present in the gradient. Determination of the particle RNA-to-protein ratio by chemical analyses of particles present in the 30 to 50 S portion of the sucrose density gradient also includes particles composed only of proteins or of proteins and snRNA which are not labeled and consequently not detected in the CsCl or Cs_2SO_4 density gradients. The isolation of Samarina particles by isotonic salt extraction at pH 8.0 in the absence of the RNase inhibitor from liver cell nuclei preserves predominantly the Samarina particles, with the particles containing the β-proteins either degraded in the nucleus or immediately during the extraction. The particles containing β-proteins are substantially more sensitive to degradation by RNase.[15] During the isolation of RNP particles by extraction of nuclei, the Samarina particles are exposed to endonuclease(s) associated with the β-proteins only during the short period of the extraction of the particles, but not during the long time required for the purification of the particles by centrifugation in sucrose density gradient. Therefore, it is probable that Samarina particles extracted from liver nuclei by isotonic salt extraction at pH 8.0 and 0 to 4° are native particles. The same could be true for Samarina particles extracted at elevated temperature but from nuclei which have a low level of endogenous RNase activity.[49,73,79,134]

Samarina particles prepared from rat liver by isotonic salt extraction at pH 8.0 sediment in a narrow band, with a sedimentation coefficient at 30 S or close to 30 S in a linear sucrose density gradient. The RNA-to-protein ratio is close to 1:4, and the same ratio is obtained by chemical analyses of particles sedimented from sucrose density gradient and by determination of buoyant density of the particles. The approximate ratio 1:4 of RNA-to-proteins corresponds to the ratio found in hnRNP particles isolated from cells in which RNA and proteins were linked together by UV irradiation in vivo.[75] These properties of the Samarina particles indicate that the particles represent entities of the nuclear RNP network which are present in vivo.

Studies on the stability of hnRNP particles in solutions of increasing ionic strength also support the conclusion that there exist two types of nuclear RNA particles. In 1973 Stévenin and co-workers[125] described a progressive release of proteins from rat brain RNPs by treatment with increasing concentrations of KCl. Treatment of 60 S particles with 0.4 *M* KCl resulted in the formation of a series of particles with the density in CsCl reaching from 1.4 to 1.7 g/cm^3. This, in the opinion of the authors, could be interpreted as the result of a progressive release of proteins from the particles. In view of our present knowledge, the results could be also explained by a difference in the release of proteins from Samarina particles and from Jacob particles in response to salt treatment. A comparison of the effects of salt on the polyparticles of different sizes has shown that the polyparticles of higher sedimentation coefficient were more resistant to the action of salt than the smaller polyparticles. Close to 70% of all pulse-labeled RNA sedimented with a CsCl density higher than 1.60 g/cm^3 after the treatment of 60 S polyparticles with 1 *M* KCl. The percentage of 110 and 160 S polyparticles which after treatment with 1 *M* KCl converted to particles with higher CsCl buoyant density than 1.60 g/cm^3 was 35 and 22%, respectively. Before the salt treatment the density in CsCl of particles of all sizes was 1.40 g/cm^3.[125] The results indicate that the Samarina particles are intermixed in the nuclear RNP network with stretches of different length of RNA associated with the β-proteins which, after breakage, will form the heterogeneous Jacob particles. The results also indicate that during the isolation of the particles there is a greater tendency to break the RNP network between the Samarina and Jacob particles than inside the stretches of RNA associated with the β-proteins.

Similar results were obtained after treatment of rat brain nuclear hnRNP particles by NaCl.[95,99] Some proteins were released from the polyparticles after treatment with 0.25 *M* NaCl. The majority of proteins which were released at this salt concentration were proteins described later by Jacob et al.[15] as α-proteins which are characteristic for Samarina particles. Approximately 75% of these proteins originally present in the polyparticles were already released at this salt concentration.[99] In addition, about 30 to 40% of the β-proteins which are characteristic for the Jacob particles were also released.[98] Treatment of the same polyparticles with 0.7 *M* NaCl released about 95% of the Samarina particle proteins and about 60% of the higher-molecular-weight β-proteins from the same type of polyparticles.[99]

Preferential release of the Samarina particle proteins after treatment of HeLa cells hnRNP particles by NaCl was also described by Pederson.[73] Recentrifugation of the 75 S hnRNP polyparticles of the HeLa cell nuclei in a linear 15 to 30% sucrose density gradient containing

0.5 *M* NaCl reduced their sedimentation velocity approximately to 50 S and their buoyant density in CsCl from 1.43 g/cm^3 to 1.75 g/cm^3. The released protein migrated as one polypeptide band (with approximate molecular weight of 40,000) in one-dimensional SDS-PAGE, but the band could comprise more than one of the major polypeptides of the Samarina particles. The proteins which were not lost from the particles by 0.5 *M* NaCl treatment were high-molecular-weight proteins belonging to the group of Jacob's β-proteins. An important result of this experiment was that even after 24 hr exposure of HeLa cell hnRNP particles, approximately 30% of the particles displayed buoyant density of native particles. This indicated to Pederson that in the particle population may exist a distinct group of different particles.[73]

The observed strong binding of the majority of heterogeneous proteins with molecular weights over 45,000 to hnRNA is in agreement with the report of LeStourgeon and co-workers.[50] Extensive dissociation of higher-molecular-weight proteins from hnRNA after the treatment with 0.7 *M* NaCl observed by Gallinaro-Matringe and co-workers[99] could be the effect of 0.2% sodium deoxycholate used for the disruption of nuclei. Treatment with higher concentrations of sodium deoxycholate leads to complete disintegration of the RNP particles.[32,124]

The existence of two types of RNPs in the nuclear RNP network is in agreement with electron microscopic observations. Two types of arrangements (fibrilar and granular) of hnRNA-nucleoproteins can be seen *in situ* or in spread chromatin.[17,18,68,71] Recent examinations of electron micrographs of hnRNP structures from the nuclei of *Drosophila* embryos have shown that the globular structures (on the average, 24 nm in diameter) are nonrandomly localized on relatively thin (4-nm wide) fibers of RNPs.[71]

Likewise the experiments in which hnRNA synthesis was inhibited by α-amanitin, or by actinomycin D support the concept of two different types of hnRNA RNPs in the cell nucleus.[135] The inhibition of RNA synthesis selectively depleted rat liver nuclei of the α-proteins of the hnRNP particles (designated in the publication as the proteins of the region C, but not identical with the C proteins of Beyer et al.[49]). The authors concluded that the α-proteins of Samarina particles are stored in the cytoplasm and enter the nucleus during transcription to associate with hnRNA.

The presence of two types of RNP complexes in the nuclear network with the resulting formation of two types of hnRNP particles after the fragmentation of this network, and the different susceptibility of these two types of particles to disintegration by RNase, could explain the variation in the proportion of the high-molecular-weight β-proteins and low-molecular-weight α-proteins in the particles prepared in different laboratories.

VII. PROTEIN COMPOSITION OF THE HNRNP PARTICLES

Heterogeneous high-molecular-weight proteins of the β-particles are not well described, and because of their overlap with nucleoplasmic and chromosomal proteins,[58] also less well defined. The molecular weight of the minor proteins is higher than the molecular weight of the main polypeptides which are characteristic for Samarina particles. In our experiments,[81] we identified in the liver hnRNP particles 19 well-expressed polypeptides which belong to this fraction of particle proteins, with many additional polypeptides present in trace amounts. About 20 predominant bands of these proteins were also observed in HeLa cells, mouse L cells,[4,73] and rat brain.[99,119] Electrophorograms have shown tissue[119] as well as species specificity[4,73] of some of these proteins. This is in contrast to the major α-proteins of Samarina particles, which are not tissue and species specific, but which differ in their quantitative distribution in different tissues.[49] In contrast to the major proteins of Samarina particles[55,122] the β-proteins are rapidly turning over,[55] contain many highly phosphorylated species, and have higher affinity for RNA.[15,50,99] In all their properties the heterogeneous high-molecular-weight proteins are similar to the proteins of the nuclear network,[26] and, as was already mentioned, they share electrophoretic mobilities with many proteins found in other nuclear fractions.[58] All these findings agree with the proposal that the Jacob particles represent a fragmented nuclear RNP network.

The Samarina particles are composed of a limited number of low-molecular-weight proteins (α-proteins). But, as was mentioned earlier, the composition of the protein moiety of the Samarina particles is likewise subject to controversy, which resulted from the fact that until recently there were no suitable techniques to study the glycine-rich proteins which represent the majority of the Samarina particle proteins.

In one of their earlier papers, only two distinct polypeptides, of 34,000 and 38,000 daltons, were reported by Martin et al.[116] to be present in the 30 S hnRNP particles. These same two polypeptides were found in particles isolated from liver of duck, chicken, rat, and monkey, as well as from Taper ascites cells. But these researchers soon realized that the number of polypeptides which can be distinguished by one-dimensional electrophoresis, and their estimated molecular weights, strongly depend on the procedure used for their electrophoresis. Inclusion of 8 *M* urea into the Laemmli SDS-polyacrylamide gels used for separation of the particles increased the apparent molecular weight of the particle polypeptides to 38,000 and 46,000, respectively. A change in the commercial source of SDS decreased the molecular weights of the same polypeptides to 35,000 and 36,000 to 37,000.[1] It is obvious that the α-proteins of the particles show similar anomalous behavior during SDS-PAGE as has been reported for some viral polypeptides.[136] Therefore, the molecular weights assigned by various authors to the major polypeptides of the Samarina particles have only descriptive values. It is perhaps of interest that, using the same organ (liver), and a similar procedure for the preparation of hnRNP particles, the authors cited in References 50, 58, 98, and 129, who used the Laemmli[137] or Waehneldt[138] procedure to analyze the Samarina particle proteins, determined the molecular weights of the major proteins to be 10 to 20% lower than those reported by authors cited in References 16, 73, 81, 108, 115, 118, and 139, who used a modification of the Shapiro et al.[140] procedure of SDS-PAGE.

Not only the molecular weight of the major polypeptides of the particles is affected by electrophoretic conditions. Extending the time of electrophoresis, Billings and Martin[1] resolved the 37,500- and 40,000-dalton components into five polypeptide bands, which suggests heterogeneity of these proteins over a very narrow molecular weight range. This is similar to the results reported by Beyer et al.[49] for liver cell hnRNP particles. These authors used the discontinuous SDS-PAGE method of Laemmli,[137] which permits the best separation of major polypeptides by one-dimensional electrophoresis. The authors studied protein composition of the 40 S hnRNP particles isolated from HeLa cells, rat liver, and Chinese hamster fibroblasts by isotonic salt extraction at pH 8.0 (see also Reference 50). It should be mentioned that after 1 hr extraction at 0° the nuclei were extracted for another 1 hr at room temperature and for 30 min at 37°. Under such conditions, the majority of Jacob particles containing the β proteins was degraded by endogenous nucleases and the obtained particles were predominantly Samarina particles. Therefore it is not surprising that only a limited amount of minor high-molecular-weight proteins was associated with the particles. In HeLa cells, the major proteins (designated as packing proteins since the authors assumed that these proteins are involved in the immediate packing of hnRNA into hnRNP particles as the hnRNA transcription proceeds) separated in the polyacrylamide gels into three groups of closely spaced doublets: A group (32,000 and 34,000 daltons), B group (36,000 and 37,000 daltons), and C group (42,000 and 44,000 daltons) (Figure 16).

In rat liver hnRNP particles the A_1 protein was observed only in traces, and the B_2 protein also was hardly detectable. In Chinese hamster lung fibroblast the B_2 protein was absent but the amount of A_1 protein was the same as in HeLa cells. Based on the observation that after a 10-min digestion with micrococcal nuclease of hnRNP particles from HeLa cells a

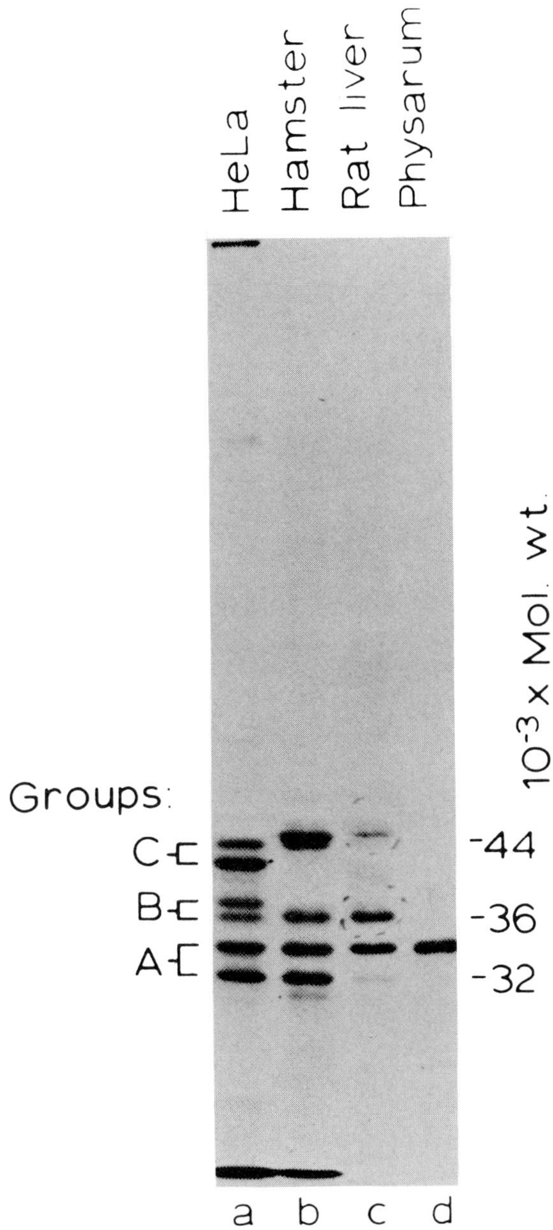

FIGURE 16. SDS-PAGE of proteins of core hnRNP particles (Samarina particles) from various species and tissues. (From LeStourgeon, W. M., Beyer, A. L., Christensen, M. E., Walker, B. W., Poupore, S. M., and Daniels, L. P., *Cold Spring Harbor Symp. Quant. Biol.*, 42, 885, 1977. With permission.)

nuclease-resistant core composed of A_2, B_1, and C_1 protein can be recovered and that the C_1, A_1, and B_2 proteins have lower resistance to digestion by trypsin, LeStourgeon considered the possibility that in rat liver and in other tissues which show simpler electrophoretic patterns of the hnRNP particle proteins the B_2, A_1, and C_1 protein could be lost as a result of the exposure of particles to higher cellular activities of nucleases and/or proteases during particle isolation.[5] The disappearance of A_1 protein could also reflect the low mitotic activity of liver cells.[50]

Despite the appeal of the concept of uniformity of Samarina particles, the outstanding electrophoretic separations of the hnRNP particle proteins obtained in LeStourgeon laboratory, and of the plausibility of the proposed explanation, this reviewer considers the question of the protein composition of Samarina particles not yet answered. The main reasons for this conclusion are the large amount of charge isomers into which the main proteins of Samarina particles can be separated by a combination of nonequilibrium pH gradient and SDS-gel electrophoresis, and the report by Peters and Comings[40] that no basic proteins which would correspond to the A_1 or B_2 proteins could be detected by the combination of the nonequilibrium pH gradient and SDS gel electrophoresis in any of the rat liver nuclear protein fractions. Proteins of the Samarina particles are also rather resistant to proteolytic degradation by endogenous proteinases and even by trypsin (in contrast to the β-proteins),[120] and the same α-proteins were identified by two-dimensional electrophoresis in the Samarina particles[40] as in the hnRNP polyparticles[120] composed of Samarina and Jacob particles.

Additionally, it is quite possible that the protein composition of Samarina particles is not uniform even in one cell type. In a nondenaturing sucrose gradient, Maundrell and Scherrer[141] isofocused 40 S hnRNP particles extracted by a modification of the Samarina procedure[12] from duck erythroblast nuclei. The particles separated in a broad zone from pH 4.2 to pH 6.8. Analysis by one-dimensional SDS-PAGE of the protein moiety of the particles from different sections of the gradient has shown a progressive enrichment of the β-proteins toward the acidic end of the gradient (which would indicate separation of the free Jacob particles from Samarina particles) and a nonuniform distribution of the different species of the α-proteins (Figure 17).

The hnRNP particle proteins which, according to the classification used in LeStourgeon et al.[49,50] would be classified as belonging to the A and B groups, had properties of a unique group of nuclear proteins. Their pI was basic, they were rich in glycyl residues, and they contained residues of an unusual amino acid N^G, N^G-dimethylarginine. In addition, the amount of glutamyl and aspartyl residues was high. Because of the basic pI of these proteins and their relatively low level of basic amino acids it is probable that many of the found glutamyl and aspartyl residues were present as amides.[8,49,50,129,142]

N^G, N^G-dimethylarginine was first identified in proteins of both subunits of HeLa cell cytoplasmic ribosomes.[143] One year later, this unusual amino acid was found in the proteins of the hnRNP particles of HeLa cells, mouse ascites cells, rat and duck liver, and in a protein of the slime mold *Physarum polycephalum* which is homologous to the major proteins of mammalian hnRNP particles.[144] The presence of N^G, N^G-dimethylarginine in the hnRNP particles of rat liver was also simultaneously described by another laboratory.[145]

Beyer and co-workers,[49] who purified the A_1 and A_2 proteins of the HeLa cells hnRNP particles to electrophoretic homogeneity by preparative SDS-gel electrophoresis determined 23 and 25 mol % of glycyl residues in A_1 and A_2 proteins, respectively. The amount of N^G, N^G-dimethylarginyl residues in these two proteins was 0.1 and 0.4 mol %, of glutaminyl residues 12.6 and 10.7 mol %, and of aspartyl residues 7.6 and 10.3 mol %, respectively. It is somewhat surprising that in A_2 proteins isolated from rat liver[50] no modified arginine was found. The amount of glycyl residues was 27.9 mol % and of glutamyl and aspartyl residues 10.8 and 11.8 mol %, respectively. In B_1 protein of rat liver the amino acids analysis has shown 1.3 mol % of N^G, N^G-arginyl, 18.6 mol % of glycyl, 8.7 mol % of glutamyl, and 9.5 mol % of aspartyl residues.

In hnRNP particles also isolated from rat liver, Patel et al.[146] could detect only small amounts of proteins which could correspond to the A_2 proteins described by LeStourgeon et al.[49,50] A protein which could correspond to the B_1 protein represented 35 to 40% of the total mass of the particle protein moiety. This protein was purified by preparative acid-urea PAGE, using the fact that in the absence of 2-mercaptoethanol the dissociation of the hnRNP particle proteins was incomplete in 0.9 *M* acetic acid, 6 *M* urea, and 10% glycerol and only

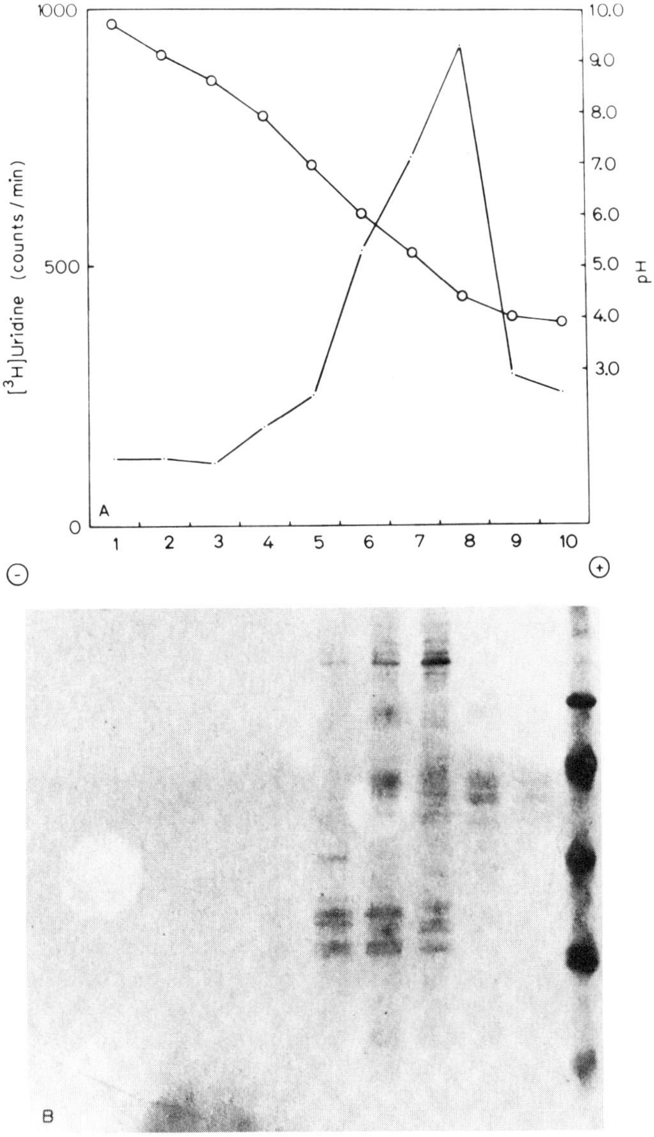

FIGURE 17. Isoelectric focusing of 40 S hnRNP particles from duck erythroblast nuclei in nondenaturing sucrose gradient. The particles were labeled with [^{3}H] uridine for 1 hr in vivo. (A) Distribution of the [^{3}H]-labeled RNA in the pH gradient; (o——o) pH, (●——●) [^{3}H] uridine. (B). SDS-PAGE of proteins recovered from corresponding fractions. The gel at the extreme right contains phosphorylase β (90,000), bovine serum albumin (68,000), ovalbumin (43,000), DNase I (31,000), and chymotrypsinogen (25,500) as molecular weight markers. (From Maundrell, K. and Scherrer, K., *Eur. J. Biochem.*, 99, 225, 1979. With permission.)

one polypeptide chain penetrated into the 10% polyacrylamide gel during electrophoresis[146] (Figure 18). The purified protein migrated as one narrow band in SDS-polyacrylamide gel. It contained 20.2 mol % of glycyl, 10.3 mol % of glutamyl and 10.6 mol % of aspartyl residues. The N^G, N^G-dimethylarginyl residues represented 0.9 mol % of all aminoacyl residues, or about 14% of the total arginyl residues (Table 2).

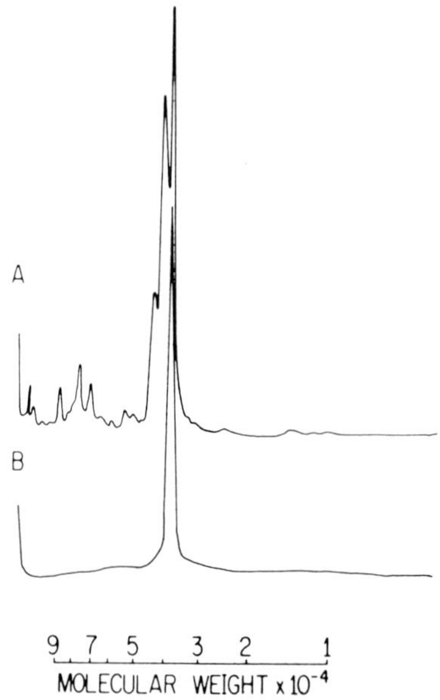

FIGURE 18. Densitometric tracings of SDS-polyacrylamide gel electrophorograms of (A) the total protein moiety of rat liver hnRNP particles, and (B) the main polypeptide purified from these particles. (From Patel, N. T., Kurosky, A., and Holoubek, V., *Biochem. Biophys. Acta,* 533, 282, 1978. With permission.)

In the same year that Beyer and co-workers[49] published their HeLa cell hnRNP particle analysis, Karn and co-workers[129] published the protein analysis of the rat liver hnRNP monoparticles. Like the particles isolated by Beyer et al.,[49] the liver hnRNP particles were also prepared by isotonic salt extraction at pH 8.0 followed by centrifugation in a linear 15 to 30% sucrose density gradient. The major proteins of the purified particles separated by one-dimensional SDS-PAGE into five bands with estimated molecular weights of 32,000, 34,000, 35,000, 41,000, and 41,500. This could represent the A_2 protein and doublets of the B and C proteins described in hnRNP particles of the HeLa cells by Beyer et al.[49] However, Karn and co-workers[129] showed that the proteins which separated on one-dimensional SDS-PAGE into five major peaks were much more heterogeneous. Two-dimensional electrophoresis using separation in an acid-urea pH 4.3 system in one direction and SDS-PAGE in the other afforded further resolution. Additional separation of these proteins could be also achieved by chromatography on phosphocellulose columns (Figure 19). By this procedure the authors identified in the 32,000- to 35,000-dalton protein group at least five major species, of which three contained N^G, N^G-dimethylarginine. All were similar in their amino acid composition to the A and B proteins of LeStourgeon and co-workers[49,50] and all contained more than 25 mol % of glycyl residues. It is of interest that additional centrifugation in metrizamide gradient of particles prepared by isotonic salt extraction at pH 8.0 and by centrifugation in sucrose density gradient almost completely removed the minor heterogeneous high-molecular-weight proteins still present in the particle fraction after centrifugation in a linear 15 to 30% sucrose density gradient[129] (Figure 20). In addition, when before the protein analysis RNA was separated from the proteins by chromatography on a DEAE-Sephadex® A-25 column, the 41,000- and 41,500-dalton polypeptides were also removed (Figure 20). These polypeptides have the highest affinity for RNA of all the major polypeptides of Samarina particle[49,50] and have metabolic turnover similar to the β-proteins.[55] They could be proteins of the nuclear matrix[48] to which the A and B proteins of Samarina particles are bound.

Table 2
THE AMINO ACID COMPOSITION OF THE PREDOMINANT (38,000-DALTON) PROTEIN OF THE RAT LIVER NUCLEI SAMARINA PARTICLES[a]

Amino acid	Residues/ mol protein[b]
Asx	34.8 ± 1 (35)
Thr	12.8 ± 1 (13)
Ser[c]	17.6 ± 1 (17—18)
Glx	36.0 ± 0.5 (36)
Pro	15.8 ± 0.1 (16)
Gly	68.4 ± 1 (68—69)
Ala	15.4 ± 0.3 (15—16)
Val	18.9 ± 0.5 (19)
Cys[c,e]	2.5 ± 0.02(2—3)
Met	2.6 ± 0.2 (2—3)
Ile	10.5 ± 1 (10—11)
Leu	16.4 ± 0.5 (16—17)
Tyr	17.0 ± 1 (17)
Phe	16.8 ± 0.3 (17)
Trp[d,e]	ND
Lys	22.6 ± 0.5 (22—23)
His	7.2 ± 0.3 (7)
N^G-Dimethylarginine	3.1 ± 0.01(3)
Arg	19.2 ± 0.5 (19)
Total	334—342

[a] Time-course hydrolysis.
[b] Calculated on the basis of 38,000 g/mol protein. (Mean values ± S.D. for 3 analysis except Ser, Cys, and Met which represent mean values ± S.E.M. for 3 analysis.) Nearest integers indicated in parenthesis.
[c] Extrapolated to time of hydrolysis.
[d] ND = not determined.
[e] insufficient protein did not permit specialized hydrolyses.

From Patel, N. T., Kurosky, A., and Holoubek, V., *Biochim., Biophys. Acta,* 533, 282, 1978. With permission.

This loss of the C-type proteins during the separation of particle proteins from RNA could explain the very simple protein composition of Samarina particles found by Samarina and other workers from the same laboratory. In our unpublished experiments we observed that the separation RNA from proteins of the rat liver Samarina particles, by particle chromatography on DEAE-cellulose column according to the procedure of Krichevskaya and Georgiev,[107] will result in the retention of not only the RNA, but also the C types of proteins. Also, precipitation of the particle RNA by 3 *M* $NaCl_2$, another procedure used Georgiev,[147] led in our laboratory to the co-precipitation of C proteins together with the RNA.

A and B proteins of the Samarina particles can be dissociated from the particles by treatment with 0.4 *M* NaCl. Fuchs and Jacob[8,142] used this method to release these proteins from the hnRNP particles of rat brain and to purify them further by chromatography on Bio-Gel® P-150. The proteins ranged in molecular weight from 28,000 to 38,000 and separated by SDS-PAGE into 10 bands, of which the bands with apparent molecular weights of 33,000, 35,000, and 37,000 were most prominent (Figure 21). Disregarding some differences in the estimated

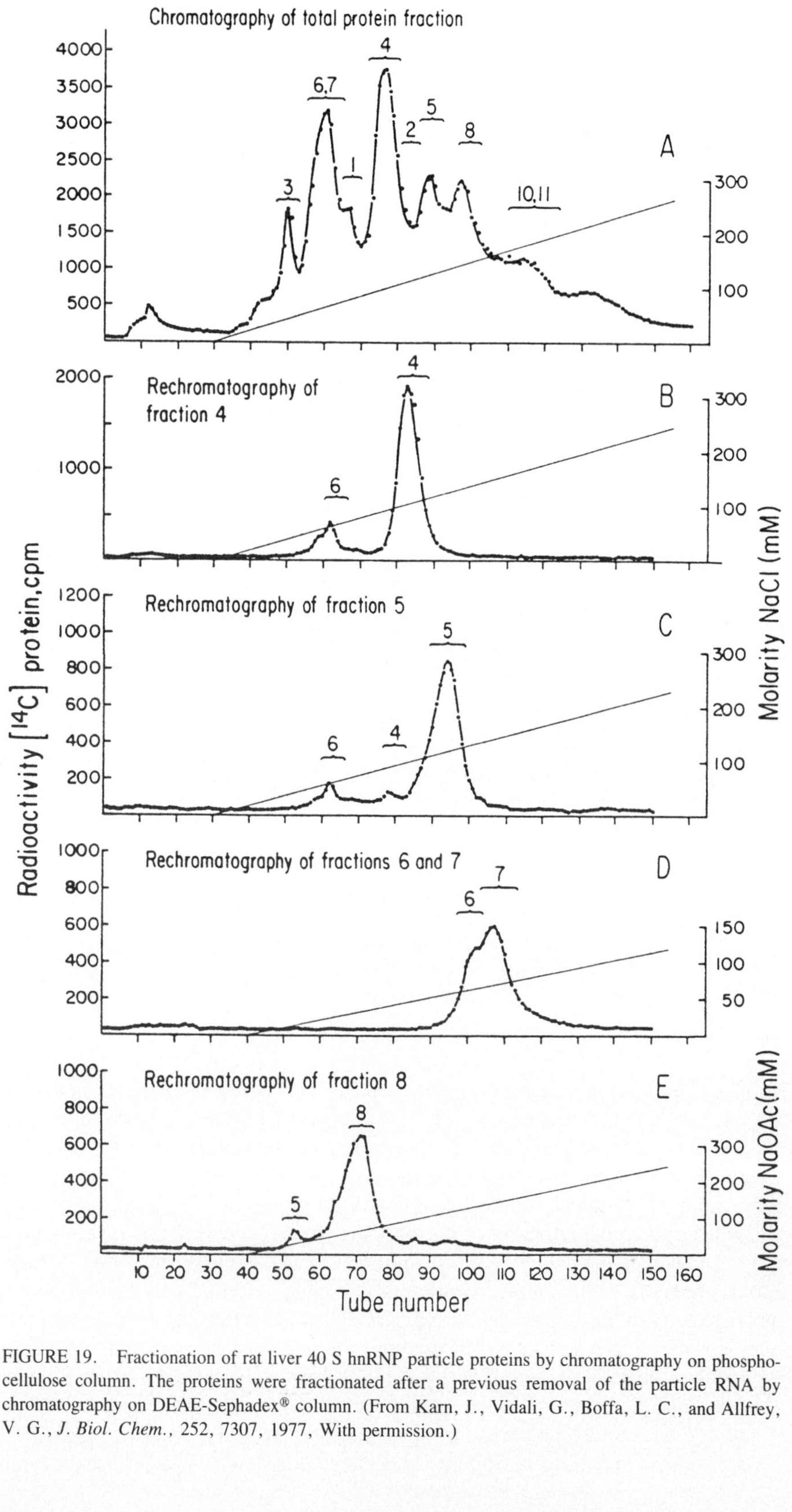

FIGURE 19. Fractionation of rat liver 40 S hnRNP particle proteins by chromatography on phosphocellulose column. The proteins were fractionated after a previous removal of the particle RNA by chromatography on DEAE-Sephadex® column. (From Karn, J., Vidali, G., Boffa, L. C., and Allfrey, V. G., *J. Biol. Chem.*, 252, 7307, 1977, With permission.)

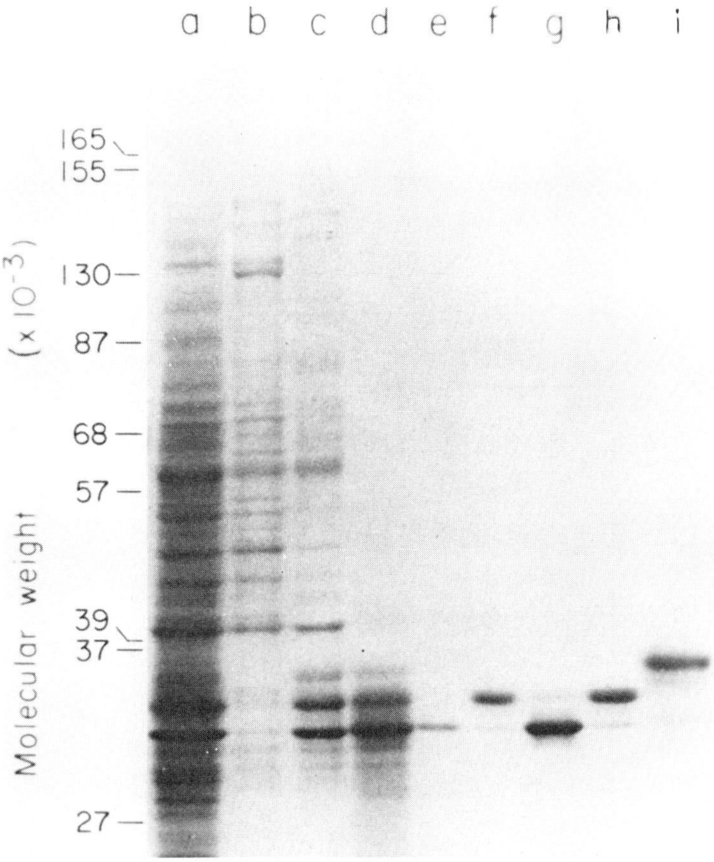

FIGURE 20. SDS-polyacrylamide gradient gel electrophoresis of protein fractions prepared by cation exchange chromatography: (a) proteins of the total 0.1 *M* NaCl, pH 8.0 extract of rat liver nuclei, (b) proteins recovered in the 12 to 20 S region of sucrose gradient after the centrifugation of the total extract on the gradient, (c) proteins of the 40 S hnRNP particles, (d) proteins of the 40 S hnRNP particles which are excluded from DEAE-Sephadex® A-25; (e to i) proteins obtained by chromatography on phosphocellulose column of particle proteins excluded from DEAE-Sephadex® A-25 (see Figure 19), (e) fraction 3, (f) fraction 6 + 7, (g) fraction 4, (h) fraction 5, and (i) fraction 8. (From Karn, J., Vidali, G., Boffa, L. C., and Allfrey, V. G., *J. Biol. Chem.*, 252, 7307, 1977. With permission.)

molecular weights, the pattern of electrophoretic separation of these polypeptides was very similar to that reported for Samarina particles of the rat liver by several authors.[81,100,108,120] Fuchs and co-workers[8] further separated a fraction obtained by chromatography on Bio-Gel® of the salt-released proteins from rat brain hnRNP monoparticles by chromatography on carboxymethyl cellulose (Figure 22). Due to the closeness in molecular weight and the existence of many charge variants of these proteins, a clear-cut separation according to charge and mass was not achieved. The amino acid composition of all eluted fractions was very similar to the amino acid composition of the A and B proteins of LeStourgeon and co-workers,[49,50] including the presence of N^G, N^G-dimethylarginine. Because of the high content of the glycyl residues (22 to 26 mol %) in these proteins, Fuchs and co-workers proposed to call these proteins glycine-rich proteins. Although the ratio of acidic to basic amino acids was close to 5 to 3, the pI of these proteins extended from 6.95 to 9.20 (Figure 23). Therefore, many of the acidic amino acids must be in the form of amides. The separation of these proteins according to the pI and molecular weights indicated the presence of more than 40 different polypeptides in the fraction of the glycine-rich hnRNP particle proteins.

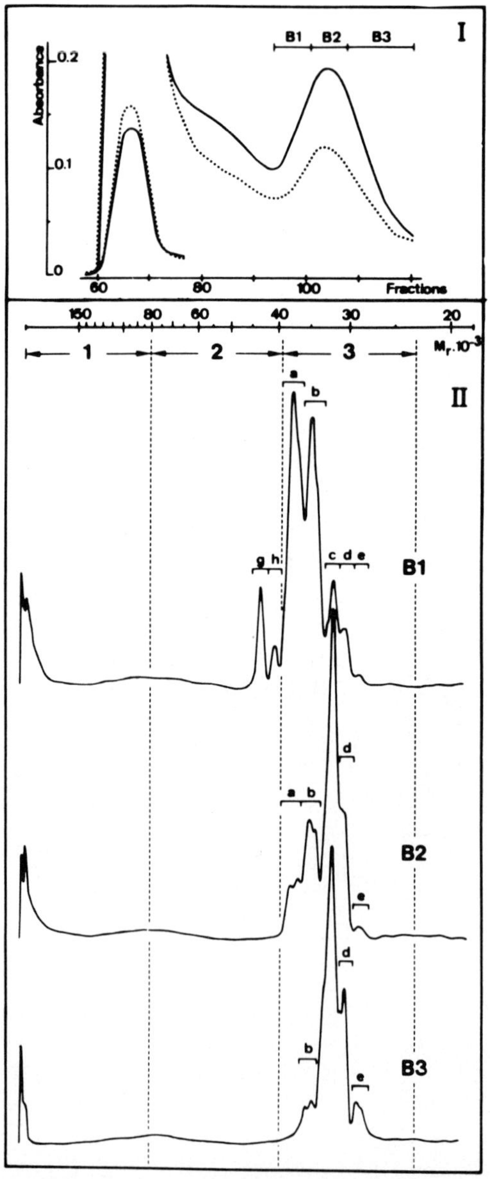

FIGURE 21. Purification of the glycine-rich hnRNA RNPs by Bio-Gel® P-150 chromatography. The glycine-rich proteins were eluted in peak B (Panel I). The fractions were pooled as indicated and analyzed by SDS-PAGE. Panel II, B1, B2, and B3 contain the densitometric tracings of the SDS-polyacrylamide gel electrophorograms; (......) absorbance at 260 nm, (——) absorbance at 280 nm. (From Fuchs, J.-P., Judes, C., and Jacob, M., *Biochemistry*, 19, 1087, 1980. With permission.)

But comparison of the trypsin and thermolysine peptide maps and the pattern of pI distribution suggested the presence of only four proteins of different primary structure in the glycine-rich protein fraction. Of these four proteins two represented more than two thirds of all glycine-rich proteins. According to the authors, the multiplicity of polypeptides which could be separated from each other was caused by extensive posttranslational modifications.

Indeed, the authors[8] have shown that peptides of the same primary structure, but modified to a different extent posttranscriptionally, differed in their apparent molecular weights up to 2000 and clearly separated into different bands in one-dimensional SDS-polyacrylamide

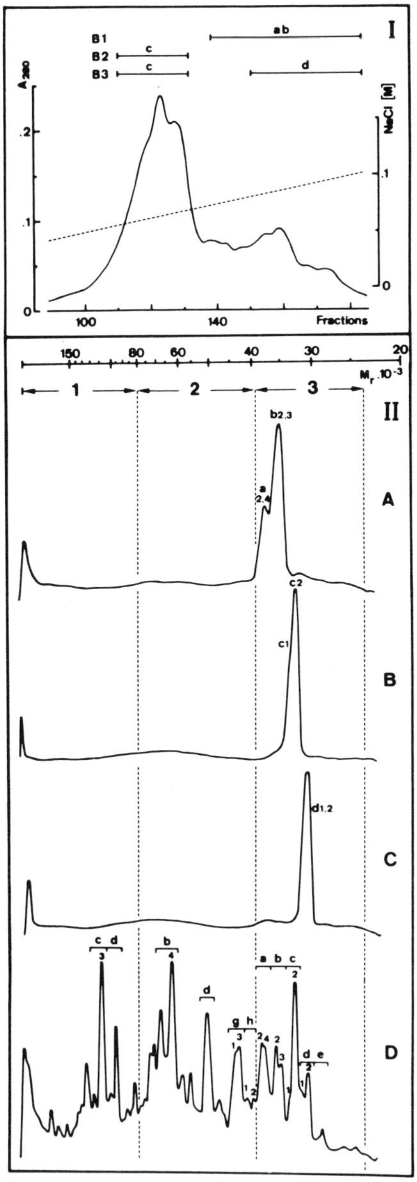

FIGURE 22. Carboxymethylcellulose chromatography of the glycine-rich hn-RNA-RNPs. The proteins obtained in the Bio-Gel® P-150 fractions B1, B2, and B3 (see Figure 21) were concentrated by $(NH_4)_2SO_4$ precipitation and applied to the carboxymethylcellulose column. All three fractions were eluted at the same concentration range (0.0 to 0.2 *M* of linear gradient of NaCl. The elution profile shown in panel I is that of B3. Panel II is showing the densitometric tracings of SDS-polyacrylamide gel electrophorograms of (A) fraction ab (from B1), (B) fraction c (from B2), and (C) fraction d (from B3). The profile of the proteins from a nuclear extract containing all the hnRNP proteins is given in (D). (From Fuchs, J.-P., Judes, C., and Jacob, M., *Biochemistry,* 19, 1087, 1980 With permission.)

gels. Therefore it is obvious that the separation of particle proteins by one-dimensional SDS-PAGE is an inadequate procedure for characterization of the protein moiety of the Samarina particles.

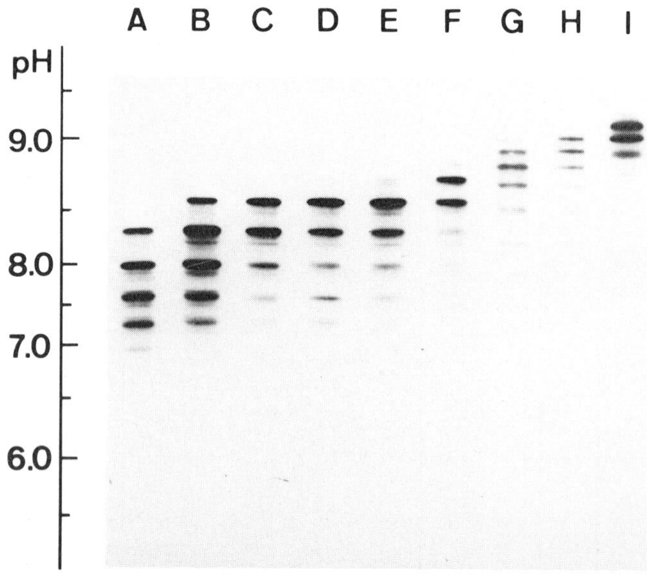

FIGURE 23. Electrofocusing of the glycine-rich hnRNA-RNPs. The proteins were fractionated by carboxymethylcellulose chromatography (see Figure 22). (A to E) correspond to fractions (in order of elution) of peak (c). (F) is enriched in b3 and c2, (G) and (H) contain a2-a4-b2, and (I) corresponds to d1-d2. (From Fuchs, J.-P., Judes, C., and Jacob, M., *Biochemistry,* 19, 1087, 1980. With permission.)

In 1975 O'Farrell[148] developed a method for two-dimensional analysis of proteins based on the combination of isoelectric focusing and SDS-PAGE which allows very detailed characterization of complex protein mixtures. This method was applied to the analysis of hnRNP particles, but since the separation of basic proteins by this procedure is inadequate, and the major proteins of hnRNP particles are basic, the results were not satisfactory.[49,149] In 1979 two French groups used simultaneously a modification of the original O'Farrell procedure[148] which replaced electrofocusing in the first dimension by nonequilibrium pH gradient[150] to analyze acidic as well as the basic proteins of the hnRNP particles. Using both the original and the modified O'Farrell procedure, Brunel and Lelay[151] analyzed the protein moiety of hnRNP particles from HeLa cells, mouse L cells, and Chinese hamster ovary cells. The authors found a high degree of evolutionary conservation among the analyzed proteins. Over a hundred individual spots were detected both in stained gels and in autoradiographs of proteins labeled with L-[^{35}S] methionine. The very abundant basic proteins of the particles separated into at least 15 peptides. Using the same techniques of two-dimensional protein separation, Maundrell and Scherrer[141] analyzed the 40 S hnRNP monoparticles prepared from chick erythroblasts. By one-dimensional SDS-PAGE the particles fractionated into approximately 50 minor polypeptides and 6 prominent bands in the molecular weight range of 30,000 to 40,000 — similar to the A, B, and C proteins observed by Beyer et al.[49] in HeLa cells. By combining both O'Farrell procedures it was possible to identify in the particles at least 65 acidic and 21 basic species of polypeptides. The proteins which separated on one-dimensional SDS-polyacrylamide gels into 6 bands were further resolved by two-dimensional separation into 15 basic and 6 neutral polypeptides. These two publications have shown heterogeneity of the major basic proteins of the hnRNP particles, a heterogeneity which was already strongly indicated by the results of Karn et al.[129] and later rigorously proven by Fuchs et al.[8]

More recently, the same two-dimensional procedures of O'Farrell[148,150] for protein analysis were used by Peters and Comings[40] to analyze proteins of rat liver cytoplasm, nuclear washes, matrix, membrane, hnRNP particles, and chromatin. The results are of special interest since all fractions of rat liver proteins were analyzed under the same conditions, and therefore it is possible to reach some conclusions about the relationship of the hnRNP particles to other nuclear structures. The hnRNP monoparticles were isolated from rat liver nuclei by isotonic salt extraction at pH 8.0 as originally described by Samarina et al.[12] The particles contained very few high-molecular-weight proteins and 91.67% of the particle proteins (determined by quantitation of proteins after separation by electrophoresis) were presented by polypeptides migrating between 34,000 and 41,000 daltons. Three major proteins estimated at 34,000, 36,000, and 41,000 daltons represented 73.6% of the particle proteins and were present in the hnRNP particles in an approximate ratio 2.3:1.7:1.0. Each of these three proteins separated into a series of isomers. All five charge isomers of each of the 34,000- and 36,000-dalton proteins were basic, and the four charge isomers of the 41,000-dalton proteins separated between pI 6.7 and 7.4. Additionally, more than 20 minor basic polypeptides in the range of 30,000 to 43,500 daltons could also be detected in the gel (Figure 24). The simpler composition, mainly representing the absence of the majority of acidic proteins observed in the liver hnRNP monoparticles by Peters and Comings[40] as compared to the more complex pattern observed in HeLa cells, mouse L cells, and Chinese hamster ovary cells by Brunel and Lelay,[151] is probably the reflection of the procedure used for particle isolation rather than of species and tissue difference. Peters and Comings[40] prepared the rat liver hnRNP particles by a procedure which results in the selective isolation of Samarina particles. Brunel and LeLay[151] used sonicated nuclei for the extraction of the hnRNP particles, a technique which results in the isolation of a mixture of Samarina and Jacob particles.

The 34,000-, 36,000-, and 41,000-dalton proteins of Peters and Comings[40] could correspond to the A_2, B_1, and C protein observed by LeStourgeon et al.[50] The observation[49] that A proteins did not resolve into charge isomers by isoelectric focusing into the O'Farrell system[148] is probably a technical problem resulting from the unsuitability of this system for the separation of basic proteins and does not contradict the conclusion concerning the identity of the proteins described by the two authors. Neither the ratio (1.3:1) of the 34,000- and 36,000-dalton proteins in the liver hnRNP particles nor the ratio (1:1) of the same proteins in the whole nuclei as reported by Peters and Comings[40] are in agreement with the ratio (4:1) observed for the A_2 and B_1 proteins in whole hnRNP particles or in the nuclease-resistant core structure of the particle by LeStourgeon et al.[49,50] The reasons for the difference are not obvious. As far as A_1 protein is concerned, it has been proposed by LeStourgeon et al.[5] that this protein, which is prominent in hnRNP particles prepared from HeLa cells but missing in hnRNP particles prepared from rat liver nuclei, is part of the rat liver nuclei hnRNP particles in vivo, but is lost because of nucleases and/or proteases activities during the isolation of hnRNP particles from rat liver. The findings of Peters and Comings[40] are not supportive of this explanation. No protein similar to the A_1 protein of HeLa cells described by Beyer et al.[49] was found in any of the studied fractions of nuclear proteins. The only protein (31,000 daltons) which could, perhaps, be homologous to this A_1 protein was found in the protein fraction extracted with 10 m*M* Tris buffer pH 7.5, but its amount in the total nuclear protein was less than 1%.[40] In contrast, the 34,000-, 36,000-, and 41,000-dalton proteins represented approximately 40% of the total nuclear proteins.[40]

The charge isomers of the 34,000-, 36,000-, and 41,000-dalton proteins were also predominant in the chromatin fraction and in the Tris wash of the nuclei[40] (Figures 25 and 26). Tight association of hnRNP proteins with chromatin was reported by many authors[53,55,152] and isotonic salt extraction at pH 8.0 was less efficient in releasing the RNPs from chromatin than procedures using sonication. In addition, the charge isomers of the 36,000-dalton protein were the most predominant polypeptides of the nuclear matrix (Figure 3) represented 25% of all the polypeptides. The charge isomers of the 41,000-dalton protein represented 2.7% of the matrix proteins.[40] These findings are very interesting, since the matrix was isolated

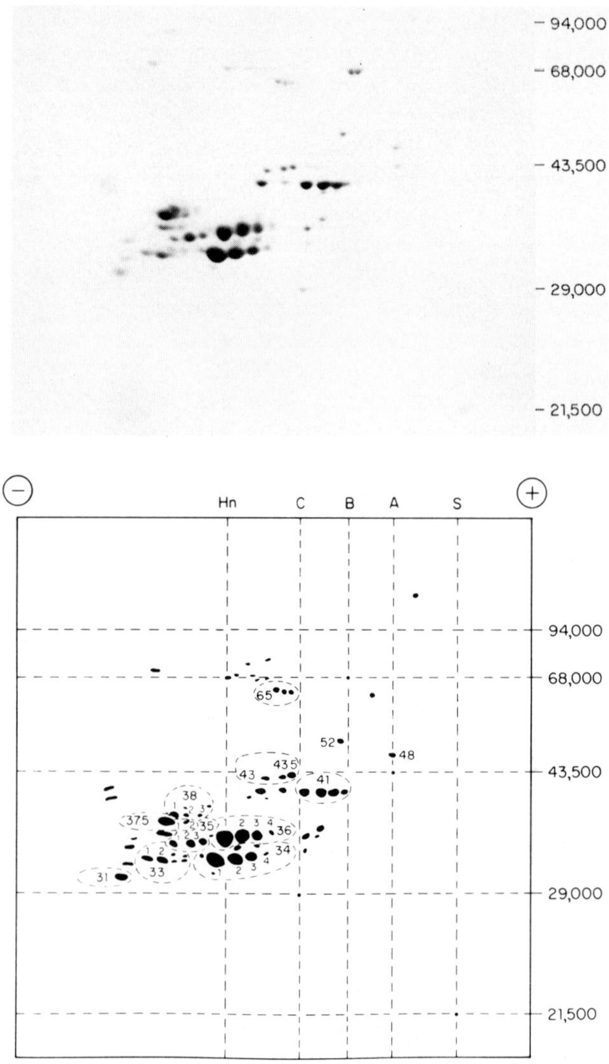

FIGURE 24. Two-dimensional gel electrophoresis of rat liver hnRNP particle proteins. The proteins were separated by nonequilibrium pH gradient electrophoresis in one direction and according to their molecular weights in the second direction (hn = position of the most basic of the charge isomers of the 36,000-dalton hnRNP particle protein, C = pH 7.4, B = pH 6.7, A = pH 6.1 and S = pH 5.5). (From Peters, K. E. and Comings, D. E., *J. Cell Biol.*, 86, 135, 1980. With permission.)

by the procedure of Berezney and Coffey[38] which includes digestion with RNase and DNase and extraction with 2 *M* CaCl. Moreover, four polypeptides of molecular weight between 65,000 and 73,000 represented 36% and one polypeptide of molecular weight 52,000 an additional 7% of the matrix proteins. These results support the notion that hnRNP particles are part of a nuclear matrix-ribonucleoprotein network.

One more result observed on the electrophorograms published by Peters and Comings[40] should be mentioned. It is the identical qualitative and quantitative separation of the charge isomers of liver hnRNP particles 34,000- and 36,000-dalton polypeptides. Peters and Comings[40] proposed that the polypeptides are coded by two genes which originated by gene duplication connected with elimination of a DNA sequence from one of the two genes which codes for

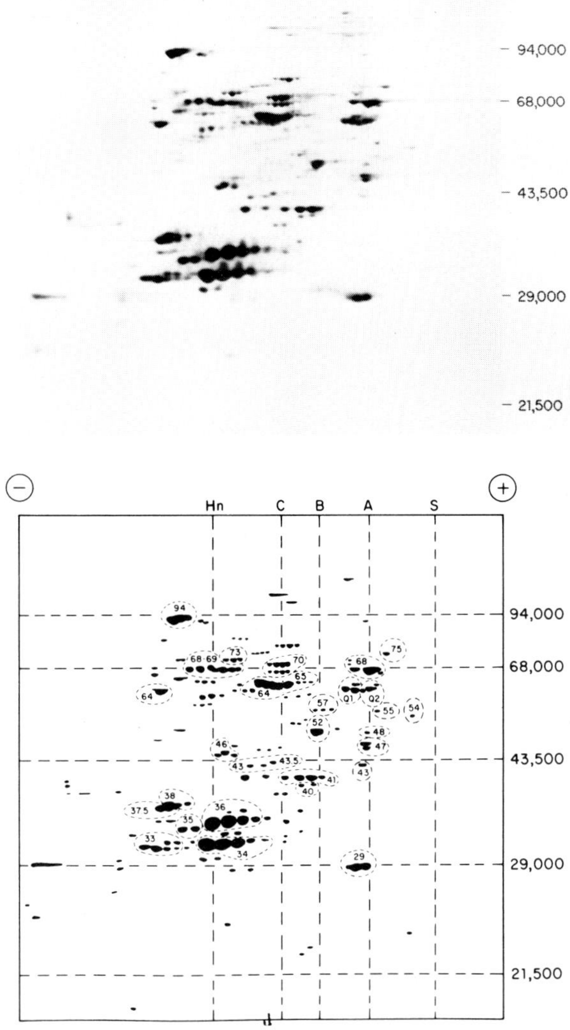

FIGURE 25. Two-dimensional gel electrophoresis of rat liver chromatin proteins. The proteins were separated by nonequilibrium pH gradient electrophoresis in one direction and according to their molecular weight in the second direction. (hn = position of the most basic of the charge isomers of the 36,000-dalton hnRNP particle protein, C = pH 7.4, B = pH 6.7, A = pH 6.1, and S = pH 5.5). (From Peters, K. E., and Comings, D. E., *J. Cell Biol.*, 86, 135, 1980. With permission.)

18 amino acids. Two such polypeptides could also originate by specific proteolytic cleavage of 18 amino acids from part of the synthesized polypeptide by some mechanism of secondary modification. Such a possibility would close the circle of considerations of the protein composition of hnRNP monoparticles (Samarina particles) and bring it back to the original proposal by Samarina and co-workers that only one major peptide is present in the Samarina particles.

Separation of the α-proteins of rat liver Samarina particles obtained by Peters and Comings[40] confirmed the earlier results of Northemann and Heinrich[120] who analyzed the protein moiety of rat liver hnRNP polyparticles by two-dimensional electrophoresis. Their results strongly support the claim that Samarina particles are a native substructure of nuclear RNPs.

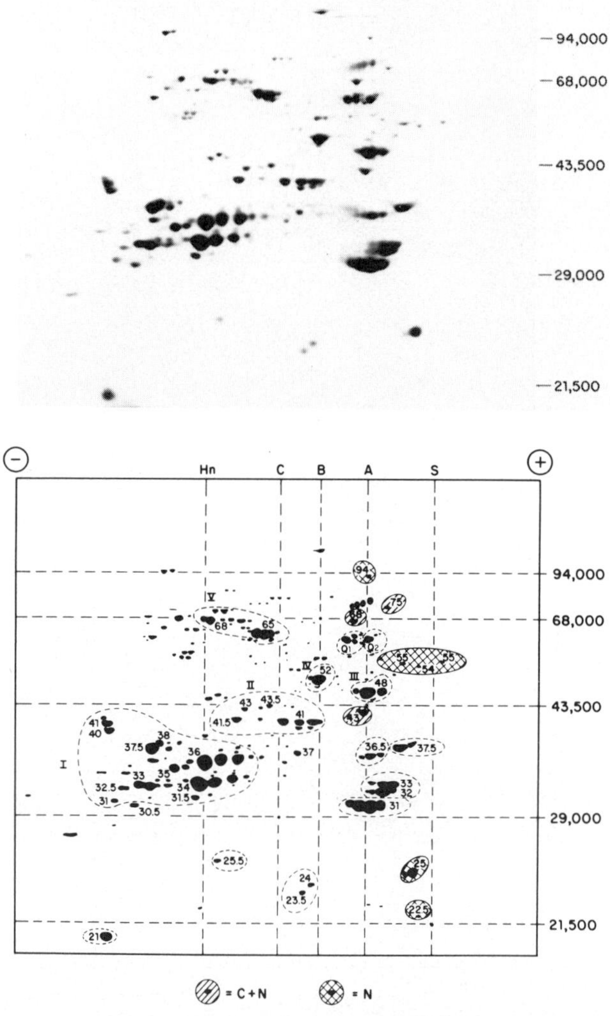

FIGURE 26. Two-dimensional gel electrophoresis of rat liver Tris wash proteins. The proteins were separated by nonequilibrium pH gradient electrophoresis in one direction and according to their molecular weight in the second direction: hn = position of the most basic of the charge isomers of the 36,000-dalton hnRNP particle protein, C = pH 7.4, B = pH 6.7., A = pH 6.1., and S = pH 5.5. C + N = proteins found in the cytoplasm, all nuclear washes, and final chromatin pellet; N = proteins not found in cytoplasm but which are present in all nuclear washes. (From Peters, K. E., and Comings, D. E., *J. Cell Biol.*, 86, 135, 1980. With permission.)

Some discrepancies in the observed pattern of the charge isomers of α-proteins on two-dimensional gels could be the result of protein carbamylation and deamination. For analysis of the charge isomers of the α-proteins of Samarina particles it is necessary to use deionized urea and to always prepare fresh urea-containing solution to avoid a carbamylation of proteins by isocyanate.[148] Also, the hnRNP particles should be prepared fresh and from fresh tissue to avoid spontaneous deamination of the glutaminyl and asparaginyl residues during the storage of cells or of hnRNP particles in frozen form.[153]

There are strong indications that the C proteins of Samarina particles differ principally from the A and B proteins. Beyer and co-workers[49] tested the stability of hnRNP monoparticles prepared from HeLa cell nuclei by low-salt extraction at pH 8.0 and further purified by centrifugation in linear 15 to 30% sucrose density gradient. According to the procedure used for their isolation and from their protein composition, the isolated 40 S hnRNP particles were predominantly Samarina particles. When the particles were exposed to an increasing salt concentration, 0.2 *M* NaCl dissociated more than 50% of the A-group proteins. At 0.4 *M* NaCl both the A and B group proteins were almost completely dissociated from RNA, but part of the C group proteins remained attached to RNA even in 0.6 *M* NaCl.

These results were confirmed by experiments of van Eekelen and van Venrooij[48] who studied the crosslinking by UV irradiation of mRNA to proteins of nuclear matrix in HeLa cells. The proteins which, based on their electrophoretic mobility, could correspond to the C group proteins of LeStourgeon and co-workers[49,50] were very efficiently cross-linked to hnRNA. The cross-linking of proteins which could correspond to the B-group proteins was much less efficient, and the authors could not demonstrate cross-linking with hnRNA of proteins which would correspond to the A proteins of LeStourgeon and co-workers.[49,50]

Similar results were obtained by Northemann and co-workers,[126] who analyzed the stability of hnRNP monoparticles prepared from rat liver nuclei in solutions of increasing ionic strength. The estimated sedimentation coefficient of the monoparticles in sucrose density gradient was 38 S. The estimated molecular weight of the proteins of the monoparticles ranged from 30,000 to 45,000 daltons. The major polypeptides in this group separated into three bands which could correspond to the A_2, B_1, and C proteins described by LeStourgeon and co-workers.[49,50] Treatment with 0.25 *M* NaCl released from the particles the A_2-like protein, and treatment with 0.5 *M* NaCl resulted in the release of B_1-like protein and in partial loss of the C-like protein. The authors assumed that the higher ionic strength disturbed the electrostatic bonds between the RNA and these proteins. The same proteins were also released by salt treatment of the 50- to 200-S hnRNP polyparticles of rat liver nuclei.[126] At the same time the majority of the minor heterogeneous proteins with molecular weights over 45,000 remained associated with hnRNA even after treatment of the RNP particles with 2 *M* NaCl.

What conclusion, if any, can be made from the various data on protein composition of hnRNP particles reported by different laboratories? In the opinion of the reviewer which, as any opinion, is biased, all presently available data can be unified into the following picture of the organization of hnRNP particles in the nucleus.

Morphological as well as biochemical data indicate that hnRNA is associated with nuclear matrix in the form of RNPs. The protein composition of the nuclear matrix-RNP network is not homogeneous. In some stretches, the RNA is associated with a heterogeneous population of predominantly acidic proteins which in the majority have molecular weights over 45,000, i.e., the β-proteins of Jacob. These fibrous stretches are interrupted by granules containing predominantly, but not exclusively, a limited number of basic proteins present as many charge isomers, i.e., the glycine-rich fraction of the α-proteins of Jacob. These proteins form a unique group of nuclear proteins characterized by high glycine content (around 25 mol %) and by the presence of an unusual amino acid, N^G, N^G-dimethylarginine.

Isolation procedures based on physical disintegration of the nuclear matrix-RNP network will break the matrix and release from chromatin a heterogeneous population of hnRNP particles composed of monoparticles containing only the α-proteins (Samarina particles with an approximate sedimentation coefficient of 30 to 35 S), particles containing predominantly the α-proteins with attached stretches of different lengths of the β-ribonucleoproteins, larger particles composed of various numbers of Samarina particles connected by β-ribonucleoproteins, and pieces of different lengths of β-ribonucleoproteins, the Jacob particles. The α-ribonucleoproteins and the β-ribonucleoproteins have very similar buoyant density and will band as one band in a density gradient. The α-ribonucleoproteins (Samarina particles) are more resistant to RNase treatment than the β-ribonucleoproteins (Jacob particles). It is also probable that activity of endogenous RNase is directly associated with the β-ribonucleoproteins. Therefore, the procedures for the isolation of the particles in which the nuclear matrix-RNP network is exposed to the activity of endogenous RNases (dependent on the cell type, length and temperature of isolation, activation, or inactivation of endogenous nucleases) result in degradation of the β-ribonucleoproteins and in the isolation of more or less pure Samarina particles. A very extensive activity of endogenous nucleases during the isolation procedure can lead to the isolation of approximately 40 S particles composed of the glycine-rich proteins only.

According to the conditions used for isolation of the hnRNP particles, including the tissue differences from which the particles are isolated, the variations in protein composition reported from different laboratories represent the variations in the proportions of Samarina, Jacob, and free-protein particles in the preparation.

The Samarina particles are composed of a limited number of glycine-rich proteins and of one or a few proteins with molecular weights around 40,000 with properties similar to β-proteins, which are probably part of the nuclear matrix proper. Discrepancy in the composition of Samarina particles which is found in the literature can be explained by the large amount of charge isomers of the glycine-rich proteins and by tight binding of the approximately 40,000-dalton protein to RNA. Procedures which are used to separate RNA from proteins of the particles before analysis of the protein moiety will also inadvertently remove the tightly bound protein together with the RNA. The charge isomers of the glycine-rich proteins will separate from each other in some systems of one-dimensional PAGE, but not in others. Two-dimensional separation of the proteins of hnRNP particles from rat liver, separating first the proteins according to their charge followed by separation according to their size, indicates the existence of only two primary glycine-rich protein species which are closely related, probably originating by gene duplication. It is even possible that only one primary species of these proteins is coded for by the genome and that some of the translated molecules of this one species are, in addition to other modifications, also cleaved by specific proteolytic enzymes.

VIII. ENZYMES AND PHOSPHOPROTEINS OF THE HNRP PARTICLES

The various reactions which are involved in the processing of hnRNA require the catalysis by many different enzymes. Some of the enzymes which could be involved in hnRNA processing were identified in the hnRNP particles. Since, until now, most of the preparative procedures did not rigorously separate the Samarina and Jacob particles from each other, it is difficult to localize precisely the individual enzymes in these particles, but in the opinion of the reviewer, the enzymes are most likely present among the β-proteins of Jacob particles.

Neissing and Sekeris[133] observed breakage of 50 to 80 S hnRNA at a limited number of sites when this RNA was incubated with the protein component of 30 S hnRNP particles isolated from rat liver. They also found Mn^{2+}-dependent poly(A) polymerase activity in the hnRNP particles.[154] Later the same authors[155] identified in hnRNP particles from rat liver nuclei two enzymatic poly(A) synthesizing activities and also a ribohomopolymer synthetase which can polymerize ATP, UTP, GTP, and CTP to the corresponding ribohomopolymers.

Autodegradation studies of RNA in hnRNP particles confirmed that endogenous ribonucleases are present among the proteins of the particles.[115,132,190] Incubation of hnRNP polyparticles isolated from rat liver at 37° for 15 min resulted in nearly complete degradation of particle hnRNA.[132] In contrast, incubation of purified rat liver hnRNP monoparticles (Samarina particles) at 37°C for 1 hr leads to a loss of only 16% of the monoparticle hnRNA.[98] These results indicate that the endogenous nucleases are probably associated with the β-proteins of the Jacob particles.

Other enzymes involved in the formation of mRNA which were reported to be present in the rat liver hnRNA particles are guanylyltransferase and N^7 and 2-o-methyltransferase.[157] These enzymes are involved in the formation of the 5′ end cap on mRNA.[158] But the same enzymes were also found to be present in the nuclear sap proteins, and a contamination of particles by these enzymes during particle purification cannot be excluded.

In addition to RNase which will hydrolyze single-stranded RNA, the particles also contain RNase attacking double-stranded RNA. Molnár et al.[159] by incubation in vitro of double- and single-stranded fragments of hnRNA with rat liver hnRNP particles have shown the presence of 5′-exonuclease and of an endonuclease specific for double-stranded RNA. Of interest are the findings of Rech and co-workers,[160] who identified ribonuclease acting against double-stranded poly r(AU) in hnRNP particles of HeLa cells. The enzymatic activity was repressed in the particles and was released by the removal of endogenous particle RNA, either by chromatography of the particles on DE52 column or by digestion of the endogenous RNA by RNase A and T1. These very interesting experiments need further exploration since Calvet and Pederson[161] could not detect degradation of HeLa cell hnRNP particle double-stranded RNA after the hydrolysis of the single-stranded RNA by RNase A and T1. The conclusion of Rech et al.[160] that hydrolysis of particle single-stranded RNA results in the activation of enzymatic activity which degrades double-stranded RNA is supported by later results from the same laboratory. Paoletti et al.[162] have found that intercalation of ethidium bromide into double-stranded RNA of HeLa hnRNP particles can be eliminated by treatment of the particles with RNases which are specific for single-stranded RNA.

Another important group of enzymes connected with hnRNP particles are the enzymes involved in the phosphorylation and dephosphorylation of proteins. Incorporation of ^{32}P into the proteins of rat brain nuclear RNP particles has been shown by Gallinaro-Matringe and Jacob in 1973.[163] The [^{32}P] radioactivity was sensitive to pronase and was found associated with RNP particles with CsCl buoyant density of 1.39 to 1.40. One third of the proteins of hnRNP particles larger than 60 S from rat brain were found to be phosphoproteins.[99] Most of these proteins had molecular weights greater than 42,000, but some migrated in SDS-PAGE with the low-molecular-weight glycine-rich proteins of Samarina particles. Also in 1973 Martin et al.[116] reported observations of efficient in vitro phosphorylation by cyclic AMP-activated protein kinase of Samarina particle proteins from Taper mouse ascites hepatoma cells, but a low incorporation of phosphates into the same proteins in vivo.

Phosphorylation of the main glycine-rich proteins of Samarina particles is controversial. Phosphorylation of the major proteins of rat liver hnRNP particles which would correspond to the glycine-rich proteins has been reported[129] but phosphorylation of the same proteins in HeLa cell hnRNP particles was not found.[50] In the opinion of the reviewer, it is probable that the glycine-rich proteins of Samarina particles do not contain any phosphorylated species, but phosphoproteins of the same molecular weight from Jacob particles migrate together with the glycine-rich proteins during SDS-PAGE. That the majority of phosphoproteins are present in the Jacob particles is shown by the fact that the phosphoprotein content of the heterogeneous complexes of rat brain RNPs is higher than that of hnRNP monoparticles.[15] According to LeStourgeon et al.,[50] only the C proteins of HeLa cell Samarina particles are phosphorylated. Schweiger and Schmidt[164] isolated from Samarina particles of rat liver nuclei a 39,000-dalton phosphoprotein which could correspond to a C protein. Phosphorylation of the C proteins and of many β-proteins of the Jacob particles supports their proposed origin from the extended nuclear matrix. The presence of many phosphorylated proteins in the matrix is well documented.[26] However, at present it cannot be rigorously excluded that some of the glycine-rich proteins are not phosphorylated, and because of their changed properties and small quantities, are lost during purification of these proteins. The phosphoproteins have higher affinity for hnRNA than the nonphosphorylated species.[50,99]

Since phosphoproteins are present in the hnRNP particles, a control of their function by phosphorylation and dephosphorylation could be anticipated. Protein kinase activity associated with hnRNP particles, which phosphorylated particle phosphoproteins, has been reported by Martin et al.[116] for hnRNP particles isolated from Taper cells nuclei, and by Schweiger and Schmidt[164] for hnRNP particles isolated from rat liver nuclei. Enzymes involved in phosphorylation and dephosphorylation of the phosphoproteins of hnRNP particles of HeLa cells have been investigated thoroughly by the Jeanteur group.[156,165] Protein kinase present in the particles was able to catalyze the incorporation of $\gamma[^{32}P]$ from ATP into phosphoserine and phosphothreonine residues of various hnRNP particle proteins. The kinase present among the particle proteins had an apparent K_m value of 0.02 m*M* for ATP — it required Mg^{2+} or Mn^{2+} for its activity — and cyclic AMP or cyclic GMP had no effect on the reaction.[156] Phosphate acceptor proteins were also associated with the particles. Predominantly two polypeptide chains (28,000 and 37,000 daltons) were labeled in vitro. Two additional polypeptides (30,000 and 52,000 daltons) of the hnRNP particles were strongly labeled in vivo after the exposure of HeLa cells to $[^{32}P]$.[166] In contrast to other phosphoproteins of the hnRNP particles which are firmly associated with RNA, the 28,000-dalton species was partly dissociated from the particles after their treatment with 0.15 *M* NaCl.[166] The dissociation of this protein species at low-salt concentration from the hnRNP particles could explain why no special attention was paid to this interesting polypeptide by other authors.

Periasamy et al.[167] purified a protein kinase from HeLa cell hnRNP particles by a combination of chromatographies on DEAE-cellulose, phosphocellulose and Sephacryl® S-200. Its molecular weight was determined by gel filtration to be approximately 48,000. Casein was a better exogenous substrate than histones. The properties of this enzyme were identical to those found for the nonpurified enzyme present in hnRNP particles.[156]

In addition to protein kinase, the HeLa cell hnRNP particles also contained a phosphoprotein phosphatase which was distinct from nonspecific alkaline phosphatase. The phosphatase was optimally active at pH 7.5, inhibited by divalent cations and *N*-ethylmaleimide, and activated by 2-mercaptoethanol.[168] The Jeanteur group[168] proposed that the activity of some of the hnRNP proteins is regulated by modification of the phosphorylation level of the particle phosphoproteins by the actions of the protein phosphatases and protein kinases.

The proposal that the extent of hnRNP particles phosphorylation could control their function is supported by our experiments in which we measured the extent of phosphorylation of the hnRNP particles in the liver of rats in which RNA processing was distorted by feeding of the carcinogen 3′-methyl-4-dimethylaminoazobenzene.[290] We have found that 48 hr after the carcinogen was administered by stomach tube, only 7% of the phosphate found in controls was still bound to proteins of the rat liver hnRNP particles. However, the specific activity of the phosphate bound to these hnRNP particle proteins increased by 800% after the treatment with the carcinogen, perhaps as a result of activation of a mechanism trying to prevent the dephosphorylation of the particle proteins triggered by the carcinogen. In contrast to Periasamy et al.[168] we could detect protein phosphatase only in the total fraction of rat liver nuclear nonhistone proteins but not in the purified hnRNP particles. The protein kinase which we found associated with rat liver hnRNP particles could be removed together with the high-molecular-weight proteins by repeated purification of the particles. The activity of the protein kinase and protein phosphatase in the fraction of nonhistone chromosomal proteins 48 hr after feeding of the carcinogen increased by 200 and 150%, respectively. Protein kinase was also detected in a crude preparation of the rat liver hnRNA particles by Karn and co-workers.[129]

In the hnRNP particle fraction these enzymes are probably not associated with Samarina particles, but with the β-proteins of the Jacob particles. This is indicated by our inability to detect any enzymatic activity in repeatedly purified Samarina particles. The analysis of the hnRNP particle proteins by SDS-PAGE by Blanchard et al.[156,165] indicates the presence of both types of particles in their particle preparation. The finding that treatment of their particles with 0.25 *M* NaCl resulted in the release of only 20% of the particle protein kinase activity[156] also agrees with the assumption that the kinase activity is associated with Jacob particles. The glycine-rich proteins of Samarina particles are almost completely dissociated from RNA after treatment with 0.3 *M* NaCl.[49]

The phosphoprotein kinase partially purified from the 0.1 *M* NaCl pH 8.0 extract of rat liver nuclei in our laboratory has properties similar to the protein kinase isolated from hnRNP particles of HeLa cells by Blanchard et al.[156] The pH optimum for its enzymatic activity is between pH 8.0 and 8.5, and the activity is cyclic AMP and cyclic GMP independent. Casein is a more efficient substrate than histones. It differs from the protein kinase isolated from HeLa cell hnRNP particles by the fact that Mn^{2+} is a more effective activator of the liver enzyme than Mg^{2+}, but our enzyme is not homogeneous. The fractionation of the 0.1 *M* NaCl pH 8.0 rat liver nuclear extract by phosphocellulose column chromatography separated the protein kinase into three peaks. The first was eluted with 0.15 *M* NaCl and could correspond to the protein kinase isolated from the HeLa cells hnRNP particles. The two other peaks were eluted close together from the column with 0.3 *M* NaCl.

IX. PROTEINS ASSOCIATED WITH POLY(A) SEQUENCES OF HNRNA

In the hnRNP polyparticles, specific proteins are associated with poly(A) and possibly also with oligo(A) sequences of hnRNA. Some hnRNA molecules contain poly(A) sequences at the 3′-OH termini[170-172] which are added to them posttranscriptionally. Presence of 3′-terminal poly(A) sequences in the hnRNP particles was first reported in 1973,[83,173] and in two laboratories small RNP particles containing poly(A) sequences were isolated which exhibited lower CsCl buoyant density than the large RNP particles.[116,174] Degradation by endogenous nucleases of the nuclear RNP polyparticles during their isolation fragmented the hnRNA and released the poly(A) sequences, resulting in the isolation of a mixture of 30 S hnRNP particles and 14 to 17 hnRNP particles.[174,175]

Quinlan and co-workers[175] extracted RNP from Taper mouse ascites hepatoma cells with 0.1 *M* NaCl at pH 9.0 and separated the small poly(A) particles by centrifugation of the extract in 5 to 30% sucrose density gradient. Particles of two sizes (15 and 17 S) containing poly(A) were found in the extract. The 17 S particles were associated with eight proteins ranging in molecular weight from 20,000 to 32,000, and the 15 S particles were associated with five proteins (57,500, 63,000, 86,000, 120,000, and 140,000 daltons) of which the 86,000-dalton protein was predominant.[1] The poly(A) sequences associated with purified 15 S RNP particles were 190- to 200-nucleotides long, and 60% of the total nuclear poly(A) sequences could be recovered from the nuclei in this form.[176] The isolated particles were RNPs as was shown by their binding to Millipore® filters at low-salt concentration and by their CsCl buoyant density of 1.36 g/cm^3. The crude 15 S RNP particles contained 70% adenylate and 20% uridylate, but the uridylate could be removed by additional purification. The authors suggest that oligo(U) sequences could be artificially associated with the poly(A) during preparation of 15-S particle RNA.[176] Since the 86,000-dalton protein was still associated with the purified particles which contained very little uridylate, the authors proposed that this protein is directly associated with the poly(A) sequences. The 86,000-dalton protein was easily displaced from the poly(A) containing particles by treatment with 0.2 *M* or higher concentration of NaCl.[1] In the cytoplasm, the poly(A) sequences of mRNA are associated by salt-stable bonds with a 78,000-dalton protein.[177] Since no 76,000- to 78,000-dalton proteins were found in salt-stable association in the 15 S RNP particles, the authors assumed that the 86,000-dalton protein associated with poly(A) RNA in the nucleus is replaced in the cytoplasm by the 78,000-dalton protein.[1]

The results and conclusions of Billings and Martin[1] differ from those reached in Pederson's

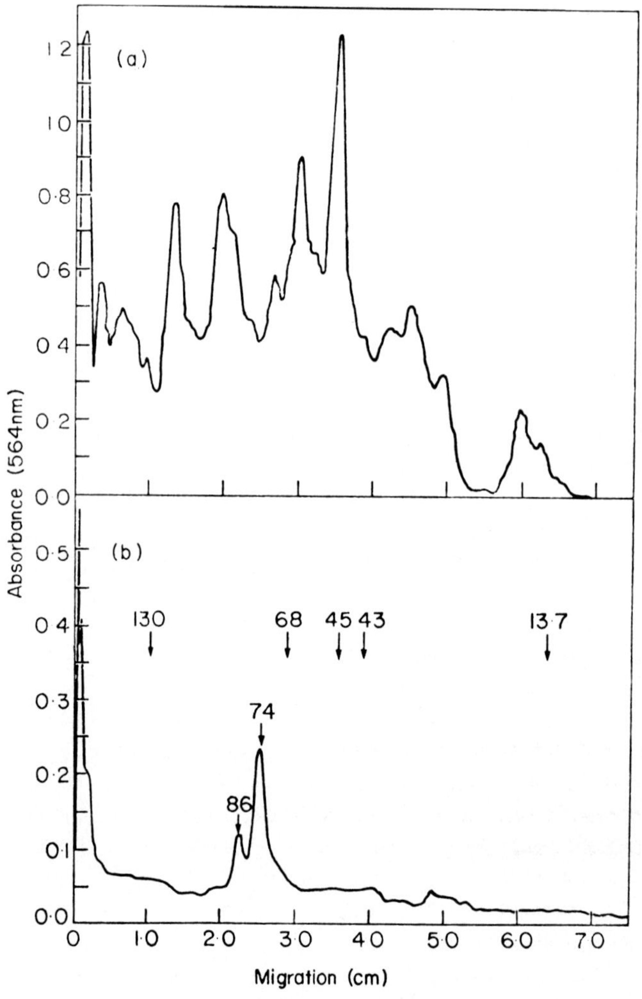

FIGURE 27. Densitometric tracing of SDS-polyacrylamide gel electrophorogram of hnRNP particle proteins associated with poly(A) sequences of hnRNA. (a) Total proteins isolated from HeLa cells hnRNP particles, and (b) proteins associated with the RNase resistant material bound to poly(U)-Sepharose®. The positions of marker polypeptides, run in a parallel gel, are indicated: β-galactosidase, 130,000; bovine serum albumin, 68,000; muscle actin, 45,000; ovalbumin, 43,000; and pancreatic RNase, 13,700. (From Kish, V. M. and Pederson, T., *J. Mol. Biol.*, 95, 227, 1975. With permission.)

laboratory: Kish and Pederson[179] isolated from HeLa cells hnRNA-protein complexes containing poly(A) and oligo(A) sequences. The basis for the isolation was the resistance of such complexes to hydrolysis by RNase A and T_1 in 0.5 *M* NaCl and their affinity for poly(U) Sepharose®. A fraction eluted from the poly(U) Sepharose® with 50% formamide was composed of 150- to 200-nucleotides-long poly(A) and 20- to 30-nucleotides-long oligo(A) sequences as well as two polypeptides with molecular weights of 74,000 and 86,000, of hich the first was the more prominent (Figure 27). The authors concluded that both poly(A) and oligo(A) sequences in HeLa cells hnRNA exist in vivo tightly complexed with polypeptides. Because of the association of a 74,000-dalton polypeptide with mRNA poly(A) sequences in cytoplasm the authors consider it possible that the protein associated with nuclear poly(A) sequences may be conserved during the transport of hnRNA from the nucleus into cytoplasm.[178] In the same laboratory a protein of similar molecular weight (72,000 to 74,000) was found to be bound to hnRNA poly(A) sequences of the slime mold *Dyctiostelium discoideum*.[180]

Billings and Martin[1] proposed as a possible explanation of the disparity between these two laboratories the difference in procedures used for purification of the poly(A)-containing complexes. Centrifugation in sucrose density gradient of hnRNP mildly digested by endogenous nucleases separated from the hnRNP particles only RNPs containing poly(A) sequences. The oligo(A) sequence remained associated with other hnRNA sequences in the 30 S hnRNP particles. On the other hand, Kish and Pederson[179] by complete enzymatic digestion of total hnRNP, prepared a mixture of ribonucleoproteins, some containing the intramolecular oligo(A) and some the 3′ terminal poly(A) sequences. It would, therefore, be possible that the 74,000- to 78,000-dalton protein is associated with the intramolecular oligo(A) sequences and not with the 3′ terminal poly(A) sequences.

Results obtained by Maxwell and co-workers[181] support the conclusions reached in Pederson's laboratory,[179] but the question will not be settled before the identity of the involved proteins is firmly established. Maxwell et al.[181] isolated by the Samarina procedure from duck erythroblast nuclei the 40, 17, 15, and 12 S RNP particles. The 40 S particles contained a mixture of α- and β-proteins, the 12 and 17 S particles were snRNA-ribonucleoproteins, and the 15 S particles contained some 200-nucleotide or longer poly(A) sequences associated with six major proteins. One of them was the 73,000-dalton protein which the authors consider identical with the protein found associated with cytoplasmic poly(A) sequences.[181]

It should be mentioned that still different results were obtained in a third laboratory. Molnár and Samarina[182] prepared by sedimentation in sucrose gradient, Cs_2SO_4 density gradient centrifugation, and affinity chromatography on a poly(dT) Sepharose® column, 14 to 15 S poly(A)-containing RNP particles. Analyses of the particles by SDS-PAGE revealed the presence of two major proteins (51,000 and 69,000 daltons) and of two or three minor components. Treatment of the particles with 0.6 *M* NaCl led to the release of the 51,000-dalton protein.

It is possible that the main poly(A)-binding protein is one of the residual nuclear matrix proteins and that the hnRNA is attached to the residual matrix through its 3′-terminal poly(A) sequences (and also through its double-stranded sequences). Herman and co-workers[22] have shown that after the removal of chromatin from HeLa cell nuclei, digestion of the RNP network with ribonuclease and elution with ammonium sulfate, only 2% of the originally present hnRNA but more than half of the poly(A) sequences and of the double-stranded regions of hnRNA remained attached to the residual nuclear matrix. The poly(A) sequences were attached to the matrix by EDTA-sensitive linkage and were released in the form of 12 S RNP particles after EDTA treatment. These results were complemented by the results of Maundrell et al.[45] who found that 73,000-dalton poly(A) binding protein is among the residual matrix proteins. According to the authors, the 73,000-dalton protein is the main poly(A) binding protein in the nucleus, as well as in the cytoplasm.[181] Therefore the presence of poly(A) ribonucleoproteins in the fraction of hnRNP polyparticles can be expected when EDTA or physical disruption of nuclear matrix is used during the polyparticles preparation.

X. RNA OF THE HNRNP PARTICLES

The presence of rapidly synthesized RNA in nuclear RNP particles was already used by the Samarina group as a marker for their isolation.[11,12] The RNA in the particles hybridized with high efficiency to intermediate repetitive sequences of DNA. The RNP complexes of rat liver contained both types of nuclear RNA sequences, those which are degraded in the nucleus and those which are transferred into the cytoplasm.[183] The presence of nucleus-restricted sequences in RNA of the 30 S RNP particles was confirmed by Martin and McCarthy,[184] who have also shown that the kinetics and turnover of RNA associated with the nuclear RNP particles were in a range expected for hnRNA.

There is good correlation between the size of the RNP polyparticles and the size of hnRNA isolated from them. RNA isolated by Samarina and co-workers[12] from rat liver Samarina particles was degraded and had a sedimentation coefficient of 4 to 8.5 S. The 60 S polyparticles contained RNA with a sedimentation coefficient around 17 S and, RNA isolated from the 120 to 130 S particles sedimented in a broad peak with a median sedimentation coefficient around 32 S. A similar dependence between the size of isolated RNA and the size of RNP polyparticles was also observed by other investigators. For example, Niessing and Sekeris[109] isolated 5 to 80 S hnRNA from 30 to 400 S hnRNP particles of rat liver nuclei. Ducamp and Jeanteur[83] prepared 9 and 12 S RNA from 30 and 60 S hnRNP polyparticles of HeLa cell nuclei, and Pederson[73] obtained 20 to 80 S RNA from 40 to 250 S hnRNP particles of the same cells.

In contrast to an earlier report by Samarina and co-workers,[12] Martin and McCarthy[184] could not quantitatively convert by mild RNase digestion labeled RNA of mouse liver polyparticles into 30 S hnRNP monoparticles, i.e., an early indication of two types of RNP structures in the polyparticles.

The hybridization techniques used in earlier experiments could detect only repetitive RNA sequences. The presence of RNA sequences copied into proteins was demonstrated for the first time in 1972 when it was shown that adenovirus-specific mRNA sequences are associated with nuclear 30 S RNP particles in adenovirus-infected cultured human FL cells.[79] Today, the presence of mRNA sequences in the nuclear hnRNP complexes is well established. Firtel and Pederson[180] have shown that 55 S RNP particles from the nuclei of cellular slime mold *Dictyostelium* contain hnRNA. Kinniburgh and Martin,[185] using a DNA probe complementary to total poly(A)$^+$ mRNA, demonstrated the presence of sequences homologous to this RNA in 30 S hnRNP monoparticles isolated from nuclei of mouse ascites cells (Figure 28). From the kinetics of hybridization the authors estimated that 10 to 15% of the RNA of the 30 S RNP particles is homologous to mRNA. Nuclear RNP particles from chick erythroblast nuclei contain RNA which coded for globin when assayed in a wheat germ cell-free system.[186] Globin mRNA sequences were also demonstrated directly in 40 S hnRNP particles prepared from duck erythroblast nuclei. Approximately 0.5% of the sequences of RNA isolated from these particles were complementary to globin mRNA.[141] Finally, Pederson and Davis[74] have shown that the 50 to 250 S hnRNP polyparticles from mouse Friend erythroleukemia cells contain covalently intact 15 S β-globin mRNA precursor and 9 S β-globin mRNA (Figures 29 and 30). The presence of both mRNA precursor and its product in hnRNP polyparticles strongly support the claim that nuclear hnRNP structures are the sites of mRNA processing. Viral RNA in adenovirus-infected cells is also associated with the nuclear hnRNP complexes.[187] During their synthesis the nascent chains of adenovirus RNA are immediately associated with proteins[188] and can be isolated in the form of 30 to 200 S RNP complexes from the infected cells. The RNA found in the hnRNP complexes is copied from all the regions of the adenovirus-2 genome.[189]

The association of hnRNA with proteins in hnRNP particles protects nucleotide sequences of mRNA from degradation by nucleases. Mild digestion of nuclear RNPs from HeLa cells by micrococcal nuclease, pancreatic RNase, or endogenous particle-bound nucleases has shown (by hybridization of the nondigested RNA with DNA complementary to HeLa cell poly(A)$^+$ mRNA) that nucleotide sequences homologous to mRNA are not present in the RNP particles in exposed stretches, and are protected from the attack by nucleases.[190] The protected mRNA sequences of the hnRNP RNA are probably present in the Samarina particles. Alonso and co-workers[191] compared hybridization of DNA, complementary to rat liver polysomal poly(A)$^+$ mRNA, to RNA isolated from the 30 S hnRNP monoparticles with the hybridization to RNA isolated from the 80 to 150 S hnRNP polyparticles (Figure 31). The proportion of mRNA sequences was higher in the RNA isolated from 30 S hnRNP monoparticles than in the RNA isolated from the 80 to 150 S hnRNP polyparticles. The

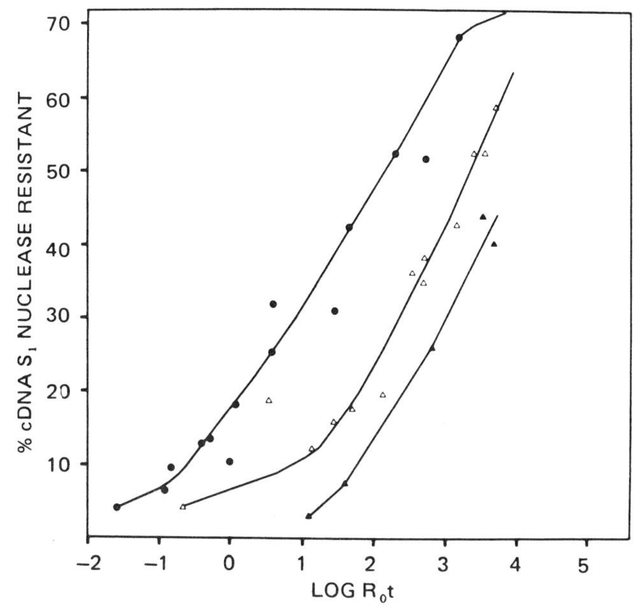

FIGURE 28. Hybridization kinetics of RNA from 30 S hnRNP particles and of poly(A)$^+$ mRNA template. with DNA complementary to poly(A)$^+$ mRNA template. The RNA was prepared from Taper ascites cells; (●——●) poly (A)$^+$ mRNA homologous reaction; (▲——▲) crude RNA from 30 S hnRNP particles; and (△——△) purified RNA from 30 S hnRNP particles. (From Kinniburgh, A. J. and Martin, T. E., *Proc. Natl. Acad. Sci. U.S.A.*, 73, 2725, 1976. With permission.)

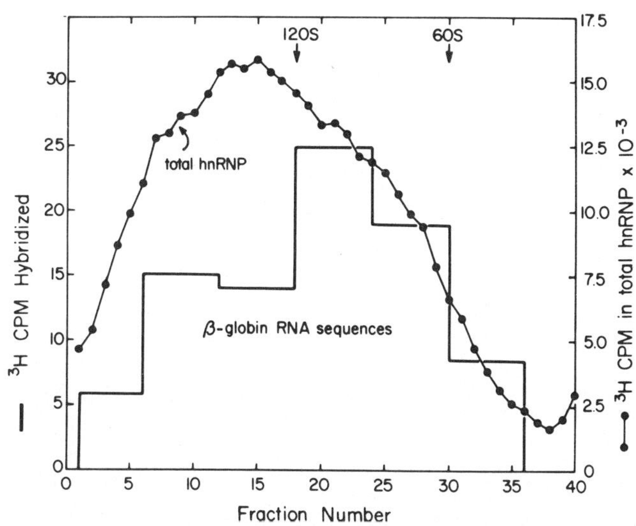

FIGURE 29. Detection of newly synthesized β-globin gene transcripts in Friend cell hnRNP particles. [^{3}H] uridine pulse-labeled hnRNP particles were isolated and centrifuged in 15 to 45% sucrose gradient. RNA was extracted from pooled gradient fractions and hybridized to nitrocellulose filters containing cloned mouse β-globin DNA; (● = Trichloroacetic acid-precipitable radioactivity in hnRNP particles. The histogram shows the total amount of β-globin RNA sequences present in each pool of gradient fractions. (From Pederson, T. and Davis, N. G., *J. Cell Biol.*, 87, 47, 1980. With permission.)

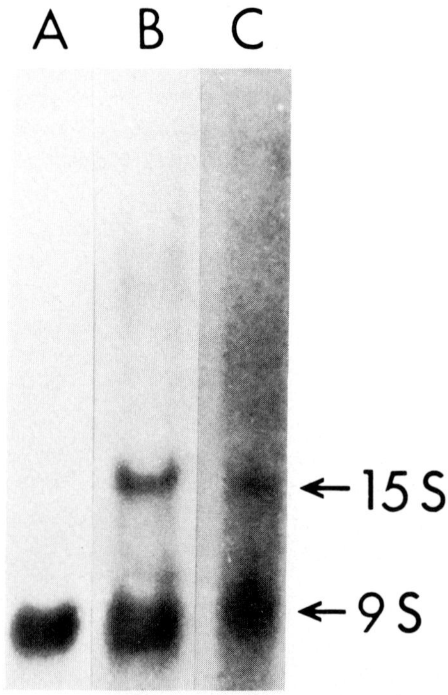

FIGURE 30. Northern blot hybridization of β-globin gene transcripts in hnRNP particles. RNA prepared from nuclei, cytoplasm, or purified hnRNP particles was electrophoresed in agarose gels after denaturation in glyoxal. After transfer to DBM paper, the RNA was hybridized with nick-translated [^{32}P] β-globin DNA. (A) Total poly(A)$^+$ cytoplasmic RNA, (B) total poly(A)$^+$ nuclear RNA, (C) RNA from purified hnRNP particles. (From Pederson, T. and Davis, N. G., *J. Cell Biol.*, 87, 47, 1980. With permission.)

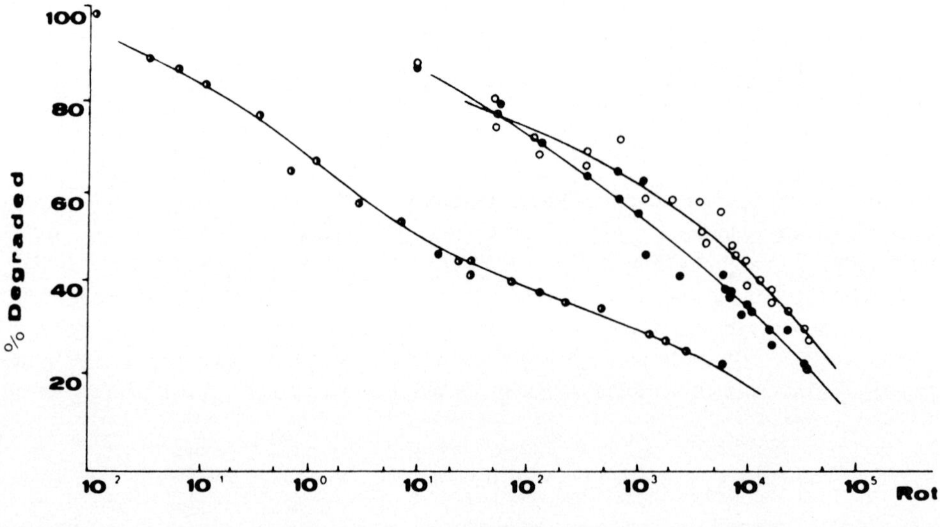

FIGURE 31. Hybridization of hnRNP particles RNA with DNA complementary to polysomal poly(A)$^+$ RNA. The RNA used in the hybridization was isolated from rat liver; (●) RNA from 30 S hnRNP monoparticles, and (●) RNA from 80 S to 150 S hnRNP polyparticles. (◑) = Poly(A)$^+$ mRNA homologous reaction. (From Alonso, A., Louis, Ch., Flytzanis, C. N., and Sekeris, C. E., *FEBS Lett.*, 93, 351, 1978. With permission.)

authors proposed that sequences of mRNA are protected during processing by association with proteins into the 30 S hnRNP particles. Severalfold enrichment of mRNA sequences in the 40 S hnRNP fraction of nuclear RNP was also reported by Maundrell and Scherrer.[141] The poly(A) sequences found at the 3′ end of many mRNA and pre-mRNA molecules are not part of the 30 S hnRNP particles. As was discussed earlier, they are associated with specific proteins and can be isolated as small RNP particles.

In addition to the sequences of mRNA, the RNA present in hnRNP particles contains many sequences which are not released from the nuclei. One type of RNA sequences found in hnRNA but not in cytoplasmic RNA are short adenylate-rich sequences (oligo(A) sequences). These sequences were identified in hnRNP particles of HeLa cells,[179] mouse ascites cells,[192] and rat liver cells.[193] The oligo(A) sequences are relatively sensitive to cleavage by nucleases, indicating that they could be present in the exposed part of the hnRNP complexes.[194]

Double-stranded stretches in hnRNA are considered as possible recognition sites for enzymes involved in RNA processing.[195] The characterization of double-stranded RNA sequences in the hnRNP particles is therefore of interest. When hnRNP particles isolated from the nuclei of HeLa cells, pulse labeled with [^{3}H] uridine for 20 min, were digested at high ionic strength with a mixture of RNases A and T_1, from 2.5 to 3.1% of the labeled nucleotide sequences were resistant to hydrolysis by the enzymes. These nucleotides were present in the RNP particles in double-stranded form.[161] However, the amount of proteins associated with the double-stranded regions of hnRNA in the particle was very small, i.e., the ratio of protein to RNA in the double-stranded RNA sequences was determined to be 0.03, only a fraction of the ratio 4.0 determined for the intact hnRNP particles.[196] Being relatively free of proteins, the double-stranded regions are accessible to digestion by nucleases attacking double-stranded structures. The accessibility in the particle of the double-stranded RNA regions was confirmed in experiments in which the double-stranded structures of hnRNA were cross-linked *in situ* by psoralen derivatives.[197,198] These experiments also definitively established that the double-stranded mRNA regions exist in living cells.[198] In addition to the double-stranded sequences which are present in the hnRNP particle, the hnRNA contains other complementary sequences which are restrained in a single-stranded form by an interaction of hnRNA with particle proteins.[161]

Since snRNAs are metabolically stable, a measurement of nucleotides present in double-stranded form in the particle, which is based on the amount of radioactivity resistant to hydrolysis by RNases A and T_1 at high ionic strength, will give lower values than other techniques. This could be the reason for a higher percentage of double-stranded RNA found in the hnRNP particles of HeLa cells by Paoletti and co-workers.[162] These authors, using intercalation of ethidium bromide as a probe for double-strandedness, estimated that 8% of total RNA in hnRNP particles is present in double-stranded form. The Scatchard plot of ethidium bromide binding to double-stranded RNA of the native hnRNP particles had a marked biphasic appearance, indicating anticooperative binding. Such binding could result if both ends of the double-stranded RNA structures were constrained in their free rotation, one relative to the other, perhaps because of the binding to the proteins on both ends of the double-stranded structures. Removal of the particle proteins by digestion with proteinase K abolished the anticooperativity of ethidium bromide binding, supporting the notion that the highly rigid structure of the double-stranded sequences is maintained through the interaction with particle proteins.

XI. SMALL NUCLEAR RNA AND HNRNP PARTICLES

The presence of discrete, metabolically stable species of small (ranging in size from about 80 to 220 nucleotides) nuclear RNA (snRNA) in the nuclei of diverse eukaryotic species has been described by many authors (for review see Reference 199). The association of small RNA with RNP structures was shown in 1971 by Rein[200] in mitotic HeLa cells.

The nomenclature of snRNA species is not yet unified. Where the authors claim an identity with species already described, the nomenclature of Ro-Choi and Busch[199] is preferred. Where exact correspondence with snRNA described in other studies is not always apparent, molecular weights or designation used in the study are used.

Association of the small nuclear RNA (snRNA) with hnRNP particles was first demonstrated by Sekeris and Niessing[201] who found in rat liver hnRNP particles low-molecular-weight RNA which differed from the hnRNA by its metabolic stability. The snRNA was found associated with hnRNP polyparticles, as well as with hnRNP monoparticles.[202] The existence of two metabolically distinguishable types of RNA in the hnRNP particles was confirmed by the study of RNA in the hnRNP particles isolated from liver of rats after the inhibition of RNA synthesis by α-amanitin or by actinomycin D.[96] In a later paper from the Sekeris laboratory,[203] the snRNA isolated from rat liver hnRNP monoparticles was separated into six electrophoretic bands with S values from 4.5 to 6.4.

The association of snRNP with hnRnP particles was confirmed by workers from many different laboratories. Zieve and Penman[204] found in HeLa cells the most abundant species (U1) of snRNA associated with 40 S RNP particles. Northemann and co-workers[291] detected several distinct RNA species, ranging in size from 5 to 8 S, associated with rat liver nuclear ribonucleoproteins. Four species of these low molecular weight RNAs were common to the ribonucleoprotein mono- and polyparticles. Guimont-Ducamp and co-workers demonstrated snRNA in hnRNP mono- and polyparticles of HeLa cells.[205] Howard[206] isolated 30 to 60 S RNP particles from mouse erythroleukemia cell nuclei containing four species of snRNA. Seifert and co-workers[132] identified six distinct species of snRNA in hnRNP polyparticles of rat liver and a similar snRNA pattern was seen in 38 S hnRNP monoparticles isolated from rat liver. It is of interest that the U1a and U1b snRNAs were present in the monoparticles in higher proportion than in the polyparticles. A similar observation was made by Gallinaro and Jacob,[207] who found that in rat brain hnRNP particles U1 snRNA represented 50% of the total snRNA in the monoparticles, but only 20% in the polyparticles, where the other two species of snRNA (the U2 and 5 S RNA) were enriched. Maxwell and co-workers[181] identified in 40 S hnRNP particles isolated from duck erythroblast nuclei eight species of snRNA. The presence of snRNA was also demonstrated in perichromatin granules.[208] The granules are probably identical with the Samarina particles as is indicated by their sedimentation coefficient of 40 S and, even more, by the presence of proteins which, when subjected to electrophoresis, exhibited migration properties similar to the basic glycine-rich proteins of the Samarina particles from liver cells.

But snRNA is also associated in the cell nucleus with structures other than the hnRNP particles. Zieve and Penman[204] found that snRNA is associated with the nuclear matrix, in addition to its association with 40 S hnRNP particles. This has been confirmed by Miller et al.,[209] who observed that all species of snRNA present in rat liver nuclei remained associated with the nuclear skeleton (matrix) when the nuclei were deprived of nucleoplasm and chromatin. It should be mentioned that under these conditions of nuclear matrix preparation, hnRNA also remains associated with the matrix. Similar results were obtained by Maundrell and co-workers[45] who found the association of all species of snRNA with the matrix of duck erythroblast nuclei. The authors prepared the matrix by extensive extraction of nuclei (already digested at 4°C with 20 μg/mℓ of DNase I) with 10 m*M* Tris pH 7.4 buffer containing, at first, only 0.2 m*M* $MgCl_2$, and in subsequent washes 2 *M* NaCl. These washes were followed by extraction with 1% Triton® X-100 and by digestion with a high concentration of DNase (200 μg/mℓ). The most abundant snRNA associated with the matrix was U1a snRNA (identified on the basis of its molecular weight). After such preparation the hnRNA also remained associated with nuclear matrix, and the concentration of globin pre-mRNA in the matrix RNA fraction was severalfold higher than in total nuclear RNA, indicating that sequences of pre-mRNA during their processing are preferentially associated with the matrix. The high concentrations of globin pre-mRNA and of U1a snRNA in the preparation of nuclear matrix indicated to the authors the role of matrix in RNA processing. Presence of snRNA species in the extracted hnRNP particles and in the remaining nuclear pellet which contained chromatin, nuclear matrix, and nuclear membranes was also observed by Gallinaro and Jacob.[207] The association of snRNA with nuclear matrix in addition to its association with hnRNP particles is not surprising if hnRNP particles are indeed broken fragments of the RNP-nuclear network complex as has been already proposed in this review.

More unexpected is the presence of free small snRNP particles in the nucleosol. Raj et al.[210] isolated small RNP particles containing U1 and U2 RNA from nuclei of Novikoff hepatoma cells by extraction with 0.01 *M* Tris-HCl buffer, pH 8.0, and by purification of the extract by chromatography on Sepharose 6B column and centrifugation in sucrose density gradient. The snRNP particles which contained U1 and U2 snRNA sedimented with the density of a nucleoprotein (1.47 g/cm^3) and, in addition to snRNAs, contained 10 polypeptides of which 2 were phosphoproteins. Howard[206] isolated two populations of RNP particles containing snRNA from mouse erythroleukemia cell nuclei, the already discussed 30 to 60 S RNP particles, and small RNP particles with a sedimentation coefficient of 10 S. In the fraction of the 10 S RNP particles he identified five molecular weight species of snRNA.

A very important discovery was the finding that RNPs containing snRNA are released from cells during some pathological processes.[211] Some patients with mixed connective tissue disease syndrome have a high titer of antibodies against specific nuclear RNPs. Douvas and co-workers[212] used IgG antibodies from two such patients to prepare an affinity column for the isolation of nuclear RNP antigenic complexes from rat liver nuclei. After the elution of the nonantigenic proteins and RNA from the column by 1 *M* NaCl, the antigenically active RNP was eluted from the column by 10 m*M* HCl. Two polypeptides (13,000 and 30,000 daltons) were identified in this RNP.

Lerner and Steitz[213] used antibodies against the small nuclear RNP particles which are present in sera of some patients with lupus erythematosus to precipitate these RNPs from the nuclear extracts of Ehrlich ascites and liver cells. In these experiments, the nuclei were broken by sonication or by shaking at 20° for 1 hr in 0.1 *M* NaCl, 1 m*M* $MgCl_2$, and 10 m*M* Tris-HCl buffer, pH 8.5. Nuclear debris was removed by centrifugation at 5300 × g for 5 min and the supernatant was used as the source of the antigen. There were two antibody systems in the sera of patients with lupus erythromatosus. One selectively precipitated 10 S particles containing U1a and U1b RNA and the other, all six molecular weight species of the snRNA. In addition to the snRNA, both precipitates contained identical polypeptides ranging in molecular weight from 12,000 to 35,000.

A different procedure for the isolation of the small RNP particles was used by Brunel et al.[214] The bonds binding snRNA to proteins are resistant to high ionic strength. It is therefore possible to isolate, by repeated equilibrium density centrifugation in CsCl, the 11 to 12 S snRNP particles from HeLa cell hnRNP polyparticles. The purification of the snRNP particles was completed by velocity sedimentation in a sucrose gradient containing 0.5 *M* CsCl and 5% sarkosyl. In the 11 to 12 S particles (density 1.43 g/cm^3) the authors identified the U1, U2, U4, U5, and U6 species of snRNA and four polypeptides ranging in molecular weight between 10,000 and 14,000. The protein composition of the particles was simpler than that reported by Lerner and Steitz,[213] but the small polypeptides found by both groups may be homologous. It is possible that not all the polypeptides are associated with snRNA by bonds resistant to high ionic strength.

Maxwell and co-workers[181] isolated particles containing snRNA from duck erythroblast nuclei. The nuclei were extracted at 37° for 45 min with isotonic pH 8.0 buffer containing 1 m*M* ATP. During such isolation the hnRNA was extensively nicked by endogenous nucleases. Centrifugation of the extract in 15 to 26% sucrose density gradient separated the RNPs into 40 S particles and smaller 12 and 17 S snRNA containing particles. The release of small snRNP particles (10 S) from hnRNP polyparticles by incubation of isotonic pH 8.0 rat liver nuclear extract at 25° for 1 hr was also reported by Seifert et al.[132] The 40 S duck erythroblasts nuclear hnRNP particles contained hnRNA (inclusive sequences of globin pre-mRNA and eight species of snRNA).[181] The 12 and 17 S particles had an approximate diameter of 15 nm and a buoyant density of 1.39 to 1.40 g/cm^3 in CsCl. They contained, in addition to all but three of the snRNA species found in hnRNP particles, two larger 150,000- and 110,000-dalton RNA species, three major species with molecular weight around 25,000, and 48,000-, 34,800-, and 27,000-dalton RNA species. The 40 S hnRNP particles differed from the 12 and 17 S particles by the presence of the major 30,000- to 40,000-dalton proteins which comprise a large fraction of the 40 S hnRNP particle proteins. In contrast, many, if not most, of the minor particle proteins were common for the 40 S hnRNP particles and for the 12 and 17 S particles (Figure 32). Protein composition of the small particles was very heterogeneous. The 12 and 17 S RNP complexes did not bind to oligo(dT) cellulose and were different in their RNA and protein composition from the 15 S RNP complexes containing the poly(A) regions of pre-mRNA.[181] On the other hand, if nuclear matrix was prepared from duck erythroblast nuclei (by digestion with DNase and by extensive extraction with salt solutions at pH 7.4 and with 1% Triton®, but with the omission of digestion with RNase), all snRNA species were retained with the matrix.[45]

The RNA and polypeptide composition of the snRNP particles prepared by Maxwell et al.[181] from duck erythroblasts was more complex than that of the snRNP particles isolated from HeLa cells by Brunel et al.[214] or from Ehrlich ascites cells by Lerner and Steitz.[213] Since the procedures used for the isolation of the particles were different and snRNP particles are part of a complex structure of nuclear RNPs, this is not surprising. It is probable that the procedure used by Brunel et al.[214] isolated snRNA only with the polypeptides directly associated with it, that is, snRNA ribonucleoproteins.

Flytzanis et al.[203] have shown that snRNA extracted from the large hnRNP polyparticles and labeled in vitro hybridized to the hnRNA extracted from the same polyparticles. This was interpreted to indicate possible participation of the snRNA in processing of the snRNA into mRNA. In 1979 Murray and Holliday[215] proposed a model for splicing of hnRNA based on the ability of snRNA to hybridize to hnRNA. Because the base composition at the 5′ end of the U1a snRNA is complementary to base sequences at the splice point of introns of pre-mRNA, U1a snRNA might align through base pairing the extremities of the intron sequences, assuring correct condition for splicing.[216,217] Rogers and Wall[217] did not find the large U1a RNA sequence complementarity to pre-mRNA present in U2 RNA or in 4.5 S RNA_I, but short sequences, which could be involved in hybridization at the splice points of pre-mRNA, were found in U2 and 4.55 RNA_I by Gallinaro et al.[218] Of course, hybrid formation by such short sequences between snRNA and two short sequences at splice points of hnRNA distant from each other may be very difficult unless helped by proteins arranging the pre-mRNA into the proper configuration. At the present time it cannot be excluded that U1a snRNA, the most abundant of all snRNA species, could be the only one involved in pre-mRNA processing. If this is indeed true, what is the function of other snRNAs in the RNP structures?

Originally it had been proposed that the function of snRNA is structural.[201] The objection against the proposed structural role of snRNA in the nuclear RNPs, raised by Gallinaro and Jacob,[207] is based on a low concentration of snRNA in nuclear RNPs. Based on their finding that snRNA represents approximately 6% of the mass of hnRNA present in rat brain nuclei, they calculated that only from one to two molecules of snRNA are present in RNPs for average-size hnRNA. In rat liver, according to Seifert et al.,[132] snRNA represents 11% of the total hnRNP polyparticle, RNA and according to Sekeris et al.[219] from 15 to 25% of all RNA of 30 S hnRNP particles. The preparation of the 30 S hnRNP particles referred to by the last authors is composed of Samarina and Jacob particles.

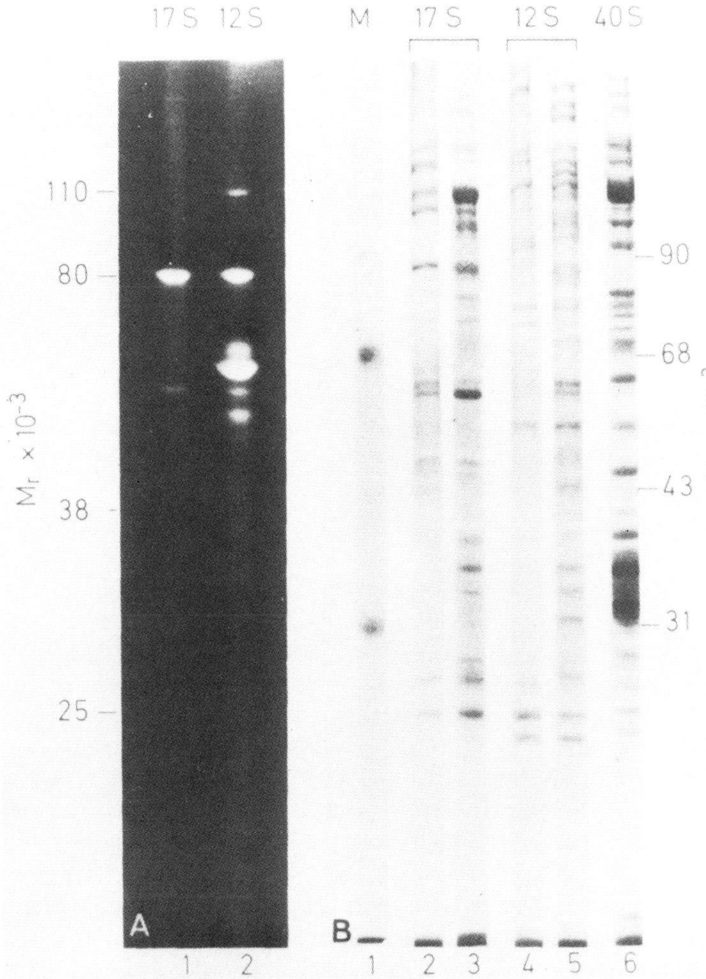

FIGURE 32. RNA and protein composition of purified 17 and 12 S RNP particles from duck erythroblast nuclei. (A). The RNA moieties of each type of particle were electrophoresed on an exponential 10 to 15% gradient polyacrylamide slab gel and stained with ethidium bromide. (B). The proteins of each type of particle were analyzed on 10% SDS-polyacrylamide gel: (1) molecular weight markers; the proteins of the purified 17 S (2) and 12 S (4) particles; the proteins of the unpurified 17 S (3) and 12 S (5) particles, before the second sucrose gradient centrifugation; (6) the proteins of the 40 S hnRNP particles. (From Maxwell, E. S., Maundrell, K., Puvion-Dutilleul, F., and Scherrer, K., *Eur. J. Biochem.*, 113, 233, 1981. With permission.)

It is probable that snRNA is present in the nuclear RNPs in a concentration too low to have a similar structural function as has rRNA in ribosomes. This is especially true since U1a snRNA is the most abundant species of snRNA and its role in RNA processing is very probable.

To the reviewer, the most likely model is in which snRNA mediates specific interactions with hnRNA of proteins involved either directly with RNA processing (as enzymes) or indirectly (as proteins determining the proper configuration of hnRNA which is required for the enzymatic recognition of specific RNA stretches). Experiments from different laboratories indicate that snRNA is hydrogen bound to pre-mRNA.[203,207,220,221] One of these publications is of additional interest. Maxwell et al.[221] used for their hybridization experiments snRNAs isolated from snRNP particles prepared from duck erythroblasts. The snRNAs were labeled in vitro at the 5′ terminus. Two principal species of the snRNA, molecular weight 27,000 and 58,000, hybridized with poly(A)$^+$ and poly(A)$^-$ pre-mRNA. The 58,000-dalton species had the same mobility as U1a snRNA. Unexpected was the finding that both snRNAs also hybridized to polysomal mRNA, demonstrating that sequences outside the intron region can be involved in the hydrogen bonding of hnRNA with snRNA.

Treatment with increasing concentrations of NaCl will lead to a preferential loss of α-proteins from hnRNP polyparticles,[99,126,219] but snRNA will not be dissociated from the particles by high-salt treatment.[126,132,219] If 30 S monoparticles which contain nicked hnRNA are exposed to high salt concentration, the Samarina particles present in the particle population will disintegrate, and some of the hnRNA will also be released.[219] Treatment of 30 S monoparticles by RNase will break Jacob particles and β-proteins and snRNA will be released. The remaining Samarina particles will contain α-proteins and some of the hnRNA.[219] In addition, metrazamide isopycnic centrifugation will separate extracted RNPs into two fractions, one enriched in snRNA and β-proteins, the other in hnRNA and α-proteins.[219] These results indicate that the snRNA is not associated directly with the α-proteins of the Samarina particles, but probably with the nuclear matrix associated with these particles. Such conclusion is in agreement with the earlier-discussed publications which found snRNA associated with nuclear matrix. That snRNA is not an integral part of Samarina particles was also shown by LeStourgeon and co-workers.[5] The authors could not find a constant quantitative relationship between the distribution of monomer 40 S hnRNP and the snRNA components recovered from fractions of sucrose density gradient, and after the dissociation of Samarina particles in 0.4 *M* NaCl, the structures containing snRNA moved throughout the gradient as in control nondissociated preparations of 40 S hnRNP particles.[5]

The presence of free snRNP particles in the cell nucleus indicates that the association of snRNA with the hnRNA-protein complex could be only temporary, i.e., characteristic for specific stages of hnRNA processing. It should be mentioned that the existence of free snRNP particles in the cell nucleus is not accepted by all workers in the field. Miller and co-workers[209] found that all nuclear snRNA remained associated exclusively with nuclear matrix when rat liver nuclei were deprived of nucleoplasm and chromatin. They explain the presence of snRNA in other nuclear fractions, reported by a different author, by experimental disintegration of the nuclear matrix. This disagreement is probably a question of the definition of nuclear matrix. If the term nuclear matrix is interpreted in a broad sense as representing the nuclear matrix proper together with the associated RNP network (from which the hnRNP particles originated by degradation) most of snRNA is indeed associated with the nuclear matrix. Nuclear matrix in this broad sense is involved actively in the processing and transport of pre-mRNA and must be a dynamic structure changing with the progress of hnRNA processing. Therefore, even in the absence of a measurable free pool of snRNA or snRNP particles, these particles could be involved in temporarily changing the interaction of hnRNA with proteins of the RNP network. The variable protein composition of snRNP particles reported in the literature supports the notion that snRNP particles are removed by the isolation procedure from large structures, and that most isolation procedures are releasing, together with the snRNAs, variable amounts of proteins with which they interact.

A key to understanding the role of some snRNP particles in hnRNA processing could be in the observations of Northemann and co-workers.[97] These researchers exposed hnRNP polyparticles (70 to 150 S, isolated from sonicated rat liver nuclei by isotonic salt extraction at pH 8.0) to increasing concentrations of urea. In the presence of 5 *M* urea the α-proteins were preferentially detached from the RNP complexes together with snRNA (Figures 33, 34, and 35). In contrast, high concentration of NaCl dissociated only the α-proteins, but not the snRNA, from polyparticles. This supports the conclusion that hydrogen bonds between hnRNA and snRNA are involved in the binding of snRNA in the hnRNP particles. That snRNA is directly hydrogen bound to hnRNA and not through a mediation of high-

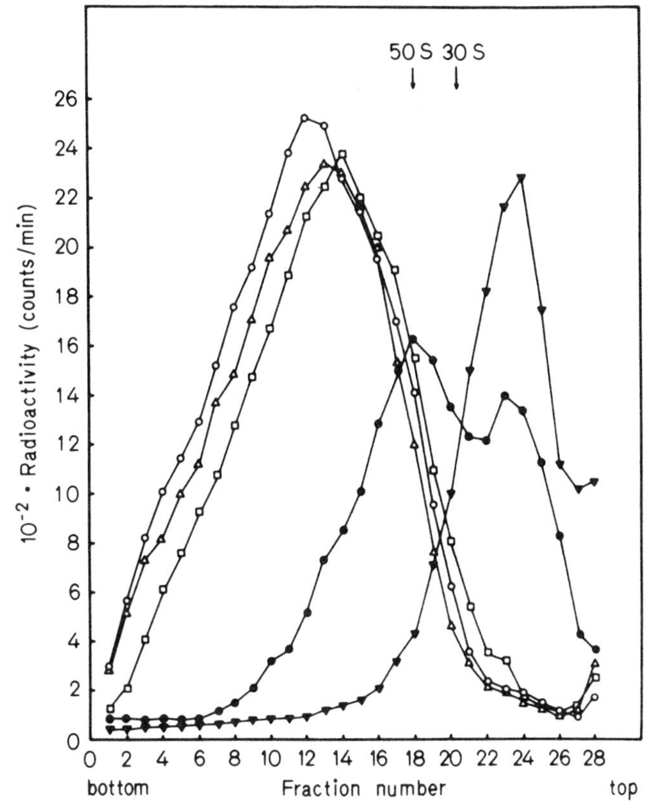

FIGURE 33. Sedimentation profiles of nuclear RNP polyparticles after incubation with urea. HnRNP polyparticles were isolated from sonicated rat liver nuclei by isotonic salt extraction at pH 8.0 in the presence of liver cytosolic RNase inhibitor. The animals were labeled for 30 min with [^{3}H]-orotic acid. The particles were purified by centrifugation on a 15 to 35% sucrose gradient, and the fractions in the range of 70 to 150 S were pooled. The polyparticles were incubated at 0° for 30 min without (o), and with 0.1 *M* (△), 0.5 *M* (□), 2.0 *M* (●) and 5.0 *M* (▼) urea. The urea-treated polyparticles were recentrifuged on a 15 to 35% sucrose gradient and trichloroacetic acid-preciptiable radio activity determined in each fraction of the gradient. (From Northemann, W., Klump, H., and Heinrich, P. C., *Eur. J. Biochem.*, 99, 447, 1979. With permission.)

molecular-weight particle proteins not released by 5 *M* urea was shown by experiments in which all particle proteins were completely degraded by trypsin. After such degradation, snRNA was found, still associated with hnRNA.[97] The authors have also shown a 41% decrease in the molar ellipticity at 264 nm of rat liver RNP polyparticles in the presence of urea, indicating a stabilization of the conformation of hnRNA by snRNA.[97] Measurements of circular dichroism of rat liver hnRNP polyparticles by Northemann et al.[97] have shown that the β-proteins are responsible for the major conformational restrain of hnRNA. Association of hnRNA with snRNA could modify this restrain allowing ionic interactions of specific hnRNA sequences with α-proteins and the formation of Samarina particles. The release of snRNA from hnRNA in the presence of urea would remove this modification, and α-proteins would be released from the particles.

Only a small amount of the bases of snRNAs take part in the formation of hybrids with hnRNA. By comparing the radioactivity in the hybrids formed by hybridization of isolated, labeled snRNA to hnRNA before and after T_1 nuclease digestion, Flytzanis and co-workers[203]

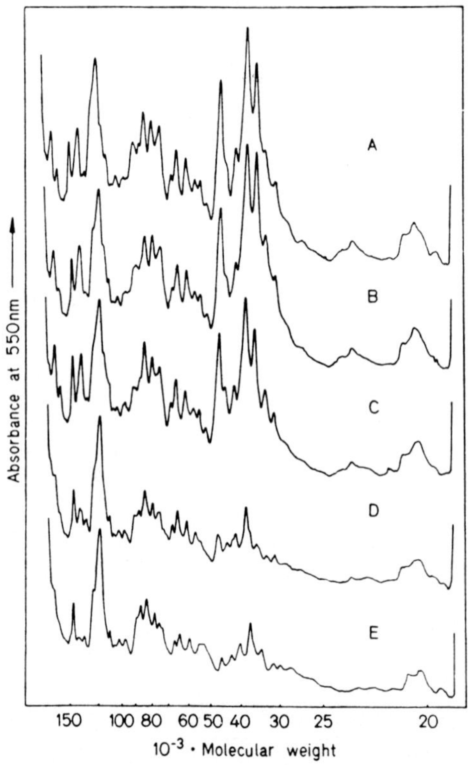

FIGURE 34. Densitometric tracings of SDS-polyacrylamide gel electrophorograms of proteins of hnRNP particles exposed to various concentrations of urea. The hnRNP particles were isolated and incubated with urea as described in the legend to Figure 33. The urea concentrations used for the treatment of the hnRNP polyparticles were none (A), 0.1 *M* (B), 0.5 *M* (C), 2.0 *M* (D), and 5.0 *M* (E). (From Northemann, W., Klump, H., and Heinrich, P. C., *Eur. J. Biochem.*, 99, 447, 1979. With permission.)

estimated that between 15 and 25 nucleotides of average snRNA are involved directly in hybrid formation. The fraction of snRNA hybridized is probably smaller for low-molecular-weight species of hnRNA than for high-molecular-weight species. It is therefore possible that parts of snRNA molecules can be involved directly in RNA-protein binding. The short hybrids formed between snRNA and hnRNA can also explain the unexpected finding that snRNP particles were separated from hnRNP particles by centrifugation in CsCl, that is, under conditions where the hydrogen bonding between hnRNA and snRNA should have been stabilized.[214] The limited amounts of formed hydrogen bonds could be weaker than the interactions between specific proteins and snRNA and the small snRNP particles would separate in CsCl from hnRNA and from hnRNA-dissociated proteins because of the difference in buoyant density.

XII. CHANGES IN THE PROTEIN COMPOSITION OF THE HNRN PARTICLES

Experiments in which RNA synthesis and processing were inhibited have shown that the formation of Samarina particles is dependent on continuous RNA processing. Low yields of hnRNP particles from liver of rats in which RNA synthesis was inhibited by α-amanitin,[96,222] actinomycin D,[96,98] or galactosamine[98] were found in different laboratories. The 30 S hnRNP particles isolated from liver in which RNA synthesis was inhibited[96] were highly depleted of hnRNA but they still contained species of snRNA. The analyzed particles were

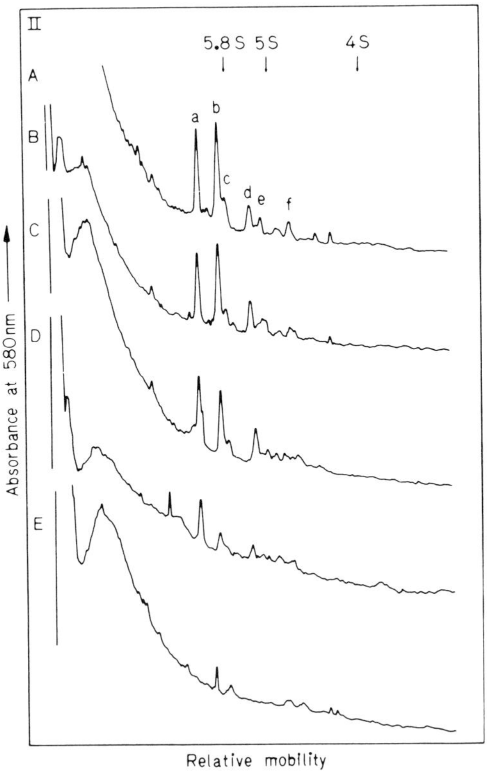

FIGURE 35. Densitometric tracings of polyacrylamide gel electrophorograms of RNA of hnRNP particles exposed to various concentrations of urea. The hnRNP particles were isolated and incubated with urea as described in the legend to Figure 33. The urea concentrations used for the treatment of the hnRNP polyparticles were none (A), 0.1 *M* (B), 0.5 *M* (C), 2.0 *M*, (D), and 5.0 *M* (E). (From Northemann, W., Klump, H., and Heinrich, P. C., *Eur. J. Biochem.*, 99, 447, 1979. With permission.)

probably a mixture of Samarina and Jacob particles. The population of the particle proteins was depleted of glycine-rich proteins of the Samarina particles.[96] In a continuation of these experiments it was observed by Stunnenberg et al.[135] that after the administration of a dose of α-amanitin or of actinomycin D which inhibited hnRNA synthesis up to 80%, the nuclei were selectively and almost completely depleted of the glycine-rich proteins of the Samarina particles. In the nucleus of liver cells from control animals the core glycine-rich proteins of the Samarina particles represented between 10 and 15% of the total nonhistone proteins. An injection of cycloheximide in a dose which inhibited 95% of the synthesis of proteins in liver cells had no effect on the yield of hnRNP particles isolated from these cells. Neither was there any significant difference in the composition of nuclear proteins.

This, together with the known metabolic stability of the particle core proteins,[55,122] excluded that depletion of the glycine-rich proteins is the result of rapid turnover of these proteins combined with an impairment of their synthesis by α-amanitin or by actinomycin D.

Results of our recent experiments[263] complement data from Sekeris laboratory. We studied the composition of 30 S hnRNP particles isolated from the liver of rats which were injected i.p. with actinomycin D (2 mg/kg of body weight) 12 hr earlier. The yield of 30 S hnRNP particles prepared from the livers of these animals was only 40% of the yield obtained from controls. The particles which were isolated after the actinomycin D treatment sedimented in a linear sucrose density gradient with the same sedimentation coefficient (30 to 35 S) as the particles isolated from the control nuclei but differed from them in composition (Figure 36). In animals treated with actinomycin D, the protein to RNA ratio of the particles increased to 32:1 from 4:1 found in the hnRNP particles isolated from livers of control rats. The most striking effect of actinomycin D was the nearly complete disappearance from the particles of all charge isomers of the glycine-rich proteins (which correspond to the A and B proteins of LeStourgeon[50]) 12 hr after the application of the drug. The most prominent protein left in the particles was a 43,000-dalton protein which corresponds to the C protein of LeStourgeon.[50] This protein has the highest affinity for RNA of the proteins of the Samarina particles[50] and could be one of the polypeptides of the nuclear matrix proper.[48]

It is of interest that actinomycin D, in addition to the inhibition of RNA synthesis, also inhibits the processing of hnRNA.[223] If the formation of double-stranded sequences on hnRNA is indeed required for the RNA processing (for review see Reference 195), the inhibition of hnRNA processing by actinomycin D could be the result of the intercalation of actinomycin D into the double-stranded stretches of hnRNA, a possibility which would be worthwhile to explore.

Based on their findings of depletion of the glycine-rich proteins of Samarina particles from rat liver nuclei, Stunnenberg et al.[135] proposed that these proteins are stored in the cytoplasm and enter the nucleus during RNA transcription. The cellular distribution of these proteins was studied by Martin and Okamura[224] by indirect immunofluorescence (Figure 37). The major, glycine-rich proteins of the Samarina particles are poorly antigenic in rabbit, but high titers of antibodies can be obtained in chickens.[225] It should be mentioned that the chicken antibodies obtained against the 30 S RNPs of Samarina particles or against the purified glycine-rich proteins of the particles had similar immunochemical and immunolocalization properties. In HeLa cells, these antibodies localized primarily within the nucleus, but not in the nucleolar and heterochromatic regions. In prophase, the antigens appeared to be retained within the nucleus until the dissolution of the nuclear envelope, at which time they become generally distributed throughout the cell. Apparently, the antigens were almost completely preserved during mitosis. During prometaphase, metaphase, and anaphase the antigens could not be seen on condensed chromosomes. During telophase, the antigens were transported back into the nucleus. The localization of antibodies in the nuclei of dormant bovine lymphocytes with a very low level of RNA synthesis indicates that the proteins of Samarina particles are stored in the cell nucleus.

It is possible that the storage of the proteins of Samarina particles may be different in liver cells and in the dormant bovine lymphocytes, but it is also possible that the inhibition of RNA synthesis by α-amanitin or by actinomycin D[135] has a different effect on the distribution of Samarina particles proteins in the cell than has the inhibition of RNA synthesis by regulatory cellular mechanisms. Despite the high metabolic stability of the major proteins of Samarina particles,[55,122] which was confirmed by their preservation during mitosis,[224] these proteins are synthesized *de novo* if required by an increase in RNA synthesis. In cells in which RNA synthesis was increased by hydrocortisone[226] or by poliovirus infection,[227] the synthesis of the major proteins of the Samarina particles was also induced.[226,227]

Another drug which affects the hnRNP particles is the hepatocarcinogen 3′-methyl-4-dimethylaminoazobenzene. Feeding of this carcinogenic azo dye to rats results in distortion of the rat liver nuclear RNA transport and processing, a distortion which is proportional to the dose of the applied hepatocarcinogen.[169,228] In rats fed the carcinogen for 4 weeks, the incorporation of ^{14}C amino acids into the proteins of a mixture of Samarina and Jacob particles was suppressed by approximately 15% in contrast to all other fractions of nuclear proteins, in which the incorporation was increased from 10 to 14%. The ^{14}C amino acids were injected 18 hr before the fractionation of the nuclei.[229] When analyzed 40 hr after the application of [^{3}H]-3′-methyl-4-dimethylaminoazobenzene by a stomach tube to the rats, the

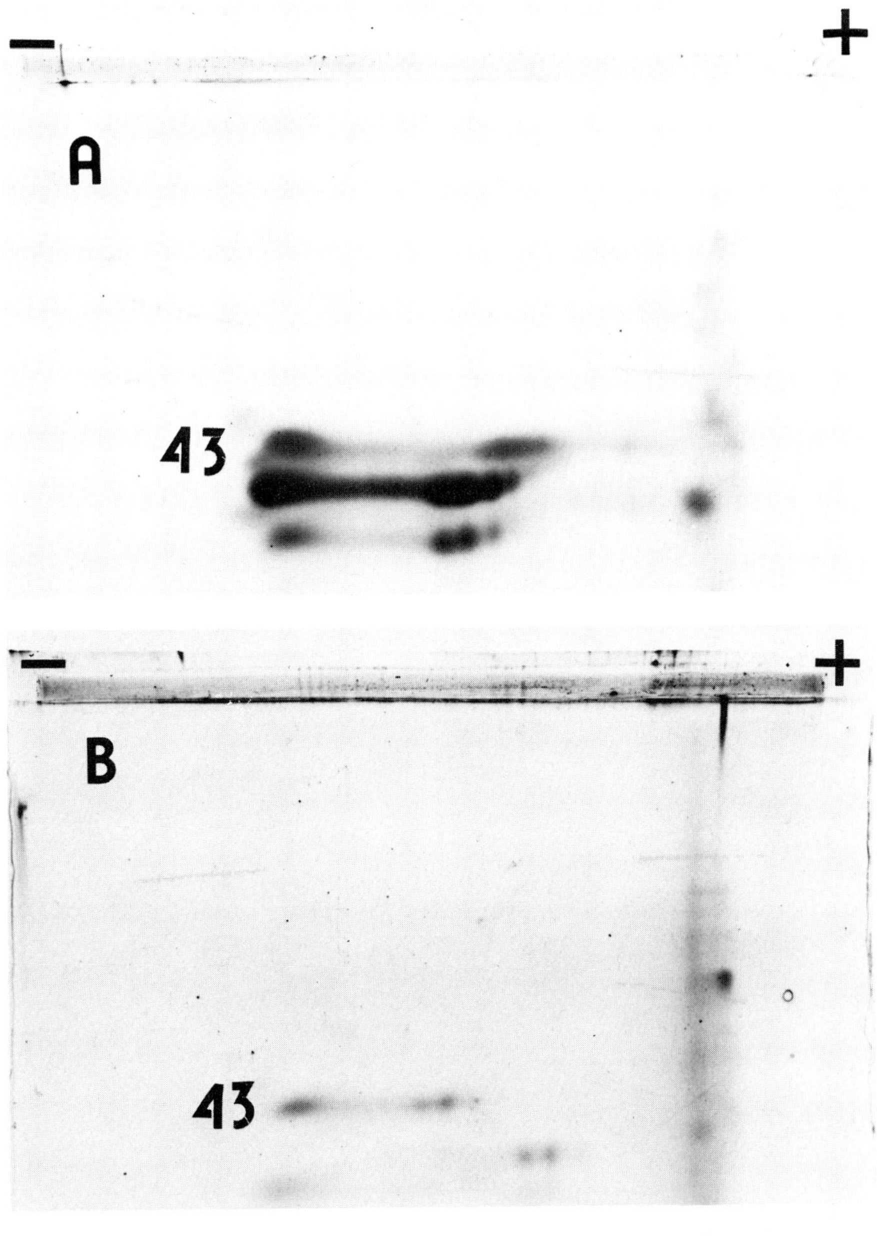

FIGURE 36. Two-dimensional gel electrophoresis of rat liver hnRNP particle proteins after the inhibition of hnRNA synthesis by actinomycin D. The proteins were separated by nonequilibrium pH gradient electrophoresis in one direction and according to their molecular weight in the second direction. (A). HnRNP particles were isolated from liver of control rat. (B). HnRNP particles were isolated from liver of rats injected i.p. with 2 mg/kg of body weight of actinomycin D 12 hr earlier.

highest binding of the labeled azodye in the rat liver nuclei was found to be to the purified Samarina particles; it has not been determined whether the binding was to the RNA or proteins of the particles.[230] The protein composition of the Samarina particles was also affected by the application of the carcinogen, though indirectly.

Analysis of the rat liver Samarina particles by SDS-PAGE has shown a decrease in the amount of one of the core proteins which was proportional to the time of azodye feeding.[231,232] The 30 S hnRNP particles isolated from animals fed the azodye for 10 weeks almost

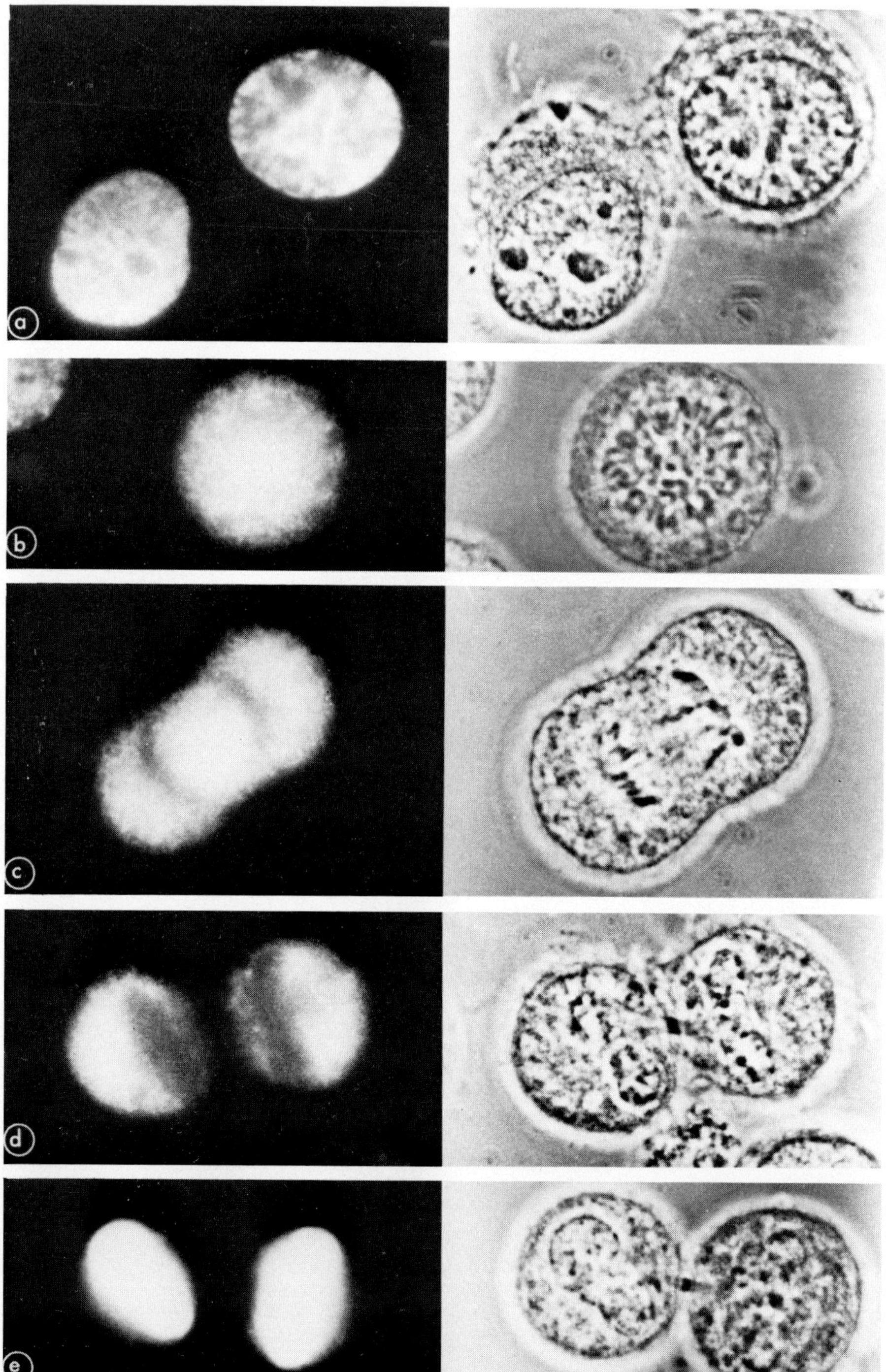

FIGURE 37. The distribution of the hnRNP particle proteins in HeLa cells during mitosis. The 30 S hnRNP particle antigens were localized by indirect immunofluorescence. The fluorescence is restricted to the nucleus in interphase (a), but becomes dispersed in metaphase (b). Condensed chromosomes (not always apparent due to overlapping cytoplasm) are not stained (c,d). Fluorescence is not diminished in mitotic stages and the antigens return rapidly to the newly formed daughter nuclei (e). Bar represents 5 μm. (From Martin, T. E. and Okamura, C. S., in *International Cell Biology (1980—1981)*, Schweiger, H. G.,Ed., Springer-Verlag, Berlin, 1981, 76. With permission.)

completely were devoid of this peptide. Surprisingly, this polypeptide was present in the 30 S hnRNP particles isolated from hepatoma induced by the same carcinogen.[232] In later experiments,[169] we observed that the toxic effect of the carcinogen resulted in temporary changes in the cell population of liver. In the first response to the feeding of 3′-methyl-4-dimethylaminoazobenzene, many hepatocytes degenerate and are replaced through the proliferation and maturation of precursor cells. In our system, the largest changes were observed after feeding the azocarcinogen for 10 weeks, when majority of the cell population consisted of transitional precursor cells. If the feeding was continued beyond 10 weeks, the precursor cells matured, and after 16 weeks the hepatocytes morphologically resembled the cells in the liver of control animals. Our recent unpublished experiments have shown that the observed changes in protein composition of the 30 S hnRNP particles are related to the changes in the cell population (Fig. 38). No difference was observed in the protein composition of particles isolated from animals fed 3′-methyl-4-methylaminoazobenzene for 16 weeks or longer.

During late stages of infection with the simian virus 40 (SV40) or adenovirus, a significant portion of the newly synthesized RNA is transcribed from the viral genome. Since the transcripts of the adenovirus and SV40 virus genomes are extensively processed, cells infected with these viruses present a suitable system to search for proteins involved in the processing of a specific pre-messenger RNA. Pagoulatos and Yaniv[149] identified in SV40-infected monkey kidney cells major new proteins, induced by the viral infection, bound to the RNA in the hnRNA particles despite the fact that the hnRNP particles contained only 7 to 10% of RNA sequences specific to SV40 DNA. In contrast, Gattoni et al.[187] found the protein composition of nuclear hnRNP polyparticles isolated from HeLa cells in which 60% of the cellular RNA was replaced by adenovirus RNA to be nearly identical to the protein composition of the same particles isolated from control cells. The only difference was in the presence of six additional minor polypeptides associated with the particles. But it could not be excluded that these polypeptides became attached to the hnRNP particles from infected cells by nonspecific adsorption.

XIII. EPILOGUE

Based on information discussed in this chapter, a model of the nuclear hnRNA-protein complexes, which is depicted in Figure 39, seems to be the most probable. This model takes into account the presence of two types of hnRNA-protein arrangement in the nuclear RNP network and the close integration of the RNPs into the nuclear matrix.

In this model hnRNA is bound to specific proteins (or a protein) of molecular weight around 40,000, which correspond to the C proteins of Beyer et al.[49] This RNP is associated with the heterogeneous proteins of the nuclear matrix forming together the nuclear ribonucleoprotein network. At specific stretches (which are nonrandomly distributed), the hnRNA is bound by hydrogen bonds to snRNA of small RNPs. Conformational changes in the hnRNA containing RNP caused by binding of the snRNA containing RNPs, result in wrapping of the segments of the RNP which contain the pre-messenger RNA sequences around protein cores composed of the glycine-rich proteins. The interactions between the RNP containing hnRNA and the glycine-rich protein containing particles are predominantly ionic and rather weak. The hnRNA containing the pre-messenger RNA (in the form of the RNP) wrapped around the cores of glycine-rich proteins is protected from unspecific degradation by endogenous nucleases and kept in the configuration required for processing. With this coiled RNP are associated various proteins involved in the processing of pre-messenger RNA (enzymes, control proteins). The association of these proteins with the pre-messenger RNA RNP could be only temporary, they could differ according to the stages of RNA processing, and could probably be easily removed during purification of nuclear RNP particles. Stretches of the hnRNA-RNP which are attached to the nuclear matrix are degraded by endogenous nucleases.

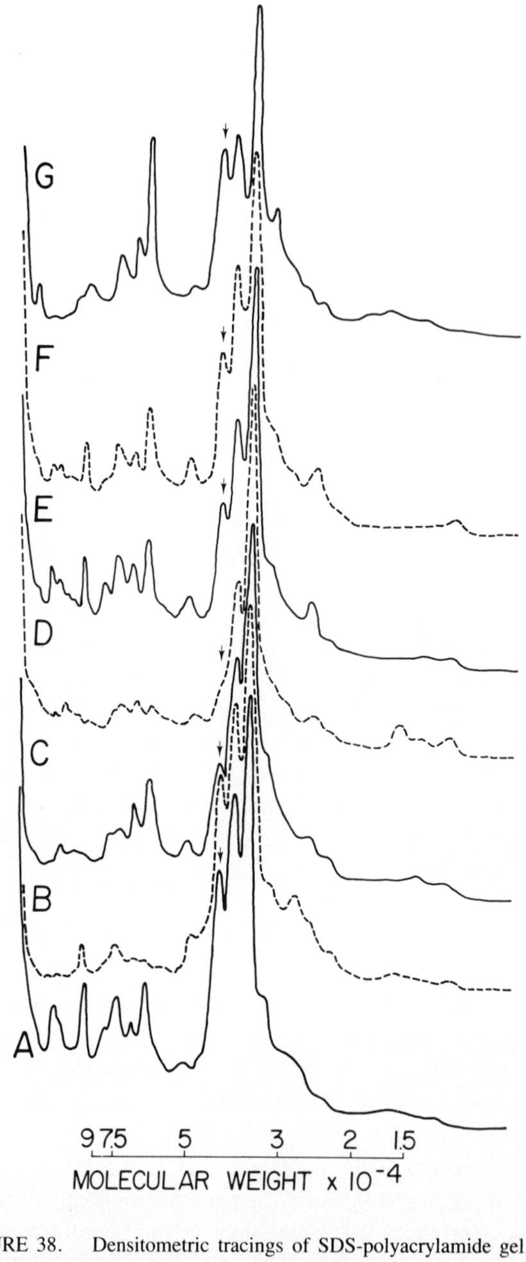

FIGURE 38. Densitometric tracings of SDS-polyacrylamide gel electrophorograms of the protein moiety of hnRNP particles isolated from liver of rats fed a hepatocarcinogen. The particles were isolated from liver of (A) control rats, (B) rats fed noncarcinogenic *p*-aminoazobenzene for 10 weeks, (C) rats fed carcinogenic 3′-methyl-4-dimethylaminoazobenzene (3′-MeDAB) for 6 weeks, (D) rats fed 3′-MeDAB for 10 weeks, (E) rats fed 3′-MeDAB for 16 weeks, and (F) rats fed 3′-MeDAB for 10 weeks followed by control diet for 6 weeks. (G) HnRNP particles were isolated from a hepatoma induced by 3′-MeDAB.

The two types of the arrangement of the hnRNA-RNP in the nuclear RNP network are broken to different extents depending on the procedure used for the isolation of nuclear hnRNP particles, and are recovered as 30 S Samarina particles (hnRNA-RNP wrapped around a protein core composed of glycine-rich proteins), as Jacob particles (hnRNA RNP associated with protein of the nuclear network), or as a mixture of both in the form of polyparticles.

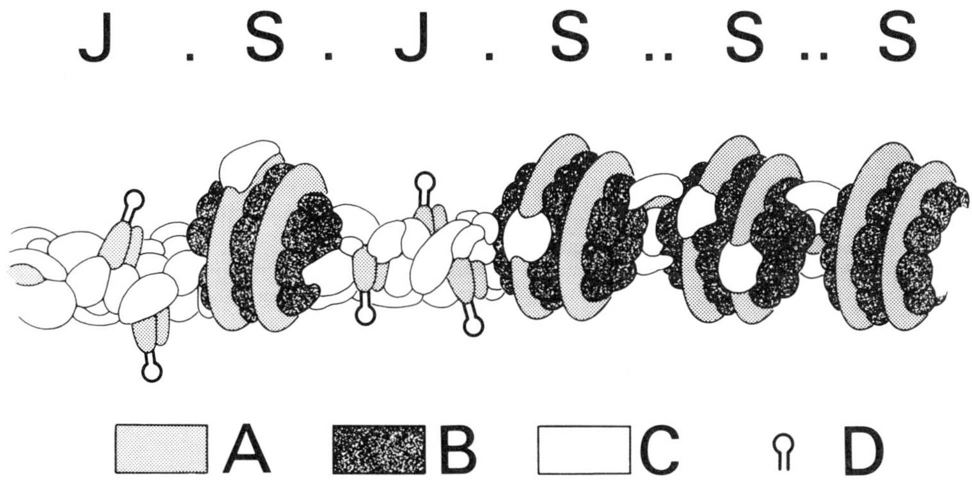

FIGURE 39. Model of the nuclear RNP network. (S) — Samarina particles, (J) — Jacob particles; (A) RNP composed of hnRNA and C type of particle proteins; (B) glycine-rich proteins; (C) proteins of the nuclear matrix, enzymes involved in hRNA processing, proteins controlling hnRNA processing, etc.; (D) double-stranded regions of hnRNA.

The dissociation of snRNA from hnRNA after the breakage of hydrogen bonds by urea will remove the restraint put on the hnRNA-RNP configuration and release the RNP from the Samarina particle core composed of the glycine-rich proteins. The same effect is achieved by an inhibition of the ionic interactions between the glycine-rich proteins and the hnRNA-RNP. An interesting detailed model of the arrangements of the glycine-rich proteins in the Samarina particles has been proposed by LeStourgeon et al.[5]

The poly(A) sequences attached at the 3′ ends of some of the hnRNA molecules are bound to unique proteins and are closely associated with the nuclear matrix in the form of small poly(A) containing RNP particles.

A model which summarizes current knowledge is useful as a hypothesis to be further tested by experiments. But we must keep in mind that a substantial uncertainty is built into any proposed model of the hnRNA-RNP by the inadequacy of our information in many crucial areas. The function of the individual species of snRNA is unknown despite the strong indication that some of these species could be involved in the processing of premessenger RNA. Also unknown is the arrangement of the RNPs containing snRNA in the nuclear RNP network. It is further uncertain whether the population of the Samarina particles is homogeneous or differs in the distribution of the charge isomers of the glycine-rich proteins. It is obvious that any final model of the Samarina particles would have to answer the question about the function of these charge isomers. Last but not least, it should be kept in mind that the Samarina and Jacob particles could be dynamic structures and that the same sequences of hnRNA could be present in the RNP network temporarily associated with different proteins in the sequence of events connected with their processing.

ACKNOWLEDGMENTS

The work of my laboratory, which is discussed in this review, was supported by a grant H-393 from The Robert A. Welch Foundation. I would like to use this occasion to acknowledge my deep gratitude to this Foundation for continuous support of my work and for their pioneering contribution to the development of basic chemical research in the state of Texas. Dr. M. Yoshida, Dr. N. T. Patel, and Dr. R. K. Upreti were involved at different times with the work done in my laboratory and their crucial contribution is gratefully acknowledged. The last two mentioned were postdoctoral fellows of the Robert A. Welch Foundation.

APPENDIX

From the time that the manuscript was submitted (April, 1981) to the time that this volume was assembled for printing, many additional papers contributing to the understanding of the chemistry and function of nuclear ribonucleoproteins have been published. This appendix is an attempt to bring at least some of these reports to the attention of the reader.

In Volumes 8 and 9 of the series *The Cell Nucleus,* seven articles directly concerned with nuclear RNPs have been published[233-247] predominantly reviewing the work done by individual research groups. Two of them were received as preprints and were already listed earlier.[3,5] These reviews are extensive and well written (even if some only from the point of view of individual research groups). Where individual research papers are concerned, many of them report results of continuing work from earlier papers which have been discussed in this review.

In earlier experiments van Eekelen and van Venrooij found by UV cross-linking of hnRNA to proteins that in HeLa cells the hnRNA is associated with 41,500- and 43,000-dalton proteins of the nuclear matrix.[48] It was assumed that these proteins corresponded to the C proteins of the Samarina particles according to the nomenclature of LeStourgeon and co-workers.[50] In addition, the two B proteins of the Samarina particles were also linked with hnRNA, but less efficiently. These results have been confirmed in experiments in which HeLa cells were infected with adenovirus and the adenoviral RNA was cross-linked by UV irradiation to nuclear proteins.[248] The adenoviral RNA isolated as RNA-protein complex by hybridization to adenovirus DNA-Sepharose®, was found cross-linked to 41,000- and 43,000-dalton proteins. These proteins comigrated on SDS gels with nuclear proteins which are cross-linked to cellular hnRNA and are probably identical with them. In experiments in which protein cross-linked viral RNA was not specifically selected by hybridization to adenovirus-DNA-Sepharose®, but the total poly(A)-containing hnRNA of infected cells was cross-linked to nuclear proteins, the authors also found the B core proteins of the RNP particles cross-linked to hnRNA, though less effectively. Since, in the experiments, the same proteins were cross-linked to poly(A)$^+$ hnRNA of infected and of noninfected cells, the authors concluded that adenoviral hnRNA is associated with the same proteins as is host hnRNA. In this last experiment, after cross-linking of proteins to hnRNA in vivo, the poly(A)-containing hnRNA was isolated from the nuclear fraction by oligo(dT)-cellulose chromatography and treated with nucleases to degrade the RNA. The proteins were separated by PAGE. In principle, the same procedure was used in all following experiments.

In an extension of their experiments van Eekelen and co-workers[249] UV-irradiated HeLa cells in which RNA was labeled either with [^{3}H] adenosine, [^{3}H] cytidine, [^{3}H] guanosine, or [^{3}H] uridine for 4 hr. After the isolation of the hnRNA-protein complexes and the degradation of the RNA, on the average, one to three nucleotides remained covalently linked per polypeptide chain. Unexpectedly, the labeling of the 41,500- and 43,000-dalton proteins was found only when the cells were labeled with [^{3}H] cytidine or [^{3}H] uridine, indicating the possibility that special spatial orientation of pyrimidines, or pyrimidine-rich stretches, can be required for RNA-protein interaction.

Earlier, Mayrand and Pederson[75] established by photochemical cross-linking of nuclear RNA to the nuclear proteins in HeLa cells in vivo that isolated nuclear RNP particles are native cellular structures. Characterization of the cross-linked proteins found that the same main proteins which are present in noncross-linked nuclear RNP polyparticles are also present in the cross-linked particles.[250] All the main core proteins of the nuclear RNP particles were also found cross-linked to hnRNA in the RNP complexes prepared from UV-irradiated nuclei of mouse Friend erythroleukemia cells[251] and from nuclear RNP polyparticles which were irradiated in vitro after their isolation from mouse L cells.[252] With the exception of the results reported by van Eekelen and co-workers,[48,248,249] all other discussed experiments have shown cross-linking of all three types of core proteins (A, B, and C) of Samarina particles to hnRNA after UV irradiation. The most efficient cross-linking of C core proteins and low efficiency cross-linking of B core proteins after a lower dose of UV irradiation,[48,248,249] and the cross-linking of all the particle core proteins to hnRNA observed in experiments which used a higher dose of UV irradiation,[250-252] agree with the differential stability of binding of the different core proteins of Samarina particles to hnRNA in solutions of increasing salt concentrations.[49,50] In addition to the cross-linking of the particle core proteins, the high-molecular-weight proteins of the polyparticles were likewise extensively cross-linked to hnRNA in HeLa cells irradiated in vivo[250] and in nuclear RNP polyparticles from mouse L cells irradiated in vitro.[252] In both experiments the proteins were labeled with [^{35}S] methionine and identified, after electrophoretic separation, by autoradiography. In an in vitro system in which RNA synthesized in isolated nuclei of Friend erythroleukemia cells was attached to the proteins by UV irradiation of the nuclei, no polypeptides larger than the particle core proteins were found cross-linked to the proteins.[251] In this experiment the nuclei were incubated with [α-^{32}P] UTP and the proteins cross-linked to newly synthesized RNA identified by the labeled uridylate which remained attached to proteins separated by electrophoresis after enzymatic hydrolysis of the cross-linked newly synthesized RNA. A cross-linking of proteins, which could be the C and B core proteins of the Samarina particles, and cross-linking of two minor 58,000- and 70,000-dalton proteins was likewise observed after UV irradiation of isolated rat liver Samarina particles in vitro.[253] The same two minor proteins banded with hnRNA when unirradiated Samarina particles were centrifuged in Cs_2SO_4 density gradient containing guanidine-HCl.[253]

By comparing the proteins cross-linked to UV-irradiated nuclear RNP polyparticles with the proteins which are cross-linked to poly(A)$^+$ mRNA by UV irradiation of polyribosomes and of free cytoplasmic messenger RNP particles, Setyono and Greenberg[252] confirmed that the proteins associated with mRNA and hnRNA are different. Also, according to these authors, the poly(A) segments of mRNAs and hnRNA are cross-linked to different proteins. The 60,000-dalton protein cross-linked with poly(A) of hnRNA differed from the 78,000-dalton protein cross-linked to the poly(A) of cytoplasmic mRNA in its partial peptide map.[252] It surely did not escape the reader that the size of the protein reported as cross-linked to poly(A) of hnRNP differs substantially from the size of most of the polypeptides reported as present in the small poly(A)-containing nuclear RNP particles[175,181] or isolated as poly(A)-protein complexes from nuclear RNPs[178-180] which were discussed earlier. Because of the diversity of procedures used for the isolation and/or characterization of these proteins the differences cannot be explained at the present time.

A different turnover of the individual core proteins of the Samarina particles and the high metabolic rate of the smaller of the A core proteins of the particles, discussed earlier,[122] were confirmed by Ivanova et al.,[254] who analyzed the proteins of Samarina particles by pulse-chase experiments in Friend erythroleukemia cells. Ivanova and co-workers[254] found that two proteins, likely corresponding to the A particle core proteins (32,000 and 34,000 daltons) turned over with a half-life of 19 and 72 hr. The 37,000- and 39,000-dalton proteins, likely corresponding to the B core proteins of the particles, turned over with a half-life of 40 and 68 hr. The difference in the half-lives of the two proteins in each of the two doublets is striking. Very unusual is the reported half-life (306 hr) of a 110,000-dalton protein which Ivanova and co-workers found as another major protein in the particles. Schweiger and Kostka[255] identified a protein of the same size in Samarina particles from rat liver nuclei. But in their experiments 3-hr labeling of the rat proteins in vivo with a mixture of [^{14}C] amino acids resulted in four to five times higher labeling of this protein than of the other proteins of the Samarina particles. If this protein indeed corresponds to the protein described by Ivanova and co-workers,[254] as indicated by the electrophorograms, the explanation of its different metabolic activity observed in the two laboratories could be an important clue to the understanding of its function. The 110,000-dalton protein described by Schweiger and Kostka is ADP-ribosylated.[255]

The experiments of Ivanova et al.,[254] furthermore, indicate the possibility of two pools of RNP particles. Mild digestion of chromatin releases particles containing newly synthesized hnRNA, which are still associated with chromatin.[256] The core proteins of Samarina particles present in this fraction of the nuclear RNPs have a half-life of only 4.5 to 8 hr. At present it is difficult to evaluate this finding. However, the assumption of homogeneity of nuclear RNP particles could be a fiction (see References 71, 73, 75, 121, and 141). The fractionation procedure of Ivanova and co-workers[254] or of Thomas and co-workers[257] could be used to look for a possible heterogeneity in the population of the RNP particles. Also a detailed analysis of Samarina particles, which can be separated in the unfixed state by electrophoresis in agarose or polyacrylamide gels,[258] could be rewarding. At present it would be possible to correlate changes in particle composition with the presence in the particles of one specific RNA transcript at different stages of RNA processing.

Thomas et al.[257] separated the hnRNP particles containing nascent hnRNA from other hnRNP particles by differential extraction of chick oviduct nuclei. The purified nuclei were first extracted with 0.15 to 0.3 *M* NaCl at pH 8.0. The use of 0.3 *M* NaCl assured better separation of pulse-labeled RNA from nonlabeled RNA. This first extraction removed, in the form of RNPs, over 90% of the ovalbumin mRNA sequences which were originally present in the nuclei. In the next step the authors released by sonication, and separated by centrifugation, hnRNP from the chromatin. The RNP contained 70% of the pulse-labeled RNA (the oviduct fragments were pulse labeled for 10 min in vitro with [^{3}H] uridine) but less than 10% of the sequences of ovalbumin mRNA which were originally present in the nuclei. The hnRNP sedimented from 30 to 350 S. Its protein composition was very heterogeneous; it contained not only all the main proteins of the Samarina particles, but also histones (with the exception of H1), and the ratio of RNA to DNA was 4:1. The report confirmed the close association of the nuclear RNP network with chromatin and also the immediate association of newly synthesized hnRNA with the proteins of the hnRNP complexes, both discussed earlier. No information about the RNP released by the 0.15 to 0.3 *M* NaCl, pH 8.0 extraction was given.

Experiments published by Pederson and Munroe[190] have shown that in nuclear RNPs of HeLa and Friend erythroleukemia cells[259] the sequences of hnRNA homologous to mRNA are protected from mild attack by nucleases. Since in the Friend erythroleukemia cells the globin mRNA sequences were more nuclease sensitive than in the nuclear RNP polyparticles, this experiment also confirmed the changes in the association of mRNA with proteins during its movement from the nucleus into the cytoplasm.

In another laboratory[260] stringent digestion with pancreatic RNase was used to identify the hnRNA sequences which are the most protected by the proteins of the RNP particles, and therefore could be directly involved in the protein-RNA interaction. The nuclear hnRNP polyparticles and Samarina particles were prepared from sonicated nuclei of HeLa cells infected with adenovirus-2. After extensive digestion with pancreatic RNase the nondigested RNA fragments were purified, 5′-end labeled, and hybridized on Southern blot with fragments of adenovirus-2 DNA which were obtained by digestion with restriction endonucleases. As in the nuclear hnRNP polyparticles, in Samarina particles 20 to 40 bases long RNA fragments were protected from digestion by pancreatic RNase. The sequences complementary to the fragment encoding the first leader and the major intron for the DNA-binding protein coded by the virus were further analyzed. The RNA sequences resistant to the hydrolysis by the RNase were tentatively located in the middle of this intron and at the leader splice junction. The same results were obtained regardless of whether the RNA fragments were derived from hnRNP polyparticles or Samarina particles. Such results would support the conclusion reached by Beyer et al.[71] from their electron-microscopic studies of the RNP morphology of nascent hnRNA in *Drosophila melanogaster* embryos using chromatin spreading technique. The authors concluded that primary structural features of nascent RNPs are nonrandom and hnRNA sequence dependent. The nascent transcript occurred in either a form of RNP fibrils or in the form of RNP particles. From the distribution of both forms on the nascent hnRNA the authors concluded that the two structural types are probably determined by the underlying RNA sequences.

Using the same technique in a more recent paper, Beyer and co-workers[261] found that the first cutting of newly synthesized RNA takes place when the RNA is still associated with chromatin. The cutting is preceded by formation of loops. From the electron-microscopic pictures it could be interpreted that the loop formation occurs by the fusion of two RNP particles (similar in size to Samarina particles) located on two different parts of a RNP fiber. Particles do not occur in the loop structure. Based on the size of the transcription units and the size of the released RNA fragments, the authors concluded that the cleavages could correspond to cleavages at the 3′end of pre-messenger RNA termini or to the release of functional domains from polycistronic transcripts, or both.

Many questions have to be answered before these results can be finally interpreted. The observation of the authors that newly synthesized RNA is mostly associated with proteins in the form of fibrils and only occasionally folded into particle-like structures does not fit well with the various results of isolation and characterization of nuclear RNP particles which were discussed in this review. Beyer and co-workers[71] proposed that this discrepancy could result from differential stability of particles subjected to treatments required for the preparation of samples for electron microscopy. Such differential stability could be based on differences in protein composition or in RNA sequences with which the proteins are associated. Despite this problem still awaiting clarification, the indication that formation of loops by coalescence of two separate RNP particles could be involved in the first step in RNA processing — the step which precedes the removal of introns — could be of extreme importance.

Since the double-stranded stretches on hnRNA are considered the possible sites of hnRNA processing[195] it is of interest that the results of the experiments of Paoletti and co-workers,[162] who used the intercalation of ethidium bromide to probe for double-strandedness of RNA in HeLa cell nuclear polyparticles, were confirmed by Borissova et al.[262] Measurements of the intercalation of ethidium bromide into double-stranded regions of RNA in rat liver Samarina particles has shown that about 6% of the total particle RNA is double-stranded. The rigid structure of the double-stranded regions which is maintained by interaction with particle proteins, reported for HeLa cells polyparticles, was also confirmed in rat liver Samarina particles. Measurements of energy transfer from fluorescamine (covalently bound to particle proteins) to ethidium bromide (adsorbed in the double-stranded RNA) indicated that the double-stranded regions stick out from the surface of the particle. This result is in agreement with the found accessibility of the particle double-stranded regions.[196-198] From energy transfer between the ethidium bromide molecules it was possible to estimate that the average length of double-stranded regions on the premessenger RNA in the Samarina particle cannot be less than 45 Å. This would correspond to 1 double-stranded stretch of approximately 18 nucleotide base pairs per Samarina particle, under the assumption of equal distribution of the double-stranded stretches between the particles. A comparison of the data obtained from polyparticles of HeLa cells nuclei[162] with the data obtained from Samarina particles of liver cell nuclei[262] imply that the double-stranded stretches on hnRNA could be restricted to Samarina particles.

In experiments discussed earlier we found nearly complete disappearance of all charge isomers of the A and B core proteins from the Samarina particles 12 hr after the application of actinomycin D (2 mg/kg of body weight).[263] In an extension of these experiments, we more recently observed that this disappearance is preceded by an inhibition of incorporation of radiomethyl into the proteins of Samarina particles, probably as the result of decreased turnover of methyl groups on the particle proteins. This decrease in the turnover of methyl groups on the particle proteins could be the result of inhibition of RNA processing by actinomycin D.[264]

We[290] observed changes in the distribution of the charged isomers of the core proteins of rat liver Samarina particles 18 hr after a partial hepatectomy — five of the charge isomers of a 43,000-dalton protein and two charge isomers of a 39,000-dalton protein were missing; 30 hr after the hepatectomy the pattern of the charge isomers of the core proteins of the Samarina particles (as determined by two-dimensional electrophoresis) returned to normal. It is possible that regenerating rat liver could be used as a model system to study the function of individual isomers of the core particle proteins in metabolism of nuclear RNA.

The main molecular-weight species of the core proteins of rat liver Samarina particles can be separated by high-performance liquid chromatography on a SynChropak® AX-300 anion-exchange column from each other (but with some cross-contamination) and from the minor higher-molecular-weight proteins of the particles.[265] It was not possible to separate, by this procedure, the charge isomers (pIs from 5.5 to 9.0) of each of the core proteins from each other, despite the heterogeneity in their charge. Probably the affinity of the individual charge isomers for each other is strong enough to prevent, under the conditions employed, their separation on an ion exchanger, and the individual core proteins eluted as aggregates of their charge isomers.

A cell-free system to study hnRNA processing would make it possible to examine the function of individual proteins of the nuclear RNP particles in the processing. In an effort to develop such a system Economidis and Pederson[251] studied the transcription of hnRNA by endogenous RNA polymerase II in isolated nuclei from Friend erythroleukemia cells. The hnRNA transcribed in the isolated nuclei was found to be immediately assembled into RNP particles with the properties of particles in which hnRNA is assembled in the cells. It appears that isolated nuclei could be a suitable system to study RNA processing. Cells in which the majority of newly synthesized RNA would be represented by a limited number of species of RNA would be the best-suited system in which processing of hnRNA could be studied in vitro. In HeLa cells infected with adenovirus-2 for 2 hr, 50 to 60% of the labeled RNA represent adenovirus-2 specific RNA. The newly synthesized virus-specific RNA sequences are found in heterogeneous RNP complexes sedimenting between 30 and 200 S, which in density and protein composition seems to be the same as the bulk of hnRNA containing RNP particles in noninfected cells. Hybridization of RNA purified from the particles to DNA restriction fragments of the viral genome bound to nitrocellulose filters have shown that all the transcript from the adenovirus-2 genome are present in the particles.[189] In an extension of these experiments it was shown that the particles contain covalently intact pre-mRNA of adenovirus DNA binding protein.[266] It is not yet possible to find the conditions under which RNP isolated from the virus-infected cells could be used as a cell-free system to study the processing of the adenovirus premessenger RNAs. When the hnRNA-containing nuclear particles isolated from the adenovirus-2-infected HeLa cells were incubated for 5 or 15 min, no processing of hnRNA occurred. The authors concluded that either some specific factor was lost during the isolation of the nuclear RNPs or that hnRNA can be processed only if the RNPs are attached to other nuclear structures (nuclear matrix). If the second eventuality will be shown to be true, construction of an in vitro system simpler than isolated nuclei would be difficult.

The preferential location of U1 snRNA in the areas of the Samarina particles,[132,204,207,267] the concentration of double-stranded stretches on hnRNA in the Samarina particles,[262] and the observations indicating that fusion of two particles resulting in formation of loops on hnRNA-protein fibers precedes the cutting of hnRNA[261] point to an important role of Samarina particles in hnRNA processing. However, notwithstanding the simpler macromolecular composition of Samarina particles (if compared with the hnRNP polyparticles), the function of even the most predominant proteins of these particles was nothing more than a proposal until now. The results of the experiments which studied the function of one isolated particle protein were published only recently. From dormant cysts of the brine shrimp *Artemia salina* can be isolated a 40,000-dalton protein which will destabilize an RNA helix.[268,269] This protein is similar in its amino acid composition to the glycine-rich core proteins of the Samarina particles. In a later publication, Thomas and co-workers[270] identified the same protein as the major protein of 30S RNP particles (Samarina particles). The particles can be isolated from intact nuclei of developed cysts of *Artemia salina*. This protein represents from 35 to 60% of the particle proteins. It can be separated into three isoelectric species with basic pIs, which appear to be posttranslational modifications of a single polypeptide chain. The other proteins found in the particles were of higher molecular weight, with two (approximately 85,000 and 110,000 daltons) being predominant. The RNA-to-protein ratio found in these particles is about 1:6. In experiments in vitro this protein will bind to RNA extracted from these particles, unstacking the RNA. At a stoichiometry of 1 protein per 10 to 12 nucleotides the secondary structure of the RNA is maximally unstacked (as determined from the circular dichroic spectrum). An addition of the protein in excess of this ratio resulted in the formation of bead-like RNA-protein complexes that are similar in appearance to the native 30 S nuclear RNP particles. No additional changes in the circular dichroic spectrum were observed. These experiments have shown for the first time, for one isolated protein of the nuclear RNP particles, that its role in the particle could be to stabilize the secondary structure of hnRNA. These results have to be integrated with the results reported earlier by Northemann et al.,[97] which have shown that particle core proteins put a specific conformational restraint on hnRNA in the particle. Probably the function of the core proteins would be, therefore, best described by the statement that these proteins change the configuration of native hnRNA into a new one. These results support the notion, expressed in many papers discussed in this review, that the function of the glycine-rich core proteins in the Samarina particles is to hold the hnRNA in proper configuration for processing.

Perhaps because of the extreme difficulties connected with the effort to characterize the function of individual parts in such complex superstructures as are hnRNP complexes, in the past year even more laboratories have turned their attention to the exploration of small RNP particles. More recently, base pairing between U1 snRNA and hnRNA has been shown *in situ*[271] and it also has been demonstrated that splicing of adenovirus early mRNA is inhibited by both anti-Sm and anti-RNP antibodies.[272] With such strong support for the proposal that U1 snRNA plays a role in the splicing of hnRNA[215-217] the attention paid to simpler structures which could have well-defined functions is not surprising. The studies of the snRNP particles were facilitated by the fact that these particles can be precipitated by antisera from patients with autoimmune diseases (for review see Reference 240).

The nuclear RNPs containing small nuclear RNA can be divided into at least two groups. One of the groups contains U-type snRNAs with 5′capped ends which are transcribed by RNA polymerase II. The RNAs present in the second group of nuclear RNP particles have uncapped 5′termini. These two groups of small nuclear RNP particles can, at least in some experiments, be distinguished by selective immunoprecipitation with different antibodies from patients with autoimmune diseases.[273] Anti-Sm serum recognizes snRNP particles containing U-type RNAs, another type of antibody designated anti-La preferentially precipitates a heterogeneous spectrum of snRNP particles containing noncapped RNA, and a third type of antibody (anti-RNP serum) selectively precipitates only snRNP particles containing U1 RNA.

The La antigen in uninfected mammalian cells consists of a group of small RNP particles containing a heterogeneous spectrum of RNAs 80 to 140 nucleotides long, with uncapped 5′termini.[273] One of these RNAs, the 4.5 S RNA exhibits similarity to the sequences of the repetitive human Alu family.[274] In virus-infected cells anti-La also reacts with RNP particles containing VA RNA encoded by the adenovirus genome,[273] or with RNP particles containing EBER RNAs encoded by the Epstein-Barr virus.[274] The controversy about the localization of these particles in the cell seems to be resolved in favor of their nuclear localization, since anti-La sera have given primarily nuclear standing by indirect immunofluorescence.[274] In contrast to snRNP particles containing U-type snRNAs, the particles which react with anti-La sera are not associated with the nuclear matrix.[275] Unfortunately, at the time of this writing little is known about their protein composition. To our knowledge, the only exception is a report that in HeLa cells infected with adenovirus the viral VA RNA can be cross-linked by UV light to a 50,000-dalton cellular protein.[276]

On the contrary, the characterization of the RNPs containing U-type of snRNAs is rapidly advancing. The following discussion is limited to this group of snRNP particles. A method for the isolation of snRNP particles, which is based on their ability to withstand the effects of a combination of high ionic strength and of moderately strong ionic detergent, has already been discussed.[214] The earlier tentative identification of snRNA species present in the mixture of the purified snRNP particles as U1, U2, U4, U5, and U6 snRNAs has been confirmed by sequencing the 3′ends of each of the five snRNA species.[277] The stability of snRNP particles at higher ionic strength and in the presence of anionic detergent was recently used in the purification of snRNP particles containing only U1 RNA.[278] U1 RNA containing snRNP particles can be selectively released from nuclear RNP polyparticles by incubation at higher temperature under isotonic conditions. RNP polyparticles purified from sonicated HeLa cell nuclei by velocity sedimentation through 15 to 30% sucrose gradient were incubated for 30 min at 25° in RSB buffer, pH 7.4, containing 0.1 *M* NaCl. The released particles containing U1 snRNA were separated from the RNP polyparticles by centrifugation in CsCl gradient which contained 0.5% sarkosyl. The final step in the purification of these particles was velocity sedimentation in sucrose gradient containing 0.5% sarkosyl and 0.5 *M* KCl. The purified particles have a buoyant density of 1.43 g/cm^3, contain only U1 or U1* snRNA, and their protein moiety consists of four major polypeptides with molecular weights between 10,000 and 14,000. The polypeptides most probably are bound to RNA by hydrophobic interactions. Since U1* snRNA is generated during the purification of the snRNP particles by specific cleavage of U1 snRNA at its 5′end[214] the obtained particles can be considered pure U1 snRNP particles. The polypeptide composition of the purified U1 snRNP particles is the same as is the polypeptide composition of the mixture of all five snRNPs. The sequences of U1, U2, U4, U5, and U6 snRNAs were determined and several homologies were found.[238] Because of these regions of homology, different snRNAs could bind to identical proteins. The same protein composition of snRNP particles also indicates that the possible specificity of function of snRNP particles containing different types of snRNA must be dependent on the RNA moiety of the particles. In contrast to these results, Wooley and co-workers,[279] by fractionation on isokinetic sucrose gradient of pH 8.0, 0.1 *M* NaCl extract of purified nuclei from *Drosophila melonogaster* embryos isolated three types of small RNP complexes. All three RNP complexes (6, 8, and 12 S) contain two major polypeptides (50,000 and 52,000 daltons). Two less-prominent polypeptides (56,000 and 75,000 daltons) are most likely limited to the 6 and 8 S RNP complexes. Yet uncharacterized RNA ranging from 80 to 600 bases in length is associated with these polypeptides.

The reports describing the composition of nuclear RNPs isolated by immunological procedures (Sm and RNP antigens) have recently been excellently reviewed,[240,280] therefore it is not necessary to analyze in this review, in detail, individual papers describing such antigens. The composition of the snRNP particles which had been purified by centrifugation in density gradients agree in principle with the composition of Sm and RNP antigens prepared from Ehrlich ascites cells,[213] Friend cells,[281] and Balb/c 3T3 cells.[275] Both the Sm and RNP antigens from these cells contain three polypeptides with molecular weight around 12,000 and, additionally, four types of polypeptides of approximately 16,000, 20,000, 26,000, and 35,000 daltons. The Sm antigens contained U1, U2, U4, U5, and U6 snRNAs, and the RNP antigens contained only U1 snRNA.[213,281] The absence of the higher-molecular-weight peptides in the particles prepared from HeLa cells by centrifugation in CsCl[214,278] could be explained by the use of a density gradient and an anionic detergent in their isolation. In *Drosophila*[282] anti-RNP serum precipitated from a pH 8.5 isotonic extract of sonicated nuclei RNPs containing U1 RNA — about 70% homologous to human U1 RNA — and 26,000- and 14,000-dalton polypeptides. The RNPs containing U1 RNA antibody sedimented at 9 to 18 S. It is interesting that the synthesis of these two proteins was not suppressed during heat shock, since most of the heat-shock genes lack intervening sequences.

The Sm and RNP antigens prepared from thymus nuclei of different species[280] also contain polypeptides with molecular weight around 13,000. The identity of at least some of these polypeptides in all types of the discussed RNPs is demonstrated by the fact that one of these polypeptides is the Sm antigenic determinant.[283,284] The amount and size of higher-molecular-weight polypeptides which were identified by different laboratories as present in the Sm and RNP antigens isolated by immunoprecipitation or immunoaffinity chromatography from the extracts of thymus cell nuclei vary considerably.

The existence of common low-molecular-weight polypeptide which is a Sm antigen determinant in all U-type snRNAs, and the requirement of additional determinants for the anti-RNP reactivity, is in agreement with the findings[214,278] that the U-RNA-containing RNP particles purified by centrifugation in CsCl can be precipitated by anti-Sm sera only, and not by anti-RNP sera.[285] At the same time, the findings that the protein composition of U1 RNA containing snRNP particles and of the particles containing other species of U-type RNA is the same,[213] strongly indicate that the expression of the antigenic sites for anti-RNP sera in the particles could be determined, at least partly, by the interactions of the proteins with the particles RNA.[213,285]

A definitive statement about the identity or diversity — possibility of proteolytic cleavage — of polypeptides found in snRNP particles prepared by different isolation procedures will be possible only after their rigorous chemical characterization. The presence of identical low-molecular-weight peptides in all studied RNPs can be expected, based on the finding that low-molecular-weight polypeptides stay associated with snRNP after centrifugation in CsCl in the presence of an anionic detergent.[214,278] The reported differences in the composition of higher-molecular-weight polypeptides are also not surprising. Any decision about what are, and what are not snRNP particles is at the present time arbitrary. The functional particles are part of the nuclear RNP network (see earlier discussion of the association of snRNAs with various nuclear structures) and different methods of isolation could remove snRNAs from this complex structure associated with different polypeptide species. Vogelstein and Hunt[275] have shown that in the nuclei of Balb/c 3T3 cells the snRNP particles reactive with anti-Sm antisera are associated with the nuclear matrix. In addition, the authors found that 21,000-, 14,500-, and 10,500- to 12,000-dalton polypeptides of these particles were significantly enriched in the proteins of the purified nuclear matrix.

A much more difficult problem is to explain the discrepancies in the published RNA composition of Sm and RNP antigens from thymus nuclei[280] and possibly also from rat.[284] With the exception of a report from one laboratory,[286] no identifiable U-type snRNAs have been found in any of the Sm and RNP antigens prepared from thymus nuclei. These antigens are RNPs, but the present RNA(s) was(were) of an unidentified small-molecular-weight species. Since the same sera, which did not precipitate any antigens containing U-type snRNA from pig thymus nuclei,[280] precipitated from HeLa cells RNP particles similar in composition to those reported by Lerner and Steitz[213] the difference could represent a difference between rapidly dividing HeLa cells and the thymus cells. Also, an identification of RNA by its size is probably not adequate because of its possible partial degradation. It has been shown by Epstein and co-workers[287] that antigen in which U1 RNA was partially cleaved by T1-RNase was precipitable by Sm-antibodies.

A different explanation for the presence of other than U-type RNAs in the Sm and RNP antigens isolated from rat liver nuclei is indicated by preliminary data published by

Douvas.[284] Under the plausible assumption that snRNP particles isolated from total nuclear extracts by immunoprecipitation are likely aggregates of specific RNA-protein complexes with nonspecific material, Douvas[284] tried to isolate a "minimal" RNP which she defined as a complex containing, in addition to RNA, only proteins which react with the antisera. As starting material she used already-purified 14 S RNP particles prepared from saline extract of sonicated rat liver nuclei. The extract contained 11% nuclear proteins and 38% nuclear RNA. The 14 S particles were separated from the Samarina RNP particles by centrifugation in 10 to 40% sucrose density gradient. The 14 S particles were further fractionated by anti-RNP immunoaffinity chromatography. The purified particles consisted only of 13,000- and 30,000-dalton polypeptides. After the transfer of these two polypeptides to nitrocellulose paper, followed by immunoautography, the anti-RNP sera reacted with the 30,000-dalton polypeptides and the anti-Sm sera reacted with the 13,000-dalton polypeptides. The 13,000-dalton polypeptides corresponded in size to the polypeptide identified as Sm antigenic determinant in other laboratories.[283,285] The identification of the 30,000-dalton polypeptide as the RNP antigenic determinant is in agreement with some of the previously published results, but differs from others.[280]

Of key interest is the preliminary characterization of the RNA present in these particles. The primary nucleic acids component of the 14 S particles is RNA larger than 20 S. All small nuclear RNA was lost during purification. As discussed earlier, 150,000- and 110,000-dalton RNA species were also found (in addition to snRNAs) in 12 and 17 S nuclear RNP particles isolated from duck erythroblast nuclei.[181] This observation could imply that the particles isolated by Douvas from rat liver, after the destruction of the nuclear RNP network by sonication and by endogenous RNase activity (as indicated by the release of free Samarina particles), contain hnRNA to which some of the proteins of the snRNP particles have been transferred.

More recently, Zieve and Penman's analysis of nuclear RNPs in HeLa cells[267] confirmed in principle earlier-discussed findings that snRNAs are associated with RNP polyparticles, the U1 snRNA being associated with the stretches of the RNP complexes forming the Samarina particles. The observed dissociation at low-salt concentration (0.1 to 0.2 *M* NaCl) of U1 RNA from the Samarina monoparticles can only be in apparent contradiction to the results from Heinrich et al.[126,132] published earlier, which indicate strong binding of snRNAs (including U1 RNA) with polyparticles. If the snRNAs are associated with the polyparticles by hydrogen bonds with hnRNA, a degradation of the hnRNA during the isolation of the 30 S Samarina monoparticles by their extraction from HeLa cell nuclei at elevated temperature could result in the dissociation at low-salt concentration of the RNPs containing U1 snRNA from the other monoparticle proteins.

Additional experimental evidence that snRNP particles can be associated with the nuclear RNP network through hnRNA has been provided by the experiments of Gallinaro and Jacob.[288] When hnRNA of the rat brain RNP particles was completely deproteinized, either by phenol extraction or by digestion with proteinase K, about 40% of snRNAs originally present in the particles stayed associated with hnRNA. At the same time, when the rat brain nuclear RNP particles were treated with 1 *M* NaCl, about half of the snRNAs originally present in the particles were released from the RNPs. Because Gallinaro and Jacob[288] found all major species of snRNA associated not only with hnRNA but also in the protein fractions released from the nuclear RNPs by 1 *M* NaCl, they proposed that the variety of types of association of snRNAs with the nuclear RNPs reflects the dynamic functional stages of snRNAs. Since snRNAs are metabolically stable, the idea that different types of association of snRNAs with the nuclear RNPs reflect the differences between their functional active form and their presence in the nuclear "pool" is attractive. The results of Gallinaro and Jacob[288] were brilliantly extended by Calvet and Pederson,[271] who after the cross-linking of RNA in nuclear RNP particles by 4′-aminomethyl-4,5′,8-trimethylpsoralen found that U1 snRNA is cross-linked to hnRNA. Inasmuch as only a fraction of the U1 sRNA is cross-linked to the hnRNA, this result is not contradictory to the proposal that U1 snRNA exists in the nuclear RNPs in different forms reflecting its function at a given time.

The last publication mentioned in this appendix illustrates a different approach used to understand the function of snRNAs. Stevens[289] described isolation of a crude preparation of 5′ to 3′ exoribonuclease from yeast. Activity of this enzyme was selectively stimulated by addition of U1 and U2 snRNA from Novikoff hepatoma cells. Addition of 5 S rRNA or tRNA from *Escherichia coli* was without effect. This stimulation probably requires formation of specific double-stranded stretches between the U1 and U2 RNA and the yeast poly(A)$^+$ RNA. Double-stranded reovirus RNA of poly(U)-poly(A) ribonucleotide were not hydrolyzed. The author speculates that this enzyme may be involved in RNA processing and that short double-stranded regions between snRNA and processed RNA may be the sites of its hydrolysis.

REFERENCES

1. **Billings, P. B. and Martin, T. E.,** Proteins of nuclear ribonucleoprotein subcomplexes, in *Methods in Cell Biology,* Vol. 17, Stein, G., Stein, J., and Kleinsmith, L. J., Eds., Academic Press, New York, 1978, 349.
2. **Heinrich, P. C., Gross, V., Northemann, W., and Scheurlen, M.,** Structure and function of nuclear ribonucleoprotein complexes, *Rev. Physiol. Biochem. Pharmacol.,* 81, 101, 1978.
3. **Jacob, M., Devilliers, G., Fuchs, J.-P., Gallinaro, H., Gattoni, R., Judes, C., and Stévenin, J.,** Isolation and structure of the ribonucleoprotein fibrils containing heterogeneous nuclear RNA, in *The Cell Nucleus,* Vol. 8, Busch, H., Ed., Academic Press, New York, 1981, 193.
4. **Kish, V. and Pederson, T.,** Isolation and characterization of ribonucleoprotein particles containing heterogeneous nuclear RNA, in *Methods in Cell Biology,* Vol. 17, Stein, G., Stein, J., and Kleinsmith, L., Eds., Academic Press, New York, 1978, 377.
5. **LeStourgeon, W. M., Lothstein, L., Walker, B. W., and Beyer, A. L.,** The composition and general topology of RNA and protein in monomer 40 S ribonucleoprotein particles, in *The Cell Nucleus,* Vol. 9, Busch, H., Ed., Academic Press, New York, 1981, 49.
6. **Martin, T. E., Pullman, J. M., and McMullen, M. D.,** Structure and function of nuclear and cytoplasmic ribonucleoprotein complexes, in *Cell Biology,* Vol. 4, Prescott, D. M. and Goldstein, L., Eds., Academic Press, New York, 1980, 137.
7. **Stévenin, J. and Jacob, M.,** Structure of premRNP. Models and pitfalls, *Mol. Biol. Rep.,* 5, 29, 1979.
8. **Fuchs, J.-P., Judes, C., and Jacob, M.,** Characterization of glycine-rich proteins from ribonucleoproteins containing heterogeneous nuclear ribonucleic acid, *Biochemistry,* 19, 1087, 1980.
9. **Samarina, O. P., Asriyan, I. S., and Georgiev, G. P.,** Vydelenie iadernykh nukleoproteidov soderzhashchikh informatsionnuiu ribonukleoinovuiu kislotu, *Dokl. Acad. Nauk S.S.S.R.,* 163, 1510, 1965.
10. **Samarina, O. P., Krichevskaya, A. A., and Georgiev, G. P.,** Nuclear ribonucleoprotein particles containing messenger ribonucleic acid, *Nature (London),* 210, 1319, 1966.
11. **Samarina, O. P., Krichevskaya, A. A., Molnar, J., Bruskov, V. I., and Georgiev, G. P.,** Nuclear ribonucleoproteins containing messenger RNA (isolation and properties) (transl.), *Mol. Biol. U.S.S.R.,* 1, 110, 1967.
12. **Samarina, O. P., Lukanidin, E. M., Molnar, J., and Georgiev, G. P.,** Structural organization of nuclear complexes containing DNA-like RNA, *J. Mol. Biol.,* 33, 251, 1968.
13. **Mant'eva, V. L., Avakyan, É. R., and Georgiev, G. P.,** The structure of nuclear D-RNA. I. The nature of the conversion of newly formed D-RNA under the action of an actinomycin block (transl), *Mol. Biol. U.S.S.R.,* 5, 256, 1971.
14. **Spirin, A. S.,** Informosomes, *Eur. J. Biochem.,* 10, 20, 1969.
15. **Stévenin, J., Gallinaro-Matringe, H., Gattoni, R., and Jacob, M.,** Complexity of the structure of particles containing heterogeneous nuclear RNA as demonstrated by ribonuclease treatment, *Eur. J. Biochem.,* 74, 589, 1977.
16. **Krichevskaya, A. A. and Georgiev, G. P.,** Nuclear ribonucleoproteins containing messenger RNA. Communication 7. Informatin, (transl.), *Mol. Biol. U.S.S.R.,* 7, 138, 1973.

17. **Bernhard, W. and Granboulan, N.,** The fine structure of the cancer cell nucleus, *Exp. Cell Res.,* Suppl. 9, 19, 1963.
18. **Monneron, A. and Bernhard, W.,** Fine structural organization of the interphase nucleus in some mammalian cells, *J. Ultrastruct. Res.,* 27, 266, 1969.
19. **Narayan, S. K., Steele, W. J., Smetana, K., and Busch, H.,** Ultrastructural aspects of the ribonucleoprotein network in nuclei of Walker tumor and rat liver, *Exp. Cell Res.,* 46, 65, 1967.
20. **Bachellerie, J.-P., Puvion, E., and Zalta, J.-P.,** Ultrastructural organization and biochemical characterization of chromatin-RNA-protein complexes isolated from mammalian cell nuclei, *Eur. J. Biochem.,* 58, 327, 1975.
21. **Faiferman, I. and Pogo, A. O.,** Isolation of a nuclear ribonucleoprotein network that contains heterogeneous RNA and is bound to the nuclear envelope, *Biochemistry,* 14, 3808, 1975.
22. **Herman, R., Weymouth, L., and Penman, S.,** Heterogeneous nuclear RNA-protein fibers in chromatin-depleted nuclei, *J. Cell Biol.,* 78, 663, 1978.
23. **Moyne, G. and Garrido, J.,** Ultrastructural evidence of mitotic perichromosomal ribonucleoproteins in hamster cells, *Exp. Cell Res.,* 98, 237, 1976.
24. **Berezney, R. and Coffey, D. S.,** Identification of a nuclear protein matrix, *Biochem. Biophys. Res. Commun.,* 60, 1410, 1974.
25. **Miller, T. E., Huang, C.-Y., and Pogo, A. O.,** Rat liver nuclear skeleton and ribonucleoprotein complexes containing hnRNA, *J. Cell Biol.,* 76, 675, 1978.
26. **Berezney, R.,** Dynamic properties of the nuclear matrix, in *The Cell Nucleus,* Vol. 7, Busch, H., Ed., Academic Press, New York, 1979, 413.
27. **Berezney, R.,** *Chromosomal Nonhistone Proteins,* Vol. IV, Hnilica, L. S., Ed., CRC Press, Boca Raton, 1983, ch. 3.
28. **Smetana, K., Steele, W. J., and Busch, H.,** A nuclear ribonucleoprotein network, *Exp. Cell Res.,* 31, 198, 1963.
29. **Steele, W. J. and Busch, H.,** Studies on the ribonucleic acid components of the nuclear ribonucleoprotein network, *Biochim. Biophys. Acta,* 129, 54, 1966.
30. **Steele, W. J., Okamura, N., and Busch, H.,** Effects of thioacetamide on the composition and biosynthesis of nucleolar and nuclear ribonucleic acid in rat liver, *J. Biol. Chem.,* 240, 1742, 1965.
31. **Faiferman, I., Hamilton, M. G., and Pogo, A. O.,** Nucleoplasmic ribonucleoprotein particles of rat liver. I. Selective degradation by nuclear nucleases, *Biochim. Biophys. Acta,* 204, 550, 1970.
32. **Faiferman, I., Hamilton, M. G., and Pogo, A. O.,** Nucleoplasmic ribonucleoprotein particles of rat liver. II. Physical properties and action of dissociating agents, *Biochim. Biophys. Acta,* 232, 685, 1971.
33. **Georgiev, G. P. and Chentsov, J. S.,** On the structural organization of nucleolo-chromosomal ribonucleoproteins, *Exp. Cell Res.,* 27, 570, 1962.
34. **Swift, H.,** Cytochemical studies on nuclear fine structure, *Exp. Cell Res.,* Suppl. 9, 54, 1963.
35. **Faiferman, I., Cornudella, L., and Pogo, A. O.,** Messenger RNA nuclear particles and their attachment to cytoplasmic membranes in Krebs tumour cells, *Nature New Biol.,* 233, 234, 1971.
36. **Mirsky, A. E. and Pollister, A. W.,** Nucleoproteins of cell nuclei, *Proc. Natl. Acad. Sci. U.S.A.,* 28, 344, 1942.
37. **Penman, S.,** RNA metabolism in the HeLa cell nucleus, *J. Mol. Biol.,* 17, 117, 1966.
38. **Berezney, R. and Coffey, D. S.,** Nuclear matrix. Isolation and characterization of a framework structure from rat liver nuclei, *J. Cell Biol.,* 73, 616, 1977.
39. **Berezney, R.,** Fractionation of the nuclear matrix. I. Partial separation into matrix protein fibrils and a residual ribonucleoprotein fraction, *J. Cell Biol.,* 85, 641, 1980.
40. **Peters, K. E. and Comings, D. E.,** Two-dimensional gel electrophoresis of rat liver nuclear washes, nuclear matrix, and hnRNA proteins, *J. Cell Biol.,* 86, 135, 1980.
41. **Hemminki, K.,** Labelling of histones and nonhistones in lung nuclear matrix and chromatin fractions, *Hoppe-Seyler's Z. Physiol. Chem.,* 358, 1125, 1977.
42. **Allen, S. L., Berezney, R., and Coffey, D. S.,** Phosphorylation of nuclear matrix proteins in isolated regenerating rat liver nuclei, *Biochem. Biophys. Res. Commun.,* 75, 111, 1977.
43. **Comings, D. E. and Okada, T. A.,** Nuclear proteins. III. The fibrillar nature of the nuclear matrix, *Exp. Cell Res.,* 103, 341, 1976.
44. **Riley, D. E., Keller, J. M., and Byers, B.,** The isolation and characterization of nuclear ghosts from cultured HeLa cells, *Biochemistry,* 14, 3005, 1975.
45. **Maundrell, K., Maxwell, E. S., Puvion, E., and Scherrer, K.,** The nuclear matrix of duck erythroblasts is associated with globin mRNA coding sequences but not with the major proteins of 40 S nuclear RNP, *Exp. Cell Res.,* 136, 435, 1981.
46. **Herlan, G. and Wunderlich, F.,** Isolation of a nuclear protein matrix from *Tetrahymena* macronuclei, *Cytobiologie,* 13, 291, 1976.
47. **Long, B. H., Huang, C.-Y., and Pogo, A. O.,** Isolation and characterization of nuclear matrix in Friend erythroleukemia cells: Chromatin and hnRNA interactions with the nuclear matrix, *Cell,* 18, 1079, 1979.

48. **van Eekelen, C. A. G. and van Venrooij, W. J.,** HnRNA and its attachment to a nuclear protein matrix, *J. Cell Biol.*, 88, 554, 1981.
49. **Beyer, A. L., Christensen, M. E., Walker, B. W., and LeStourgeon, W. M.,** Identification and characterization of the packaging proteins of core 40 S hnRNP particles, *Cell,* 11, 127, 1977.
50. **LeStourgeon, W. M., Beyer, A. L., Christensen, M. E., Walker, B. W., Poupore, S. M., and Daniels, L. P.,** The packaging proteins of core hnRNP particles and the maintenance of proliferative cell states, *Cold Spring Harbor Symp. Quant. Biol.*, 42, 885, 1977.
51. **Augenlicht, L. H., McCormick, M., and Lipkin, M.,** Digestion of RNA of chromatin and nuclear ribonucleoprotein by Staphylococcal nuclease, *Biochemistry,* 15, 3818, 1976.
52. **Miller, O. L., Jr. and Hamkalo, B. A.,** Visualization of RNA synthesis on chromosomes, *Int. Rev. Cytol.*, 33, 1, 1972.
53. **Bhorjee, J. S. and Pederson, T.,** Chromatin: Its isolation from cultured mammalian cells with particular reference to contamination by nuclear ribonucleoprotein particles, *Biochemistry,* 12, 2766, 1973.
54. **Tata, J. R. and Baker, B.,** Sub-nuclear fractionation. II. Intranuclear compartmentation of transcription *in vivo* and *in vitro, Exp. Cell Res.*, 83, 125, 1974.
55. **Augenlicht, L. H. and Lipkin, M.,** Appearance of rapidly labeled, high molecular weight RNA in nuclear ribonucleoprotein, *J. Biol. Chem.*, 251, 2592, 1976.
56. **Kimmel, C. B., Sessions, S. K., and MacLeod, M. C.,** Evidence for an association of most nuclear RNA with chromatin, *J. Mol. Biol.*, 102, 177, 1976.
57. **Liew, C. C. and Chan, P. K.,** Identification of nonhistone chromatin proteins in chromatin subunits, *Proc. Natl. Acad. Sci. U.S.A.*, 73, 3458, 1976.
58. **Stevenin, J., Gattoni, R., Gallinaro-Matringe, H., and Jacob, M.,** Nuclear ribonucleoprotein particles contain specific proteins and unspecific non-histone nuclear proteins, *Eur. J. Biochem.*, 84, 541, 1978.
59. **McKnight, S. L. and Miller, O. L., Jr.,** Ultrastructural patterns of RNA synthesis during early embryogenesis of *Drosophila melanogaster, Cell,* 8, 305, 1976.
60. **Finch, J. T. and Klug, A.,** Solenoidal model for superstructure in chromatin, *Proc. Natl. Acad. Sci. U.S.A.*, 73, 1897, 1976.
61. **Worcel, A. and Benyajati, C.,** Higher order coiling of DNA in chromatin, *Cell,* 12, 83, 1977.
62. **Grebanier, A. E. and Pogo, A. O.,** Cross-linking of proteins in nuclei and DNA-depleted nuclei from Friend Erythroleukemia cells, *Cell,* 18, 1091, 1979.
63. **Smith, H. C. and Berezney, R.,** DNA polymerase α is tightly bound to the nuclear matrix of actively replicating liver, *Biochem. Biophys. Res. Commun.*, 97, 1541, 1980.
64. **Miller, O. L., Jr. and Bakken, A. H.,** Morphological studies of transcription, in *Gene Transcription in Reproductive Tissue,* Diczfalusy, E., Ed., Karolinska Institutet, Stockholm, 1972, 155.
65. **Girard, M. and Baltimore, D.,** The effect of HeLa cell cytoplasm on the rate of sedimentation of RNA, *Proc. Natl. Acad. Sci. U.S.A.*, 56, 999, 1966.
66. **Zawislak, R., Stevenin, J., and Jacob, M.,** Artificial complexes of rat brain RNA and nuclear soluble proteins, *Biochemie,* 56, 91, 1974.
67. **Stevens, B. J. and Swift, H.,** RNA transport from nucleus to cytoplasm in *Chironomus* salivary glands, *J. Cell Biol.*, 31, 55, 1966.
68. **Malcolm, D. B. and Sommerville, J.,** The structure of chromosome-derived ribonucleoprotein in oocytes of *Triturus cristatus carnifex* (Laurenti), *Chromosoma,* 48, 137, 1974.
69. **NcKnight, S. L. and Miller, O. L., Jr.,** Post-replicative nonribosomal transcription units in *D. melanogaster* embryos, *Cell,* 17, 551, 1979.
70. **Busby, S. and Bakken, A.,** A quantitative electron microscopic analysis of transcription in sea urchin embryos, *Chromosoma,* 71, 249, 1979.
71. **Beyer, A. L., Miller, O. L., Jr., and McKnight, S. L.,** Ribonucleoprotein structure in nascent hnRNA is nonrandom and sequence-dependent, *Cell,* 20, 75, 1980.
72. **Wilt, F. H., Anderson, M., and Ekenberg, E.,** Centrifugation of nuclear ribonucleoprotein particles of sea urchin embryos in cesium sulfate, *Biochemistry,* 12, 959, 1973.
73. **Pederson, T.,** Proteins associated with heterogeneous nuclear RNA in eukaryotic cells, *J. Mol. Biol.*, 83, 163, 1974.
74. **Pederson, T. and Davis, N. G.,** Messenger RNA processing and nuclear structure: Isolation of nuclear ribonucleoprotein particles containing β-globin messenger RNA precursors, *J. Cell Biol.*, 87, 47, 1980.
75. **Mayrand, S. and Pederson, T.,** Nuclear ribonucleoprotein particles probed in living cells, *Proc. Natl. Acad. Sci. U.S.A.*, 78, 2208, 1981.
76. **Lund-Larsen, T. R.,** Effects of lead ions and Triton X-100 on release and degradation of rapidly labelled ribonucleoprotein from rat liver nuclei, *Int. J. Biochem.*, 6, 657, 1975.
77. **Martin, T. E. and McCarthy, B. J.,** Synthesis and turnover of RNA in the 30-S nuclear ribonucleoprotein complexes of mouse ascites cells, *Biochim. Biophys. Acta,* 277, 354, 1972.
78. **Köhler, K. and Arends, S.,** Particles of nuclear origin carrying rapidly labelled RNA, *Eur. J. Biochem.*, 5, 500, 1968.

79. **Lukanidin, E. M., Zalmanzon, E. S., Komaromi, L., Samarina, O. P., and Georgiev, G. P.,** Structure and function of informofers, *Nature New Biol.*, 238, 193, 1972.
80. **Molnár, J. and Komáromy, L.,** Isolation of nuclear ribonucleoprotein particles containing D-RNA from rat liver by gel-filtration, *Acta Biochim. Biophys. Acad. Sci. Hung.*, 7, 151, 1972.
81. **Patel, N. T. and Holoubek, V.,** Dependence of the composition of the protein moiety of nuclear ribonucleoprotein particles on the extent of particle purification as studied by electrophoresis including a two-dimensional procedure, *Biochim. Biophys. Acta*, 474, 524, 1977.
82. **Kulguskin, V. V., Krichevskaya, A. A., Lukanidin, E. M., and Georgiev, G. P.,** Studies on dissociation and reconstitution of nuclear 30-S ribonucleoprotein particles containing pre-mRNA, *Biochim. Biophys. Acta*, 609, 410, 1980.
83. **Ducamp, Ch. and Jeanteur, Ph.,** Characterization of nuclear RNP particles from HeLa cells. Analysis of protein and RNA constituents. Presence of poly(A), *Biochimie*, 55, 1235, 1973.
84. **Schumm, D. E. and Webb, T. E.,** Transport of informosomes from isolated nuclei of regenerating rat liver, *Biochem. Biophys. Res. Commun.*, 48, 1259, 1972.
85. **Schumm, D. E., Morris, H. P., and Webb, T. E.,** Cytosol-modulated transport of messenger RNA from isolated nuclei, *Cancer Res.*, 33, 1821, 1973.
86. **Smuckler, E. A. and Koplitz, M.,** Thioacetamide-induced alterations in nuclear RNA transport, *Cancer Res.*, 34, 827, 1974.
87. **Patel, N. T., Yoshida, M., and Holoubek, V.,** *In vitro* incorporation of 3′-methyl-4-dimethylaminoazobenzene into liver nuclei and release of RNA from the nuclei, *Cancer Res.*, 41, 743, 1981.
88. **Sauermann, G.,** Studies on ribonucleic acid metabolism using nuclear columns. Release of rapidly labeled RNA from rat liver nuclei, *Hoppe-Seyler's Z. Physiol. Chem.*, 357, 1117, 1976.
89. **Spirin, A. S. and Nemer, M.,** Messenger RNA in early sea-urchin embryos: Cytoplasmic particles, *Science*, 150, 214, 1965.
90. **Spohr, G., Granboulan, N., Morel, C., and Scherrer, K.,** Messenger RNA in HeLa cells: An investigation of free and polyribosome-bound cytoplasmic messenger ribonucleoprotein particles by kinetic labelling and electron microscopy, *Eur. J. Biochem.*, 17, 296, 1970.
91. **Ishikawa, K., Kuroda, C., and Ogata, K.,** Release of ribonucleoprotein particles containing rapidly labeled ribonucleic acid from rat liver nuclei. Effect of adenosine-5′-triphosphate and some properties of the particles, *Biochim. Biophys. Acta*, 179, 316, 1969.
92. **Ishikawa, K., Kuroda, C., and Ogata, K.,** Messenger ribonucleoprotein complexes released from rat liver nuclei by ATP. II. Chemical and metabolic properties of the protein moiety of messenger ribonucleoprotein complexes, *Biochim. Biophys. Acta*, 213, 505, 1970.
93. **McCarty, K. S., Parsons, J. T., Carter, W. A., and Laszlo, J.,** Protein-synthetic capacities of liver nuclear subfractions, *J. Biol. Chem.*, 241, 5489, 1966.
94. **Matringe, H. and Jacob, M.,** Brain nuclear and cytoplasmic ribonucleoproteins carrying DNA-like RNA: Comparison of their proteins with those from soluble nuclear and cytoplasmic supernatants by polyacrylamide gel electrophoresis, *Biochimie*, 54, 1169, 1972.
95. **Stévenin, J. and Jacob, M.,** Effects of sodium chloride and pancreatic ribonuclease on the rat-brain nuclear particles: The fate of the protein moiety, *Eur. J. Biochem.*, 47, 129, 1974.
96. **Louis, Ch. and Sekeris, C. E.,** Isolation of informofers from rat liver. Effects of α-amanitin and actinomycin D, *Exp. Cell Res.*, 102, 317, 1976.
97. **Northemann, W., Klump, H., and Heinrich, P. C.,** The effect of urea on the structure of nuclear ribonucleoprotein particles from rat liver, *Eur. J. Biochem.*, 99, 447, 1979.
98. **Gross, V., Weiss, E., Northemann, W., Scheurlen, M., and Heinrich, P. C.,** Comparative studies of the effects of galactosamine and actinomycin D on nuclear ribonucleoprotein particles from rat liver, *Exp. Cell Res.*, 109, 331, 1977.
99. **Gallinaro-Matringe, H., Stévenin, J., and Jacob, M.,** Salt dissociation of nuclear particles containing DNA-like RNA. Distribution of phosphorylated and nonphosphorylated species, *Biochemistry*, 14, 2547, 1975.
100. **Northemann, W., Gross, V., Scheurlen, M., and Henrich, P. C.,** Studies on the preparation and properties of ribonucleoprotein particles from rat liver nuclei, *Biochim. Biophys. Acta*, 519, 406, 1978.
101. **Moulé, Y. and Chauveau, J.,** Particules ribonucléoprotéiques 40 s des noyaux de foie de rat, *J. Mol. Biol.*, 33, 465, 1968.
102. **Stévenin, J., Gallinaro-Matringe, H., and Jacob, M.,** Preparation of nuclear particles and nucleosol without cross-contamination, *Biochimie*, 57, 1099, 1975.
103. **Reddi, K. K.,** Action of micrococcal phosphodiesterase on tobacco mosaic virus nucleic acid, *Nature (London)*, 182, 1308, 1958.
104. **Reddi, K. K.,** Degradation of tobacco mosaic virus nucleic acid with micrococcal phosphodiesterase, *Biochim. Biophys. Acta*, 36, 132, 1959.
105. **Webb-Walker, B., Lothstein, L., Baker, C. L., and LeStourgeon, W. M.,** The release of 40 S hnRNP particles by brief digestion of HeLa nuclei with micrococcal nuclease, *Nucleic Acids Res.*, 8, 3639, 1980.

106. **Stévenin, J., Gattoni, R., Divilliers, G., and Jacob, M.,** Rearrangements in the course of ribonuclease hydrolysis of pre-messenger ribonucleoproteins, *Eur. J. Biochem.*, 95, 593, 1979.
107. **Krichevskaya, A. A. and Georgiev, G. P.,** Further studies on the protein moiety in nuclear DNA-like RNA containing complexes, *Biochim. Biophys. Acta*, 194, 619, 1969.
108. **Niessing, J. and Sekeris, C. E.,** The protein moiety of nuclear ribonucleoprotein particles containing DNA-like RNA: Presence of heterogeneous and high molecular-weight polypeptide chains, *FEBS Lett.*, 18, 39, 1971.
109. **Niessing, J. and Sekeris, C. E.,** Further studies on nuclear ribonucleoprotein particles containing DNA-like RNA from rat liver, *Biochem. Biophys. Acta*, 247, 391, 1971.
110. **Parsons, J. T. and McCarty, K. S.,** Rapidly labeled messenger ribonucleic acid-protein complex of rat liver nuclei, *J. Biol. Chem.*, 243, 5377, 1968.
111. **Sarasin, A.,** Particules ribonucléoprotéiques 40 S des noyaux de foie de rat. Proprietes des proteines de ces particules, *FEBS Lett.*, 4, 327, 1969.
112. **Olsnes, S.,** Characterization of protein bound to rapidly labelled RNA in polyribosomes from rat liver, *Eur. J. Biochem.*, 15, 464, 1970.
113. **Morel, C., Kayibanda, B., and Scherrer, K.,** Proteins associated with globin messenger RNA in avian erythroblasts: Isolation and comparison with the proteins bound to nuclear messenger-like RNA, *FEBS Lett.*, 18, 84, 1971.
114. **Gallinaro-Matringe, H. and Jacob, M.,** Comments on the method of analysis of the proteins from nuclear particles, *FEBS Lett.*, 41, 339, 1974.
115. **Molnar, J. and Samarina, O. P.,** Autodegradation of pre-mRNA containing nuclear ribo-nucleoprotein particles. The effect of autodegradation on the double-stranded RNA sequences and on the protein composition of particles, *Mol. Biol. Rep.*, 3, 195, 1976.
116. **Martin, T., Billings, P., Levey, A., Ozarslan, S., Quinlan, T., Swift, H., and Urbas, L.,** Some properties of RNA: protein complexes from the nucleus of eukaryotic cells, *Cold Spring Harbor Symp. Quant. Biol.*, 38, 921, 1973.
117. **Schweiger, A. and Kostka, G.,** Preparation in undenatured form of the main protein bound to heterogeneous nuclear RNA in liver and hepatoma cells, *Mol. Biol. Rep.*, 3, 353, 1977.
118. **Albrecht, C. and van Zyl, I. M.,** A comparative study of the protein components of ribonucleoprotein particles isolated from rat liver and hepatoma nuclei, *Exp. Cell Res.*, 76 ,8, 1973.
119. **Gallinaro-Matringe, H., Stévenin, J., and Jacob, H.,** The nuclear particles from rat brain and liver are similar but not identical, *Differentiation*, 9, 147, 1977.
120. **Northemann, W. and Heinrich, P. C.,** Proteinase as probes of the structure of rat liver ribonucleoprotein particles carrying heterogeneous nuclear RNA, *Biochim. Biophys. Acta*, 564, 67, 1979.
121. **Kumar, A. and Pederson, T.,** Comparison of proteins bound to heterogeneous nuclear RNA and messenger RNA in HeLa cells, *J. Mol. Biol.*, 96, 353, 1975.
122. **Martin, T., Jones, R., and Billings, P.,** HnRNP core proteins: Synthesis, turnover and intracellular distribution, *Mol. Biol. Rep.*, 5, 37, 1979.
123. **Gattoni, R., Stévenin, J., Devilliers, G., and Jacob, M.,** Size heterogeneity of monoparticles from nuclear ribonucleoproteins containing premessenger RNA. *FEBS Lett.*, 90, 318, 1978.
124. **Stévenin, J. and Jacob, M.,** Étude de la dissociation, par le désoxycholate de sodium, des polyparticules ribonucléoprotéiques nucléaires du cerveau de rat, *Eur. J. Biochem.*, 29, 480, 1972.
125. **Stévenin, J., Zawislak, R., and Jacob, M.,** Etude de l'action de sels de cations monovalents sur les particules ribonucléoprotéiques de cerveau de rat, *Eur. J. Biochem.*, 33, 241, 1973.
126. **Northemann, W., Seifert, H., and Heinrich, P. C.,** The effect of sodium chloride on the structure of ribonucleoprotein particles from rat liver nuclei, *Hoppe-Seyler's Z. Physiol. Chem.*, 360, 877, 1979.
127. **Kulguskin, V. V., Lukanidin, E. M., and Georgiev, G. P.,** Free informofers and reconstitution of 30 S RNP, *Cold Spring Harbor Symp. Quant. Biol.*, 42, 911, 1977.
128. **Martin, T., Billings, P., Pullman, J., Stevens, B., and Kinniburgh, A.,** Substructure of nuclear ribonucleoprotein complexes, *Cold Spring Harbor Symp. Quant. Biol.*, 42, 899, 1977.
129. **Karn, J., Vidali, G., Boffa, L. C., and Allfrey, V. G.,** Characterization of the non-histone nuclear proteins associated with rapidly labeled heterogeneous nuclear RNA, *J. Biol. Chem.*, 252, 7307, 1977.
130. **Gross, A. J. and Sizer, I. W.,** The oxidation of tyramine, tyrosine and related compounds by peroxidase, *J. Biol. Chem.*, 234, 1611, 1959.
131. **Prosvirnin, V. V., Ruzidic, S., and Samarina, O. P.,** Cross-linked informofers, *Nucleic Acids Res.*, 7, 1649, 1979.
132. **Seifert, H., Scheurlen, M., Northemann, W., and Heinrich, P. C.,** Low molecular weight RNAs as components of nuclear ribonucleoprotein particles containing heterogeneous nuclear RNA, *Biochim. Biophys. Acta*, 564, 55, 1979.
133. **Niessing, J. and Sekeris, C. E.,** Cleavage of high-molecular-weight DNA-like RNA by a nuclease present in 30-S ribonucleoprotein particles of rat liver nuclei, *Biochim. Biophys. Acta*, 209, 484, 1970.

134. **Köhler, K. and Arends, S.**, Particles of nuclear origin carrying rapidly labelled RNA, *Eur. J. Biochem.*, 5, 500, 1968.
135. **Stunnenberg, H., Louis, Ch., and Sekeris, C. E.**, Depletion in nuclei of proteins associated with hnRNA, as a result of inhibition of RNA synthesis, *Exp. Cell Res.*, 112, 335, 1978.
136. **Swaney, J. B., vande Woude, G. F., and Bachrach, H. L.**, Sodium dodecylsulfate-dependent anomalies in gel electrophoresis: Alterations in the banding patterns of foot-and-mouth disease virus polypeptides, *Anal. Biochem.*, 58, 337, 1974.
137. **Laemmli, U. K.**, Cleavage of structural proteins during assembly of the head of bacteriophage T4, *Nature (London)*, 227, 680, 1970.
138. **Waehneldt, T. V.**, Preparative isolation of membrane proteins by polyacrylamide gel electrophoresis in the presence of ionic detergent (SDS): Proteins of rat brain myelin, *Anal. Biochem.*, 43, 306, 1971.
139. **Kul'guskin, V. V.**, Nuclear ribonucleoprotein particles containing messenger RNA. 12. Study of the dissociation and reconstruction of 30 S particles, (Transl.), *Mol. Biol. U.S.S.R.*, 11, 474, 1977.
140. **Shapiro, A. L., Viñuela, E., and Maizel, J. V., Jr.**, Molecular weight estimation of polypeptide chains by electrophoresis in SDS-polyacrylamide gels, *Biochem. Biophys. Res. Commun.*, 28, 815, 1967.
141. **Maundrell, K. and Scherrer, K.**, Characterization of pre-messenger-RNA-containing nuclear ribonucleoprotein particles from avian erythroblasts, *Eur. J. Biochem.*, 99, 225, 1979.
142. **Fuchs, J.-P. and Jacob, M.**, Fractionation of constituents of ribonucleoproteins containing heterogeneous nuclear ribonucleic acid, *Biochemistry*, 18, 4202, 1979.
143. **Chang, F. N., Navickas, I. J., Chang, C. N., and Dancis, B. M.**, Methylation of ribosomal proteins in HeLa cells, *Arch. Biochem. Biophys.*, 172, 627, 1976.
144. **Christensen, M. E., Beyer, A. L., Walker, B., and LeStourgeon, W. M.**, Identification of N^G,N^G-dimethylarginine in a nuclear protein from the lower eukaryote *Physarum polycephalum* homologous to the major proteins of mammalian 40 S ribonucleoprotein particles, *Biochem. Biophys. Res. Commun.*, 74, 621, 1977.
145. **Boffa, L. C., Karn, J., Vidali, G., and Allfrey, V. G.**, Distribution of N^G,N^G-dimethylarginine in nuclear protein fractions, *Biochem. Biophys. Res. Commun.*, 74, 969, 1977.
146. **Patel, N. T., Kurosky, A., and Holoubek, V.**, Isolation and characterization of the predominant protein in nuclear ribonucleoprotein particles from rat liver, *Biochim. Biophys. Acta*, 533, 282, 1978.
147. **Georgiev, G. P.**, Precursor of mRNA (pre-mRNA) and ribonucleoprotein particles containing pre-mRNA, in *The Cell Nucleus*, Vol. 3, Busch, H., Ed., Academic Press, New York, 1974, 67.
148. **O'Farrell, P. H.**, High resolution two-dimensional electrophoresis of proteins, *J. Biol. Chem.*, 250, 4007, 1975.
149. **Pagoulatos, G. N. and Yaniv, M.**, Proteins bound to heterogeneous nuclear RNA of simian-virus-40-infected cells, *Eur. J. Biochem.*, 91, 1, 1978.
150. **O'Farrell, P. Z., Goodman, H., and O'Farrell, P. H.**, High resolution two-dimensional electrophoresis of basic as well as acidic proteins, *Cell*, 12, 1133, 1977.
151. **Brunel, C. and Lelay, M.-N.**, Two-dimensional analysis of proteins associated with heterogeneous nuclear RNA in various animals cell lines, *Eur. J. Biochem.*, 99, 273, 1979.
152. **Kimmel, C. B., Sessions, S. K., and McLeod, M. C.**, Evidence for an association of most nuclear RNA with chromatin, *J. Mol. Biol.*, 102, 177, 1976.
153. **Robinson, A. B.**, Evolution and the distribution of glutaminyl and asparaginyl residues in proteins, *Proc. Natl. Acad. Sci. U.S.A.*, 71, 885, 1974.
154. **Niessing, J. and Sekeris, C. E.**, A homoribopolynucleotide synthetase in rat liver nuclei associated with ribonucleoprotein particles containing DNA-like RNA, *FEBS Lett.*, 22, 83, 1972.
155. **Niessing, J. and Sekeris, C. E.**, Synthesis of polynucleotides in nuclear ribonucleoprotein particles containing heterogeneous RNA, *Nature New Biol.*, 243, 9, 1973.
156. **Blanchard, J.-M., Brunel, C., and Jeanteur, Ph.**, Characterization of an endogenous protein kinase activity in ribonucleoprotein structures containing heterogeneous nuclear RNA in HeLa cell nuclei, *Eur. J. Biochem.*, 79, 117, 1977.
157. **Bajszár, G., Szabó, G., Simoncsits, A., and Molnár, J.**, Methylated cap formation by enzymes bound to nuclear informofer particles, *Mol. Biol. Rep.*, 4, 93, 1978.
158. **Furuichi, Y., Muthukrishnan, S., Tomasz, J., and Shatkin, A. J.**, Mechanism of formation of reovirus mRNA 5′-terminal blocked and methylated sequence m^7GpppGmpC, *J. Biol. Chem.*, 251, 5043, 1976.
159. **Molnár, J., Bajszár, G., Marczinovits, I., and Szabó, G.**, Cleavage of pre-mRNA sequences by ribonucleases bound to nuclear RNP particles of rat liver, *Mol. Biol. Rep.*, 4, 157, 1978.
160. **Rech, J., Brunel, C., and Jeanteur, Ph.**, HnRNP from HeLa cells contain a ribonuclease active on double-stranded RNA, *Biochem. Biophys. Res. Commun.*, 88, 422, 1979.
161. **Calvet, J. P. and Pederson, T.**, Secondary structure of heterogeneous nuclear RNA: Two classes of double-stranded RNA in native ribonucleoprotein, *Proc. Natl. Acad. Sci. U.S.A.*, 74, 3705, 1977.
162. **Paoletti, J., Rech, J., Brunel, C., and Jeanteur, Ph.**, Constrained configuration of double-stranded ribonucleic acid in HeLa hnRNP and its relaxation by ribonuclease D, *Biochemistry*, 19, 5223, 1980.

163. **Gallinaro-Matringe, H. and Jacob, M.,** Nuclear particles from rat brain: complexity of the major proteins and their phosphorylation *in vivo*, *FEBS Lett.*, 36, 105, 1973.
164. **Schweiger, A. and Schmidt, D.,** Isolation of RNA-binding proteins from rat liver nuclear 30 S-particles, *FEBS Lett.*, 41, 17, 1974.
165. **Blanchard, J.-M., Ducamp, Ch., and Jeanteur, Ph.,** Endogenous protein kinase activity in nuclear RNP particles from HeLa cells, *Nature (London)*, 253, 467, 1975.
166. **Blanchard, J.-M., Brunel, C., and Jeanteur, Ph.,** Phosphorylation *in vivo* of proteins associated with heterogeneous nuclear RNA in HeLa cell nuclei, *Eur. J. Biochem.*, 86, 301, 1978.
167. **Periasamy, M., Brunel, C., and Jeanteur, Ph.,** Protein kinase associated with hnRNP from HeLa cell nuclei, *Biochimie*, 61, 823, 1979.
168. **Periasamy, M., Brunel, C., Blanchard, J.-M., and Jeanteur, Ph.,** Characterization of a phosphoprotein phosphatase activity in ribonucleoprotein particles from HeLa cell nuclei, *Biochem. Biophys. Res. Commun.*, 79, 1077, 1977.
169. **Patel, N. T., Folse, D. S., and Holoubek, V.,** Release of repetitive nuclear RNA into the cytoplasm in liver of rats fed 3′-methyl-4-dimethylaminoazobenzene, *Cancer Res.*, 39, 4460, 1979.
170. **Greenberg, J. R. and Perry, R. P.,** Relative occurrence of polyadenylic acid sequences in messenger and heterogeneous nuclear RNA of L cells as determined by poly(U)-hydroxylapatite chromatography, *J. Mol. Biol.*, 72, 91, 1972.
171. **Edmonds, M., Vaughan, M. H., Jr., and Nakazato, H.,** Polyadenylic acid sequences in the heterogeneous nuclear RNA and rapidly labeled polyribosomal RNA of HeLa cells: possible evidence for a precursor relationship, *Proc. Natl. Acad. Sci. U.S.A.*, 68, 1336, 1971.
172. **Darnell, J. E., Philipson, L., Wall, R., and Adesnik, M.,** Polyadenylic acid sequences: role in conversion of nuclear RNA into messenger RNA, *Science*, 174, 507, 1971.
173. **Cornudella, L., Faiferman, I., and Pogo, A. O.,** Polyadenylic acid sequences in ascites cells nuclear particles and membrane-bound messenger RNA, *Biochim. Biophys. Acta*, 294, 541, 1973.
174. **Samarina, O. P., Aitkhozhina, N. A., and Besson, J.,** Poly A blocks in the nuclear ribonucleoprotein complexes containing pre-mRNA, *Mol. Biol. Rep.*, 1, 193, 1973.
175. **Quinlan, T. J., Billings, P. B., and Martin, T. E.,** Nuclear ribonucleoprotein complexes containing polyadenylate from mouse ascites cells, *Proc. Natl. Acad. Sci. U.S.A.*, 71, 2632, 1974.
176. **Quinlan, T. J., Kinniburgh, A. J., and Martin, T. E.,** Properties of a nuclear polyadenylate-protein complex from mouse ascites cells, *J. Biol. Chem.*, 252, 1156, 1977.
177. **Blobel, G.,** A protein of molecular weight 78,000 bound to the polydenylate region of eukaryotic messenger RNAs, *Proc. Natl. Acad. Sci. U.S.A.*, 70, 924, 1973.
178. **Kish, V. M. and Pederson, T.,** Poly(A)-rich ribonucleoprotein complexes from HeLa cell messenger RNA, *J. Biol. Chem.*, 251, 5888, 1976.
179. **Kish, V. M. and Pederson, T.,** Ribonucleoprotein organization of polyadenylate sequences in HeLa cell heterogeneous nuclear RNA, *J. Mol. Biol.*, 95, 227, 1975.
180. **Firtel, R. A. and Pederson, T.,** Ribonucleoprotein particles containing heterogeneous nuclear RNA in the cellular slime mold *Dictyostelium discoideum*, *Proc. Natl. Acad. Sci. U.S.A.*, 72, 301, 1975.
181. **Maxwell, E. S., Maundrell, K., Puvion-Dutilleul, F., and Scherrer, K.,** Distribution of low-molecular-weight RNAs in avian erythroblast nuclear ribonucleoprotein complexes associated with pre-messenger RNA, *Eur. J. Biochem.*, 113, 233, 1981.
182. **Molnár, J. and Samarina, O. P.,** Protein composition of nuclear 14 S ribonucleoprotein particles containing poly(A), *Mol. Biol. Rep.*, 2, 1, 1975.
183. **Mant'eva, V. L., Avakyan, É. R., and Georgiev, G. P.,** Nuclear ribonucleoproteins containing messenger RNA. 6. Classes of nuclear D-RNA bound to informofers, (transl.), *Mol. Biol. U.S.S.R.*, 3, 428, 1969.
184. **Martin, T. E. and McCarthy, B. J.,** Synthesis and turnover of RNA in the 30-S nuclear ribonucleoprotein complexes of mouse ascites cells, *Biochim. Biophys. Acta*, 277, 354, 1972.
185. **Kinniburgh, A. J. and Martin, T. E.,** Detection of mRNA sequences in nuclear 30 S ribonucleoprotein subcomplexes, *Proc. Natl. Acad. Sci. U.S.A.*, 73, 2725, 1976.
186. **Hendrick, D. and De Kloet, S. R.,** Globin messenger sequences in nuclear ribonucleoprotein particles of avian erythroblasts, *Mol. Biol. Rep.*, 2, 225, 1975.
187. **Gattoni, R., Stévenin, J., and Jacob, M.,** Comparison of the nuclear ribonucleoproteins containing the transcripts of adenovirus-2 and HeLa cell DNA, *Eur. J. Biochem.*, 108, 203, 1980.
188. **Blanchard, J.-M.,** HnRNP proteins interact with nascent adenovirus RNA, *Biochem. Biophys. Res. Commun.*, 97, 524, 1980.
189. **Blanchard, J.-M. and Weber, J.,** Nuclear ribonucleoprotein particles from adenovirus infected HeLa cells, *Mol. Biol. Rep.*, 7, 107, 1981.
190. **Munroe, S. H. and Pederson, T.,** Messenger RNA sequences in nuclear ribonucleoprotein particles are complexed with protein as shown by nuclease protection, *J. Mol. Biol.*, 147, 437, 1981.
191. **Alonso, A., Louis, Ch., Flytzanis, C. N., and Sekeris, C. E.,** Distribution of coding sequences within the nuclear hnRNA-protein complexes, *FEBS Lett.*, 93, 351, 1978.

192. **Kinniburgh, A. J. and Martin, T. E.,** Oligo(A) and oligo(A)-adjacent sequences present in nuclear ribonucleoprotein complexes and mRNA, *Biochem. Biophys. Res. Commun.*, 73, 718, 1976.
193. **Alonso, A., Winter, H., and Krieg, L.,** Poly(A) and oligo(A)-adjacent sequences in RNA from the large hnRNP complexes of rat liver, *Biochim. Biophys. Acta,* 652, 72, 1981.
194. **Kinniburgh, A. J., Billings, P. B., Quinlan, T. J., and Martin, T. E.,** Distribution of hnRNA and mRNA sequences in nuclear ribonucleoprotein complexes, *Prog. Nucleic Acid Res. Mol. Biol.*, 19, 335, 1976.
195. **Naora, H.,** Some aspects of double-stranded hairpin structures in heterogeneous nuclear RNA, *Int. Rev. Cytol.*, 56, 255, 1979.
196. **Calvet, J. P. and Pederson, T.,** Nucleoprotein organization of inverted repeat DNA transcripts in heterogeneous nuclear RNA-ribonucleoprotein particles from HeLa cells, *J. Mol. Biol.*, 122, 361, 1978.
197. **Calvet, J. P. and Pederson, T.,** Photochemical cross-linking of secondary structure in HeLa cell heterogeneous nuclear RNA *in situ, Nucleic Acids Res.*, 6, 1993, 1979.
198. **Calvet, J. P. and Pederson, T.,** Heterogeneous nuclear RNA double-stranded regions probed in living HeLa cells by crosslinking with psoralen derivative aminomethyltrioxsalen, *Proc. Natl. Acad. Sci. U.S.A.*, 76, 755, 1979.
199. **Ro-Choi, T. S. and Busch, H.,** Low-molecular-weight nuclear RNAs, in *The Cell Nucleus,* Vol. 3, Busch, H., Ed., Academic Press, New York, 1974, 151.
200. **Rein, A.,** The small molecular weight monodisperse nuclear RNA's in mitotic cells, *Biochim. Biophys. Acta,* 232, 306, 1971.
201. **Sekeris, C. E. and Niessing, J.,** Evidence for the existence of a structural RNA component in the nuclear ribonucleoprotein particles containing heterogeneous RNA, *Biochem. Biophys. Res. Commun.*, 62, 642, 1975.
202. **Deimel, B., Louis, Ch., and Sekeris, C. E.,** The presence of small molecular weight RNAs in nuclear ribonucleoprotein particles carrying hnRNA, *FEBS Lett.*, 73, 80, 1977.
203. **Flytzanis, C., Alonso, A., Louis, Ch., Krieg, L., and Sekeris, C. E.,** Association of small nuclear RNA with hnRNA isolated from nuclear RNP complexes carrying hnRNA, *FEBS Lett.*, 96, 201, 1978.
204. **Zieve, G. and Penman, S.,** Small RNA species of the HeLa cell: Metabolism and subcellular localization, *Cell,* 8, 19, 1976.
205. **Guimont-Ducamp, Ch., Sri-Widada, J., and Jeanteur, Ph.,** Occurrence of small molecular weight RNAs in HeLa nuclear ribonucleoprotein particles containing HnRNA (HnRNP), *Biochimie,* 59, 755, 1977.
206. **Howard, E. F.,** Small nuclear RNA molecules in nuclear ribonucleoprotein complexes from mouse erythroleukemia cells, *Biochemistry,* 17, 3228, 1978.
207. **Gallinaro, H. and Jacob, M.,** An evaluation of small nuclear RNA in hnRNP, *FEBS Lett.*, 104, 176, 1979.
208. **Daskal, Y., Komaromy, L., and Busch, H.,** Isolation and partial characterization of perichromatin granules, *Exp. Cell Res.*, 126, 39, 1980.
209. **Miller, T. E., Huang, C.-Y., and Pogo, A. O.,** Rat liver nuclear skeleton and small molecular weight RNA species, *J. Cell Biol.*, 76, 692, 1978.
210. **Raj, N. B. K., Ro-Choi, T. S., and Busch, H.,** Nuclear ribonucleoprotein complexes containing U1 and U2 RNA, *Biochemistry,* 14, 4380, 1975.
211. **Notman, D. D., Kurata, N., and Tan, E. M.,** Profiles of antinuclear antibodies in systemic rheumatic diseases, *Ann. Int. Med.*, 83, 464, 1975.
212. **Douvas, A. S., Stumph, W. E., Reyes, P., and Tan, E. M.,** Isolation and characterization of nuclear ribonucleoprotein complexes using human anti-nuclear ribonucleoprotein antibodies, *J. Biol. Chem.*, 254, 3608, 1979.
213. **Lerner, M. R. and Steitz, J. A.,** Antibodies to small nuclear RNAs complexed with proteins are produced by patients with systemic lupus erythematosus, *Proc. Natl. Acad. Sci. U.S.A.*, 76, 5495, 1979.
214. **Brunel, C., Sri Widada, J., Lelay, M.-N., Jeanteur, Ph., and Liautard, J.-P.,** Purification and characterization of a simple ribonucleoprotein particle containing small nucleoplasmic RNAs (snRNP) as a subset of RNP containing heterogeneous nuclear RNA (hnRNP) from HeLa cells, *Nucleic Acids Res.*, 9, 815, 1981.
215. **Murray, V. and Holliday, R.,** Mechanism for RNA splicing of gene transcripts, *FEBS Lett.*, 106, 5, 1979.
216. **Lerner, M. R., Boyle, J. A., Mount, S. M., Wolin, S. L., and Steitz, J. A.,** Are snRNPs involved in splicing? *Nature (London),* 283, 220, 1980.
217. **Rogers, J. and Wall, R.,** A mechanism for RNA splicing, *Proc. Natl. Acad. Sci. U.S.A.*, 77, 1877, 1980.
218. **Gallinaro, H., Lazar, E., Jacob, M., Krol, A., and Branlant, C.,** Small RNA in hnRNP fibrils and their possible function in splicing, *Mol. Biol. Rep.*, 7, 31, 1981.
219. **Sekeris, C. E., Prüsse, A., Alonso, A., and Louis, Ch.,** HnRNA-snRNA-protein complexes: possible structure and function, *Period. Biol.*, 82, 309, 1980.

220. **Jelinek, W. and Leinwand, L.,** Low molecular weight RNAs hydrogen-bonded to nuclear and cytoplasmic poly(A)-terminated RNA from cultured Chinese hamster ovary cells, *Cell,* 15, 205, 1978.
221. **Maxwell, E. S., Maundrell, K., and Scherrer, K.,** Two low molecular weight nuclear RNAs, isolated from avian erythroblast nuclear ribonucleoprotein complexes, hybridize to duck pre-messenger and globin messenger RNA, *Biochem. Biophys. Res. Commun.,* 97, 875, 1980.
222. **Kostka, G. and Schweiger, A.,** Composition of hnRNA-associated proteins in rat liver is specifically altered after cycloheximide treatment, *Mol. Biol. Rep.,* 6, 57, 1980.
223. **Levis, R. and Penman, S.,** The metabolism of poly(A)$^+$ and poly(A)$^-$ hnRNA in cultured Drosophila cells studied with a rapid uridine pulse-chase, *Cell,* 11, 105, 1977.
224. **Martin, T. E. and Okamura, C. S.,** HnRNP protein distribution in various differentiated vertebrate cells, in *International Cell Biology (1980—1981),* Schweiger, H. G., Ed., Springer-Verlag, Berlin, 1981, 76.
225. **Jones, R. E., Okamura, C. S., and Martin, T. E.,** Immunofluorescent localization of the proteins of nuclear ribonucleoprotein complexes, *J. Cell Biol.,* 86, 235, 1980.
226. **Pederson, T.,** Gene activation in eukaryotes: Are nuclear acidic proteins the cause or the effect? *Proc. Natl. Acad. Sci. U.S.A.,* 71, 617, 1974.
227. **Stanton, G. J. and Holoubek, V.,** Stimulation of synthesis of the proteins of 30-S nuclear ribonucleoprotein particles in human amnion U cells by viral infection, *Biochim. Biophys. Acta,* 477, 151, 1977.
228. **Patel, N. T. and Holoubek, V.,** Azocarcinogen-induced release of chromatin-associated RNA into liver cytoplasm: The possible role of reiterated RNA sequences in the control of RNA processing, *Chem. Biol. Interactions,* 35, 129, 1981.
229. **Yoshida, M. and Holoubek, V.,** Early effects of carcinogenic aminoazo dyes on the protein patterns and metabolism in rat livers, *Int. J. Biochem.,* 7, 259, 1976.
230. **Yoshida, M. and Holoubek, V.,** Binding of 3′-methyl-4-dimethylaminoazobenzene to nuclear ribonucleoprotein particles, *Life Sci.,* 18, 49, 1976.
231. **Patel, N. T. and Holoubek, V.,** Protein composition of liver nuclear ribonucleoprotein particles of rats fed carcinogenic aminoazo dyes, *Biochem. Biophys. Res. Commun.,* 73, 112, 1976.
232. **Patel, N. T. and Holoubek, V.,** Protein composition of nuclear ribonucleoprotein particles isolated from liver of rats in the early stages of feeding of 3′-methyl-4-dimethylaminoazobenzene and from hepatoma induced by the same carcinogen, *Chem. Biol. Interactions,* 19, 303, 1977.
233. **Sommerville, J.,** Immunolocalization and structure organization of nascent RNP, in *The Cell Nucleus,* Vol. 8, Busch, H., Ed., Academic Press, New York, 1981, 1.
234. **Puvion, E. and Moyne, G.,** *In situ* localization of RNA structures, in *The Cell Nucleus,* Vol. 8, Busch, H., Ed., Academic Press, New York, 1981, 59.
235. **Daskal, Y.,** Perichromatin granules, in *The Cell Nucleus,* Vol. 8, Busch, H., Ed., Academic Press, New York, 1981, 117.
236. **Maclean, N. and Gregory, S. P.,** Transcription in isolated nuclei, in *The Cell Nucleus,* Vol. 8, Busch, H., Ed., Academic Press, New York, 1981, 139.
237. **Sekeris, C. E. and Guialis, A.,** Low-molecular-weight nuclear ribonucleoprotein particles, in *The Cell Nucleus,* Vol. 8, Busch, H. Ed., Academic Press, New York, 1981, 247.
238. **Reddy, R. and Busch, H.,** U snRNA's of nuclear snRNP's, in *The Cell Nucleus,* Vol. 8, Busch, H. Ed., Academic Press, New York, 1981, 261.
239. **Pogo, O. A.,** Heterogeneous nuclear RNA-protein complexes and nuclear matrix, in *The Cell Nucleus,* Vol. 8, Busch, H. Ed., Academic Press, New York, 1981, 331.
240. **Douvas, A. S. and Tan, E. M.,** Human antibodies to RNA-containing particles, in *The Cell Nucleus,* Vol. 8, Busch, H., Ed., Academic Press, New York, 1981, 369.
241. **Samarina, O. P. and Krichevskaya, A. A.,** Nuclear 30 S RNP particles, in *The Cell Nucleus,* Vol. 9, Busch, H., Ed., Academic Press, New York, 1981, 1.
242. **Comings, D. E. and Peters, K. E.,** Two-dimensional gel electrophoresis of nuclear particles, in *The Cell Nucleus,* Vol. 9, Academic Press, New York, 1981, 89.
243. **Martin, T. E. and Okamura, C. S.,** Immunocytochemistry of nuclear hnRNP complexes, in *The Cell Nucleus,* Vol. 9, Busch, H., Ed., Academic Press, New York, 1981, 119.
244. **Jeanteur, Ph.,** Enzymatic activities associated with hnRNP, in *The Cell Nucleus,* Vol. 9, Busch, H., Ed., Academic Press, New York, 1981, 145.
245. **Augenlicht, L. H.,** Digestion products of nuclear ribonucleoprotein, in *The Cell Nucleus,* Vol. 9, Busch, H., Ed., Academic Press, New York, 1981, 171.
246. **Webb, T. E., Schumm, D. E., and Palayoor, T.,** Nucleocytoplasmic transport of mRNA, in *The Cell Nucleus,* Vol. 9, Busch, H., Ed., Academic Press, New York, 1981, 199.
247. **Roy, R. K., Sarkar, S., Guha, C., and Munro, H. N.,** RNP particles involved in release of *in vitro* synthesized poly(A)-containing RNA in isolated nuclei, in *The Cell Nucleus,* Vol. 9, Busch, H., Ed., Academic Press, New York, 1981, 289.
248. **van Eekelen, C. A. G., Mariman, E. C. M., Reinders, R. J., and van Venrooij, W. J.,** Adenoviral heterogeneous nuclear RNA is associated with host cell proteins, *Eur. J. Biochem.,* 119, 461, 1981.

249. **van Eekelen, C. A., Riemen, T., and van Venrooij, W. J.,** Specificity in the interaction of hnRNA and mRNA with proteins as revealed by *in vivo* cross linking, *FEBS Lett.*, 130, 223, 1981.
250. **Mayrand, S., Setyono, B., Greenberg, J. R., and Pederson, T.,** Structure of nuclear ribonucleoprotein: Identification of proteins in contact with poly(A)$^+$ heterogeneous nuclear RNA in living HeLa cells, *J. Cell Biol.*, 90, 380, 1981.
251. **Economidis, I. V. and Pederson, T.,** Assembly of nuclear ribonucleoprotein particles during *in vitro* transcription, *Proc. Natl. Acad. Sci. U.S.A.*, 79, 1469, 1982.
252. **Setyono, B. and Greenberg, J. R.,** Proteins associated with poly(A) and other regions of mRNA and hnRNA molecules as investigated by cross-linking, *Cell*, 24, 775, 1981.
253. **Marczinovits, I. and Molnár, J.,** Ultraviolet light-induced cross-linking of HnRNA to informofer proteins *in vitro*, *Mol. Biol. Rep.*, 8, 111, 1982.
254. **Ivanova, E., Pironcheva, G., and Djondjurov, L.,** Turnover of the major polypeptides of 40-S monomer particles, *Eur. J. Biochem.*, 113, 569, 1981.
255. **Schweiger, A. and Kostka, G.,** Properties of a 110,000 molecular weight protein in rat liver hnRNP particles, *Mol. Biol. Rep.*, 7, 47, 1981.
256. **Djondjurov, L., Ivanova, E., and Tsanev, R.,** Two chromatin fractions with different metabolic properties of non-histone proteins and newly synthesized RNA, *Eur. J. Biochem.*, 97, 133, 1980.
257. **Thomas, P. S., Shepherd, J. H., Mulvihill, E. R., and Palmiter, R. D.,** Isolation of a nuclear ribonucleoprotein fraction from chick oviduct containing ovalbumin messenger RNA sequences, *J. Mol. Biol.*, 150, 143, 1981.
258. **Volkova, I. V. and Gauze, L. N.,** Homogeneity of 30 S fraction of nuclear ribonucleoprotein particles. Electrophoresis of intact particles, (transl.), *Mol. Biol. U.S.S.R.*, 16, 100, 1982.
259. **Pederson, T. and Munroe, S. H.,** Ribonucleoprotein organization of eukaryotic RNA, XV. Different nucleoprotein structures of globin messenger RNA sequences in nuclear and polyribosomal ribonucleoprotein particles, *J. Mol. Biol.*, 150, 509, 1981.
260. **Ohlsson, R. I., van Eekelen, C., and Philipson, L.,** Non-random localization of ribonucleoprotein (RNP) structures within an adenovirus mRNA precursor, *Nucleic Acids Res.*, 10, 3053, 1982.
261. **Beyer, A. L., Bouton, A. H., and Miller, O. L., Jr.,** Correlation of hnRNP structure and nascent transcript cleavage, *Cell*, 26, 155, 1981.
262. **Borissova, O. F., Krichevskaya, A. A., and Samarina, O. P.,** Structural investigation of nuclear RNP particles containing pre-mRNA by different fluorescence techniques, *Nucleic Acids Res.*, 9, 663, 1981.
263. **Upreti, R. K., and Holoubek, V.,** The effect of inhibition of RNA synthesis by actinomycin D on the population of basic polypeptides of the 30 S nuclear ribonucleoprotein particles, *Biochimie*, 64, 247, 1982.
264. **Upreti, R. K. and Holoubek, V.,** Methylation of proteins of the nuclear ribonucleoprotein particles in liver and kidney of rats injected with actinomycin D, *Biochimie*, 64, 435, 1982.
265. **Upreti, R. K. and Holoubek, V.,** Separation of proteins of nuclear ribonucleoprotein particles by high-performance liquid chromatography, *Anal. Chim. Acta*, 131, 239, 1981.
266. **Kucherer, C., Marty, L., and Blanchard, J. -M.,** Presence of the pre-mRNA for the 72K DNA binding protein in hnRNP from early adenovirus-2 infected HeLa cells, *Biochem. Biophys. Res. Commun.*, 105, 603, 1982.
267. **Zieve, G. and Penman, S.,** Subnuclear particles containing a small nuclear RNA and heterogeneous nuclear RNA, *J. Mol. Biol.*, 145, 501, 1981.
268. **Marvil, D. K., Nowak, L., and Szer, W.,** A single-stranded nucleic acid-binding protein from *Artemia salina*. I. Purification and characterization, *J. Biol. Chem.*, 255, 6466, 1980.
269. **Nowak, L., Marvil, D. K., Thomas, J. O., Boublik, M., and Szer, W.,** A single-stranded nucleic acid-binding protein from *Artemia salina*. II. Interaction with nucleic acids, *J. Biol. Chem.*, 255, 6473, 1980.
270. **Thomas, J. O., Raziuddin, Sobota, A., Boublik, M., and Szer, W.,** A RNA helix-destabilizing protein is a major component of *Artemia salina* nuclear ribonucleoproteins, *Proc. Natl. Acad. Sci. U.S.A.*, 78, 2888, 1981.
271. **Calvet, J. P. and Pederson, T.,** Base-pairing interactions between small nuclear RNAs and nuclear RNA precursors as revealed by psoralen cross-linking *in vivo*, *Cell*, 26, 363, 1981.
272. **Yang, V. W., Lerner, M. R., Steitz, J. A., and Flint, S. J.,** A small nuclear ribonucleoprotein is required for splicing of adenoviral early RNA sequences, *Proc. Natl. Acad. Sci. U.S.A.*, 78, 1371, 1981.
273. **Lerner, R. M., Boyle, J. A., Hardin, J. A., and Steitz, J. A.,** Two novel classes of small ribonucleoproteins detected by antibodies associated with lupus erythematosus, *Science*, 211, 400, 1981.
274. **Lerner, M. R., Andrews, N. C., Miller, G., and Steitz, J. A.,** Two small RNAs encoded by Epstein-Barr virus and complexed with protein are precipitated by antibodies from patients with systemic lupus erythematosus, *Proc. Natl. Acad. Sci. U.S.A.*, 78, 805, 1981.
275. **Vogelstein, B. and Hunt, B. F.,** A subset of small nuclear ribonucleoprotein particle antigens is a component of the nuclear matrix, *Biochem. Biophys. Res. Commun.*, 105, 1224, 1982.

276. **van Eekelen, C., Buijtels, H., Linné, T., Ohlsson, R., Philipson, L., and van Venrooij, W.,** Detection of a cellular polypeptide associated with adenovirus-coded VA RNA using in vitro labeling of proteins crosslinked to RNA, *Nucleic Acids Res.*, 10, 3039, 1982.
277. **Sri-Widada, J., Liautard, J. P., Assens, C., and Brunel, C.,** Primary structure identification of snRNAs present in highly purified snRNPs from HeLa cells, *Mol. Biol. Rep.*, 8, 29, 1981.
278. **Sri-Widada, J., Assens, C., Liautard, J. P., Jeanteur, Ph., and Brunel, C.,** Isolation of a pure U_1 snRNP from HeLa cells, *Biochem. Biophys. Res. Commun.*, 104, 457, 1982.
279. **Wooley, J. C., Cone, R., and Chung, S. Y.,** Isolation and characterization of nuclear ribonucleoprotein complexes from *Drosophila melanogaster, Mol. Biol. Rep.*, 7, 5, 1981.
280. **MacGillivray, A. J., Carroll, A. R., Dahi, S., Naxakis, G., Sadaie, M. R., Wallis, C. M., and Jing, T.,** The composition of the nuclear antigens Sm and RNP of human rheumatic and connective tissue diseases and the relevance of their autoantibodies as probes for RNA processing mechanisms, *FEBS Lett.*, 141, 139, 1982.
281. **Barque, J. P., Yeni, P., Peraudeau, L., Danon, F., and Larsen, C. J.,** Nuclear ribonucleoproteins recognized by human antinuclear antibodies in retrovirus-infected cells, *Biochem. Biophys. Res. Commun.*, 99, 284, 1981.
282. **Wieben, E. D., and Pederson, T.,** Small nuclear ribonucleoproteins of *Drosophila:* Identification of U1 RNA-associated proteins and their behavior during heat shock, *Mol. Cell. Biol.*, 2, 914, 1982.
283. **White, P. J. and Hoch, S. O.,** Definition of the antigenic polypeptides in the Sm and RNP ribonucleoprotein complexes, *Biochem. Biophys. Res. Commun.*, 102, 365, 1981.
284. **Douvas, A. S.,** Antibodies occurring in two different rheumatic diseases react with the same nuclear ribonucleoprotein particle, *Proc. Natl. Acad. Sci. U.S.A.*, 79, 5401, 1982.
285. **Assens, C., Liautard, J. -P., Sri-Widada, J., Brunel, C., and Jeanteur, Ph.,** Highly purified snRNPs retain antigenic determinants towards anti Sm antibodies, *Biochem. Biophys. Res. Commun.*, 106, 953, 1982.
286. **White, P. J., Gardner, W. D., and Hoch, S. O.,** Identification of the immunogenically active components of the Sm and RNP antigens, *Proc. Natl. Acad. Sci. U.S.A.*, 78, 626, 1981.
287. **Epstein, P., Reddy, R., and Busch, H.,** Site-specific cleavage by T1 RNase of U-1 RNA in U-1 ribonucleoprotein particles, *Proc. Natl. Acad. Sci. U.S.A.*, 78, 1562, 1981.
288. **Gallinaro, H. and Jacob, M.,** The status of small nuclear RNA in the ribonucleoprotein fibrils containing heterogeneous nuclear RNA, *Biochem. Biophys. Acta,* 652, 109, 1981.
289. **Stevens, A.,** A yeast endoribonuclease stimulated by Novikoff hepatoma small nuclear RNAs U1 and U2, *Biochem. Biophys. Res. Commun.*, 107, 642, 1982.
290. **Holoubek, V.,** unpublished data.
291. **Northemann, W., Scheurlen, M., Gross, V., and Heinrich, P. C.,** Circular dichroism of ribonucleoprotein complexes from rat liver nuclei, *Biochem. Biophys. Res. Commun.*, 76, 1130, 1977.

Chapter 3

ORGANIZATION AND FUNCTIONS OF THE NUCLEAR MATRIX

Ronald Berezney

TABLE OF CONTENTS

I. INTRODUCTION

The isolation and characterization of the nuclear ground substance or *nuclear matrix* has provided a new stimulus for investigating the relationships of nuclear form and function in the eukaryotic cell. The nuclear matrix is currently an exploding area of research and investigators are exploiting the matrix for studying nuclear functions. Perhaps most exciting are the generally positive associations which have been found to date between a variety of functionally related properties and isolated matrix structures. These associations include: DNA loop attachment sites,[1-3] in vivo replicating DNA,[1,4-8] the major replicative enzyme, DNA polymerase alpha,[9] endogenous template driven in vitro DNA synthesis in isolated nuclear matrix,[10,11] possible enrichment of at least certain gene sequences at the DNA attachment sites of the matrix,[12-14] hn-RNA and hn-RNP (ribonucleoprotein) particles including primary transcripts of specific gene sequences and later spliced intermediates,[15-23] ribosomal precursor RNA,[24] sn-RNA,[25,26] steroid receptor binding sites,[27-31] viral proteins such as the T-antigen,[32-34] viral transcripts,[20,22] carcinogen binding sites,[35-37] reversible in vitro expansion of the isolated matrix,[38] association of heat shock proteins in *Drosophila melanogaster,*[39] and protein phosphorylation.[40,41]

It is conceivable that the above mentioned list of matrix-associated properties will be outdated by the time this chapter appears in print. This is an inevitable property of an ever-burgeoning field of investigation. Another inevitable consequence is the confusion and ambiguity often created by "too much, too soon". While the studies reported so far are very exciting, it will become more important and clearly more difficult to define the true significance of these findings and the subsequent precise mechanisms of nuclear matrix associations and functions. A good example of this is the current controversy concerning active gene sequences enriched on the nuclear matrix. Clearly, additional experiments are required to elucidate the meaning of these possible associations. Designing and performing more critical experiments in the future should, in turn, lead to an increasing knowledge and insight into matrix-related phenomena.

This review is, therefore, written with the primary goal of providing a critical perspective of current developments in the matrix field. Readers interested in more detailed accounts of various aspects of nuclear matrix organization, isolation, and biochemistry are referred to a number of previous reviews on this topic.[5,42-48]

II. THE *IN SITU* MATRIX

A. Chromatin Domains

Classical light microscopy studies of the late 19th and early 20th centuries revealed three main structural components in the eukaryotic nucleus: (1) the chromatin, which actually corresponded to only the dense or heterochromatinic regions of chromatin, (2) the nucleolus, and (3) nuclear sap or karyolymph whose transparent appearance led to the belief that this was a clear fluid or transparent gel. This light microscopic picture has dominated until recently our general conception of the cell nucleus.

Clearly, this simple view of the nucleus as a homogeneous bag containing chromatin and the specialized nucleolar structure has served as a reasonable starting point for investigation. For one thing, it put the emphasis on the genetic material: the chromatin and the precise chromosome structures visible during mitosis and meiosis. Indeed, cytological observations in the early 1900s in conjunction with genetic studies provided the first firm evidence that the cell nucleus is the depository of the genetic information which is contained within the chromosomes. It was, therefore, natural to consider that the chromatin network observed in interphase nuclei represented a more amorphous, unraveled form of the chromosomes.

The model of the interphase nucleus as a bag of chromatin immersed or embedded within a homogeneous, unstructured nucleoplasm was radically changed with the application of electron microscopy to the cell nucleus. These studies revealed elaborate structure in the interchromatinic regions between condensed chromatin[49,50] and led Fawcett[51] to conclude, in 1966, that rather than nuclear sap or nucleoplasm, the terms *nuclear matrix* or *nuclear ground substance* were more appropriate designations for this complex nuclear structure.

Figure 1 is a thin-sectioned electron micrograph which illustrates the basic morphology of the *in situ* rat liver nucleus. Dense chromatin structures are found along the nuclear periphery in close association with the surrounding nuclear envelope, along the periphery of the nucleolus, and at other locations in the nuclear interior. Between the dense chromatin areas are the interchromatinic regions which contain at least two main structural components: the diffuse or *euchromatin* of the nucleus and the nonchromatin *nuclear matrix*. It is here that the distinction becomes vague. Although early electron microscopic studies reported the presence of presumptive nonchromatin granular and fibrillar structures in this region,[49,50] it was generally very difficult to distinguish between diffuse chromatin fibers sectioned in various directions and so-called nonchromatin granules and fibers.

An important breakthrough occurred with the development of the EDTA regressive staining procedure by Bernhard and Monneron.[52,53] With this technique it was possible to preferentially bleach the DNA-containing chromatin structures while maintaining significant contrast of the nonchromatin structure. A remarkably new view of the cell nucleus was revealed. As shown in Figure 2, an extensive nonchromatin fibrogranular material or matrix is observed within the interchromatinic regions under conditions wherein the dense chromatin regions show little or no staining. The maintenance of distinct chromatin regions (which appear as white or slightly gray in the micrograph) around the nuclear periphery and nucleolus suggests that at least the major spatial relationships between the dense chromatin and interchromatinic regions are preserved during this staining procedure. Moreover, restaining the sections with uranyl acetate resulted in the typically intense staining of the dense chromatin and the nucleus appeared similar to those prepared without the EDTA regressive staining.[52]

The nuclear pore complexes which serve as major passageways for macromolecules between the nucleus and cytoplasm[54] appear particularly dense with this staining procedure (see arrows in Figure 3). An important observation in this regard is that *in situ* matrix structures in the nuclear interior appear continuous with the nuclear pore complexes (Figures 2 and 3). This was also suggested from the earlier studies of Monneron and Bernhard[53] and Franke and Falk.[55] As shown in Figure 3, channels of matrix material radiate from the interior of the nucleus to the nuclear pore complexes.

Puvion and Bernhard[56] studied *in situ* nuclear structure by combining cryoultramicrotomy with EDTA regressive staining and other staining procedures. In this manner they were able to observed nuclear structures in both thin and thick (up to 2 to 3 μm) unembedded sections using conventional accelerating voltages on a standard transmission electron microscopy.

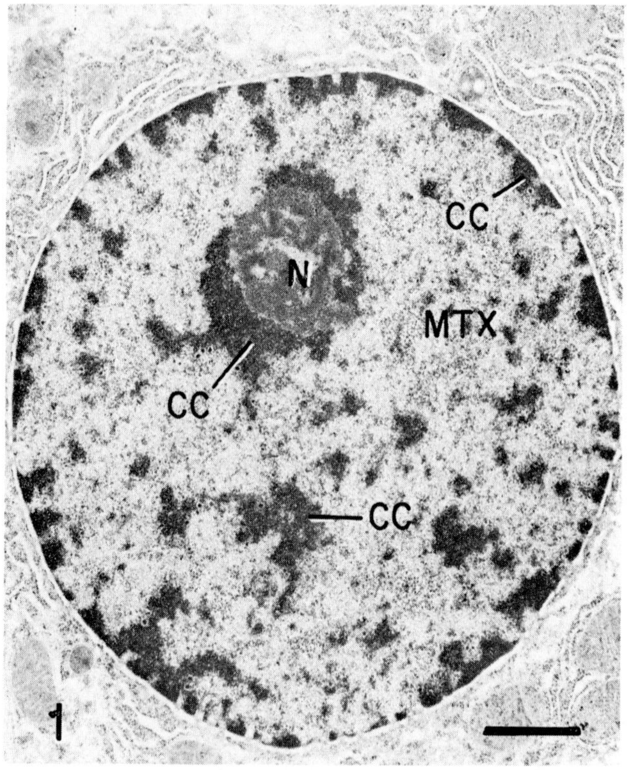

FIGURE 1. Thin-sectioned electron micrograph of a rat liver nucleus *in situ* by standard fixation with glutaraldehyde and staining with uranyl acetate and lead citrate: CC, condensed chromatin; MTX, matrix; N, nucleolus. Bar = 1 μm.

Figure 4 shows a contiguous matrix structure in a 2- to 3-μm frozen section of rat liver. The three-dimensional continuity of this network was confirmed by stereoscopic analysis.[56,57]

It is important to recall that the interchromatinic regions of the nucleus contain, aside from the nonchromatin matrix material, diffuse chromatin. Since diffuse chromatin is believed to contain the actively functioning euchromatin of the genome,[58] this immediately suggests a close *in situ* association between actively replicating and transcribing chromatin and nonchromatin matrix components. It is conceivable that components of the matrix actually become part of the diffuse chromatin structure or, vice versa, domains of the chromatin structure become part of the matrix. Moreover, these postulated associations may be dynamic and related to the functional state of the chromatin. Clearly, much more study of isolated chromatin and matrix systems is needed to understand the possible interrelationships between these two nuclear components. For example, recent reports indicate that the nuclear matrix contains the binding sites for supercoiled DNA loops[1-3] and that isolated chromatin and matrix preparations might share a discrete fraction of common polypeptides.[42,59,60]

B. Ribonucleoprotein Domains

Aside from domains of interaction with the chromatin, the *in situ* matrix also contains ribonucleoprotein (RNP) domains. Indeed, the earlier work on analysis of interchromatinic structures in nuclei *in situ* stressed the RNP nature of these components[49,50] and the EDTA regressive staining pattern of the *in situ* nucleus is generally interpreted as an enhancement

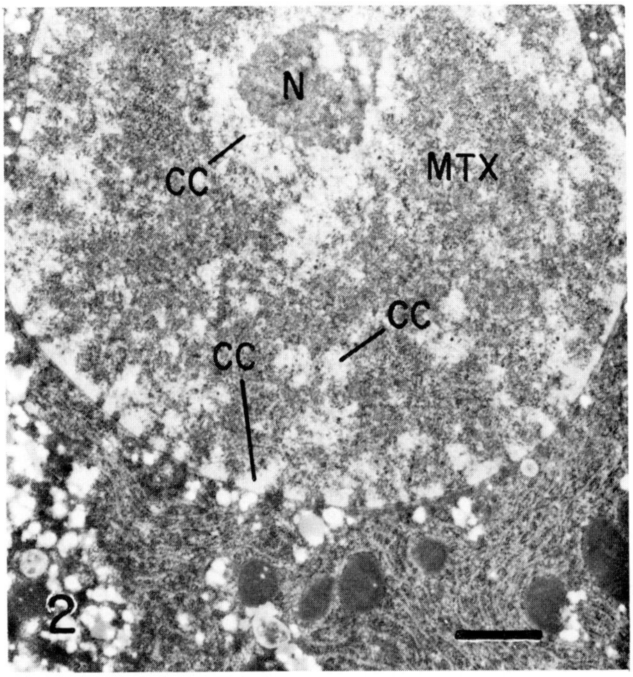

FIGURE 2. Thin-sectioned electron micrograph of a rat liver nucleus *in situ* by the EDTA regressive staining procedure of Bernhard.[52] Under these conditions the DNA-containing chromatin structure is only very lightly stained revealing an elaborate *in situ* matrix structure throughout the nuclear interior. CC, condensed chromatin; MTX, matrix; N, nucleolus. Bar = 1 μm.

of these presumptive RNP domains.[52,53,61] High resolution electron microscopic autoradiography in correlation with EDTA regressive staining supports this interpretation. Newly transcribed extranucleolar RNA is localized nearly exclusively along the borders and within the interior of the interchromatinic regions.[61,62] Pulse-chase experiments have led to the idea that once released from the chromatin template, the newly transcribed RNA migrates through the *in situ* matrix structure as RNP particles or fibrils.[61,63-65] This suggests that the *in situ* matrix contains a network of RNP particles extending from the sites of chromatin transcription to final release through the nuclear pore complexes. The strands or channels of *in situ* matrix structure which lead from the interior to the nuclear pore complex structures (see Figures 2 and 3) may, therefore, represent dynamic assembly lines for the transport of RNP particles. Since the RNA generally undergoes numerous splicing and other processing events following transcription, it is conceivable that these posttranscriptional events occur in close association with *in situ* matrix components (e.g., along the postulated "assembly lines"). Thus, the *in situ* matrix may provide a structural template for both the organization and functioning of processing events on discrete RNA molecules.

C. Dynamic Properties

The dynamic properties of nuclear structure have been widely documented.[42] Suffice it to say that the cell nucleus responds to physiological changes be they cell cycle related, or induced externally or internally via hormones, drugs, or other environmental cues — by a variety of specific morphological alterations. One of the most common alterations involves change in size of the nucleus. This has been most impressively observed in nuclear trans-

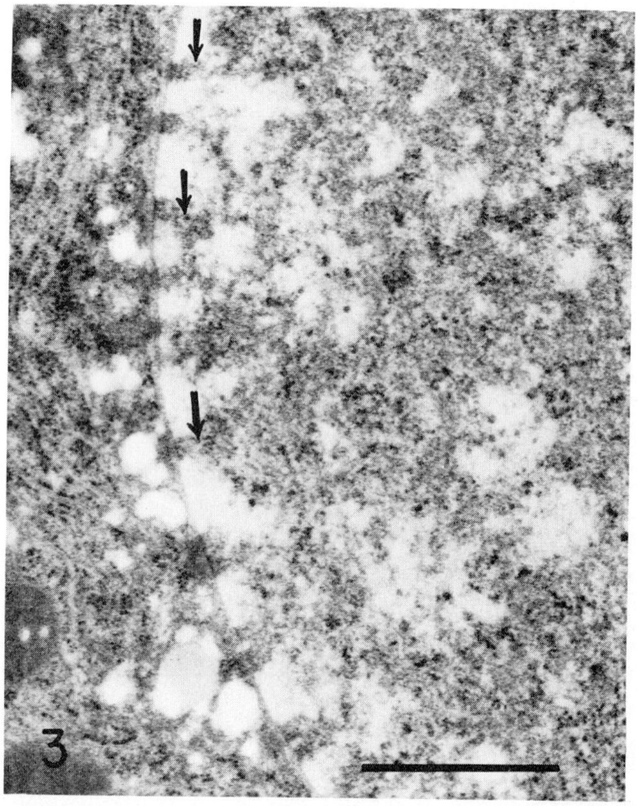

FIGURE 3. Thin-sectioned electron micrograph of an *in situ* rat liver nucleus in the area of the nuclear pore complex by the EDTA regressive staining procedure of Bernhard.[52] In this micrograph the interior *in situ* matrix appears contiguous with the nuclear pore complexes of the surrounding nuclear envelope. Arrows indicate the "channels" of matrix material radiating towards the nuclear pore complexes. Bar = 1 μm.

plantation and cell fusion experiments where enormous enlargements have been measured in cell nuclei activated to synthesize DNA.[66-68] It appears, however, that the cell nucleus may be constantly changing size during the cell cycle. The HeLa cell nucleus, for example, doubles in volume between G_1 and the end of S phase.[69]

Associated with increases in nuclear size is a decrease in the relative volume occupied by the condensed chromatin and a corresponding increase in the *in situ* matrix. This is significant since, as discussed in the previous two subsections, the matrix appears to contain actively replicating and transcribing euchromatin as well as transporting RNP particles. Increase in the size of the overall matrix may, therefore, be important for activation of these fundamental nuclear processes. Moreover, these observations suggest that changes in overall nucleus size may be largely accounted for by corresponding changes in the *in situ* matrix volume. In some cases of nuclear activation there is also a massive decrease in the condensed chromatin regions. Although this decrease may not directly account for the increase in nuclear size, it leads to a corresponding massive increase in the amount of euchromatin in the matrix regions which in turn may "activate" the matrix to increase in volume.

In addition to size changes, treatment of cells with drugs which inhibit various biosynthetic processes often lead to striking perturbations in nuclear shape and internal morphology.[42] Many of these drugs are carcinogens (e.g., aflatoxin, dimethylnitrosamine, ethionine, lasiocarpine, and thioacetamide). Indeed, it has been known since the pioneering studies by

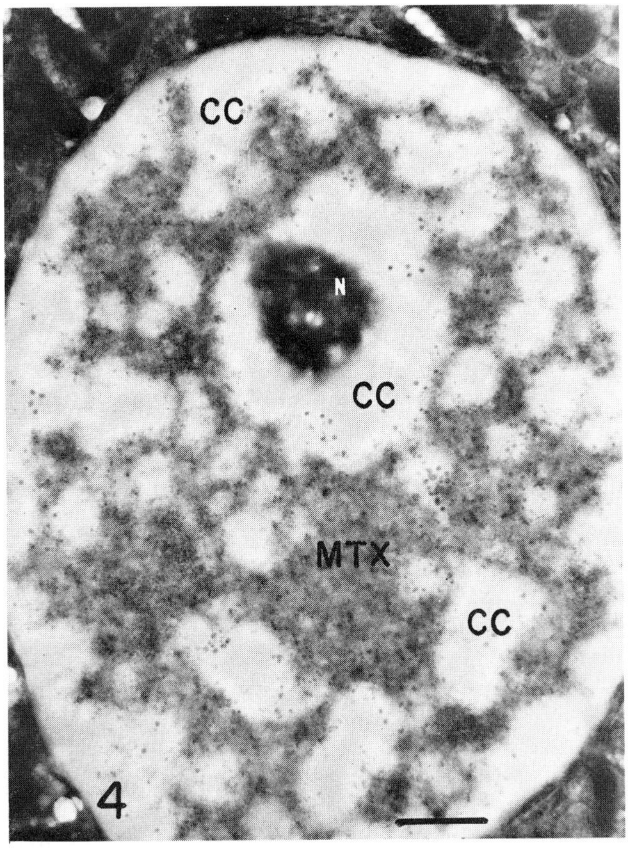

FIGURE 4. Thick-frozen section (2 to 3 μm) of an *in situ* rat liver nucleus using a modification of the Bernhard procedure to reveal a continuous nonchromatin matrix structure in the nuclear interior: CC, condensed chromatin; MTX, matrix; N, nucleolus. Bar = 1 μm. (From Puvion, E. and Bernhard, W., *J. Cell Biol.*, 67, 200, 1975. With permission.)

the 19th century cytopathologists that cancer cells are often characterized by the bizarre shape and internal morphology of their cell nuclei. This is dramatically illustrated in Figure 5 which shows the irregular shape and morphology of nuclei in Zajdela ascites hepatoma cells. Although alterations in nucleolar and chromatin structure have been most widely studied, there have also been considerable reports of bizarre changes in the *in situ* matrix regions of rapidly proliferating neoplastic cells including the appearance of intranuclear, membrane-like structures.[42]

D. Molecular Organization and Functions

Our current knowledge of *in situ* nuclear structure, therefore, leads us to a view of the eukaryotic nucleus as a complex three-dimensional structural mosaic of condensed chromatin and interchromatinic areas. This model view is schematically illustrated in Figure 6. The nonchromatin nuclear matrix is depicted as a dynamic but continuous ground substance in the interchromatinic region which interacts with both diffuse euchromatin and RNP particles. At least two distinct structural domains are postulated within the *in situ* matrix: the chromatin domains and the RNP domains. This dual aspect of nuclear matrix organization has important implications. It immediately suggests heterogeneity of the matrix at both the organizational and functional levels. The chromatin domains of the matrix, for example, may be quite

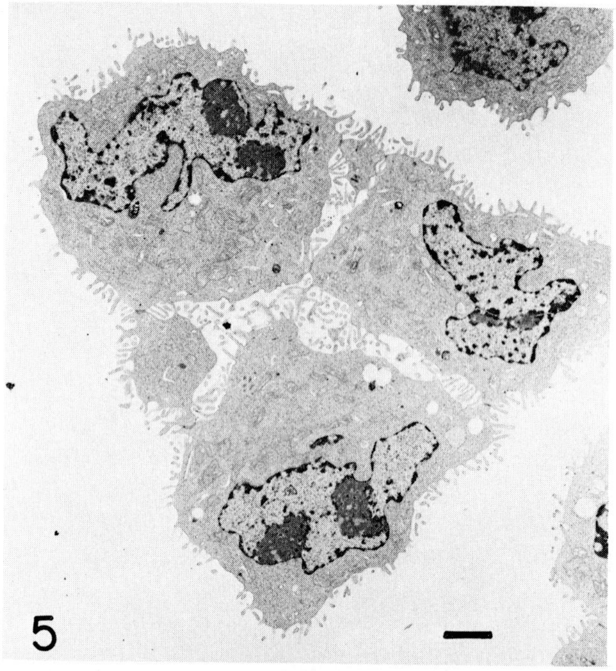

FIGURE 5. Thin-sectioned electron micrograph through Zajdela ascites hepatoma cells. Note the bizarre shape of the nuclei. Bar = 1 μm.

different from the RNP domains with respect to both individual matrix components and their organization. In studying nuclear matrix organization it is, therefore, important to define the specific type(s) of association under consideration.

From a functional point of view the matrix can be divided into domains of replication and/or transcription of chromatin vs. domains of posttranscriptional processing. Since the *in situ* matrix appears to be a continuous structure, it is tempting to consider that the matrix has a fundamental role in the structural integration of these processes. At the level of posttranscriptional processes the matrix may, therefore, provide a continuous assembly line or "laboratory bench" for the organization of the processing events and the movement of the processed RNP structures through the nucleus and final transit into the cytoplasm at the nuclear pore complexes (see Figure 6). At the level of replication and transcription the matrix may provide a structural template for the expression of functional loops of DNA and perhaps the organization of active replicational and/or transcriptional complexes.

The continuous nature of the matrix system also suggests that these two distinct structural and functional domains (e.g., chromatin and RNP particles) may be interlinked. In this manner the matrix may serve an important integrative and regulatory role for cascade and assembly line events. For example, events occurring at the level of the nuclear pore complex, e.g., interaction with factors on the cytoplasmic side of the pore complex may result in changes in RNA processing, which in turn lead to other changes which may ultimately affect transcriptional events. While these considerations are highly speculative, growing evidence of functional associations with isolated matrix systems (see Introduction) indicate their relevance as working hypotheses for future studies.

A very controversial and often confusing issue with respect to the matrix is whether or not there exists a matrix structure free of chromatin and RNP domains. In other words, does an *independent, nonchromatin, non-RNP, in situ matrix exist?* In a way this is a loaded question because it is conceivable that the matrix exists as a collage of chromatin and RNP

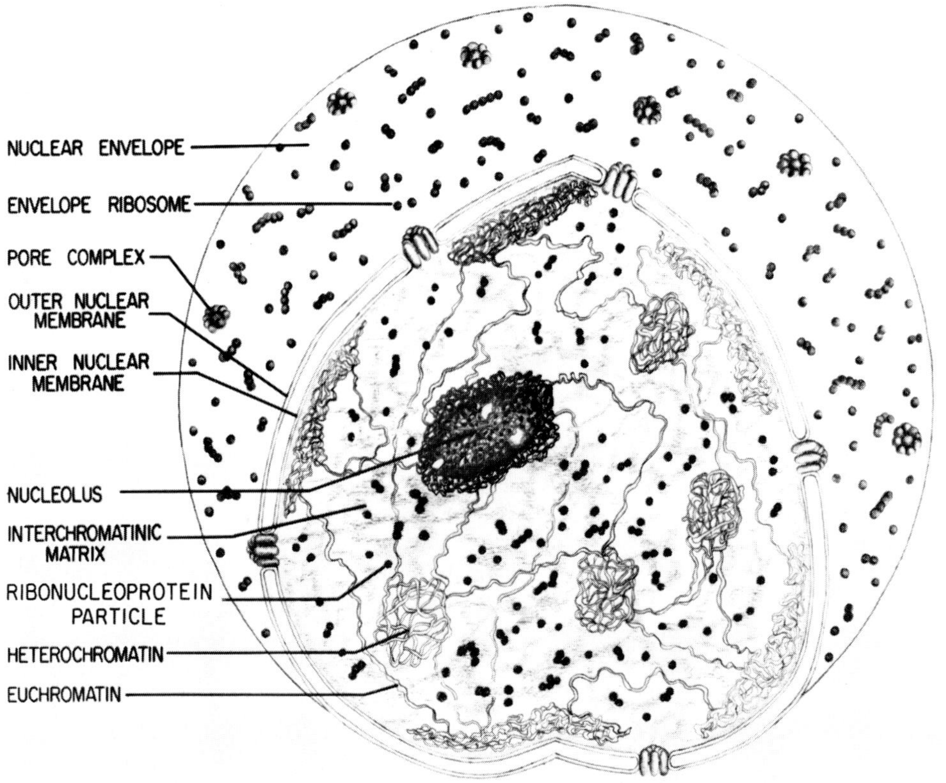

FIGURE 6. Schematic model of a typical eukaryotic cell nucleus. The nucleus is surrounded by a double-membraned nuclear envelope containing nuclear pore complexes. Ribosome-like structures are found on the surface of the outer membrane as individual particles and "polysome-like" arrays. The chromatin in the nuclear interior is interpreted as a continuous system of condensed and diffused regions. The structure of the interchromatinic matrix (see text) is simplified to contain only RNP particles and a more diffuse matrix network. This may represent an oversimplification as the RNP structures may be more diffusely spread throughout the matrix in both fibrous and granular configurations. The overall matrix is visualized as a continuous structure, associated with the nuclear pore complexes at the periphery, as well as with the chromatin and the nucleolus in the interior. The peripherally located matrix may correspond to the lamina-dense layer often observed in close association with the inner nuclear membrane. This model is obviously a static interpretation of nuclear structure. The *in situ* nuclear matrix may actually be a dynamic structure in which specific connections between various regions of the matrix and associations with other nuclear structures (e.g., nuclear pore complexes and chromatin) may be subject to change. The model was drawn by L. A. Buchholtz. (From Berezney, R., in *The Cell Nucleus*, Vol. 7, Busch, H., Ed., Academic Press, New York, 1979, 413. With permission.)

domains with no "free matrix" in between. While it is possible to isolate nuclear protein matrix (see Section III) free of DNA and RNA this has no direct bearing on this question. The protein matrix may only be prepared following extensive digestions with DNases and RNases to digest away tightly bound DNA and RNA. Thus, rather than supporting the existence of "free protein matrix" structure in the nucleus, these studies suggest that tightly bound chromatin and RNP are integral components of the matrix structure. In other words, the maintenance of matrix structure following RNA and DNA digestion merely emphasizes the structural role of the matrix proteins and not necessarily an *in situ* existence free of chromatin and RNP domains.

Clearly, more detailed molecular studies of matrix organization are required to determine if the matrix is simply a tight mosaic of chromatin and RNP domains or contains extensive regions of "free protein matrix". It should be mentioned, however, that the laboratories of Brasch and Sinclair[70] and Ghosh et al.[71,72] have independently demonstrated an extensive protein-rich, fibrillar nuclear matrix in cells treated with α-amanitin to inhibit RNA synthesis. Following drug treatment there is an extensive condensation or "margination" of the chromatin so that little or no euchromatin is present in the interchromatinic regions. The α-amanitin also markedly reduces the RNA levels within the nucleus so that the RNP domains of the matrix are also depleted. The presence of an elaborate, presumably *in situ* protein matrix structure supports the view that extensive "free matrix regions" can exist and that the chromatin and RNP domains are added to this basic infrastructure. However, these studies cannot directly address the question as to what extent these "free matrix regions" exist in uninhibited normal cells.

E. Universality

Nuclear matrix structure appears to be a universal feature of the eukaryotic nucleus. It has been observed *in situ* in a wide variety of cells and organisms in both the animal and plant kingdoms from primitive unicellular eukaryotes to specialized mammalian cells including man.[42] A particularly interesting example are the very primitive dinoflagellates, such as *Gyrodinium cohni*. These unicellular organisms have eukaryotic organelles in the cytoplasm and a true nucleus with a surrounding nuclear envelope. The DNA in the nucleus, however, is not organized into a chromatin structure with significant amounts of histones.[73,74] Rather, the DNA is arranged in bundles of highly coiled DNA similar in general appearance to the prokaryotic pronucleus.[75-77] Most significantly, fibrogranular nuclear matrix structure is evident throughout the nuclear structure between the dense DNA bundles.[75] Since these organisms first appeared on the earth approximately 2 billion years ago, the nuclear matrix may have preceded the evolution of chromatin in the cell nucleus. This is significant as it suggests that chromatin may have evolved special properties to enable it to effectively interact with the preexisting matrix structure.

III. NUCLEAR MATRIX ISOLATION

A. Nuclear Isolation

Electron microscopic analysis of *in situ* cellular structure is invaluable for defining the basic organization of specific cellular components. When combined with other techniques, such as cytochemistry and autoradiography, additional information can be obtained on the localization of specific molecules and molecular processes. Electron microscopy is equally important for study of isolated cellular components. It is the electron microscope which allows the investigator to identify the isolated organelle by comparison with *in situ* structures. This analysis has two main purposes: it demonstrates the purity and degree of intactness and it reveals the *structural correspondence* between the isolated and the *in situ* structures. The word *correspondence* is critical because it is virtually inevitable that at least some modification of *in situ* structure will occur during isolation. The goal of a successful isolation, therefore, is to minimize structural alterations while preserving the biological activities which the particular organelle performs. The power of a successful isolation is self evident for it enables a detailed molecular dissection of the structure and its associated molecular processes.

This type of structure-function in vitro approach has revolutionized our understanding, over the past 30 years, of many basic cellular and molecular events. Moreover, the revolution is still in progress. In vitro structural systems which maintain functional activities represent a major strategy used to decipher the molecular details of specific processes ranging from protein synthesis on polyribosomes to oxidative phosphorylation within mitochondria. The cell nucleus is no exception to this type of in vitro analysis which is particularly valuable due to the complexity of nuclear structure. Fortunately, procedures have been developed for isolating nuclei with a high degree of integrity and purity. Moreover, the major structural features of the nucleus are preserved under the proper isolation conditions. This is demonstrated by thin section electron microscopy for the isolated rat liver nucleus (Figure 7).

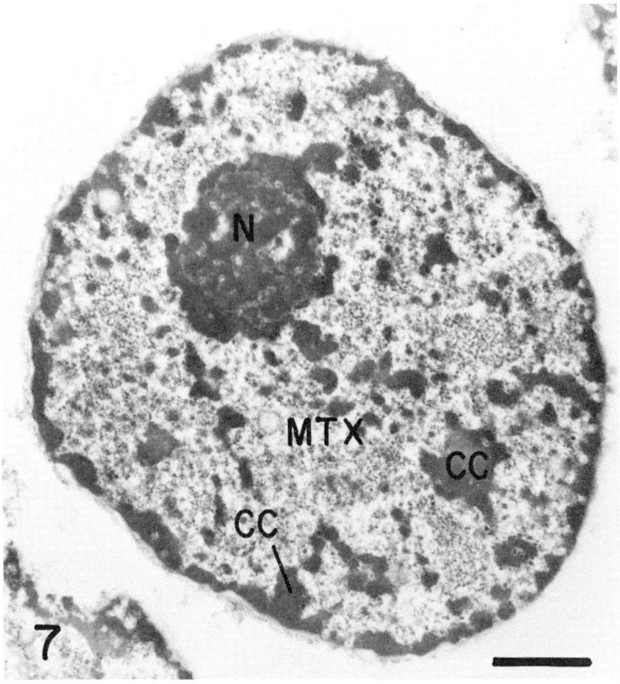

FIGURE 7. Thin-sectioned electron micrograph of an isolated rat liver nucleus. Many of the major structural features of the *in situ* nucleus are preserved during nuclear isolation: CC, condensed chromatin; MTX, matrix; N, nucleolus. Bar = 1 μm.

Figure 8 demonstrates nuclear matrix in isolated nuclei, as revealed by the EDTA regressive staining procedure, similar to the *in situ* structure. The major visible difference is that the matrix in isolated nuclei appears somewhat more aggregated than in the *in situ* state (compare Figures 2 and 8).

B. Residual Nuclear Structures

A number of early studies have indicated that nuclei maintain their overall structure after extraction of chromatin and other components under high-salt conditions, e.g., $\geq 1\ M$ NaCl (see References 5 and 42 for reviews). Particularly noteworthy were the pioneering studies of Mirsky and Ris[78] in their analysis of "residual chromosomes" in interphase nuclei and the reported persistence of residual nuclear structures following extraction of both histones and DNA from isolated nuclei.[79] The first electron microscopy of these residual nuclear structures was reported by Georgiev and Chentsov in 1960[80] who termed these structures "nucleonems". Significantly, Georgiev and Chentsov suggested the association of nucleonema structure with both chromatin and RNP domains.[80] The studies of Georgiev and Chentsov[80] were followed by a number of other electron and light microscopic investigations.[81-84] The laboratory of Busch et al.[83,84] stressed the RNP nature of these structures.[83,84]

C. DNA-Depleted Nuclear Matrix

Following these early observations, Berezney and Coffey[85,86] reported a more comprehensive structural and biochemical analysis of nuclei following consecutive treatments designed to remove chromatin (low- and high-salt extractions) and nuclear envelope components (Triton® X-100). The nuclear structures were monitored during and after treatment by phase contrast and Nomarski interference microscopy.[86] Figure 9 is a summary of a standard nuclear

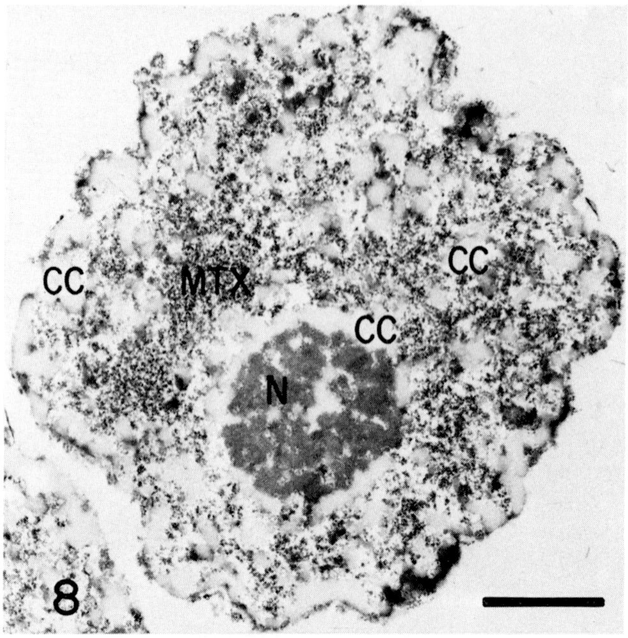

FIGURE 8. Thin sectioned electron micrograph of an isolated rat liver nucleus using EDTA regressive staining procedure of Bernhard.[52] A matrix structure is observed similar to *in situ* matrix although somewhat more condensed: CC, condensed chromatin; MTX, matrix; N, nucleolus. Bar = 1 μm.

matrix isolation scheme from rat liver which is currently in use in our laboratory. The electron micrographs in this figure illustrate the maintenance of overall nuclear structure during the fractionation. The first step involves cleaving the nuclear DNA by the endogenous endonucleases in rat liver[87,88] or by exogenous nucleases such as DNase I or micrococcal nuclease. This step is critical for successful isolation of nuclear matrix structures by this technique. (Other procedures reported in Section III.D have been developed for isolating matrix which still contain high-molecular-weight DNA.) The digested nuclei are then extracted with low ionic strength media which releases a bulk chromatin fraction containing 70 to 80% of the total nuclear DNA. This chromatin fraction is depleted in nonhistone proteins and consists predominantly of oligonucleosomes.[89] The remaining 20 to 30% of the total nuclear DNA is associated with the nuclear structures previously termed nuclear matrix I or *low-salt matrix*.[42] Electron microscopy reveals that the characteristic condensed chromatin regions of the nucleus are no longer visible in the low-salt matrix (Figure 9).[42,86] EDTA regressive staining of this fraction shows a structure similar to the isolated nucleus indicating preservation of the nonchromatin matrix structure.[5] Since the interior fibrogranular structure in the standard uranyl acetate-lead citrate staining technique (Figure 9) is similar to that after EDTA regressive staining, the remaining chromatin appears to be closely associated with the nonchromatin matrix structure. Indeed, these morphological relationships suggest that the low-salt matrix is enriched in the chromatin domains which may be closely associated with the *in situ* matrix.

The low-salt matrix is then extracted with high ionic strength media (2 *M* NaCl). This removes most of the remaining chromatin, termed the high-salt soluble chromatin, leaving nuclear matrix II structures[42] containing 1 to 3% of the total DNA. This DNA is very resistant to salt as determined by additional extractions with 2 *M* NaCl. The nuclear matrix II structures

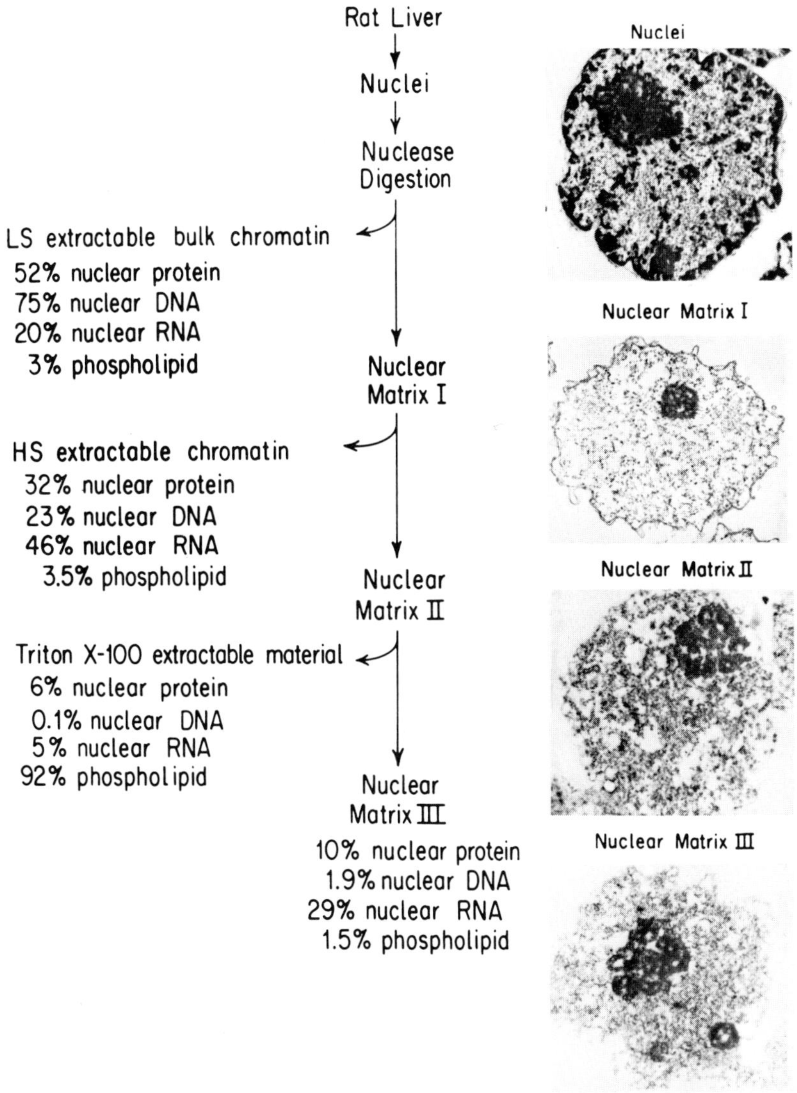

FIGURE 9. Outline for the preparation of nuclear matrix rat liver. For more details see the text and a previous review.[42] This scheme was composed by H. C. Smith.

are also called *high-salt matrix*. However, in many studies nuclear envelope components are also removed by treatment of nuclear matrix II with nonionic detergents such as Triton® X-100. This treatment removes little DNA ($<0.1\%$) but significant amounts of protein (5 to 7%,) RNA (5 to 10%), and phospholipid ($\geqslant 95\%$). Electron microscopy suggests that the major structural effect is on the nuclear envelope. The double-membraned envelope is no longer visible. Instead, a single lamina layer is seen devoid of phospholipid and containing residual nuclear pore complexes. This nuclear structure is termed nuclear matrix III or the *high-salt-resistant nuclear matrix*. It contains 10% of the total nuclear protein, 30% of the total nuclear RNA, 1 to 3% of the total nuclear DNA, and 2 to 5% of the total nuclear phospholipid. The RNA and DNA associated with the high-salt-resistant nuclear matrix is tightly bound but can be effectively removed by exhaustive treatment with RNase and DNase I. This fraction is termed the *nuclear protein matrix* or nuclear matrix IV and consists largely of protein with only trace amounts of RNA, phospholipid and carbohydrate.[42,86]

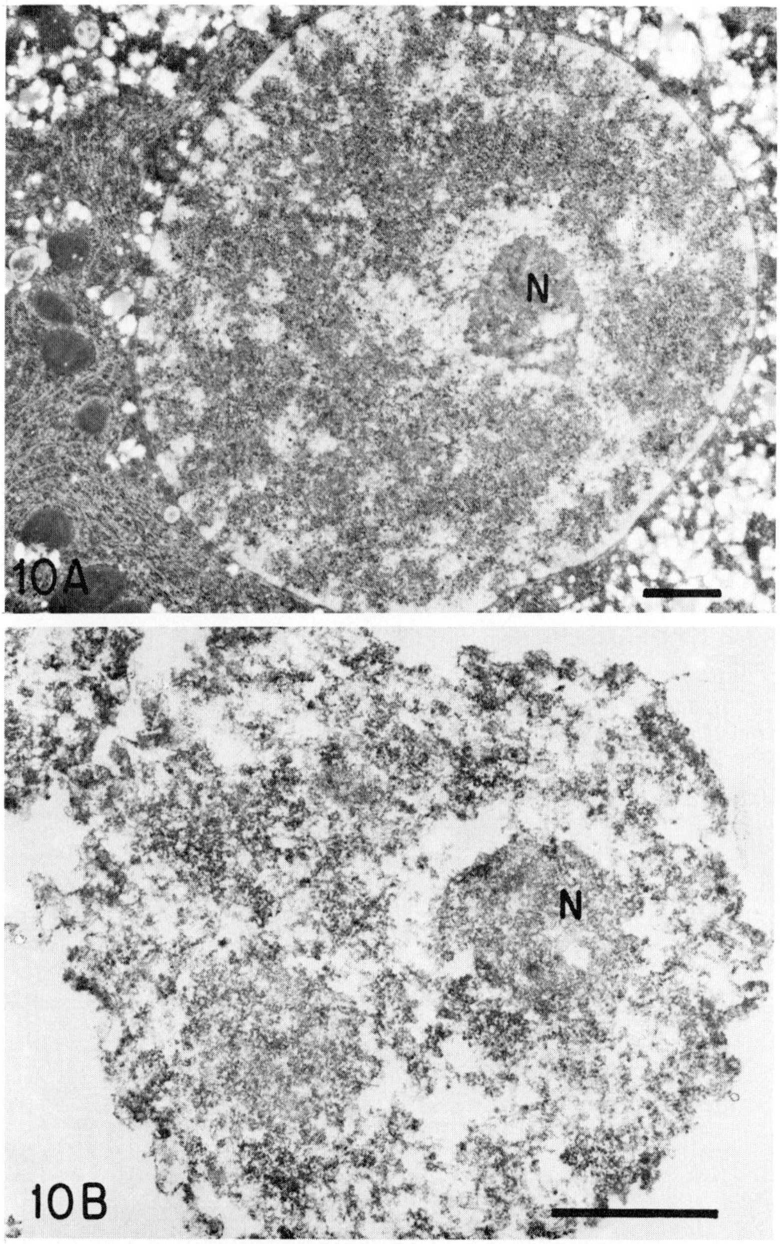

FIGURE 10. Comparison of the *in situ* and isolated rat liver nuclear matrix by thin-sectioned electron microscopy using the EDTA regressive staining procedure of Bernhard:[52] A, *in situ* nucleus; B, isolated nuclear matrix (nuclear matrix III[42]); N, nucleolar structure. Bar = 1 μm.

Figure 10 shows the thin-sectioned appearance of the *in situ* matrix in intact cells compared to the high-salt-resistant nuclear matrix (nuclear matrix III) employing the EDTA regressive staining procedure. It appears that many of the major structural features of the intact nuclear matrix are preserved under these conditions of matrix isolation. However, there are also structural alterations which occur during isolation. Most notable is the apparent absence of the strands of matrix material which, *in situ*, often radiate to the nuclear pore complexes

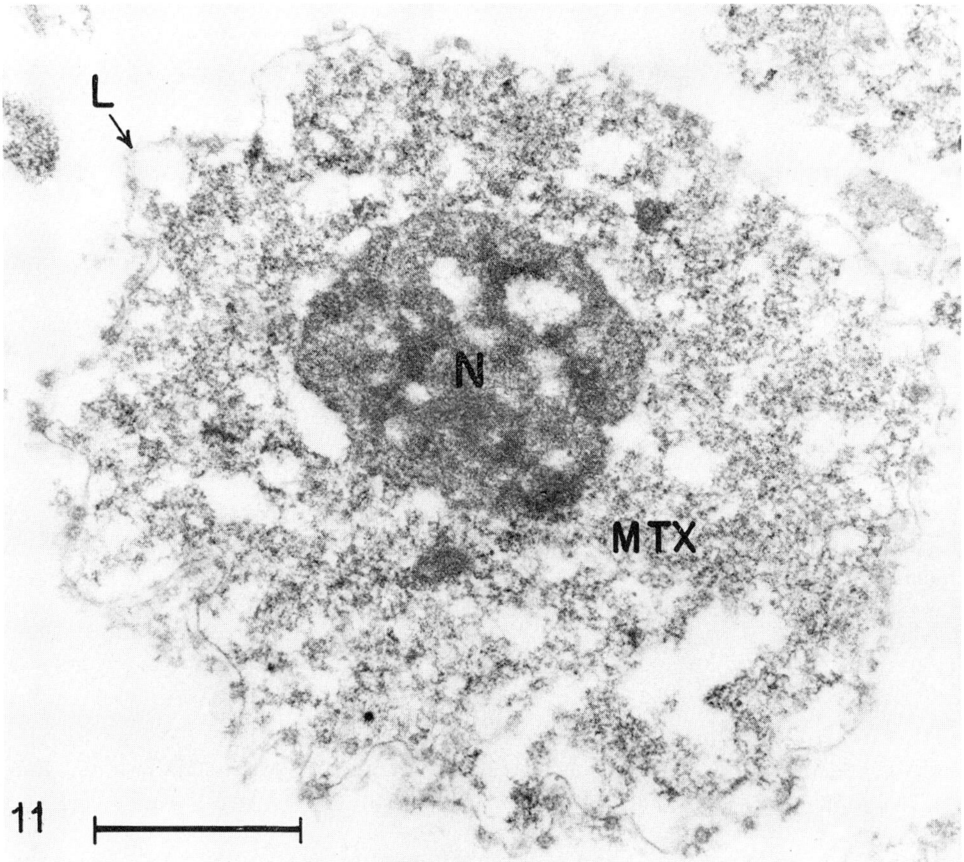

FIGURE 11. Thin-sectioned electron micrograph of an isolated nuclear matrix (nuclear matrix III) using conventional glutaraldehyde fixation and staining with uranyl acetate and lead citrate: MTX, internal matrix; N, nucleolar matrix; L, pore complex lamina. Bar = 1 μm. (From Berezney, R. and Buchholtz, L. A., *Biochemistry*, 20, 4995, 1981. With permission.)

(see Figures 2 and 3). Instead, large aggregates of matrix structures are generally observed in close apposition to the remnants of the surrounding nuclear envelope layer (Figures 10 and 11). Finally, much of the nuclear envelope is removed during the detergent treatment including over 95% of the phospholipid. Remarkably, however, a surrounding *pore complex lamina* remains which contains morphologically recognizable nuclear pore complexes (Figure 11).

The three main structural domains of the isolated nuclear matrix (in all further discussion the term *nuclear matrix* will refer to high-salt-resistant nuclear matrix, i.e., nuclear matrix III or similarly prepared matrix structures) are clearly visualized by thin-sectioning electron microscopy employing standard glutaraldehyde fixation and staining procedures (Figure 11). These domains are the surrounding *pore complex lamina, residual nucleoli* and the *fibrogranular matrix* in the interior (Figure 11). As summarized above, there are structural perturbations in all three domains during isolation from intact cells. However, there is also striking preservation of morphology including the general size and shape of the overall nuclear structures and of the nuclear pore complexes. The three-dimensional electron microscopic studies of Bouvier et al.[90] support this conclusion and reveal continuity among the structural domains of the isolated matrix. Moreover, previous studies revealed a close resemblance of the fibrogranular internal matrix structures of isolated matrices to the fibrogranular structures of the *in situ* matrix.[86] Interestingly, this structural correspondence is also apparent in the isolated nuclear protein matrix (nuclear matrix IV) following removal of tightly bound RNA and DNA from the nuclear matrix with RNase and DNase I.[86] This is significant since the removal of the tightly bound RNA and DNA results in a massive decrease of the overall matrix size.[86] Despite this contraction, there is preservation of fine structure. One exception is the residual nucleoli, which now appear as highly electron-dense, tightly packed structures[86] (see also Section V.F).

A number of investigators have demonstrated a similar isolated nuclear matrix structure in both liver[91-93] and a wide variety of eukaryotic cells and organisms throughout the phylogenetic scale.[7,14,16-18,21,23,24,28,32,34,38,42,43,59,94-101] It is important, however, to note that although the published isolation procedures generally employ high-salt solutions to extract the chromatin, other aspects of the matrix isolation procedure often vary enormously. It is apparent that different cell types require special considerations. For example, the matrix isolation summarized in this chapter for rat liver (see Figure 9) results in highly disrupted matrix structures when applied directly to HeLa cell nuclei.[102] However, this problem can be largely overcome by slight modifications in the isolation incuding the presence of 2 to 5 m*M* $MgCl_2$ during all steps of the fractionation.[102] Most isolated nuclei also require exogenous digestion with DNase I, micrococcal nuclease, or restriction enzymes to prevent trapping of matrix structures in the DNA gel which forms after exposure to high-salt solutions. Rat liver nuclei also appear to be unusually stable to centrifugation. Thus, more gentle centrifugation conditions are often required to isolate intact matrix structures from other cell types.

A generally applicable nuclear matrix isolation procedure will hopefully be developed as more is learned about the nuclear matrix. At present, however, ''matrix making'' is very much an art that requires constant monitoring at all stages of the preparation with both light and electron microscopic techniques. Without this careful monitoring there is no guarantee that isolated matrix preparations actually are composed of intact matrix structures. Moreover, the word intact is always a relative term when dealing with isolated cellular components. Considerable variations in matrix structures have been reported in individual preparations. This matter will be further developed in Section IV.

D. DNA-Rich Nuclear Matrix

The DNA-depleted nuclear matrix usually contains 1 to 3% of the total nuclear DNA. By inhibiting endogenous nuclease activity during isolation it is possible to prepare nuclear matrix structures from rat liver which contain a major portion of the total nuclear DNA.[2] Moreover, most of this associated DNA is of very high molecular weight (≥500kb). Figure 12 demonstrates the ''nuclear-like'' appearance of the DNA-rich nuclear matrix preparation examined under phase contrast microscopy. The matrices appear intact and unaggregated with a well-developed internal nuclear structure including nucleoli and other dense regions interspersed with more diffuse regions. Thin-sectioning electron microscopy confirmed the interior location of at least the bulk of the matrix-attached high-molecular-weight DNA (Figure 13). A prominent dense-staining network of DNA fibrils were resistant to extensive digestion with RNase but were completely removed by DNase I.[2] In addition, DNA-depleted matrix clearly lack this network of dense fibrils (Figure 11).

There is also a remarkable clustering of DNA fibrils along the periphery, the borders of the residual nucleoli, and in a number of interior regions (Figure 13). A similar pattern is characteristic of condensed chromatin *in situ* (see Figure 11). This raises an intriguing possibility: might the original deposition of DNA into tightly packed (heterochromatin) and diffuse (euchromatin) regions be at least partially preserved in the DNA-rich nuclear matrix? This is important for it suggests that the attachment of nuclear DNA to isolated nuclear matrix is not simply due to random, artifactual binding of DNA during the matrix isolation.

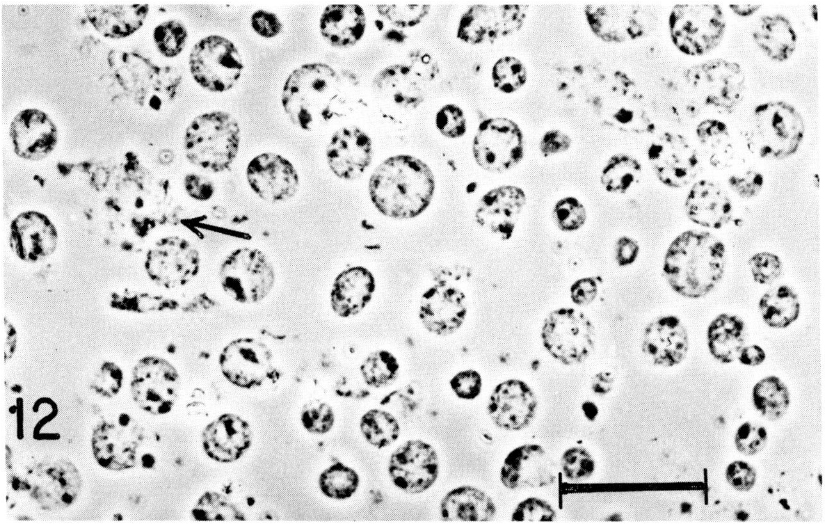

FIGURE 12. Phase contrast light microscopy of DNA-rich nuclear matrix isolated from rat liver. Arrow indicates a disrupted matrix structure. Bar = 40 μm. (From Berezney, R. and Buchholtz, L. A., *Biochemistry*, 20, 4995, 1981. With permission.)

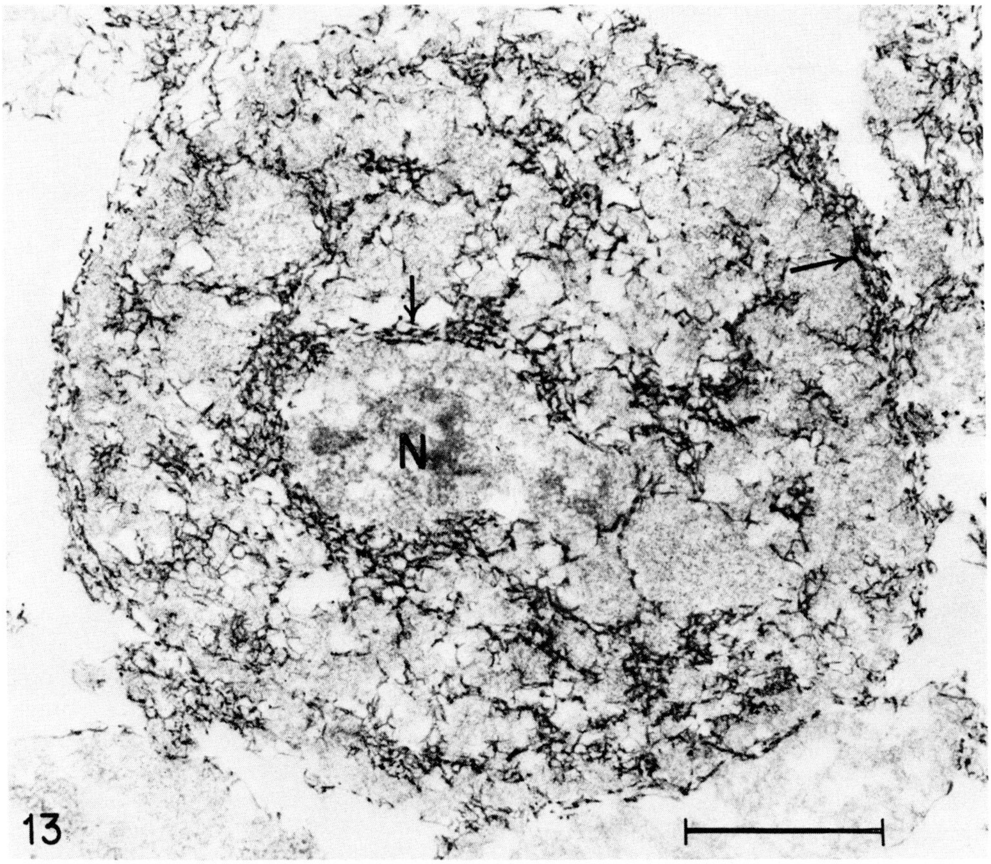

FIGURE 13. Thin-sectioned electron micrograph of a DNA-rich nuclear matrix isolated from rat liver. Note the perinucleolar and peripherally localized DNA fibrils (arrows). Bar = 1 μm. (From Berezney, R. and Buchholtz, L. A., *Biochemistry*, 20, 4995, 1981. With permission.)

Further evidence that this association is not an artifact comes from reconstitution experiments in which there was no detectable high-salt-resistant binding of high-molecular-weight DNA to DNA-free, DNA-depleted, or DNA-rich matrices.[2] Similar reconstruction experiments have been performed by Vogelstein and colleagues on matrices isolated from 3T3 fibroblasts, with virtually identical results.[7,103]

Cook and Brazell[104] previously reported that residual nuclear structures from HeLa cells termed "nucleoids", or more recently "cages", contain 70 to 75% of the total nuclear DNA. Most importantly the DNA attached to the "nucleoids" was highly supercoiled. This led Cook and Brazell to suggest that eukaryotic DNA was arranged in a series of topologically constrained domains which was due to the attachment of the DNA to the nucleoid structure in a series of repeating loops. Comparison of the results of Cook and Brazell with those of our laboratory suggests that the "nucleoids" or "cages" may be a form of nuclear matrix. Consistent with this interpretation, Vogelstein et al.[105] reported that the DNA attached to the nuclear matrix is highly supercoiled and our preliminary results with rat liver DNA-rich nuclear matrix also indicates a significant degree of supercoiling.[106] The residual nuclear structures of Keller and colleagues,[107-111] termed "nuclear ghosts" or "nucleoskeletons", may also be a form of isolated nuclear matrix. The absence of thin-section electron microscopic analysis of both the "nucleoid" and the "nucleoskeleton" structure make it difficult to evaluate their actual structural relationship to typical nuclear matrix preparations. Recently, however, Ševaljević and Petrović[60] reported that sea urchin "nuclear ghosts" prepared by the method of Riley et al.[107] were very similar to isolated nuclear matrices in composition, polypeptide profile, and thin-sectioning electron microscopy.

E. Chromosome Scaffold

Laemmli and colleagues[112,113] have reported that the DNA in isolated chromosomes is bound to a core structure termed the *chromosome scaffold.* By removing the histones from chromosomes with either high-salt or polyelectrolytes such as heparin or dextran sulfate, they observed a halo of highly convoluted DNA loops surrounding a central scaffold structure.[112] Cleavage of the DNA loops with nucleases results in DNA-depleted scaffold structures analogous to the DNA-depleted matrix isolated from interphase nuclei.[113,114] It is interesting to consider whether the chromosome scaffold represents a discrete subset of the interphase nuclear matrix. For example, the scaffold may be enriched in the DNA attachment sites or chromatin domain of the interphase matrix, but is presumably depleted in the RNP domain of the matrix. In this regard, preliminary results suggest that the chromosome scaffold proteins might be a distinct subset of the total nuclear matrix proteins[115] (see also Section V.E).

F. The Reality of the Isolated Nuclear Matrix

A number of earlier critics of the nuclear matrix have called these structures "artifacts of isolation." In particular, the high-salt extractions have been criticized for inducing aggregation and rearrangement of nuclear components. From the ultrastructural studies it is clear, however, that the isolated matrix is not simply a random aggregation of preexisting soluble components. Aside from an overall maintenance of nuclear morphology, there is also considerable preservation of nuclear pore complex, internal matrix, and nucleolar structure in the isolated matrix (Section III. C). Berezney and Coffey.[86] for example, reported a striking resemblance between the fine structure of the *in situ* matrix and that of the isolated internal matrix (Section III. C). Consistent with a general correspondence between the structure of the isolated nuclear matrix and the *in situ* matrix, Herlan et al.[116] found that actinomycin D-induced alterations in the *in situ* matrix of *Tetrahymena pyriformis* are maintained in the isolated matrix. Also Bekers et al.[117] described a striking reorganization of internal matrix structure in isolated nuclear matrices which closely paralleled the *in situ* movement of chromosomes during the "closed" mitotic stages of *Physarum polycephalum.*

However, there are also structural alterations. Many of these alterations relate to the overall pattern of the internal matrix (see Section III. C) and might be due to the removal of chromatin components which, in turn, results in shifts in the topographical distribution of internal matrix structure.

While we cannot rule out the possibility that the observed macroscopic shifts in structural topography are the results of major changes in molecular associations of the individual matrix components, the high-resolution electron microscopic studies indicate a similar fibrogranular structure in the isolated state as *in situ* (Section III. C). Future studies employing cross-linking agents should help resolve this issue. For example, it has been recently demonstrated that RNP particles are bound to isolated nuclear matrix by similar proteins as RNP particles *in situ* [21] (see Section VIII. A).

With regard to high salt aggregation it should be noted that high-salt solutions are often employed to prevent aggregation of proteins in solution rather than to induce it. This includes the nonhistone proteins of which the nuclear matrix proteins are a major class.[47,136] In addition, it is possible to isolate similar nuclear matrices at ionic strengths much lower than 2 *M* NaCl, e.g., 0.25 *M* $(NH_4)_2SO_4$[118] and with polyelectrolytes instead of high salt.[119] While major aggregation of nuclear matrix components may not occur in high salt it is conceivable that certain associations do shift. This may be an inevitable consequence of salt perturbation.

A reasonable position at present with regard to the reality of the isolated nuclear matrix is that there is a general *structural correspondence* of the isolated matrix to the *in situ* matrix. There are probably also alterations during isolation which, in turn, may be significant for both the molecular organization and functional properties of the isolated matrices. For example, it is conceivable that the high-salt-resistant nature of the nuclear matrix may be a property induced by the high-salt treatment itself. If this were the case, then the isolated matrix might represent a more "locked-in" or static state of what is *in situ* a more dynamic structure.[42] Further studies of nuclear matrix organization and function should ultimately clarify these matters.

IV. NUCLEAR MATRIX VS. NUCLEAR LAMINA

The isolated nuclear matrix typically consists of three main structural domains: a surrounding pore-complex lamina, residual nucleoli, and a fibrogranular internal matrix. There is no doubt, however, that specific components of the matrix can be separated. This must occur, for example, in the isolation of nucleoli or the nuclear envelope. The separation of nucleoli from the matrix structure is readily understood since the procedures usually involve sonication to disrupt overall nuclear structure followed by purification of the released nucleoli by centrifugation techniques.[120] Thus, it is anticipated that the matrix would be highly fragmented by this technique and nucleoli readily separated from chromatin, nuclear envelope, and internal matrix components.

The situation is more complex with regard to the nuclear envelope. Although some procedures do involve initial sonication and separation of nuclear envelope on sucrose gradients,[121-126] others appear remarkably similar to the initial steps in matrix isolation which involve treatment of isolated nuclei with nuclease and salt solutions[127-134] (for review see Wunderlich et al.).[54] Indeed, in the development of the first nuclear matrix isolation procedure,[85] this author slightly modified the nuclear envelope isolation of Kay et al.[134] In the following year Aaronson and Blobel[135] used the same procedure of Kay et al.[134] to isolate the pore-complex lamina from rat liver presumably free of internal matrix components. A controversy thus developed in which the significance of both the nuclear matrix and nuclear lamina preparations were questioned. For example, it was considered that the internal matrix of the isolated nuclear matrix preparations may have resulted from "artifactual aggregations" created during the isolation procedure. In this sense the nuclear lamina preparations could be considered "cleaner" preparations. In opposition, it was suggested that the nuclear lamina preparations were missing internal matrix due to specific steps in the isolation which induced this separation. Consistent with this possibility several laboratories reported that more intact matrix preparations resulted when prepared in the presence of protease inhibitors such as phenylmethylsulfonyl fluoride (PMSF) or chloride (PMSC).[16,100,101,136,137] Preliminary studies from our laboratory with isolated nuclear matrix further indicate that the internal portion of the nuclear matrix is more sensitive to removal following mild protease digestion compared to the surrounding nuclear lamina.[138]

An important development in this controversy resulted from the studies of Kaufmann et al.[139] These investigators found that isolation of "empty matrix" from rat liver is dependent on the digestion of nuclei with significant levels of RNase A. When the RNase digestion was performed before high-salt extraction relatively "empty matrices" or nuclear lamina ghosts were observed. In contrast, typical matrix structures with significant amounts of internal matrix structure were seen when the RNase digestion occurred after high-salt extraction. These results suggest a role for RNA (either directly or indirectly) in maintaining internal matrix structure. Thus, nuclear lamina free of internal matrix and nucleoli structures can be effectively isolated when the RNA is significantly degraded. Preliminary results from our laboratory indicate that at least 95% of the total nuclear RNA must be released by combined RNase and high-salt extraction to effectively release the internal matrix.[140]

While the RNase effect of Kaufmann et al.[139] has been confirmed in our laboratory for rat liver[140] it may not be applicable to other cells. For example, van Venrooij et al.[137] reported no visible effect of pretreatment with RNase on the interior structure of nuclear matrices isolated from HeLa cells. Fisher et al.[101] reported similar results in matrices isolated from *Drosophila melanogaster*. Adolph,[119] however, did observe the RNase effect in HeLa cells. Perhaps subtle differences in isolation procedures will ultimately explain these variable findings.

Another parameter in stabilizing internal matrix structure in isolated nuclear matrix may be disulfide bonds. Kaufmann et al.[139] reported that the release of internal matrix material by RNase treatment is depressed when the nuclei are exposed to conditions which promote disulfide bond formation. Treatment of nuclei with sulfhydryl blocking agents such as iodoacetamide or *N*-ethylmaleimide before RNase treatment resulted in uniformly empty matrix structures. Importantly, sulfhydryl oxidizing agents were no longer capable of stabilizing matrix structure following prior treatment with sulfhydryl blocking agents.[139] While experiments in our laboratory with rat liver[140] confirm the finding of Kaufmann et al.,[139] other investigations have not observed a disruptive effect on internal matrix integrity when sulfhydryl blockers or reducing agents were added before cell breakage.[101,137,141] Therefore, the potential role of protein disulfides in internal matrix structure remains to be clarified.

In summary, the earlier controversy between the nuclear matrix and nuclear lamina appears to be largely resolved. Two of the major laboratories who were earlier in favor of "empty nuclear matrix" have now reported isolation of matrix with considerable internal structure.[101,118,141,142] What remains to be resolved, however, are the conditions which allow a clear separation between the lamina and internal matrix. Both RNA and protein disulfide bonds may be involved, but the situation appears to be more complex than initially suggested by Kaufmann et al.[139]

V. NUCLEAR MATRIX PROTEINS

A. Polypeptide Composition

Earlier studies indicated that the proteins comprising "residual nuclear structures" extracted with high-salt solutions (1*M*NaCl) consisted predominantly of acidic nonhistone proteins.[79,143,144] These high-salt-resistant nonhistone proteins can be further fractionated into

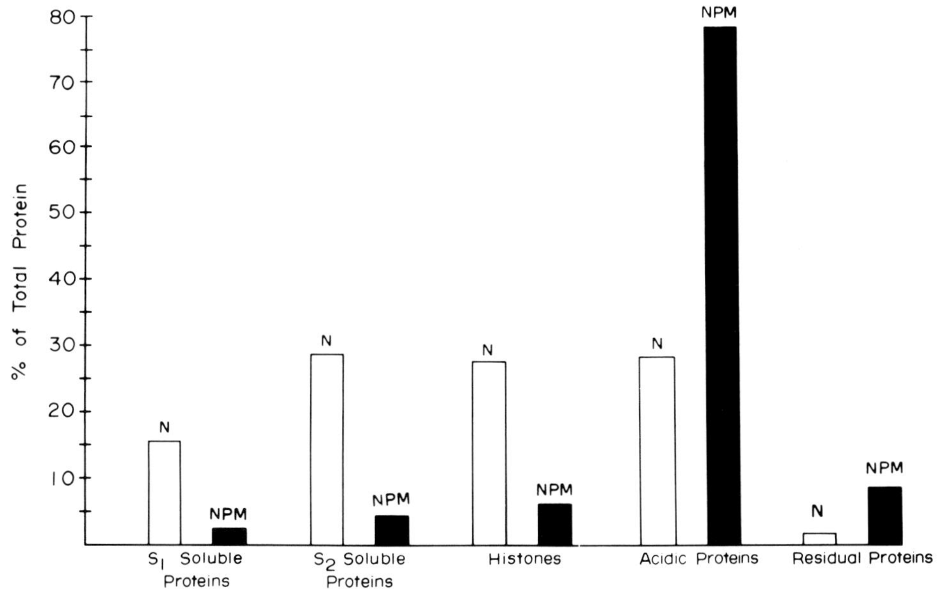

FIGURE 14. Chemical fractionation of nuclear matrix protein (black bars, NPM) compared to total nuclear protein (white bars, N). The general classes of nuclear proteins were fractionated by the extraction procedures of Steele and Busch.[144] (From Berezney, R. and Coffey, D. S., *J. Cell Biol.*, 73, 616, 1977. With permission.)

the NaOH-soluble acidic proteins and the residual proteins insoluble in NaOH. Berezney and Coffey[86] exposed isolated rat liver nuclei and nuclear matrix to a series of extractions classically used to separate total nuclear proteins into soluble nucleoplasmic proteins (S1 and S2), histone, acidic nuclear proteins, and residual proteins (insoluble in NaOH).[144] The results shown in Figure 14 demonstrated that the matrix proteins consisted predominantly of acidic nonhistone proteins (~80% of total protein) with a smaller amount of residual proteins (~10% of total protein) and were depleted in the soluble proteins and histones which are major components of total nuclear protein. The possible presence of small amounts of histones in the nuclear matrix is suggested in Figure 14 and will be discussed in Section V.D. Amino acid analysis of the total matrix proteins yielded an acidic-to-basic amino acid ratio of 1.46:1 and is consistent with their designation as acidic nonhistone proteins.[86] A similar acidic-to-basic ratio was measured in total nuclear matrix proteins from *Tetrahymena pyriformis* by Wunderlich and Herlan.[38] However, analysis of matrix proteins by two-dimensional electrophoresis demonstrates considerable heterogeneity in charge, ranging from acidic to basic (see Figure 17).

The polypeptide profile of the total matrix proteins on one-dimensional SDS-acrylamide gels is shown in Figure 15. Although the matrix proteins are heterogeneous in size, they are somewhat less complex than total nuclear protein. In particular, the matrix is relatively devoid of polypeptides with molecular weights below 45,000. The majority of the matrix polypeptides range between 50,000 to 150,000 daltons. Most prominent are a series of three bands between 60,000 to 70,000 daltons as well as several higher and lower molecular weight components.

The polypeptide profile of rat liver DNA-depleted matrix is virtually identical to the DNA-rich matrix containing high-molecular-weight DNA (Figure 15). Moreover, it was found that the recovery of total nuclear protein in DNA-depleted and DNA-rich nuclear matrices was nearly identical.[2] These results suggest that the DNA-rich matrices are not significantly contaminated with chromosomal protein which might be directly attached to the long DNA fibrils and, thus, not be directly attached to the matrix structure.

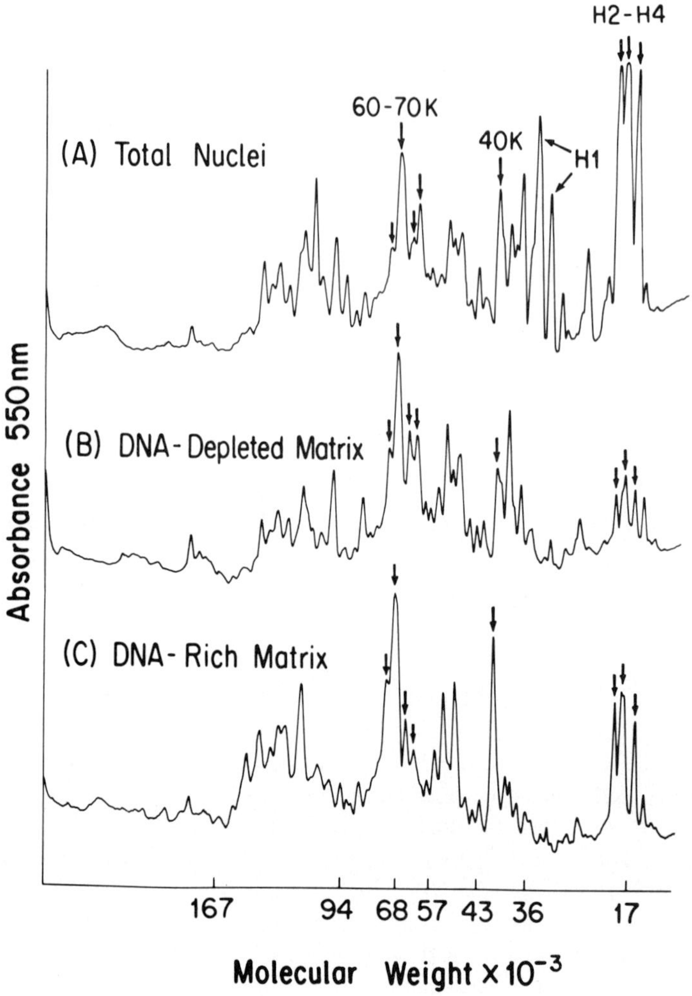

FIGURE 15. SDS-acrylamide gel electrophoresis of total nuclear protein and nuclear matrix protein prepared from both DNA-rich and DNA-depleted (nuclear matrix III) nuclear matrices.

B. Comparison of Matrix Polypeptide Profiles

Previous studies of nuclear matrix or "matrix-like" preparations suggested considerable differences in the polpeptide profiles depending on cell type and/or the preparative procedures. This matter was discussed in a previous review[42] and it was concluded that it was difficult to evaluate the meaning of reported variations in the molecular weights of matrix polypeptides in lieu of the different procedures employed for matrix isolation and SDS-acrylamide gel electrophoresis. For example, Berezney[136] and Berezney et al.[97] reported large quantitative variations of matrix polypeptide profiles which were dependent on the presence or absence of the protease inhibitors PMSF and/or sodium tetrathionate in the isolation solutions. Sodium tetrathionate is also a mild oxidative agent which can oxidize protein sulfhydryl groups to disulfides. Indeed, recent results indicate that the major effect of sodium tetrathionate on isolated nuclear matrix is to stabilize internal matrix structure via disulfide bond formation.[139] Thus, the degree of autooxidation of matrix protein sulfhydryl groups during isolation is another parameter which can vary among isolation procedures and lead to corresponding differences in polypeptide profiles.

Perhaps the most variable factor among matrix preparations is the degree of cytoplasmic contamination. This aspect is often overlooked by investigators. It is becoming clichéd for authors to state, for example, that the "nuclear preparations used in these studies were essentially free of cytoplasmic debris." In some cases, the basic criterion is phase contrast or Nomarski interference microscopy. Although interference microscopy is preferable to phase contrast, both are only indicative of massive cytoplasmic contamination. Contaminants which may be tightly bound to the nuclear structures might not be visible. However, if they are resistant to the high-salt and detergent treatments, minor contaminants at the nuclear level would become major components at the matrix level.

It is, therefore, imperative to routinely monitor preparations by electron microscopy and modify nuclear isolations to improve their purity. This requires special considerations. For example, actin-containing microfilaments are generally disrupted during normal fixation and embedding procedures for electron microscopy. Therefore, it might be incorrectly concluded that actin found in isolated matrix was a part of the matrix structure unless precautions were taken to preserve the actin cables (e.g., tannic acid fixation) which might be adventitiously bound to the exterior surfaces of the matrix structures.

Similarly, certainly intermediate filaments, at least, are resistant to extraction with high-salt solutions and could also represent a variable contaminant of nuclear matrix preparations. Consistent with the above possibilities, Capco et al.[118] have recently demonstrated that HeLa cell nuclei prepared by the Penman double detergent technique[145] are surrounded by a corona of filaments. The corresponding nuclear matrices contain major components which comigrate with actin and vimentin on two-dimensional gels.[118] The HeLa cell "nucleoids" or "cages" first characterized by Cook and Brazell[104] represent an extreme case of nuclear matrix structures highly contaminated by cytoplasmic components.[146] This is reasonable since the nucleoids are isolated directly from cell lysates without previous purification of nuclei.[104] While these structures are useful for studies of matrix-attached DNA loops and replication,[6] they are clearly not suited for direct studies of natively associated matrix proteins.

Shelton and associates[147] have stressed that the three major matrix polypeptides first reported by Berezney and Coffey on one-dimensional gels[85] show significant variability in apparent molecular weights depending on the acrylamide gel system. For example, the highest molecular weight component, designated lamin A, has been reported with molecular weights varying from 68,000 to 80,000. More importantly, in a given acrylamide gel system lamin A as well as lamin C from rat liver, HeLa cells, and chicken erythrocytes migrated identically.[147] These results stress the importance of comparing SDS-acrylamide gel systems before drawing conclusions about the polypeptide profiles from different matrix preparations.

Starting from highly purified nuclei, our laboratory has found that the polypeptide patterns of nuclear matrix proteins from rat liver and Chinese hamster DON cells are virtually identical on both one-dimensional and two-dimensional acrylamide gels.[148] Similar results were reported by Cobbs and Shelton[149] for HeLa and rat liver nuclear matrix proteins run on one-dimensional gels. Although this result appears at variance with the findings of Hodge et al.[32] who reported the major HeLa nuclear matrix polypeptides between 49,000 to 57,000 daltons, the studies of Riley and Keller,[108] van Eekelen and van Venrooij,[21] and our laboratory[218] all support the results of Cobbs and Shelton.[149]

Taken together, the above results suggest that the major matrix polypeptide components, at least, may be largely conserved in mammalian cells. Interestingly, nuclear matrix from the invertebrate *Drosophila melanogaster* also have a one-dimensional polypeptide profile similar to mammalian cells.[101] Lower in the phylogenetic scale, however, the situation is unclear. While nuclear matrix from the ciliate protozoan *Tetrahymena pyriformis* have a polypeptide profile similar to mammalian cells,[38,94] the profiles from sea urchin *(Paracentrotus lividus)*[59,60] and slime mold (Physarum polycephalum)[95] are quite different, with lower-molecular-weight polypeptides (15,000 to 40,000) as major components. Interestingly, Wunderlich and Herlan[38] reported the appearance of an 18,000-dalton polypeptide as a major component in *Tetrahymena* when the isolation conditions were changed to obtain nuclear matrix capable of reversible expansion and contraction.

C. Nuclear Matrix and Lamina Proteins

An important aspect of the relationship between nuclear matrix and lamina centers on the identification and localization of the proteins. While the structural aspects of this controversy have been largely resolved (see Section IV), there are still important questions to answer concerning the proteins. Immunocytochemical studies of Gerace et al.,[150] Krohne et al.,[151] and Ely et al.[152] clearly demonstrated that polyclonal antibodies to each of the three 60,000- to 70,000-dalton polypeptides from isolated rat liver lamina preparations only react with the periphery of the nucleus (e.g., the lamina) and not the interior matrix in both whole cells and isolated nuclei (see also Stick and Hausen[153]). Appropriate quantitation of the various fractions obtained during nuclear lamina isolation further revealed the absence of significant amounts of immunologically cross-reacting components in the released material.[150] This led to the reasonable conclusion that most, if not all, of these three polypeptides were specific to the peripherally localized nuclear lamina and were not significant components of the internal matrix. Unfortunately, this also led to the premature conclusion that the three 60,000- to 70,000-polypeptides found in intact nuclear matrix preparations were, therefore, entirely accountable by the lamina localized polypeptides. Quantitation of the total protein as well as of the polypeptide profiles of isolated lamina and matrix suggested that the content of the three 60,000- to 70,000-dalton polypeptides in the total nuclear matrix must be at least severalfold greater than in isolated nuclear lamina.[154] This indicated that there might be 60,000- to 70,000-dalton ''nonlamina polypeptides'' in the nuclear matrix which have approximately the same apparent molecular weights on one-dimensional SDS-acrylamide gels as the 60,000- to 70,000-dalton lamina-specific polypeptides, but which are antigenically distinct.[154]

The resolution of this problem illustrates the importance of two-dimensional electrophoresis and the cautious manner in which results based only on one-dimensional electrophoresis should be interpreted. While the 60,000- to 70,000-dalton polypeptides in isolated lamina preparations contain three main clusters on two-dimensional gels termed lamin A, B, and C[155] (Figure 16), the corresponding 60,000- to 70,000-dalton polypeptides in isolated nuclear matrix are much more complex on two dimensions. This was first demonstrated by Peters and Comings[156] and is illustrated in Figure 17. Aside from the lamin A, B, and C clusters, the total matrix contains several other more basic clusters of similar molecular weight which are missing from isolated lamina preparations (compare Figures 16 and 17). Based on Coomassie® blue staining, we estimate that the nonlamina, matrix polypeptides represent as much as 50 to 70% of the total 60,000- to 70,000-dalton polypeptides in the nuclear matrix fraction. Moreover, we have recently discovered that a monoclonal antibody (J-16), which reacts with all three lamina specific polypeptides in rat liver, does not react with the nuclear matrix specific 60,000- to 70,000-dalton polypeptides.[157] Another monoclonal antibody (J-9) reacts with several of the more basic, matrix-specific polypeptides as well as lamins A and C.[157] Immunofluorescence reveals a peripheral nuclear location for J-16 while J-9 stains both the nuclear interior and periphery in intact cells.[157] In this regard, Fisher et al.[101] have recently reported that polyclonal antibodies to a major polypeptide (74,000 daltons) of the *Drosophila melanogaster* nuclear matrix stains both the peripheral rim and the interior of the *in situ* nucleus, and Werner et al.[158] have demonstrated that antibodies to a 68,000-dalton, tightly associated DNA binding protein also stain the periphery and interior of the nucleus in intact cells (see Section V.D for further discussion of the DNA binding proteins).

Densitometric analysis of the two-dimensional polypeptide profile in Figure 16 suggests that the three lamin polypeptide clusters represent the majority of total protein ($\geq$70%) in the isolated nuclear lamina preparation. In contrast, the corresponding profile for total nuclear matrix (Figure 17) is much more heterogeneous and the lamins represent 10 to 20% of the total stained protein. These results are different from previous reports which suggested the relative amounts of 60,000- to 70,000-dalton polypeptides in the isolated nuclear matrix and

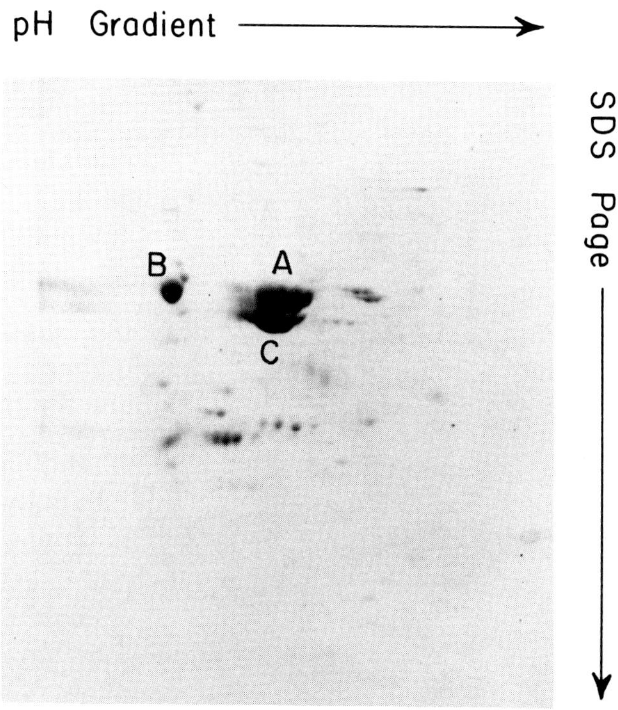

FIGURE 16. Two-dimensional electrophoresis of proteins from isolated rat liver nuclear lamina. Lamins A, B, and C are indicated. The pH gradient ranged fron approximately 4.5 (left side) to 8.5 (right side). The polypeptides were migrated from top to bottom in the second dimension in a 10% SDS-acrylamide gel.

lamina were similar (approximately 30 to 40% of total stained protein).[86,150,154] Several factors may account for this difference. Firstly, the nuclear matrices used for the analyses of Figure 17 were isolated by a rapid procedure in the presence of PMSF and sodium tetrathionate. This optimizes the recovery of internal matrix material and leads to a corresponding decrease in the relative amount of 60,000- to 70,000-dalton polypeptide in the total matrix protein.[136] Secondly, a major portion of the 60,000- to 70,000-dalton polypeptides in nuclear matrix are not lamins A, B, or C. Thirdly, the lamina preparations in this study were highly purified and isolated in the presence of PMSF. This suggests that previous data for nuclear lamina may have been based on preparations significantly contaminated with nonlaminar components. Also, earlier nuclear lamina preparations were performed in the absence of PMSF and included two consecutive incubations at room temperature in the presence of DNase I.[135,159] Since commercial preparations of DNase I are often contaminated with slight amounts of proteases, proteolytic degradation may have been a factor in these isolations.

In summary, current studies now suggest that the total 60,000- to 70,000-dalton polypeptides in the eukaryotic cell nucleus can be divided into at least two main structural domains: the lamins-specific domain (lamins A, B, and C) and the interior matrix-specific domain. These two domains may be distinguished immunologically. Drawing on the studies of Peters and Comings,[156] it is likely a third domain can be thought of as representing those 60,000- to 70,000-dalton polypeptides not associated with either high-salt-resistant lamina or matrix.

Finally, it appears that the two-dimensional polypeptide profile of isolated nuclear lamina is distinct from that of nuclear matrices isolated under conditions which retain extensive

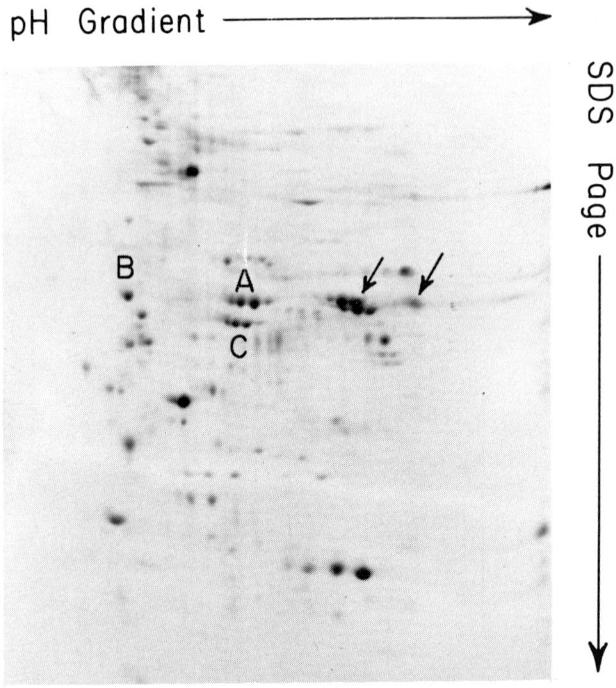

FIGURE 17. Two-dimensional electrophoresis of proteins from isolated rat liver nuclear matrix (nuclear matrix III). Lamins A, B, and C are indicated. More basic matrix polypeptides in the 60,000 to 70,000 molecular weight range are indicated with arrows. The pH gradient ranged from approximately 4.5 (left side) to 8.5 (right side). The polypeptides were migrated from top to bottom in the second dimension in a 10% SDS-acrylamide gel.

amounts of internal matrix structure. The lamina proteins appear to represent a relatively small subset of the total matrix polypeptide population and are highly enriched in lamin proteins A, B, and C.

D. Nuclear Matrix and Chromatin Proteins

The possible close *in situ* association of nuclear matrix components with chromatin was reviewed in Section II of this chapter. Significantly, this association appeared to be predominantly with the diffuse euchromatin of the nucleus and led to the consideration of specific structural domains of functioning chromatin closely associated with the matrix. Studies with isolated nuclear matrix support this possibility. In vivo incorporation experiments suggest that the DNA replicational sites are bound to the nuclear matrix (see Section VII.A for details). Appropriate DNA loop models (see Section VI.A and B) further suggest that the small DNA fragments tightly bound to the nuclear matrix represent the basal attachment sites for repeating DNA loops or chromatin domains in the nucleus and that these or analogous matrix-DNA attachment sites function as matrix bound replicational sites and possibly transcriptional sites.[1,6-8,19] DNA attachment sites, therefore, represent a potentially important structural and functional domain of interaction with chromatin.

It is interesting to consider the proportion of total matrix protein that is potentially involved in binding DNA at the putative matrix attachment sites. We estimate a minimum of 10^8 copies of matrix protein per isolated rat liver nucleus, based on an average protein content per matrix of 10 pg and an average molecular weight of 60,000. The number of DNA binding sites associated with the matrix is estimated at 125,000, based on an average of 80 kb per matrix-attached DNA loop.[8] Thus, only 1 out of 1000 or 0.1% of the total matrix proteins may be directly involved in DNA binding, assuming one matrix protein per attachment site. If a complex of approximately 10 matrix proteins is involved (~ 600,000 daltons) this still represents only a small proportion of the total matrix protein (~ 1%).

Identifying the putative proteins involved in DNA binding at matrix attachment sites might, therefore, be another case of finding the proverbial needle in a haystack. Despite this possible difficulty, Razin et al.[3] have made an initial attempt at fractionating and identifying these DNA binding proteins. Comings and associates[47,160] have also studied the DNA binding properties of nuclear matrix protein complexes dispersed by a freeze-thawing procedure in high-salt-urea solutions. Their results indicate a general affinity of matrix proteins for DNA which is not species specific but shows strong preference for A-T-rich sequences.[160] Unfortunately, the DNA binding was not resistant to high-salt solutions (⩾0.5*M*NaCl) and, therefore, the relationship of this in vitro binding to the high-salt-resistant binding at the putative matrix-DNA attachment sites is unclear. Similarly, Bowen et al.[161] have developed protein blotting techniques to study DNA binding to specific nuclear matrix polypeptides. While interesting patterns of DNA binding were observed, most of this binding was not resistant to 0.5 *M*NaCl).The small amount of binding resistant to salt extraction (1 to 2 % of total binding in 0.05 *M*NaCl), however, may be significant. Further applications of this protein blotting technique could be useful in characterizing proteins at the DNA attachment sites.

A potentially important observation is the presence of small but reproducible amounts of the nucleosomal core histones H2A, H2B, H3, and H4 in both DNA-depleted and DNA-rich matrices (Figure 15). It is, therefore, important to determine whether DNA is anchored to the matrix via a matrix-bound core nucleosomal structure. Approximately 2 to 5% of the total nuclear histone is recovered in the matrix fraction based on quantitative densitometric measurements.[2] This provides $\sim 10^6$ to 10^7 copies of histones per matrix structure. Since each nucleosome consists of an octamer of the four core histones, there are clearly enough matrix-bound core histones to account for one or more nucleosomes at each of the 10^5 putative matrix DNA attachment sites.

While it remains to be determined whether the matrix-bound histones are actually involved in the anchoring of DNA to the matrix structure, it is unlikely that the tight association of histones with the matrix is an "isolation artifact" since both the DNA-rich and DNA-depleted matrices contain similar levels of bound histones. Moreover, these histones are not removed by repeated extractions with 2*M*NaCl and/or low ionic strength solutions (10m*M* Tris, pH 7.4). In addition, we have not detected 2 *M* NaCl-resistant binding of exogenously added histone to isolated matrices.[2] Finally, Grebanier and Pogo[162] have demonstrated a specific association of histone H3 with a small proportion of matrix proteins after chemical cross-linking of intact nuclei. This strongly argues for a close interaction between histone H3 and certain matrix proteins. The inability to clearly identify histones in earlier nuclear matrix studies may have been due to several factors, such as high sensitivity of histones to proteolytic degradation and use of SDS-acrylamide gel systems which do not completely resolve histones.

Another approach to studying the interaction of chromatin with nuclear matrix components is to first identify polypeptide species common to both isolated fractions. This strategy is designed to remove much of the "hay" from the "needle" by concentrating further analysis on the common polypeptides. Based on one-dimensional SDS-acrylamide electrophoresis, we found that a substantial amount of the total chromatin nonhistone proteins (⩾20%) migrate in the same molecular weight range as the total matrix polypeptides.[42] Ševaljević and Poznanović[59] have reported a similar finding in isolated sea urchin matrix and residual chromatin proteins. The comprehensive two-dimensional electrophoretic analysis of Peters and Comings[156] reveal major differences but also some apparent similarities between chromatin and nuclear matrix nonhistone proteins. Further studies in our laboratory[163] indicate that the high-salt-resistant (2*M*NaCl) chromatin proteins are relatively devoid in lamins A, B, and C but contain significant amounts of the more positively charged, presumably internal matrix-specific 60,000- to 70,000-dalton polypeptides, as well as certain higher-molecular-weight matrix polypeptides (see Section V.C and Figures 16 and 17). This finding is consistent with the view that the matrix-DNA attachment sites are localized predominantly in the nuclear interior.

It will be interesting in future investigations to determine to what extent the proteins found tightly bound to DNA fragments after extraction of histones and high-salt-extractable nonhistone proteins from isolated chromatin[164] are also components of the nuclear matrix. This could be a direct approach to the elusive "needle". In this regard, Hnilica and associates [165-168] have employed immunological approaches to study the proteins present in the high-salt-resistant chromatin fractions. Interestingly, the immunological response of these nuclear antigens is dependent on the presence of small amounts of DNA associated with the residual chromatin proteins. Since the immunoactivity is also destroyed by proteases, it appears that the associated DNA maintains the protein antigens in a more "immunoreactive state." Cambell et al.[169] have further reported that these immunoreactive protein-DNA complexes are highly enriched in chromosome scaffolds isolated from HeLa cells (see also Section V.E). Whether isolated nuclear matrices are similarly enriched in these nuclear antigens is an important question which remains to be answered.

E. Nuclear Matrix, Mitotic Apparatus, and Chromosome Scaffold Proteins

The nuclear matrix is a ubiquitous structural element of the somatic cell nucleus. While the fate of the matrix during mitosis is unclear, it is conceivable that certain components of the matrix become associated with the mitotic spindle apparatus. Consistent with this possibility Bekers et al.[117] have demonstrated that interphase nuclear matrix components are rearranged during the "closed" mitotic events in *Physarum polycephalum.* A remarkable relationship was found between structural rearrangements of isolated nuclear matrix from mitotic cells and the positioning of the chromosomes in prophase, metaphase, and anaphase. This led to the suggestion that matrix structure is involved in the positioning and perhaps movement of the chromosomes.[117] The close association of mitotic chromosomes with the nuclear matrix of *Physarum polycephalum* is consistent with the postulated role of the nuclear matrix in the organization and function of chromatin domains including chromatin condensation-decondensation events.[42]

While the involvement of the nuclear matrix in the open mitotic events of higher eukaryotic cells remains to be determined, it appears that certain components of the interphase matrix, namely, the DNA-attached site domains, persist in the chromosomes as the chromosomal scaffold structure.[112-114] Recent results suggest that the chromosome scaffold proteins are a distinct subset of the total chromosomal proteins in Chinese hamster DON cells.[115] Interestingly, the scaffold proteins are depleted in lamins A and C but contain small amounts of the nonlamina 60,000- to 70,000-dalton polypeptides.[115,170] The chromosome scaffolds also contain a series of higher-molecular-weight polypeptides, and a major component at 57,000 daltons which cross-reacts with vimentin antibodies.[170] Since the scaffold may be concentrated in the DNA attachment site complexes and is probably devoid of the RNP domains of the matrix, it may represent a less ambiguous system for investigating these DNA binding sites. Hnilica and co-workers,[169] for example, have taken an immunological approach to demonstrate the concentration of certain DNA-binding proteins in chromosome scaffolds isolated from HeLa cells (see Section V.D).

F. Nucleolar Matrix Proteins

A major structural domain of the isolated nuclear matrix is the residual nucleolar structure. Since procedures are available for isolating morphologically intact nucleoli with a high degree of purity, it is possible to directly analyze the properties of this presumptive *nucleolar matrix* or *core*.

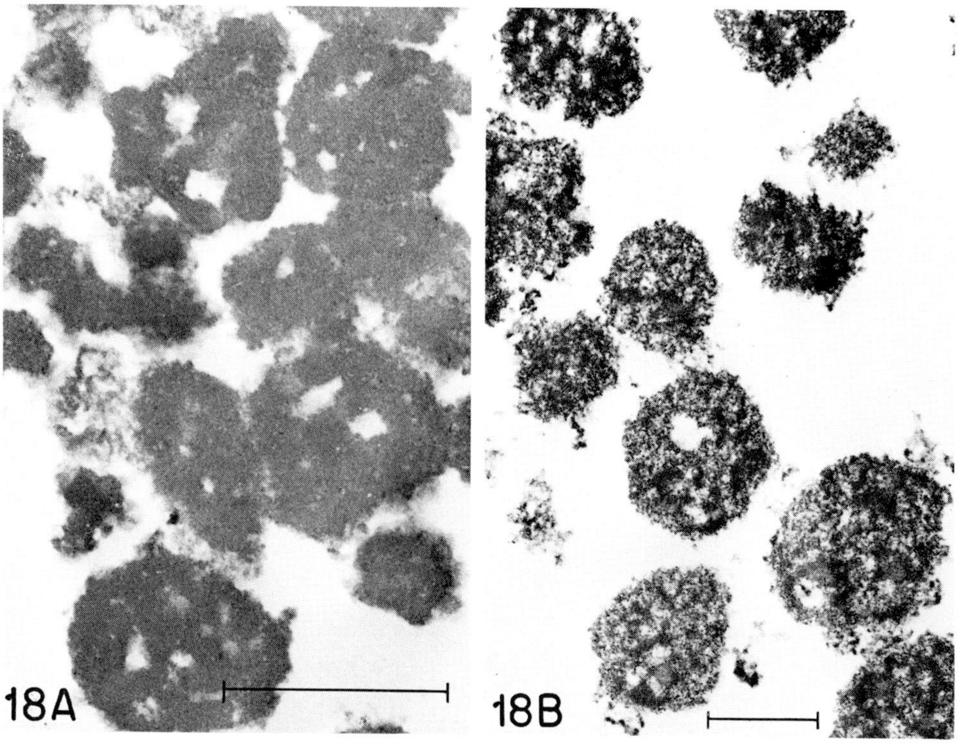

FIGURE 18. Survey thin-sectioned electron micrographs of (A) isolated nucleolar protein matrix. Average diameter of the protein cores is 0.8 μm based on 100 separate measurements. (B) Isolated nucleoli. Average diameter is 1.5 μm (100 measurements). Bar = 1.0 μm.

The isolation of a nucleolar protein matrix was first reported in 1976 by Berezney and Coffey.[5] Isolated nucleoli were subjected to the same series of treatments as isolated nuclei for isolation of the nuclear protein matrix (nuclear matrix IV, see Section III.C). The final nucleolar protein core consisted of 94.8% protein, 0.8% DNA, and 3.8% RNA. Approximately 30% of the total nucleolar protein, 5% of the total RNA, and 2% of the total DNA were recovered in the protein core.

A diminished size of the nucleolar protein matrix compared to the isolated nucleoli was indicated by both phase contrast and electron microscopic analysis (Figure 18). The highly condensed and osmiophilic protein cores are approximately one half the diameter of the corresponding untreated nucleoli. Isolated nucleoli demonstrated considerable preservation of *in situ* structure (Figure 19A) including a thread-like reticulum or nucleolonema containing dense fibrillar (F) and granular-fibrillar (G) regions. The latter regions contain electron-dense RNP granules (150 to 250 Å) embedded in a more diffuse fibrillar structure. The condensed fibrillar areas are occasionally interrupted by more diffuse areas termed fibrillar centers. Between the nucleolonemal reticulum are more open regions or "lacunae."

The isolated nucleolar protein matrix (Figures 18B and 19B) is a highly condensed fibrous structure with staining characteristics resembling the compact nucleolonemal fibrillar regions of the intact nucleolus. The possible derivation of this protein core structure from the fibrillar nucleolonema is further supported by the presence of distinct fibrillar centers in the final core structure (Figure 19B). The smaller size of the core could conceivably be due to collapse or condensation of the fibrillar nucleolonema after removal of the granular RNP components. Presence of nucleolar lacunae within the core structure, however, suggests the preservation of at least a portion of the nucleolonemal reticulum. Lining the lacunae, as well as along the periphery of the nucleolar cores, are less electron-dense fibrillar structures. Similarly localized fibrillar structures are seen in the isolated nucleoli.

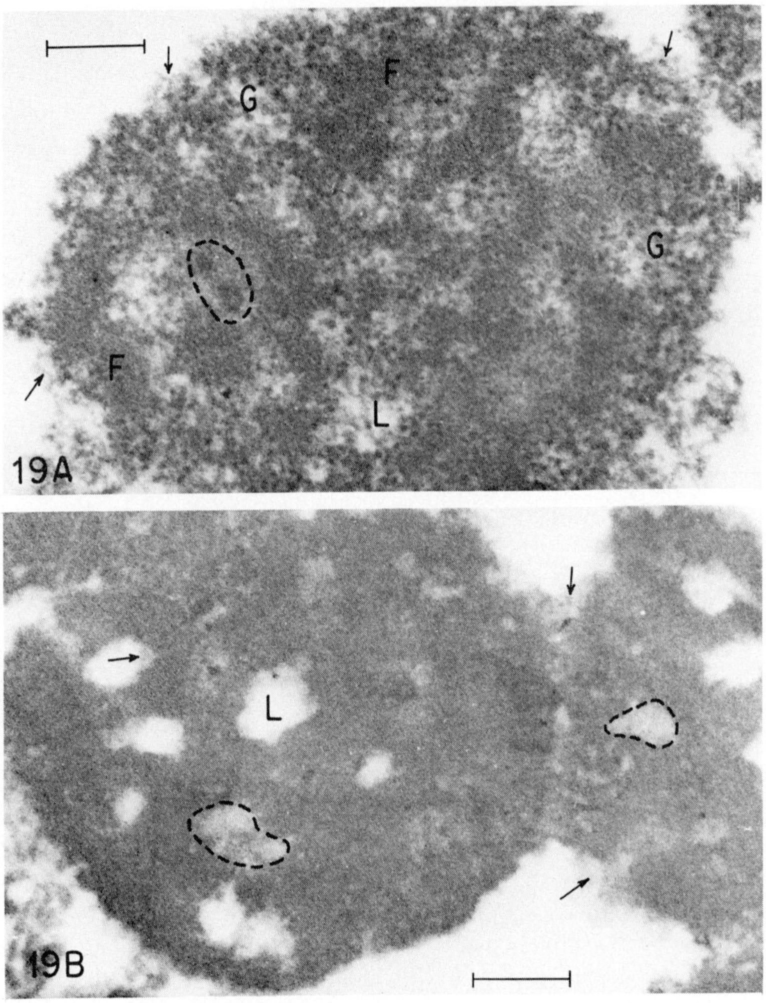

FIGURE 19. Thin-sectioned electron micrographs of isolated nucleoli and nucleolar protein matrix. (A) Isolated nucleolus. Characteristic *in situ* structural features are preserved in the isolated nucleolus. These include the fibrillar nucleolonemal reticulum (F), the granular regions of the nucleolonema (G), fibrillar centers (encircled dashed line) and more open areas or lacunae (L). Note the diffuse nature of the fibrillar components along the periphery (arrows). Bar = 0.2 μm. (B) Nucleolar protein matrix. A highly condensed fibrous structure is apparent with electron-dense staining characteristics similar to the fibrillar nucleolonema of the intact nucleolus. Distinct fibrillar centers (encircled dashed line) and lacunae (L) interrupt the otherwise homogeneous structure. Note the more diffuse fibrillar structure along the periphery and lining the lacunae (arrows). Bar = 0.2 μm.

The structure of the nucleolar matrix protein core is similar to the residual nucleolus previously identified in the total nuclear protein matrix structure (see Reference 86). In the nuclear matrix, however, the residual nucleolus appears continuous with the internal matrix structure. While it is not clear whether a nucleolar core-internal matrix continuity exists *in situ* it is interesting to note that fragments of internal matrix structures are often detected along the periphery of both isolated nucleoli and the nucleolar matrix cores.

Resolution of the total nucleolar proteins on one-dimensional SDS-acrylamide gels reveals a heterogeneous distribution of polypeptides with molecular weights ranging from 13,000 to 200,000 (Figure 20). The nucleolar matrix core proteins have an overall polypeptide

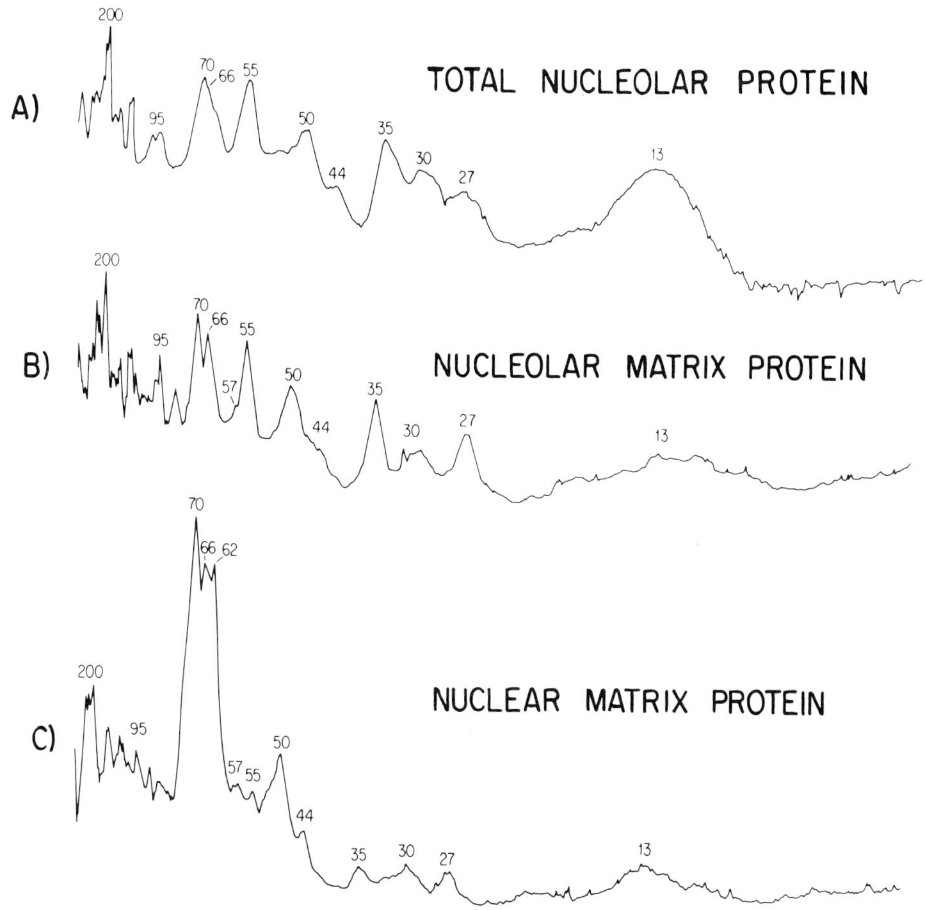

FIGURE 20. SDS-acrylamide gel electrophoresis of the total nucleolar protein (A), nucleolar matrix protein (B), and the total nuclear matrix protein (C) The SDS-solubilized polypeptides were run on 10% acrylamide gels as previously described (Berezney and Coffey[86]). Approximately 50 μg of protein was placed on each gel. Direction of electrophoresis on the densitometric tracings is from left to right. The numbers above the polypeptide peaks indicate the estimated apparent molecular weights × 10^{-3}. (Taken in part from Berezney, R. and Coffey, D. S., *J. Cell Biol.*, 73, 616, 1977. With permission.)

profile similar to the total nucleolar proteins with prominent peptide regions at approximately 200,000, 95,000, 70,000, 66,000, 55,000, 50,000, 44,000, 35,000, 30,000, and 27,000 daltons (Figure 20). The core proteins are enriched in the lower-molecular-weight polypeptides (27,000 to 44,000) compared to the total nuclear matrix and are relatively devoid of the characteristic 60,000- to 70,000-dalton polypeptides. Comings and Peters[47] used two-dimensional electrophoresis to demonstrate that these prominent low-molecular-weight nucleolar matrix core proteins are basic in charge. In conclusion, both one- and two-dimensional electrophoresis demonstrates that the nucleolar matrix is composed of a discrete set of prominent polypeptides and has an overall profile distinct from both the nuclear lamina and total nuclear matrix.

G. Protein Domains in the Nuclear Matrix

Polypeptide analysis of isolated nuclear lamina (Section V.C) and nucleolar matrix (Section V.F) suggest that discrete populations of polypeptides or "protein domains" correspond to discrete structural domains of the matrix. To further fractionate the nuclear matrix, Berezney[154] employed a mild sonication procedure to separate the total matrix into a protein fibril and

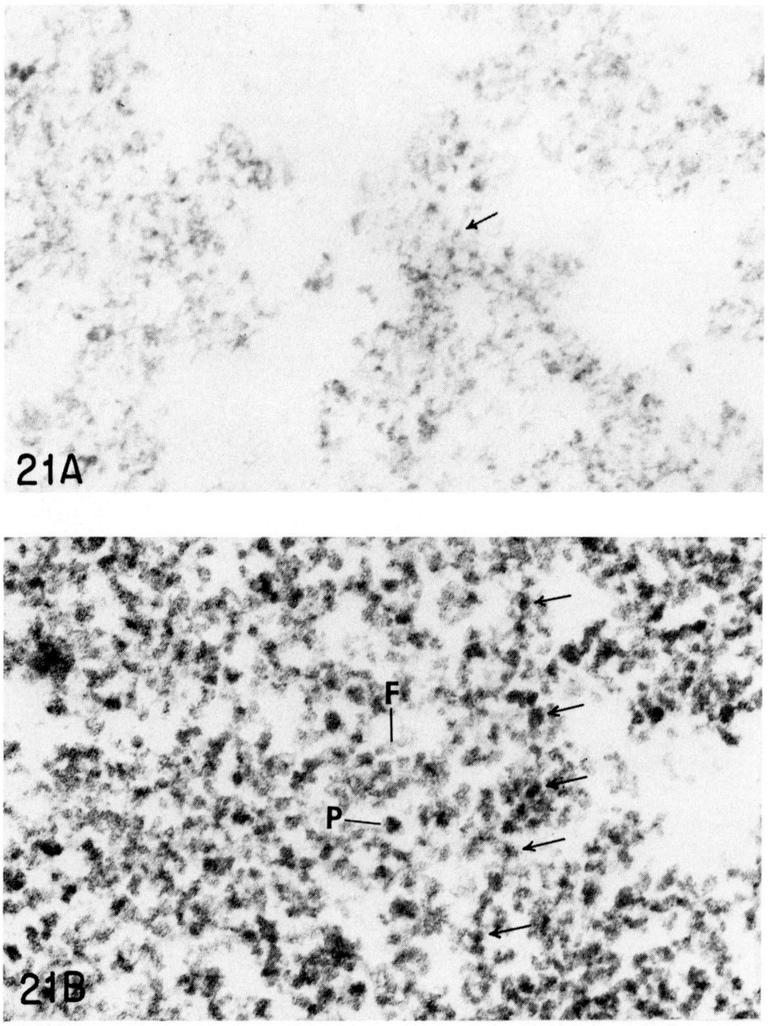

FIGURE 21. Thin-sectioned electron microscopy of (A) matricin fraction. The isolated fibrils resemble structures observed in the intact nuclear matrix and form a diffuse reticular structure. Fibrils as thin as 30 to 50 Å are visible. Many globular structures are arranged along the length of the fibrils (arrow). Bar = 0.2 μm. (B) Residual RNP fraction: although this fraction is rich in matrix particles (P), fibrous matricin can also be detected (F). Chainlike arrays of particles appear to be interlinked by matricin (arrows). Bar = 0.2 μm. (From Berezney, R., *J. Cell Biol.*, 85, 641, 1980. With permission.)

a RNP-particle fraction. The protein fibril fraction termed *matricin*[91,154] is shown in Figure 21A. It contained approximately 25% of the total matrix protein and was over 90% protein in composition with a smaller amount of DNA (8%) and only trace amounts of RNA.[154] The RNP fraction (Figure 21B) contained nearly 99% of the total matrix-bound RNA and was composed of 80% protein, 19% RNA, and 1% DNA.

SDS-acrylamide gel electrophoresis revealed a striking difference in polypeptide profiles between the two subfractions (Figure 22). While the matricin was composed almost entirely of the three 60,000- to 70,000-dalton major matrix polypeptides, the RNP fraction was very heterogeneous, although it also included the 60,000- to 70,000-dalton proteins as major components.[154] In accordance with the preponderance of immunocytochemical

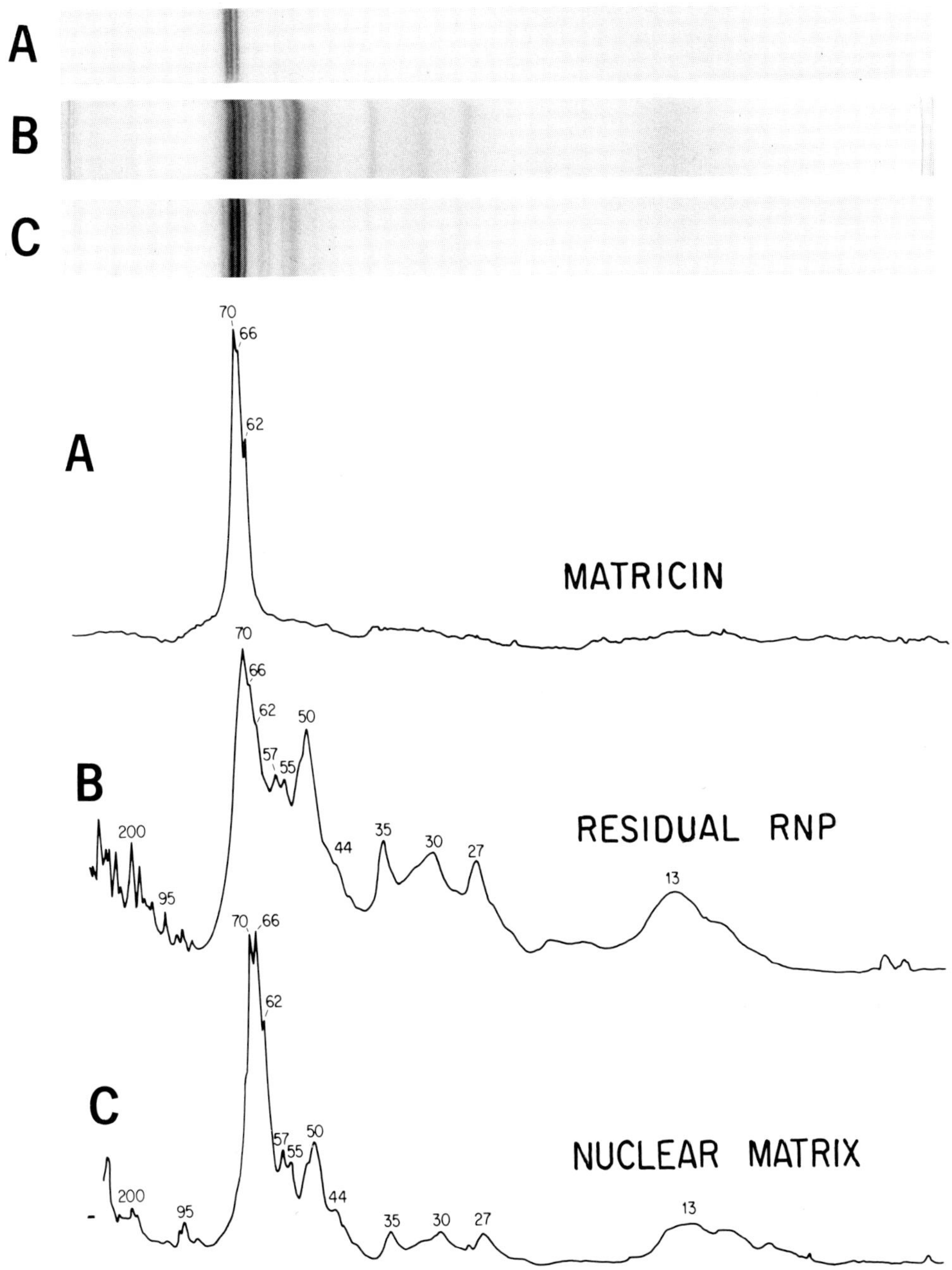

FIGURE 22.. SDS-acrylamide gel electrophoresis of matrix proteins. The SDS-solubilized polypeptides were run on 10% acrylamide gels as previously described.[86] Electrophoresis if from left to right. Densitometric tracings are shown below the photographed gels. The numbers above the polypeptide peaks indicate the estimated apparent molecular weights $\times 10^{-3}$. (A) Matricin, 30 μg protein. (B) Residual RNP fraction, 50 μg protein. (C) Total nuclear matrix, 40 μg protein. (From Berezney, R., *J. Cell Biol.*, 85, 641, 1980. With permission.)

evidence[150-153] and the recent profiles of isolated nuclear lamina on two-dimensional electrophoresis (see Figure 16), it is likely that the ''matricin'' fraction is at least highly enriched in nuclear lamina components. Indeed, these results are consistent with the view that the lamina may be composed solely of lamins A, B, and C. Of course this identification is tentative since nuclear pore complexes were not visible in this dispersed fraction.

Assignment of the matricin fraction to the nuclear lamina is also consistent with the distribution of RNA in the two subfractions as the isolated lamina is depleted in RNA.[29,135] The RNP fraction may, therefore, be a more purified fraction of internal matrix. Electron microscopy of the RNP fraction showing discrete RNP particles enmeshed in a fibrous matrix (Figure 21B) is consistent with this view. In any case, it can be concluded that mild perturbation of the overall matrix structure can lead to release of components with discrete polypeptide compositions or protein domains. Such approaches might be useful for the coordinated elucidation of specific functional domains associated with the matrix.

H. Synthesis and Turnover of Nuclear Matrix Proteins

Evidence has accumulated which indicates that nonhistone proteins undergo extensive changes in synthesis and turnover in cells stimulated to proliferate as well as in various stages of the cell cycle.[171-175] Since nuclear matrix proteins constitute a major portion of the total nuclear nonhistone protein,[47] it is important to determine whether these proteins undergo dynamic alterations in composition, synthesis and/or turnover.

Hodge et al.[32] in their examination of HeLa matrix proteins during the cell cycle, reported a similar polypeptide profile in G_1, S, and late S/G_2. Densitometric evaluations of stained SDS-acrylamide gels indicated differences in the relative amounts of the major matrix polypeptides (45,000 to 75,000 daltons). Only mid-S phase cells presented a pattern similar to randomly dividing cells which included a predominance of the 49,000-, 53,000-, and 57,000-dalton polypeptides over the 60,000- to 70,000-dalton polypeptides. Similar results were obtained by Riley and Keller[110] for HeLa "nuclear ghost" polypeptides.

Hodge et al.[32] studied amino acid incorporation into matrix proteins during the HeLa cell cycle and found that the highest relative rate of incorporation occurred during late G_1 stages. Substantial incorporation of amino acids was measured during the S phase but declined to much lower levels during late S/G_2. SDS-acrylamide gel patterns of the newly synthesized matrix proteins revealed two main groups of labeled polypeptides in randomly dividing cells of 45,000 to 75,000 daltons and 90,000 to 130,000 daltons. Further studies by Hodge et al.[32] investigated the fate of matrix proteins during mitosis. These data suggest that some matrix polypeptides, particularly those of 50,000 to 75,000 daltons, survive mitosis and are present in the next G_1 stage. Other matrix polypeptides, for example the group at 90,000 to 130,000 daltons, may not survive mitosis.

Heimminki[176] has studied the amino acid labeling of histones and nonhistone proteins in lung tissue. The matrix proteins incorporated label at a relatively high rate exceeding the next highest nonhistone protein fraction by threefold. SDS-acrylamide gel electrophoresis of the labeled matrix proteins indicated significant labeling of the major matrix polypeptides between 45,000 to 95,000 daltons. However, a significant amount of label appeared in a large-molecular-weight polypeptide of greater than 100,000 daltons which is not a major polypeptide component as determined by densitometry of stained gels.[176]

Rat liver matrix proteins are also synthesized rapidly. The incorporation of [^{3}H]-leucine into total liver matrix protein was approximately 3.5 times greater than the total nuclear proteins.[42]

Though still in an early stage of development it is already apparent that the matrix proteins are synthesized in vivo at an unusually high rate. The results of further more comprehensive studies of matrix protein synthesis and turnover in proliferating and nonproliferating cells should prove to be very interesting indeed.

I. Phosphorylation of Nuclear Matrix Proteins

Allen et al.[40] previously demonstrated that the nuclear matrix proteins of rat liver are preferentially phosphorylated in an in vitro system consisting of isolated nuclei (Figure 23). Moreover, the level of matrix protein phosphorylation was enhanced threefold in a period during liver regeneration (12 hr post partial hepatectomy) which just precedes the onset of DNA synthesis (14 to 16 hr).

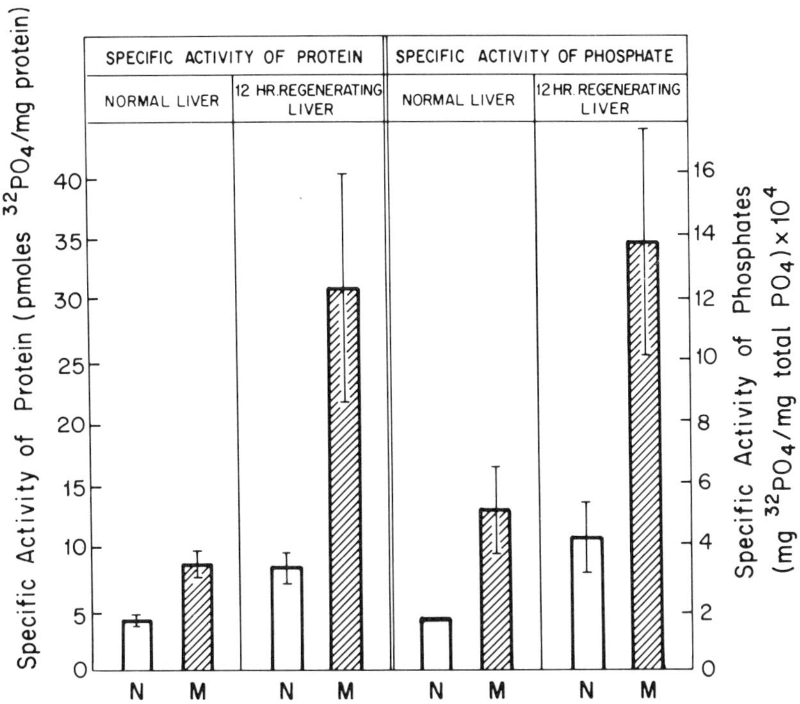

FIGURE 23. [γ-³²P]-ATP phosphorylation of isolated nuclei and the resulting specific activity of the total nuclear proteins (N) and the isolated nuclear matrix proteins (M). Data are presented for normal liver nuclei and nuclei isolated 12 hr after partial hepatectomy (From Allen, S. L., Berezney. R., and Coffey, D. S., *Biochem. Biophys. Res. Commun.*, 75, 111, 1977. With permission.)

Analysis on SDS-acrylamide gels revealed a heterogeneous distribution of phosphorylated polypeptides in nuclear matrix from quiescent liver (Figure 24). The corresponding gel patterns from 12-hr regenerating liver nuclear matrix demonstrated a significant enhancement in the phosphorylation of a discrete set of high-molecular-weight polypeptides between approximately 60,000 to 200,000 daltons (Figure 24). The laboratory of Ševaljević and co-workers[41] has extended these studies to examine phosphorylation in isolated nuclear matrix. They found that isolated matrices contain an active protein kinase activity and reported stage-specific, differential phosphorylation of nuclear matrix proteins during sea urchin development. Moreover, an apparent inverse relationship was observed between the level of phosphorylation in isolated nuclear matrix vs. residual chromatin proteins (high-salt-resistant) and the developmental stage of the sea urchin embryos.[41]

Ševaljević's group[177] has recently examined the localization of protein kinase activity following fractionation into matricin fibrils and RNP particles. In rat liver they found that the in vitro protein phosphorylation was considerably higher in the RNP fraction than in the matricin fraction. This is consistent with the finding of protein kinase activities associated with isolated RNP particles.[178] Moreover, isolated nuclear envelopes and nuclear lamina have been reported to have protein kinase activities.[179-182] It is, therefore, likely that the nuclear lamina is a major component responsible for the protein kinase activity in isolated nuclear matrix. Finally, it is important to note that the matrix-bound protein kinase phosphorylates a variety of high-molecular-weight polypeptides (e.g., 50,000 to 200,000),[41,177] whereas, at least one of the nuclear envelope-associated enzymes is more specific for the lamin proteins.[179] In this later regard it is interesting to note that the lamin polypeptides are

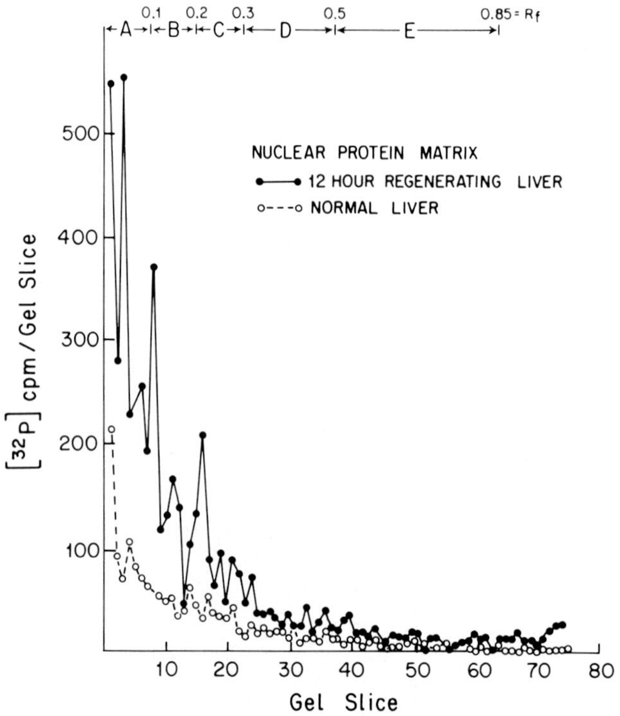

FIGURE 24. [^{32}P] radioactivity associated with the nuclear matrix proteins isolated from normal liver and 12 hr after partial hepatectomy. The matrix proteins were phosphorylated in vitro in isolated nuclei using [γ-^{32}P]-ATP. SDS-acrylamide gel electrophoresis was performed on 100 μg of protein. (From Allen, S. L., Berezney, R., and Coffey, D. S., *Biochem. Biophys. Res. Commun.*, 75, 111, 1977. With permission.)

differentially phosphorylated in the mitotic stage compared to interphase cells.[155] It was speculated that phosphorylation of nuclear lamin polypeptides may be involved in the reversible disassembly-assembly of the nuclear lamina and envelope during mitosis.[155]

VI. DNA LOOP ORGANIZATION

A. Attachment of DNA Loops to the Nuclear Matrix

It is often an exciting period in science when two fields merge into one. Such seems to be the case with respect to the nuclear matrix and the organization of DNA loops. Several lines of evidence now indicate that the huge molecules of eukaryotic DNA are arranged in the cell nucleus in repeating domains or loops with an average size of 50 to 200 kb.[104,112,183,184] These putative DNA loops are highly supercoiled, indicating that they are constrained or fixed at both ends. Moreover, the domains of supercoiled DNA persist and are of similar size throughout the cell cycle including mitosis.[185]

Independent of the above studies on DNA loops were the investigations into the isolated nuclear matrix initiated by Berezney and Coffey in 1974.[85] While comprehensive studies from a number of different laboratories suggested that the isolated matrix was basically a nonchromatin framework structure of the nucleus, it was also apparent that a small but reproducible amount of the total nuclear DNA remained tightly bound to the matrix following exhaustive treatments with high-salt solutions and nonionic detergents.[32,86,91,94,95] The possible relevance of this *residual matrix DNA* for chromatin structure and function was brought to the forefront by the discovery of Berezney and Coffey[4] that newly replicated DNA was initially associated with the residual matrix DNA in regenerating rat liver (see Section VII.A). This finding led to the concept that the DNA replicational sites are bound to the isolated nuclear matrix.

A further important development in these studies was the report from Paulson and Laemmli[112] on the association of DNA loops with chromosome scaffolds. This represented the first clear electron microscopic visualization of the DNA loops postulated to exist by other investigators via DNA supercoiling measurements[104,184] and nuclease digestion experiments.[183] The small amount of residual DNA which remained as an integral constituent of the chromosome scaffold following nuclease digestion and histone removal[113,114] immediately suggested that this ''residual scaffold DNA'' might be the mitotic equivalent of the nuclear matrix residual DNA.

Indications that a discrete set of nonhistone proteins having migration rates in SDS-acrylamide gels similar to matrix and scaffold proteins was involved in constraining the DNA loops in a supercoiled state,[186] gave credence to the emerging view that the supercoiled DNA loops were constrained by their attachment to components of the interphase matrix and the mitotic chromosome scaffold, respectively (see also Razin et al.[3]). If the DNA loops were actually anchored to the nuclear matrix, then it should be possible to isolate nuclear matrices with intact DNA loops. This was reported by Berezney and Buchholtz[2] in structures termed *DNA-rich nuclear matrices* (see Section III.D). It appears that these rat liver DNA-rich matrices are similar to the HeLa cell ''nucleoid'' or ''cage'' structures which Cook and Brazell[104] isolated to first demonstrate supercoiling in eukaryotic DNA. Direct comparisons of the polypeptide composition between ''nucleoids'' and DNA-rich nuclear matrix are hindered by the greater degree of cytoplasmic contamination present in preparation of ''nucleoids''.[146]

When the DNA-rich matrices are examined by whole-mount electron microscopy, an extensive DNA halo is observed which radiates as far as 20 to 30 μm from the center of a dense matrix core (Figure 25). At higher magnification the DNA in this halo appears as a series of highly folded loops. This is most evident along the periphery of the halo where individual loops are often distinguishable (Figure 26). Importantly, analogous DNA halo structures were previously observed attached to the mitotic chromosome scaffold.[112] Also, similar loop structures have been demonstrated by McCready et al. in HeLa ''nucleoids''.[187]

Matrix fibers can be seen extending from the actual matrix core to various positions in the DNA halo where they often terminate within specific DNA loops (Figures 25 and 26). It is intriguing to consider the possibility that these associations represent attachment sites between the DNA loops and the matrix fibrils. While further studies are required to rule out spurious associations created during the electron microscopic preparation, preliminary enzymatic digestion experiments have indicated some role for matrix proteins in maintaining the intricate DNA loop structures. For example, DNase I treatment destroys the DNA loops but leaves the matrix core with radiating matrix fibers intact (Figure 27). RNase has no visible effect on either the DNA loops or the matrix fibril structure. In contrast, proteinase K destroys both the matrix fibrils and the convoluted DNA loops. This digestion leaves the DNA spread randomly on the electron microscopic grids similar to high-molecular-weight linear DNA (Figure 27).

B. DNA Sequences at the Matrix Attachment Sites

A fundamental question to consider is whether there are specific attachment sequences which anchor the DNA at the putative matrix attachment sites. Such a family of sequences could provide a mechanism for the propagation of an exact spatial ordering of DNA loops from one generation to another. Moreover, if DNA loops represent functional as well as structural units of eukaryotic DNA, then these hypothetical anchoring sequences could be

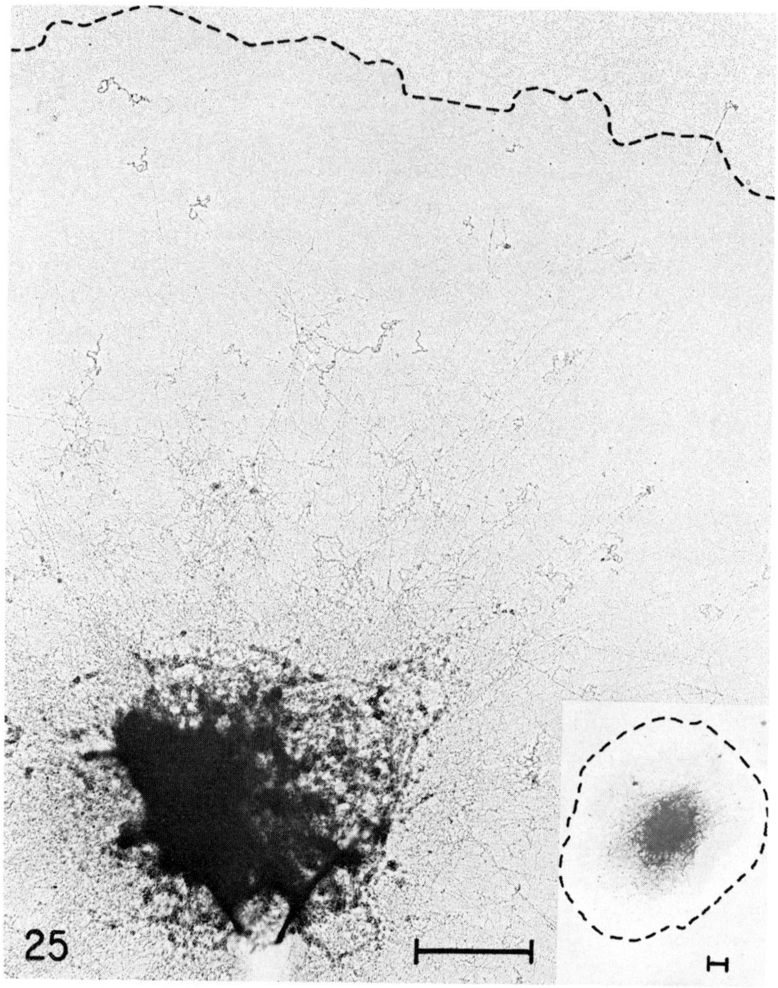

FIGURE 25. Whole-mount spread electron micrograph of a DNA-rich nuclear matrix. Broken lines denote the border of DNA loop structures which extend up to 30 μm and form a DNA "halo" around the electron-dense central matrix structure. Shadowed with platinum-palladium. Bar = 2.0 μm.

involved in the expression or regulation of replication and/or transcription. Since estimates of the number of matrix-DNA attachment sites vary from approximately 5×10^4 to 1×10^5 per nucleus,[8,188] attachment site sequences might represent a subclass of the repetitive or middle repetitive DNA families which are characteristic of eukaryotic cells.[189-191]

A number of laboratories have investigated this possibility in recent years using a similar experimental strategy. Since the purpose of these studies was to determine if any class of sequences was enriched at or near the matrix attachment site, it was necessary to examine DNA which contained sequences closer to and farther away from the matrix. Importantly, it was found that the level of nuclease digestion prior to matrix isolation determines both the size and relative recovery of DNA in the matrix.[188,192,193] This is illustrated in Figure 28 which shows a linear relationship between the amount of and average fragment size of the matrix-attached DNA. Significantly, the average fragment size approaches zero as the amount of DNA on the matrix approaches zero.

Although earlier studies suggested an enrichment in repetitive DNA at the basal attachment sites,[188,194,195] more recent results are contradictory. For example, Bowen[196] reported that

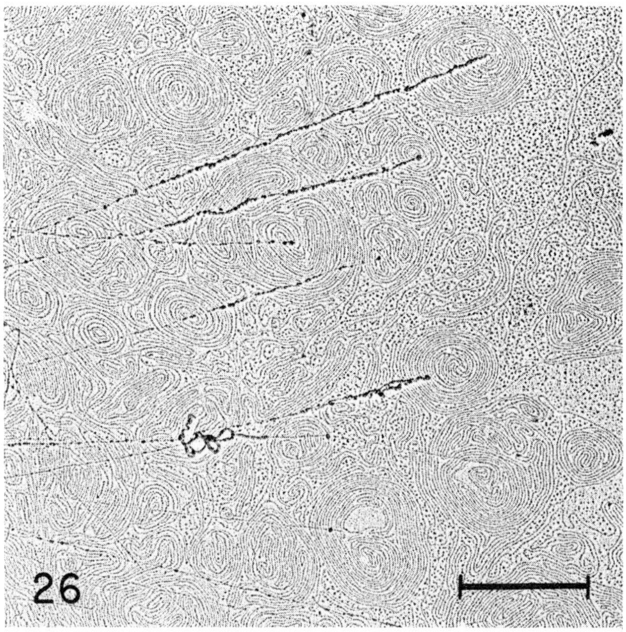

FIGURE 26. Whole-mount spread electron micrograph of a DNA-rich nuclear matrix in a region along the periphery of the DNA halo (see Figure 25). The DNA loops are wound into highly convoluted "fingerprint" patterns. Matrix fibrils extend from the central matrix core structure (see Figure 25) to the convoluted DNA loops. Bar = 0.5 μm.

digestion of HeLa metaphase chromosomes with the restriction enzymes Hae III and Eco RI resulted in a putative enrichment in human satellite II and III sequences. However, digestion with micrococcal nuclease revealed no enrichment in either highly repetitive or middle repetitive sequences. Earlier, Razin et al.[188] reported a similar phenomenon with respect to the enrichment of satellite sequences by using restriction enzymes. However, these investigators still found an enrichment in middle repetitive DNA after use of micrococcal nuclease. In contradiction, recent studies from our laboratory[192,193] demonstrate that small fragments of DNA located near both the mitotic scaffold and interphase matrix attachment sites have reassociation kinetics virtually indistinguishable from total chromosomal DNA (Figure 29). In support of these results, Kuo[197] has recently reported that the DNA at the scaffold is not a unique sequence class of genomic DNA. Importantly, in arriving at this conclusion, Kuo used two different methods for both scaffold isolation and nuclease digestion (Eco RI or micrococcal nuclease).

We have also examined the reassociation kinetics of DNA fragments from both rat and mouse liver interphase matrices prepared with several different nucleases including micrococcal nuclease, DNase I, and the endogenous nuclease of liver cells.[192] The results illustrated in Figure 30 revealed, regardless of the nuclease employed, no reproducible enrichment in repetitive sequences. In fact, no enrichment in repetitive sequences was detected in any of the DNA-containing fractions generated during nuclear matrix isolation.[192] Moreover, we have screened a series of different-sized matrix DNA fragments from mouse liver with a libraryof cloned repetitive mouse DNA sequences using the Grunstein-Hogness blot technique[192] This clone library included the major mouse satellite sequences as well as repetitive sequences repeated only about 500 times per genome. No reproducible enrichment in specific repetitive sequences was detected (Figure 31). In contrast, Matsumoto[198] has reported that nuclear matrices from bovine cells are enriched in bovine satellite DNA.

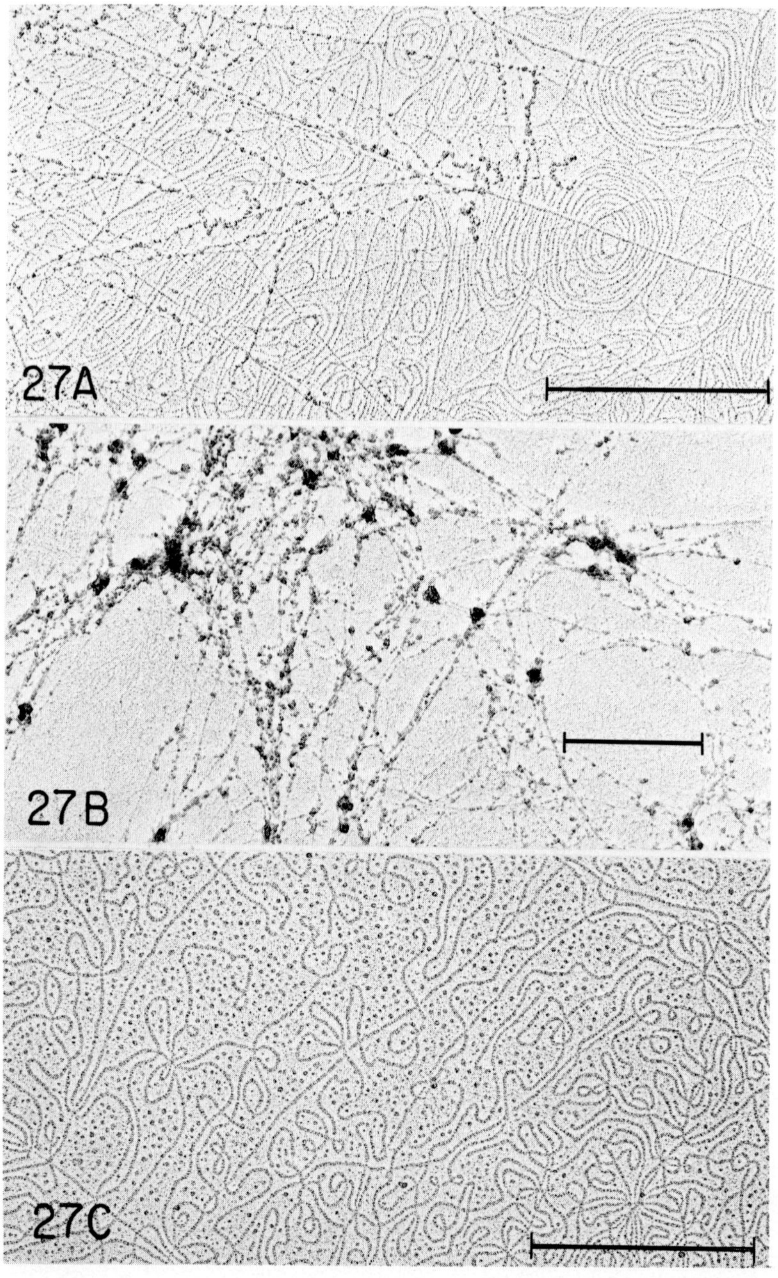

FIGURE 27. Effect of DNase and protease treatment on matrix-attached DNA loops. (A) Control DNA-rich nuclear matrix from rat liver; matricin fibrils and highly convoluted DNA loops are seen. Bar = 1.0 μm. (B) DNase treated matrix. Isolated DNA-rich matrices were treated with DNase I (200 units DNase I/mg DNA, 30 min, 30°C). The DNA loops are removed leaving a distinct network of matricin fibrils. Bar = 1.0 μm. (C) Protease treated matrix. DNA-rich matrices were treated with proteinase K (Boehringer-Mannheim; 100 μg proteinase K/mg matrix protein, 30 min, 30°C). The matricin protein fibrils are disintegrated; the characteristic folded looped structure is removed leaving the DNA randomly arranged. Bar = 1.0 μm.

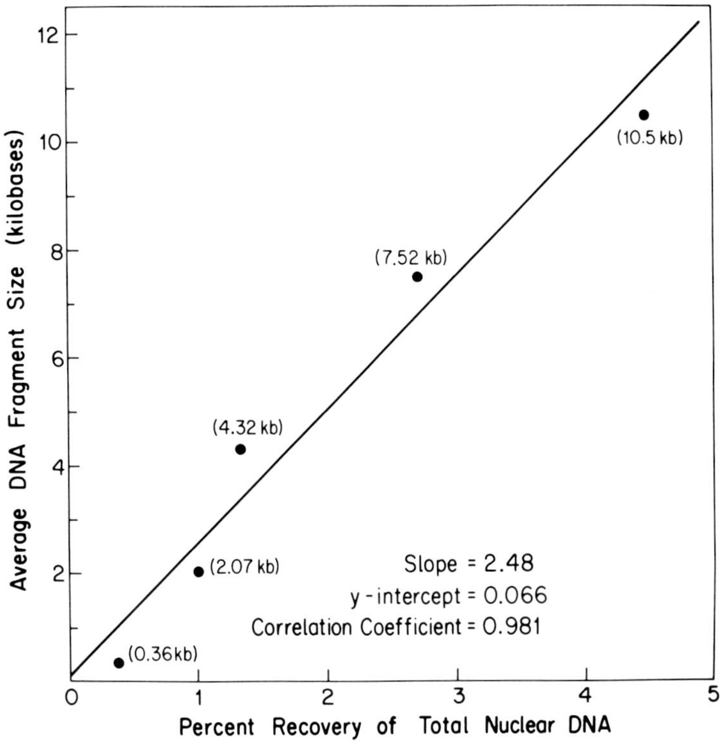

FIGURE 28. Relationship between the average length and amount of DNA in isolated nuclear matrix. Matrices with different amounts of attached DNA were generated by endogenous digestion of rat liver nuclei between 10 and 120 min at 37°C. The matrix-attached DNA fragments were purified and the weighed average fragment size determined in nondenaturing 1% agarose gels with λ-DNA, PM-2-DNA, and Hae III-digested PM-2 as molecular weight markers. (From Berezney, R., Basler, J., Buchholtz, L. A., Smith, H. C., and Siegel, A. J., in *The Nuclear Envelope and the Nuclear Matrix,* Maul, G. G., Ed., Alan R. Liss, New York, 1982, 113. With permission.)

In considering DNA loops and their attachment sites it is natural to view these as "static loop structures" (see e.g., Cook and Brazell[199]). However, evidence is growing which suggests that the DNA loops are dynamic structures at least during DNA replication (see Section VII.A). In an attempt to resolve this issue in nonreplicating cells, we have performed matrix-DNA-driven annealing experiments with matrix DNA fragments isolated from quiescent liver cells.[192] If the DNA is fixed at permanent sites, then the DNA closest to these sites should represent a very small proportion of the sequence complexity of total genomic DNA (<1%) and should be unable to significantly drive a total nuclear DNA probe. The results of this matrix-driven reassociation revealed that the small matrix DNA fragments (≤350 bp) drove the total genomic probes with virtually identical kinetics and to the same extent as the homologous matrix DNA probe (Figure 32). This led us to propose that the DNA loops may be dynamic rather than statically fixed structures.[192,193] In this manner, the number of attachment sites is preserved while the actual sites of DNA interaction are in a state of rapid exchange (see Figure 33).

This rapid exchange model has important implications for the possible regulation of specific DNA loop structures in relation to replicational and/or transcriptional events. It is conceivable, for example, that dynamic *DNA loop programs* serve as the basis for specific transcriptional and developmental events. In this regard, recent findings concerning actively

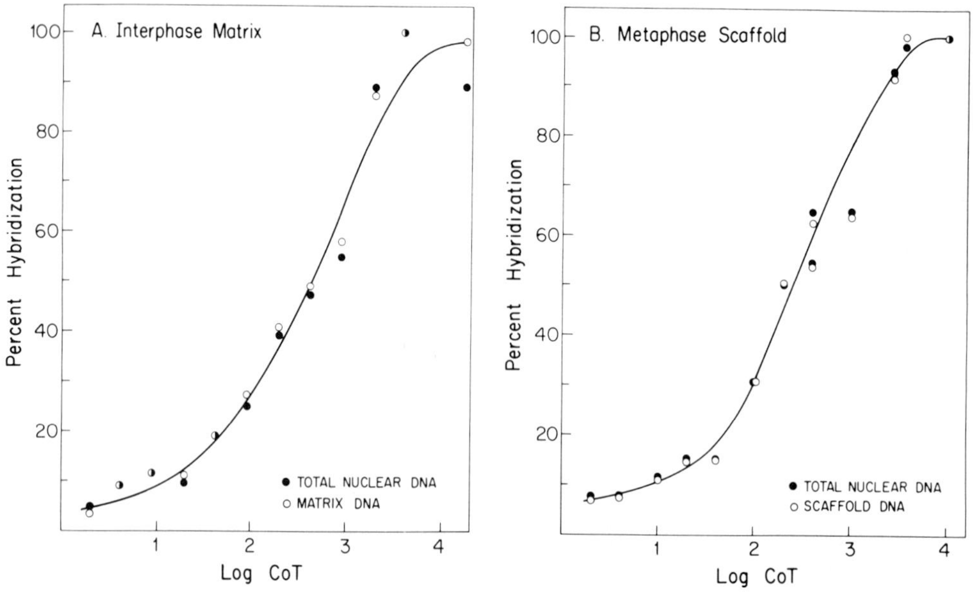

FIGURE 29. Total genomic DNA-driven annealing of Chinese hamster DON cell interphase matrix (A), and metaphase scaffold (B), DNA fragments. Matrix and scaffold DNA fragments (≤600 bp,o) were isolated by elution from nondenaturing agarose gels, nick-translated with $[^{32}P]$ label and reassociated in the presence of an excess of total nuclear DNA driver (approximately 500 bp) and a homologous total DNA probe (approximately 400 bp,●) nick-translated with $[^{3}H]$ label. Hybridization was measured by the S_1 method. All values were normalized such that the maximum extent of hybridization of each probe represented 100%. (From Berezney, R., Basler, J., Buchholtz, L. A., Smith, H. C., and Siegel, A. J., in *The Nuclear Envelope and the Nuclear Matrix,* Maul, G. G., Ed., Alan R. Liss, New York, 1982, 183. With permission.)

transcribed gene sequences at the matrix attachment sites are of great interest[12,13] and will be considered in further detail in Section VIII. While Cook and Brazell[199] have presented evidence in favor of static DNA loops, their primary observations of enrichment or depletion of specific gene sequences at the base of the DNA loops have not been reproduced by Kuo[197] who also favors a dynamic DNA loop model. The possible role of matrix attached DNA loops in DNA replication is discussed in the following section (Section VII.A).

VII. DNA REPLICATION

A. In Vivo DNA Replication and DNA Loop Models

Aside from an obvious role for the three-dimensional arrangement and packaging of chromatin, the organization of eukaryotic DNA into repeating loops may also order the DNA into distinct functional units. For instance, in the dynamic loop model proposed in Figure 33, the rapid exchange of DNA with a fixed number of attachment sites could provide a basis for the regulation of specific loop formation during DNA replication and transcription. Since eukaryotic replication occurs in discontinuous replicon units with a size similar to the matrix attached loops,[1,7,8] it is intriguing to consider whether replicons represent matrix-attached DNA loops in the process of replication.

Electron microscopic autoradiographic studies have suggested a close association of DNA replicational sites with the *in situ* matrix structure.[200] Isolated nuclear matrices have enabled a direct quantitation of newly replicated DNA associated with the matrix. The earlier studies of Berezney and Coffey[4,5] demonstrated an initial enrichment of newly replicated DNA associated with the nuclear matrix following a 1 min in vivo pulse in regenerating rat liver. With increasing pulse time there was a rapid and progressive decrease in labeled DNA attached to the matrix and a corresponding increase in the bulk chromatin.[4,5] These studies

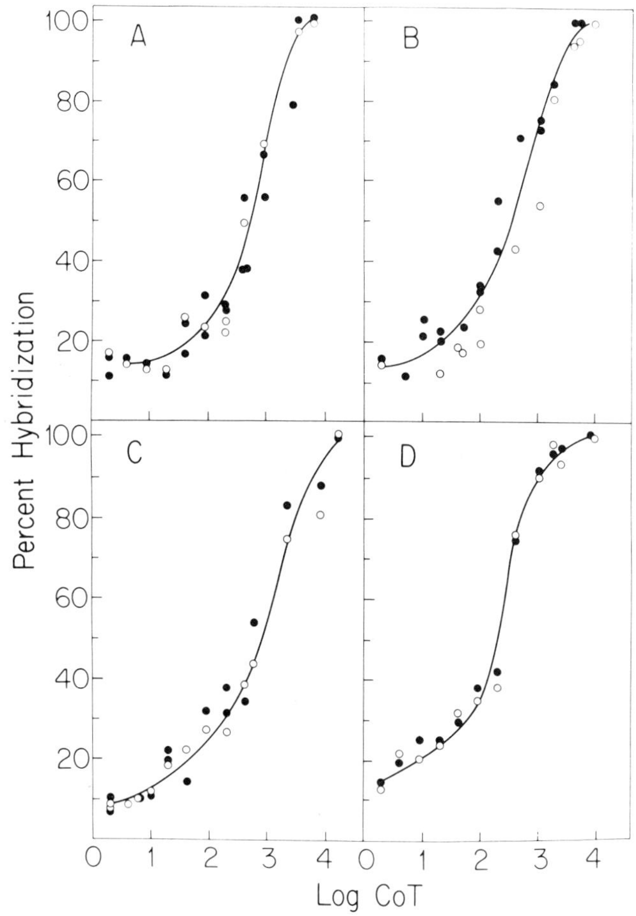

FIGURE 30. Total genomic DNA-driven reassociation of matrix-attached and total nuclear DNA probes. (A) Rat liver nuclear matrix-attached DNA (≤350 bp, prepared by the endogenous nuclease method) and total rat liver nuclear DNA (~400 bp), probes annealing in the presence of an excess of total rat liver nuclear DNA (≤ 1200 bp). (B) Mouse liver matrix-attached DNA (≤ 350 bp, prepared by the endogenous nuclease method) and total mouse liver DNA (~400 bp) annealing in the presence of an excess of total mouse liver nuclear DNA(~800 bp). (C) Rat liver matrix-attached DNA (~800 bp), prepared by DNase I method, and total rat liver DNA (~800 bp), driven by total rat liver nuclear DNA (≤1200 bp). (D) Rat liver matrix-attached DNA (≤350 bp, prepared by the micrococcal nuclease method) and total rat liver nuclear DNA (~400 bp) probes driven by an excess of total nuclear DNA (≤1200 bp). Total probes (O); matrix probes (●). (From Basler, J., Hastie, N. D., Pietras, D., Matsui, S.-I., Sandberg, A. A., and Berezney, R., *Biochemistry*, 20, 6921, 1981. With permission.)

were the first to suggest that the DNA replicational sites are fixed to the matrix and that previously replicated DNA moves progressively away from the matrix replicational sites. Recently, we repeated these experiments with improved matrix isolation techniques and confirmed the original observations.[8] The logic behind this kinetic model is illustrated in Figure 34. If DNA replication occurs only at matrix-bound replicational sites, then newly replicating DNA should move progressively away from the matrix-bound sites and into more distal regions of the DNA loops. Most importantly, this ''decay'' of labeled DNA from the nuclear matrix should be closely related to the in vivo rate of replicational fork movement and the average fragment size of the matrix-attached DNA.

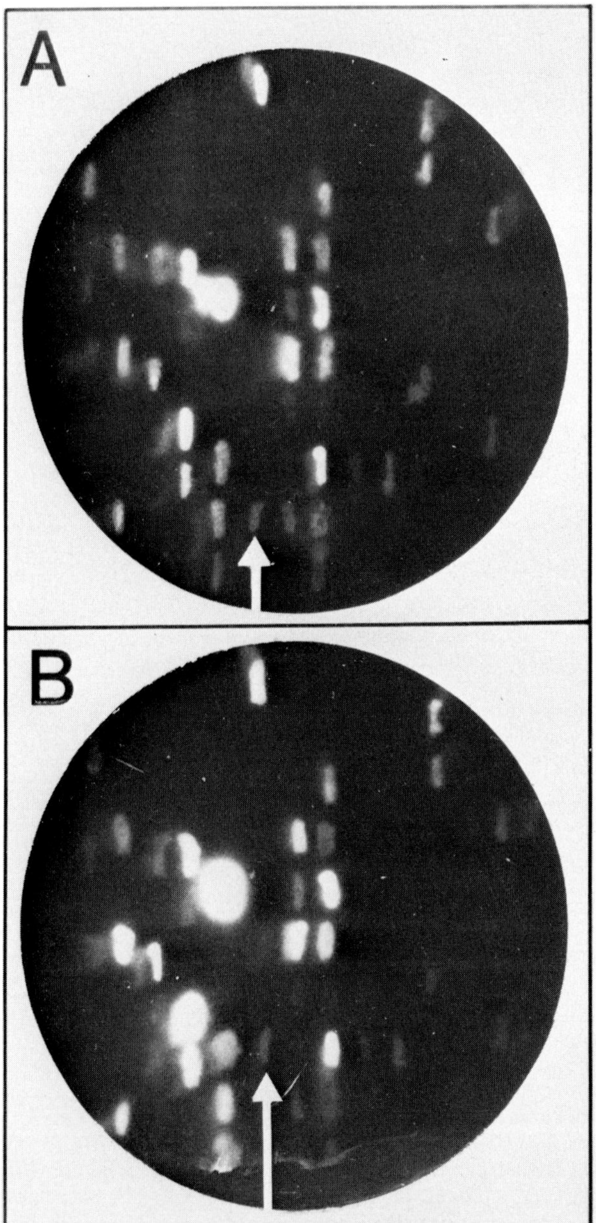

FIGURE 31. Grunstein-Hogness hybridization of (A) total nuclear mouse DNA probe (~400 bp) and (B) small nuclear matrix mouse DNA (~350 bp) to cloned repetitive mouse DNA fragments. [^{32}P]-labeled DNA probes [(5 to 10) × 10^7 cpm/μg] prepared by nick translation and hybridization were performed as described previously.[192] Arrow indicates one clone whose position was changed from filter A to filter B to facilitate identification during handling. This figure is a direct positive print of the autoradiograph. (From Basler, J., Hastie, N. D., Pietras, D., Matsui, S.-I., Sandberg, A. A., and Berezney, R., *Biochemistry,* 20, 6921, 1981. With permission.)

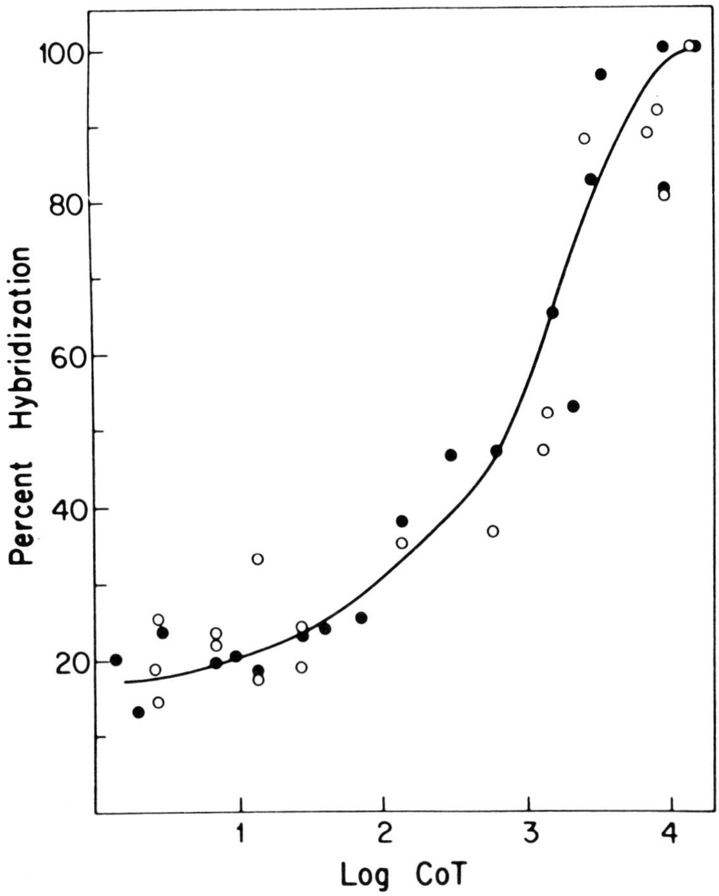

FIGURE 32. Rat liver nuclear matrix DNA-driven annealing of total nuclear and matrix DNA. Matrix probe, ≤350 bp (O), and matrix DNA driver (800 bp) were prepared via the endogenous nuclease method. Total nuclear DNA probe, ~400 bp (●), was from nick-translated total sheared rat liver DNA. (From Basler, J., Hastie, N. D., Pietras, D., Matsui, S.-I., Sandberg, A. A., and Berezney, R., *Biochemistry*, 20, 6921, 1981. With permission.)

With these two parameters as a basis, we estimated the theoretical decay curve for replicating DNA associated with isolated nuclear matrices from regenerating liver.[8] The actual experimental data, based on in vivo pulse labeling and sizing of the labeled DNA fragments, were strikingly similar to the theoretical curve (Figure 35). It should be noted that the curve should change in relation to the average fragment size of DNA attached to the matrix. This has been confirmed in our laboratory for several fragment sizes ranging from 0.8 to 4.5 kb. [201] Other investigators have reported similar results in several different cell types.[1,6,7,202] Vogelstein et al.,[105] for example, used light autoradiography to directly visualize the migration of newly replicated DNA from the central core of the matrix into the DNA halo structure with increasing pulse time or chase. Appropriate reconstruction experiments also suggest that the high-salt-resistant binding of both replicating and nonreplicating DNA to the matrix is not an artifact induced during isolation.[2,7,8,103]

The aforementioned investigations all suggest that the DNA replicational sites are bound to the nuclear matrix. Appropriate DNA loop models have been proposed which emphasized the dynamic interaction of DNA loops with the nuclear matrix.[1,6-8] Figure 36 illustrates two alternative views for replication of matrix attached DNA loops. Although other studies of

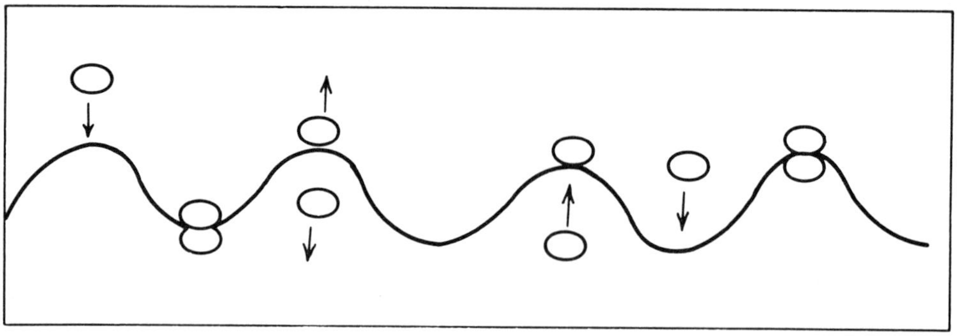

FIGURE 33. Dynamic loop model of DNA. DNA attachment sites (O) are depicted in different states of association and dissociation with the DNA strand. (From Berezney, R., Basler, J., Buchholtz, L. A., Smith, H. C., and Siegel, A. J., in *The Nuclear Envelope and the Nuclear Matrix*, Maul, G. G., Ed., Alan R. Liss, New York, 1982, 183.)

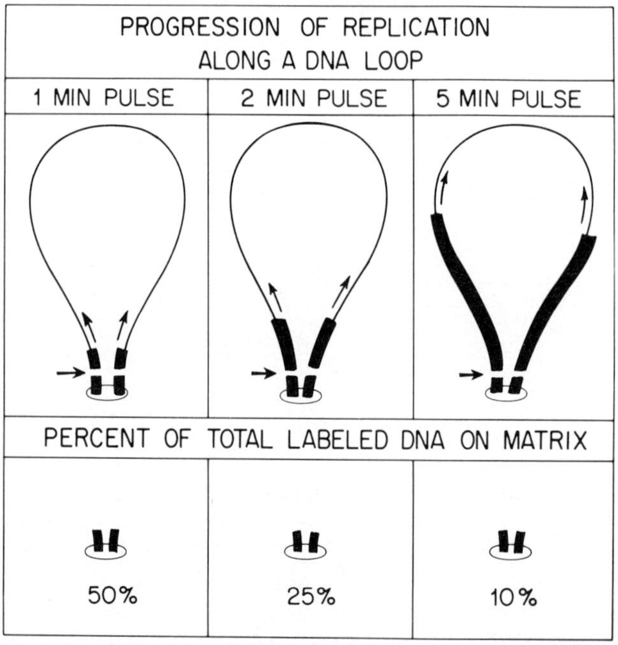

FIGURE 34. Progression of replication along a matrix-attached DNA loop. Knowing the average fragment size of the nuclear matrix-attached DNA and the in vivo rate of fork elongation, the percentage of total newly replicated DNA associated with the matrix-attached DNA fragments can be calculated with the assumption that DNA replication occurs at matrix attachment sites. In these calculations the average size of the matrix-attached DNA fragments is 1.6 kilobases and the average rate of in vivo fork movement is 3.2 kilobases. This diagram gives a pictorial representation of the kinetic decay curves shown in Figure 35.

matrix-associated replication have favored the "fixed matrix" model,[1,7] it is likely that more detailed information is needed concerning matrix replicational sites before alternative views such as the "sliding matrix" model can be ruled out.[8]

B. Matrix-Bound DNA Polymerase Alpha

Implicit in the above loop models (Figure 36) is the view that functional replicational complexes or *replisomes* are bound to the matrix during replication of the DNA loops. With

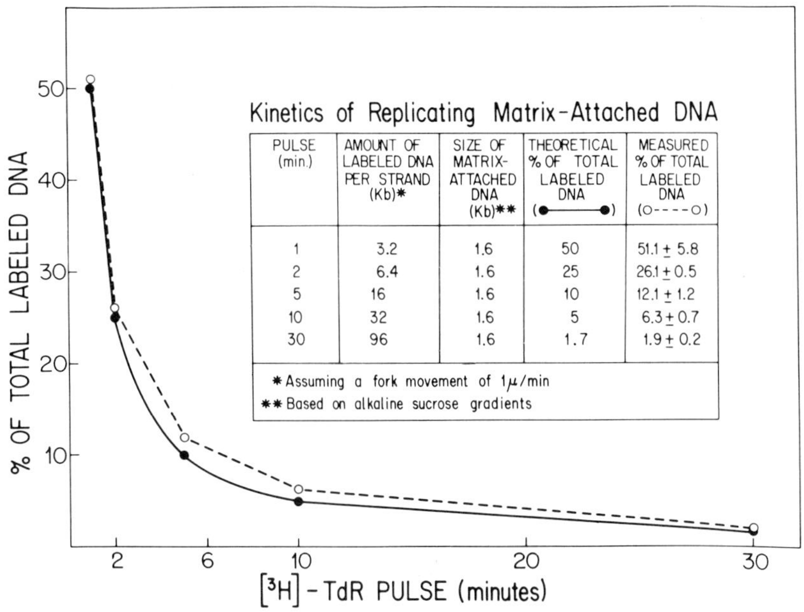

PULSE (min.)	AMOUNT OF LABELED DNA PER STRAND (Kb)*	SIZE OF MATRIX-ATTACHED DNA (Kb)**	THEORETICAL % OF TOTAL LABELED DNA (●——●)	MEASURED % OF TOTAL LABELED DNA (o----o)
1	3.2	1.6	50	51.1 ± 5.8
2	6.4	1.6	25	26.1 ± 0.5
5	16	1.6	10	12.1 ± 1.2
10	32	1.6	5	6.3 ± 0.7
30	96	1.6	1.7	1.9 ± 0.2

* Assuming a fork movement of 1μ/min
** Based on alkaline sucrose gradients

FIGURE 35. Kinetics of association of replicating DNA with the nuclear matrix. Experimental data which described the kinetic association of replicating DNA attached to the matrix structure are compared with a theoretical kinetic model. In this model it is assumed that replication occurs in continuous association with the matrix structure and that previously replicated DNA is found progressively further away from the "matrix replicational sites". (See Figure 34 for direct visualization of this phenomenon.) This can be envisioned as the result of the movement of the matrix replicational sites along the replicating DNA strands or vice versa. In the theoretical calculations, the average rate of fork movement is estimated as 1 μm or 3.2 kb/min. The average size of the 1-min-labeled matrix-attached DNA fragments is estimated as 1.6 kb. (From Berezney, R. and Buchholtz, L. A., *Exp. Cell Res.*, 132, 1, 1981. With permission.)

this in mind our laboratory is currently investigating the possible association of the DNA replicational machinery with the nuclear matrix. In 1980 Smith and Berezney[9] reported that isolated nuclear matrices from actively replicating regenerating rat liver contained significant (10 to 25% of total nuclear activity) DNA polymerase alpha activity (Figure 37) which is widely believed to represent the major replicative enzyme in mammalian cells. The matrix-bound alpha polymerase was completely inhibited by aphidicolin, a specific inhibitor of DNA polymerase alpha, but was resistant to dideoxy TTP an inhibitor of DNA polymerases beta and gamma, but not alpha. Moreover, DNA polymerase beta (a presumptive DNA repair enzyme) was detected in only trace quantities on the regenerating liver matrices.[9]

If the matrix-bound alpha polymerase is part of a functional replicational complex, then the alpha polymerase should continue synthesis of the matrix-attached replicating DNA fragments. This was recently demonstrated by Smith and Berezney[11] who characterized an endogenous in vitro synthesis system in nuclear matrix isolated from regenerating rat liver. Figure 38 shows linear kinetics for endogenous DNA synthesis on the matrix for at least 60 min at 37°C. Moreover, approximately 90% of the total DNA synthesis on the matrix is driven by the alpha polymerase (Figure 38). Alkaline sucrose gradients of the matrix-synthesized DNA demonstrated one prominent peak at 4S, i.e., 100 to 200 nucleotides (Figure 39), which is the characteristic size of single-stranded Okazaki replicational fragments in mammalian cells.

Smith and Berezney[11,203] further reported striking intranuclear redistributions of DNA polymerase alpha activity preceding and during in vivo replication in regenerating rat liver. The most marked induction of in vitro enzyme ectivity occurred coincident with the onset

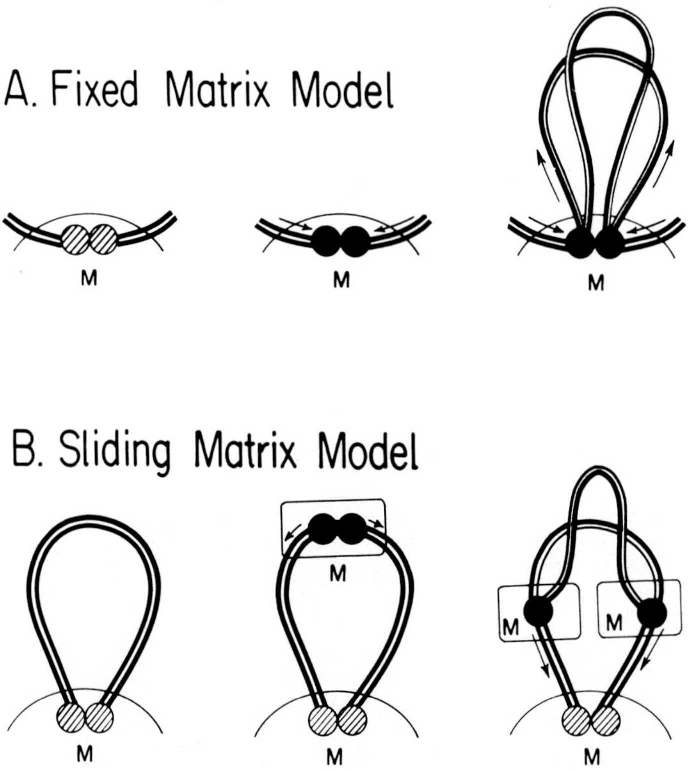

FIGURE 36. Diagrammatic representation of hypothetical structural models for matrix-bound replication. In both models the thick lines of the DNA refer to the newly replicated daughter strands. (A) Fixed matrix model. The diagram on the left indicates a region of the DNA between matrix-attached loops which is closely associated with matrix (hatched circles). Whether this corresponds to a defined ''interloop matrix attachment site'' is unclear. In the middle diagram, a fixed bidirectional replicational complex (solid circles) becomes associated with this interloop attachment site. DNA from the two adjacent loops then moves in opposite directions through the replicational complex. This forms a new looped structure containing the newly synthesized DNA (right diagram). The DNA of the newly replicated loop is therefore derived from each of two adjacent loops. (B) Sliding matrix model. In the left diagram is shown a loop of unreplicated DNA attached to the matrix (hatched circles). In this model, two adjacent matrix-bound replicational complexes become attached to the DNA loop at a site other than the initial DNA attachment site (solid circles, middle diagram). Replication then proceeds in both directions by the sliding of the matrix-attached replicational complexes through the DNA in the direction of the original matrix attachment sites (right diagram). Replication of the loop terminates at these sites resulting in the formation of two copies of the original loop. (From Berezney, R. and Buchholtz, L. A., *Exp. Cell Res.*, 132, 1, 1981. With permission.)

of in vivo replication and was seen in the high-salt-resistant domain of the nuclear matrix. Moreover, DNA polymerase alpha activities redistributed *pre-replicatively* from bulk chromatin to the high-salt-resistant matrix and reached a maximal percent of the total nuclear activity (~30%) during in vivo replication. At the conclusion of replication, DNA polymerase alpha activities shifted back into the bulk chromatin and were no longer high-salt resistant. In contrast, DNA polymerase beta activities were randomly distributed in the nuclear subfractions throughout the proliferative response.[203] We concluded that DNA polymerase alpha was dynamically assembled or ''activated'' on the nuclear matrix and that this process was initiated before the onset of in vivo replication. Since this activation includes DNA synthesis on the matrix-attached DNA, these results further suggest that functional replicational complexes, or replisomes, are *''pre-replicatively activated''* on the matrix. Although it is not known whether other replicative factors are assembled on the nuclear matrix, Reddy and Pardee[204] demonstrated in Chinese hamster fibroblasts that in addition to a DNA polymerase

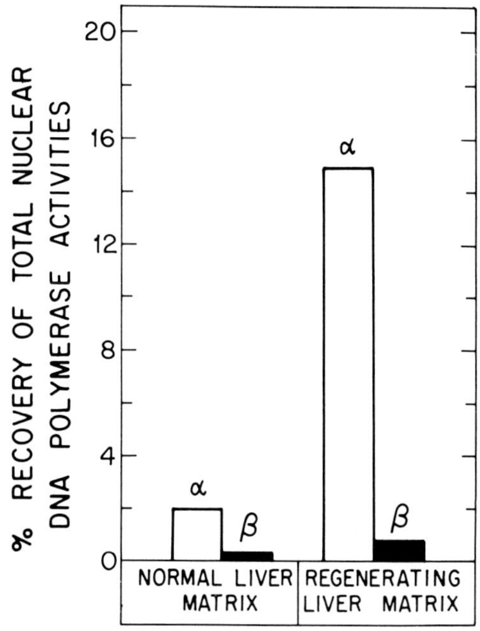

FIGURE 37. Recovery of total nuclear DNA polymerase alpha and beta activities in isolated nuclear matrices. Recoveries are based on the incorporation of [^{3}H] label into acid-precipitable counts on activated calf thymus DNA. It is important to note that the recovery of total alpha polymerase in normal liver (2%) is a maximal estimate since no alpha activity was detected in this fraction. (From Smith, H. C. and Berezney, R., *Biochem. Biophys. Res. Commun.*, 97, 1541, 1980. With permission.)

activity, several enzymes involved in DNA precursor metabolism redistribute prereplicatively from the cytoplasm to the nucleus and are associated in a rapidly sedimenting nuclear complex.

Recently, we have localized the matrix-bound DNA polymerase alpha activity in the matrix interior.[205] The nuclear lamina isolated from regenerating rat liver according to Kaufmann et al[139] was deficient (3 to 15% of total matrix activity) in both endogenous template-engaged DNA polymerase alpha and alpha polymerase activity measured with exogenous DNA template. In contrast, the lamina contained approximately 50% of the corresponding matrix-bound beta polymerase activities. This localization was confirmed by high-resolution electron microscopic autoradiographic analysis following in vitro DNA synthesis in isolated nuclear matrices.[205] The recent immunocytochemical studies of Bensch et al.[206] using monoclonal antibodies to DNA polymerase alpha also suggest that the polymerase is largely localized in the *in situ* nuclear matrix regions of intact cells.

VIII. TRANSCRIPTION AND THE RNP DOMAIN OF THE NUCLEAR MATRIX

A. RNA Associated with the Nuclear Matrix

Electron microscopic autoradiographic studies have demonstrated the localization of newly transcribed RNA within the *in situ* nuclear matrix.[61-65] Pulse-chase experiments further revealed that the newly transcribed RNA "remains" in the *in situ* matrix regions during its stay in the nucleus.[61-65] These *in situ* localization studies form the basis for the view that RNP domains represent an important functional component of the nuclear matrix (see Section II.B and D).

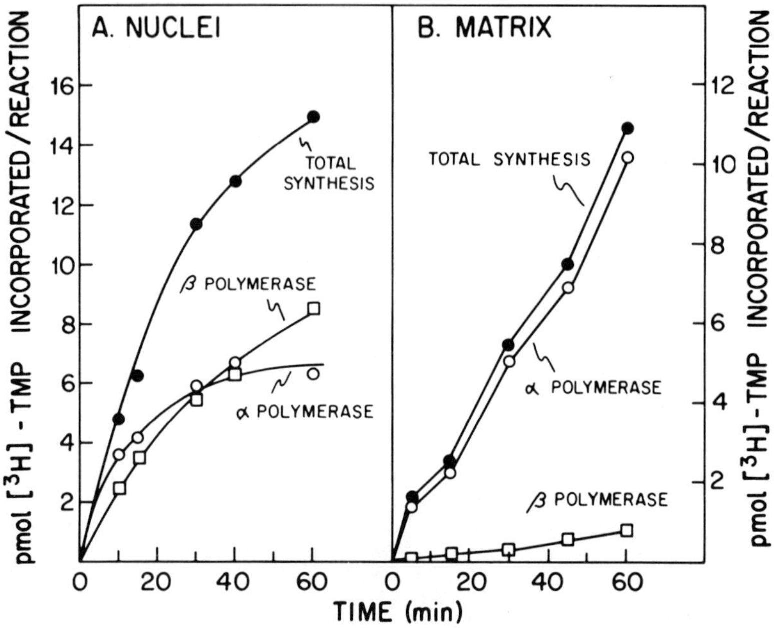

FIGURE 38. Kinetics of endogenous in vitro DNA synthesis. Isolated nuclei (A) and nuclear matrix (B) were prepared from regenerating liver 22 hr after partial hepatectomy and assayed for DNA polymerase activities on the endogenous DNA template: total polymerase activity (●—●); alpha polymerase activity (○—○); beta polymerase activity (□—□). DNA polymerase alpha and beta were discriminated by the inhibition of alpha polymerase with *N*-ethylmaleimide. (From Smith, H.C. and Berezney, R., *Biochemistry*, 21, 6751, 1982. With permission.)

It is not surprising, therefore, that RNA is a major component of the isolated high-salt-resistant nuclear matrix. This was first recognized by Steele and Busch[207,208] who performed an extensive analysis of the RNA in the nuclear matrix. Both ribosomal RNA and hn-RNA (heterogeneous nuclear RNA) were found in significant amounts. Much of the ribosomal RNA was probably localized in the residual nucleoli. Moreover, these pioneering studies were the first to demonstrate hn-RNA in isolated nuclear matrix.[207]

Unfortunately, it was nearly 10 years before Faiferman and Pogo[209] confirmed this finding and further demonstrated that the hn-RNA was not predominantly a component of the intact nuclear envelope. With the development of nuclear matrix isolation technology in the 1970s, several laboratories extended these earlier observations. For example, Miller et al.[16,26] concluded that nearly all the hn-RNA and the nuclear small-molecular-weight RNAs (sn-RNA) are bound to isolated nuclear matrix. Importantly, Miller et al.[16] also found that the association of hn-RNA with the matrix required the presence of protease inhibitors (PMSF or PMSC) during nuclear and matrix isolation. When isolated in the absence of PMSC, most of the rapidly labeled RNA was released in the form of RNP particles. This suggested that RNP particles are natively associated with the nuclear matrix.

Further insight into the association of hn-RNP particles with the nuclear matrix have come from the studies of van Eekelen and van Venrooij.[21] These investigators used *in situ* cross-linking to demonstrate the association of hn-RNA with two major polypeptides (41,500 and 43,000 daltons) in the nuclear matrix. While these represented only minor components of the total matrix proteins, they correspond to major components of isolated RNP particles.[210] Thus, it appears that the hn-RNP particles may be bound to the matrix at specific sites. Importantly, the hn-RNA in isolated nuclear matrices were cross-linked to the same 41,500- and 43,000-dalton polypeptides.[21]

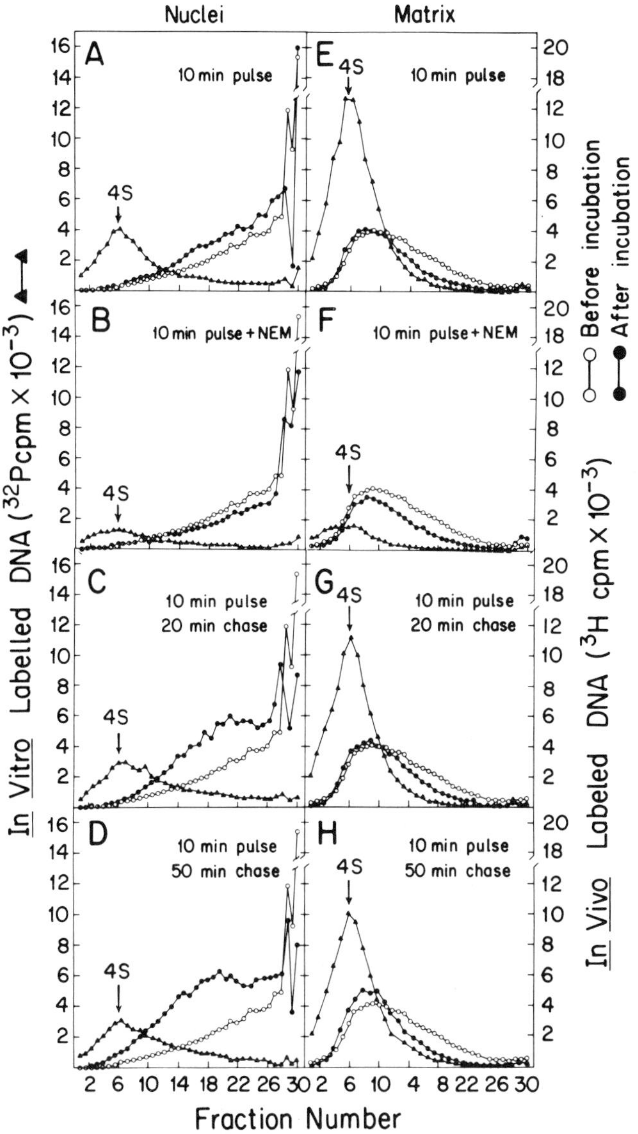

FIGURE 39. Alkaline sucrose gradients in vitro synthesized DNA in nuclei and matrix. Nuclei and nuclear matrix were isolated from actively replicating regenerating rat liver which had been long-term labeled in vivo with [^{3}H]-thymidine. In vitro DNA synthesis was then performed in the isolated nuclei and matrix with or without *N*-ethylmaleimide with [^{32}P]-dCTP as the labeled nucleotide The DNA was purified following the in vitro DNA synthesis and centrifuged on 5 to 20% alkaline sucrose gradients (see Berezney and Buchholtz[8]). The direction of migration is from left to right. In vivo [^{3}H]-labeled DNA before the DNA synthesis incubation (O-O); in vivo [^{3}H]-labeled DNA after the DNA synthesis incubation (●—●); in vitro [^{32}P]-labeled DNA (▲—▲). (From Smith, H. C. and Berezney, R., *Biochemistry*, 21, 6751, 1982. With permission.)

Maundrall et al.[18] reported that only two major polypeptides of hn-RNP particles remain associated with nuclear matrices isolated from duck erythroblasts. One of these, a 43,000-dalton polypeptide, may correspond to the "RNP-matrix binding protein" of van Eekelen and van Venrooij.[21] The second polypeptide (73,000 daltons), comigrated on two-dimensional gels with the presumptive poly(A) binding protein of globin RNP.[18] This suggests that hn-RNP may also be bound to the matrix via the poly(A) binding protein. Consistent with this view, Herman et al.[15,25] reported that a significant proportion of the poly(A) tracts of hn-RNA (approximately two thirds) remain bound to the matrix following RNase digestion of nuclei. This led Herman et al.[15,25] to propose that hn-RNP particles are bound to the nuclear matrix via the poly(A) tails as well as by double-stranded RNA regions. More recent results from Basler and Berezney,[211] however, suggest that while poly(A) tracts may be involved, other single-stranded regions of the RNA molecules are more critical for maintaining a high-salt-resistant binding to the matrix.

B. The Matrix as a Gene-Splicing Apparatus

Progress in the eukaryotic gene splicing field suggests that a multitude of molecular events must be coordinated to perform this complex process. The matrix with its three-dimensional structural organization may potentially serve as an "active work bench" for gene splicing and other RNA processing events. Thus, aside from providing structural pathways for the migration of RNP particles from the sites of transcription to final release into the cytoplasm via the nuclear pore complexes (see Section II and Figure 6), the matrix may provide a functional assembly line whereby RNA processing and intranuclear transport occur in concert.

While the above discussion is highly speculative, future studies employing in vitro nuclear matrix systems should enable a direct test of these possibilities. In this regard Ross et al.[23] and Maundrell et al.[18] have reported that globin gene transcripts are enriched in nuclear matrices isolated from erythroblast cells. This included both primary transcripts and later processed intermediates. Similar results have been recently reported by Mariman et al.[22] for the association of adenoviral transcripts with isolated nuclear matrix. It is also important to note that the nuclear sn-RNAs are tightly bound to the nuclear matrix where they may function in hn-RNA processing events.[48]

The nuclear matrix may also play an important role in ribosomal RNA transcription and processing. For example, Pardoll and Vogelstein[13] reported an approximately sixfold enrichment in ribosomal RNA genes in rat liver nuclear matrix. Presumably, these sequences are tightly bound to the residual nucleoli of the isolated matrix. Wunderlich and associates[24] have reported a comprehensive study of the kinetics of association of ribosomal RNA precursors with the nuclear matrix of *Tetrahymena pyriformis*. Interestingly, their results suggest that earlier transcripts are associated in significantly greater proportions with the high-salt-resistant matrix than later processed molecules. This suggests differences in the RNP-matrix associations during processing. These differences in matrix associations may be functionally significant. For example, a "looser" matrix association might be critical for effective transport of the fully processed transcripts into the cytoplasm. The isolated nucleolar matrix reported in Section V.F is a potentially important in vitro system to further explore ribosomal RNA transcriptional-processing events.

C. Transcription

Herman et al.[15] reported a "steady state" association of total labeled hn-RNA with the nuclear matrix of approximately 60%. This percent of association was independent of time from a 15-sec pulse to a 3-hr chase. Recently Basler and Berezney[211] demonstrated that essentially all of the labeled nuclear RNA ($\geqslant$95%) is bound to the matrix between 1-min and 4-hr pulses, if vanadium nucleoside complexes (a specific RNase inhibitor) are added during matrix isolation. Since most of the labeled RNA in pulses $\leqslant$1 min represent nascent RNA molecules, these results can be explained by one of two possibilities: (1) nascent RNA chains are bound to isolated nuclear matrix via a matrix-bound transcriptional complex, or (2) nascent chains are bound to isolated nuclear matrix via posttranscriptional association of the "nascent RNP" with the nuclear matrix. The inability to clearly distinguish, at the moment, between these two possibilities has resulted in much confusion. Jackson et al.,[19] for example, have interpreted their pulse-labeling experiments as direct evidence for the association of functional transcriptional complexes with the matrix. Clearly, this is only one of two possibilities.

Enrichment in active gene sequences in the matrix-attached DNA fragments has also been reported.[12-14,19] However, these enrichments are generally slight when compared to the recovery of DNA in the matrix and suggest that most of the "active gene sequences" are actually not close to the basal attachment sites (see Section VI.B). For example, Jackson et al.[19] found approximately 2-fold enrichment in total actively transcribed DNA sequences in a matrix containing approximately 5% of total nuclear DNA and Robinson et al.[14] reported a 3.4- to 3.6-fold enrichment of ovalbumin gene sequences in matrices having 12 to 14% of the total DNA. Moreover, this enrichment phenomenon of actively transcribed genes on the matrix is still controversial, since Basler et al.[192] found no enrichment in actively transcribed DNA sequences in rat liver nuclear matrices and Ross et al.[23] have been unable to demonstrate enrichment of globin genes in erythroblast nuclear matrix DNA fragments despite the fact that the globin gene transcript is nearly quantitatively associated with the matrix.

This latter finding suggests that the situation may be more complex than originally meets the eye. It had been assumed by some critics of active gene sequences on the matrix that the phenomenon was an artifact of nascent RNA bound to the matrix "pulling" its own gene sequence along with it.[197,212] The results of Ross et al.[23] demonstrate this is, at least, not a general explanation of the active gene phenomenon.

It is conceivable that the enrichment of active gene sequences on the matrix is related to transcription. If the DNA loops are actually in a state of flux (see Figure 33), then perhaps only slight enrichments would be observed at the matrix-attachment sites. What is needed is a direct test to determine whether functional transcriptional complexes are bound to nuclear matrices during active transcription. Preliminary results from our laboratory suggest that small but reproducible amounts of RNA polymerase II-driven in vitro transcription occur in isolated nuclear matrix. Moreover, there is a severalfold enhancement of activity in cells with higher rates of in vivo transcription.

The virtually instantaneous association of nascent RNA with the nuclear matrix is very intriguing. It supports the concept of a direct and possibly continuous linkage system between the functional chromatin domain of transcription and the RNP domain (see Section II.D). It will be important to decipher the nature of these interactions in future studies.

IX. VIRUSES, HORMONES, CARCINOGENS, AND THE NUCLEAR MATRIX

A number of other properties of isolated nuclear matrix systems are consistent with the postulated role of the matrix as a structural milieu for nuclear function. In particular, several major components which have a profound influence on DNA replication, transcription, and ultimately on cell growth and expression have been found to be tightly bound to the salt-resistant matrix. These include DNA viruses, steroid hormones, and chemical carcinogens. With regard to DNA viruses, the concept is emerging that the key to the success of the virus is its ability to "infect" and to "take over" the replicational and transcriptional apparatus of the nucleus. Central to this thesis are the studies of Pigiet and associates[34] demonstrating the association of a discrete, highly phosphorylated form of T-antigen associated with the nuclear matrix of polyoma-infected cells. Since the T-antigen is a major replicational and transcriptional regulatory protein of the virus, infection of the matrix with T-antigen may represent an important controlling element in viral replication, transcription, and propagation. Evidence has already accumulated that certain viral-specific proteins[32] (see, however, Younghusband and Maundrell[213]), viral DNA,[34,213] and viral-specific RNP particles[22] are tightly bound to the matrix. It also appears that the viral transcripts are anchored to the matrix by the same hn-RNP proteins as host transcripts.[22]

The finding of hormone-responsive high-affinity binding sites for estradiol and dihydrotestosterone in the nuclear matrix of the appropriate target cells[27-30] lends support to the concept that the matrix may be directly involved in transcriptional regulation. In this context, the possible enrichment of ovalbumin gene sequences in the nuclear matrix from oviduct takes on additional significance.[14] Future studies with isolated matrix systems should be helpful in identifying the relationship of the ovalbumin gene DNA binding sites and the estradiol binding sites on the matrix. It is anticipated, for example, that at least some of the estradiol binding sites are also ovalbumin gene DNA binding sites. Moreover, it is conceivable that appropriate in vitro systems can be developed to elucidate the functional events of hormone binding and gene activation.

Several investigations have shown that the carcinogenic polycyclic aromatic hydrocarbon, benzo [α] pyrene, is preferentially bound to the nuclear matrix following exposure to the carcinogen in intact animals,[36] minced tissue systems,[35,36] and in vitro in isolated nuclei.[37] In all cases, approximately 50% of the total covalent-bound benzo [α] pyrene derivative in isolated nuclei was recovered in the high-salt matrix. Significantly, the bulk chromatin fraction obtained by the low-salt extractions of Berezney and Coffey[86] (see Section III.C) was depleted in bound carcinogen.[37] Ueda and associates[214] further demonstrated that the covalent binding of benzo [α] pyrene to the nuclear matrix is substantially inhibited by presumptive anticarcinogenic vitamin A compounds, such as retinol. In contrast, benzo [α] pyrene bound to the bulk chromatin was unaffected.

Evidence is also accumulating that the catalytic components cytochrome P-448 and epoxide hydrolase, which are responsible for the metabolic conversion of benzo [α] pyrene to the active carcinogenic form, are found in the nuclear interior, in addition to being components of the endoplasmic reticulum and nuclear envelope.[215,216] While it is presently unclear whether the cytochrome P-448 and epoxide hydrolase are associated with the nuclear matrix, chromatin, or both, Bresnick et al.[217] have recently demonstrated both of these components in isolated nucleoli, which generally are a constituent of the isolated nuclear matrix. This finding is consistent with earlier immunocytochemical localization which revealed major deposits of stain over regions resembling nucleoli in size and distribution.[216] Interestingly, Bresnick et al.[216] also observed numerous microdeposits of stain throughout the nuclear interior. This stippled pattern of staining might be indicative of discrete domains of cytochrome P-448 over the chromatin and/or internal matrix regions.

It is intriguing to consider the nuclear matrix as an integrative structural machinery for the initial binding, metabolic conversion, and ultimate expression of benzo [α] pyrene and potentially, other carcinogens and drugs. In this regard, it will be important in future studies to determine the precise binding sites for benzo [α] pyrene. The carcinogen, for example, may bind at or close to the matrix DNA attachment sites where it could potentially affect replicational and transcriptional processes as well as sequences on the DNA loops.

As previously emphasized, the nuclear matrix is a potentially important system for studying cancer.[42] As a step in this direction, our laboratory has initially characterized nuclear matrices isolated from Zajdela hepatoma cells.[97]

X. NUCLEAR FUNCTION AND THE DYNAMIC MATRIX

An earlier review stressed the dynamic nature of the nuclear matrix.[42] These dynamic aspects may be a property of matrix function. Thus, the *raison d'etre* of the matrix may be to perform and perhaps to orchestrate the dynamics of nuclear functions. This is, in turn, expressed in the apparent transient nature of the DNA loops, the dynamic association of replicational complexes, and the flux of RNA from transcriptional sites of the nucleus to final release into the cytoplasm for genetic expression. All of the above are exciting possibilities worthy of future exploration.

ACKNOWLEDGMENTS

I am extremely grateful to Dr. Joseph Basler, Dr. Harold C. Smith, Ms. Linda A. Buchholtz, and Mr. Alan J. Siegel who performed many of the experiments reported from our laboratory in this review. Special thanks to Dr. Lubomir S. Hnilica for his kind assistance above and beyond the call of duty. The preparation of this chapter was aided by U.S.P.H.S. grant GM 23922.

REFERENCES

1. **Dijkwel, P. A., Mullenders, L. H. F., and Wanka, F.,** Analysis of the attachment of replicating DNA to a nuclear matrix in mammalian interphase nuclei, *Nucleic Acids Res.*, 6, 219, 1979.
2. **Berezney, R. and Buchholtz, L. A.,** Isolation and characterization of rat liver nuclear matrices containing high molecular weight deoxyribonucleic acid, *Biochemistry*, 20, 4995, 1981.
3. **Razin, S. V., Chernokhvostov, V. V., Roodyn, A. V., Zbarsky, I. B., and Georgiev, G. P.,** Proteins tightly bound to DNA in the regions of DNA attachment to the skeletal structures of interphase nuclei and metaphase chromosomes, *Cell*, 27, 65, 1981.
4. **Berezney, R. and Coffey, D. S.,** Nuclear protein matrix: association with newly synthesized DNA, *Science*, 189, 291, 1975.
5. **Berezney, R. and Coffey, D. S.,** The nuclear protein matrix: isolation, structure and functions, *Adv. Enzyme Regul.*, 14, 63, 1976.
6. **McCready, S. J., Godwin, J., Mason, D. W., Brazell, I. A., and Cook, P. R.,** DNA is replicated at the nuclear cage, *J. Cell Sci.*, 46, 365, 1980.
7. **Pardoll, D. M., Vogelstein, B., and Coffey, D. S.,** A fixed site of DNA replication in eucaryotic cells, *Cell*, 19, 527, 1980.
8. **Berezney, R. and Buchholtz, L. A.,** Dynamic association of replicating DNA fragments with the nuclear matrix of regenerating liver, *Exp. Cell Res.*, 132, 1, 1981.
9. **Smith, H. C. and Berezney, R.,** DNA polymerase α is tightly bound to the nuclear matrix of actively replicating liver, *Biochem. Biophys. Res. Commun.*, 97, 1541, 1980.
10. **Berezney, R. and Smith, H. C.,** In vitro DNA synthesis in isolated nuclear matrix, *J. Cell Biol.*, 91, 67a, 1981.
11. **Smith, H. C. and Berezney, R.,** Nuclear matrix-bound DNA synthesis: an *in vitro* system, *Biochemistry*, 21, 6751, 1982.
12. **Nelkin, B. D., Pardoll, D. M., and Vogelstein, B.,** Localization of SV-40 genes within supercoiled loop domains, *Nucleic Acids Res.*, 8, 5623, 1980.
13. **Pardoll, D. M. and Vogelstein, B.,** Sequence analysis of nuclear matrix associated DNA from rat liver, *Exp. Cell Res.*, 128, 466, 1980.
14. **Robinson, S. I., Nelkin, B. D., and Vogelstein, B.,** The ovalbumin gene is associated with the nuclear matrix of chicken oviduct cells, *Cell*, 28, 99, 1982.
15. **Herman, R., Weymouth, L., and Penman, S.,** Heterogeneous nuclear RNA-protein fibers in chromatin-depleted nuclei, *J. Cell Biol.*, 78, 663, 1978.
16. **Miller, T. E., Huang, C. Y., and Pogo, A. O.,** Rat liver nuclear skeleton and ribonucleoprotein complexes containing HnRNA, *J. Cell Biol.*, 76, 675, 1978.
17. **Long, B. H., Huang, C. Y., and Pogo, A. O.,** Isolation and characterization of nuclear matrix in Friend erythroleukemia cells: chromatin and hnRNA interactions with the nuclear matrix, *Cell*, 18, 1079, 1979.
18. **Maundrell, K., Maxwell, E. S., Puvion, E., and Scherrer, K.,** The nuclear matrix of duck erythroblasts is associated with globin mRNA coding sequences but not with the major proteins of 40S nuclear RNP, *Exp. Cell Res.*, 136, 435, 1981.
19. **Jackson, D. A., McCready, S. J., and Cook, P. R.,** RNA is synthesized at the nuclear cage, *Nature (London)*, 292, 552, 1981.
20. **Jackson, D. A., Caton, A. J., McCready, S. J., and Cook, P. R.,** Influenza virus RNA is synthesized at fixed sites in the nucleus, *Nature (London)*, 296, 366, 1982.

21. **van Eekelen, C. A. G. and van Venrooij, W. J.,** hnRNA and its attachment to a nuclear protein matrix, *J. Cell Biol.*, 88, 554, 1981.
22. **Mariman, E. C. M., van Eekelen, C. A. G., Reinders, R. J., Berns, A. J. M., and van Venrooij, W. J.,** Adenoviral heterogeneous nuclear RNA is associated with the host nuclear matrix during splicing, *J. Mol. Biol.*, 154, 103, 1982.
23. **Ross, D. A., Yen, R. W., and Chae, C. B.,** Association of globin ribonucleic acid and its precursors with the chicken erythroblasts nuclear matrix, *Biochemistry*, 21, 764, 1982.
24. **Herlan, G., Eckert, W. A., Kaffenberger, W., and Wunderlich, F.,** Isolation and characterization of an RNA-containing nuclear matrix from *Tetrahymena macronuclei*, *Biochemistry*, 18, 1782, 1979.
25. **Herman, R., Zieve, G., Williams, J., Lenk, R., and Penman, S.,** Cellular skeletons and RNA messages, *Prog. Nucleic Acid Res. Mol. Biol.*, 19, 379, 1976.
26. **Miller, T. E., Huang, C. Y., and Pogo, A. O.,** Rat liver nuclear skeleton and small molecular weight RNA species, *J. Cell Biol.*, 76, 692, 1978.
27. **Barrack, E. R., Hawkins, E. F., Allen, S. L., Hicks, L. L., and Coffey, D. S.,** Concepts related to salt resistant estradiol receptors in rat uterine nuclei: nuclear matrix, *Biochem. Biophys. Res. Commun.*, 79, 829, 1977.
28. **Barrack, E. R. and Coffey, D. S.,** The specific abinding of estrogens and androgens to the nuclear matrix of sex hormone responsive tissues, *J. Biol. Chem.*, 255, 7265, 1980.
29. **Agutter, P. S. and Birchall, K.,** Functional differences between mammalian nuclear protein matrices and pore lamina complex laminae, *Exp. Cell Res.*, 124, 453, 1979.
30. **Clark, J. H. and Markaverich, B. M.,** Heterogeneity of estradiol-binding sites and the nuclear matrix, in *The Nuclear Envelope and the Nuclear Matrix*, Maul, G. G., Ed., Alan R. Liss, New York, 1982, 259.
31. **Vollmer, G., Haase, A., and Eisele, K.,** Androgen receptor complex binding in murine skeletal muscle nuclei, *Biochem. Biophys. Res. Commun.*, 105, 1554, 1982.
32. **Hodge, L. D., Mancini, P., Davis, F. M., and Heywood, P.,** Nuclear matrix of HeLa S3 cells. Polypeptide composition during adenovirus infection and phases of the cell cycle, *J. Cell Biol.*, 72, 194, 1977.
33. **Deppert, W.,** Simian virus 40 (SV40)-specific proteins associated with the nuclear matrix isolated from adenovirus type 2-SV40 hybrid virus-infected HeLa cells carry SV-40 U-antigen determinants, *J. Virol.*, 26, 165. 1978.
34. **Buckler-White, A. J., Humphrey, G. W., and Pigiet, V.,** Association of polyoma T antigen and DNA with the nuclear matrix from lytically infected 3T6 cells, *Cell*, 22, 37, 1980.
35. **Blazsek, I., Vauhkonen, M., and Hemminki, K.,** Binding of benzo(α)pyrene into lung and thymocyte nuclear fractions, *Res. Commun. Chem. Pathol. Pharmacol.*, 23, 611, 1979.
36. **Hemminki, K. and Vainio, H.,** Preferential binding of benzo(α)pyrene into nuclear matrix fraction, *Cancer Lett.*, 6, 167, 1979.
37. **Ueyama, H., Matsuura, T., Nomi, S., Nakayasu, H., and Ueda, K.,** Binding of benzo(α)pyrene to rat liver nuclear matrix, *Life Sci.*, 29, 655, 1981.
38. **Wunderlich, F. and Herlan, G.,** A reversibly contractile nuclear matrix, *J. Cell Biol.*, 73, 271, 1977.
39. **Sinibaldi, R. M. and Morris, P. W.,** Putative function of *Drosophila melanogaster* heat shock proteins in the nucleoskeleton, *J. Biol. Chem.*, 256, 10735, 1981.
40. **Allen S. L., Berezney, R., and Coffey, D. S.,** Phosphorylation of nuclear matrix proteins in isolated regenerating rat liver nuclei, *Biochem. Biophys. Res. Commun.*, 75, 111, 1977.
41. **Ševaljević , L., Poznonović, G., Petrović, M., and Krtolica, K.,** Protein kinase activity and lectin binding ability of the nuclear matrix, *Biochem. Int.*, 2, 77, 1981.
42. **Berezney, R.,** Dynamic properties of the nuclear matrix, in *The Cell Nucleus*, Vol. 7, Busch, H., Ed., Academic Press, New York, 1979, 413.
43. **Shaper, J. H., Pardoll, D. M., Kaufmann, S. H., Barrack, E. R., Vogelstein, B., and Coffey, D. S.,** The relationship of the nuclear matrix to cellular structure and function, *Adv. Enzyme Regul.*, 17, 213, 1979.
44. **Agutter, P. S. and Richardson, J. C. W.,** Nuclear non-chromatin proteinaceous structures: their role in the organization and function of the interphase nucleus, *J. Cell Sci.*, 44, 395, 1980.
45. **Berezney, R.,** The nuclear matrix: a structural milieu for the intranuclear attachment and replication of eucaryotic DNA, in *International Cell Biology (1980—1981)*, Schweiger, H. G., Ed., Springer-Verlag, Berlin. 1981, 214.
46. **Barrack, E. R., Hawkins, E. F., and Coffey, D. S.,** The specific binding of estradiol to the nuclear matrix, in *Steroid Hormone Receptor Systems*, Leavitt, W. W. and Clark, J. H., Eds., Plenum Press, New York, 1979, 47.
47. **Comings, D. E. and Peters, K. E.,** Two-dimensional gel electrophoresis of nuclear proteins, in *The Cell Nucleus*, Vol. 9, Busch, H.,Ed., Academic Press, New York, 1981, 89.
48. **Pogo, A. O.,** Heterogeneous nuclear RNA-protein complexes and nuclear matrix, in *The Cell Nucleus*, Vol. 8, Busch, H., Ed., Academic Press, New York, 1981, 331.

49. **Bernhard, W. and Granbonlan, N.,** The fine structure of the cancer cell nucleus, *Exp. Cell Res.*, Suppl. 9, 19, 1963.
50. **Swift, H.,** Cytochemical studies on nuclear fine structure, *Exp. Cell Res.*, Suppl. 9, 54, 1963.
51. **Fawcett, D. W.,** *An Atlas of Fine Structure. The Cell Its Organelles and Inclusions*, W. B. Saunders, Philadelphia, 1966, 3.
52. **Bernhard, W.,** A new staining procedure for electron microscopical cytology, *J. Ultrastruct. Res.*, 27, 250, 1969.
53. **Monneron, A. and Bernhard, W.,** Fine structural organization of the interphase nucleus in some mammalian cells, *J. Ultrastruct. Res.*, 27, 266, 1969.
54. **Wunderlich, F., Berezney, R., and Kleinig, H.,** The nuclear envelope: an interdisciplinary analysis of its morphology, composition and functions, in *Biological Membranes*, Vol. 3, Chapman, D. and Wallach, D. F. H., Eds., Academic Press, New York, 1976, 241.
55. **Franke, W. W. and Falk, H.,** Appearance of nuclear pore complexes after Bernhard's staining procedure, *Histochemie*, 24, 266, 1970.
56. **Puvion, E. and Bernhard, W.,** Ribonucleoprotein components in liver cell nuclei as visualized by cryoultramicrotomy, *J. Cell Biol.*, 67, 200, 1975.
57. **Puvion, E.,** personal communication, 1982.
58. **Frenster, J. H.,** Ultrastructure and function of heterochromatin and euchromatin, in *The Cell Nucleus*, Vol. 1, Busch, H., Ed., Academic Press, New York, 1974, 565.
59. **Poznanović, G. and Ševaljević, L.,** The isolation and characterisation of the nuclear matrix from sea urchin embryos, *Cell Biol. Int. Rep.*, 4, 701, 1980.
60. **Ševaljević L. and Petrović, M.,** On the similarity between the nuclear network and chromatin nonhistone proteins of sea urchin embryos, *Cell Mol. Biol.*, 27, 147, 1981.
61. **Puvion, E. and Moyne, G.,** *In situ* localization of RNA structures, in *The Cell Nucleus*, Vol. 8, Busch, H., Ed., Academic Press, New York, 1981, 59.
62. **Fakan, S. and Bernhard, W.,** Localization of rapidly and slowly labeled nuclear RNA as visualized by high resolution autoradiography, *Exp. Cell Res.*, 67, 129, 1971.
63. **Fakan, S. and Bernhard, W.,** Nuclear labeling after prolonged ^{3}H-uridine incorporation as visualized by high resolution autoradiography, *Exp. Cell Res.*, 79, 431, 1973.
64. **Nash, R. E., Puvion, E., and Bernhard, W.,** Perichromatinic fibrils as components of rapidly labeled extranucleolar RNA, *J. Ultrastruct. Res.*, 53, 395, 1975.
65. **Fakan, S., Puvion, E., and Spohr, G.,** Localization and characterization of newly synthesized nuclear RNA in isolated rat hepatocytes, *Exp. Cell Res.*, 99, 155, 1976.
66. **Graham, C. F., Arms, K., and Gurdon, J. B.,** The induction of DNA synthesis by egg cytoplasm, *Dev. Biol.*, 14, 349, 1966.
67. **Harris, H.,** *Nucleus and Cytoplasm*, Clarendon Press, Oxford, 1968.
68. **Harris, H.,** *Cell Fusion*, Clarendon Press, Oxford, 1970.
69. **Maul, G. G., Maul, H. M., Scogna, J. E., Lieberman, M. W., Stein, G. S., Hsu, B. Y. I., and Borun, T. W.,** Time sequence of nuclear pore formation in phytohemagglutinin-stimulated lymphocytes and HeLa cells during the cell cycle, *J. Cell Biol.*, 55, 433, 1972.
70. **Brasch, K. and Sinclair, G. D.,** The organization, composition and matrix of hepatocyte nuclei exposed to α-amanitin, *Virchows Arch. B Cell Pathol.*, 27, 193, 1978.
71. **Ghosh, S., Paweletz, N., and Ghosh, I.,** Cytological identification and characterization of the nuclear matrix, *Exp. Cell Res.*, 111, 363, 1978.
72. **Ghosh, S. and Ghosh, I.,** Ultrastructural characterization of collapsed interchromatin material in α-amanitin treated cells, *Indian J. Exp. Biol.*, 18, 1082, 1980.
73. **Rizzo, P. J. and Nooden, L. D.,** Chromosomal proteins in the dinoflagellate alga *Gyrodinium cohnii*, *Science*, 176, 796, 1972.
74. **Rizzo, P. J. and Nooden, L. D.,** Isolation and chemical composition of dinoflagellate nuclei, *J. Protozool.*, 20, 666, 1973.
75. **Kubai, D. F. and Ris, H.,** Division in the dinoflagellate *Gyrodinium cohnii* (Schiller). A new type of nuclear reproduction *J. Cell Biol.*, 40, 508, 1969.
76. **Livolant, F. and Bouligand, Y.,** New observations on the twisted arrangement of dinoflagellate chromosomes, *Chromosoma*, 68, 21, 1978.
77. **Livolant, F. and Bouligand, Y.,** Double helical arrangement of spread dinoflagellate chromosomes, *Chromosoma*, 80, 97, 1980.
78. **Mirsky, A. E. and Ris, H.,** The composition and structure of isolated chromosomes, *J. Gen. Physiol.*, 34, 475, 1951.
79. **Mirsky, A. E. and Osawa, S.,** The interphase nucleus, in *The Cell*, Vol. 2, Brachet, J. and Mirsky, A. E., Eds., Academic Press, New York, 1961, 677.
80. **Georgiev, G. P. and Chentsov, Y. S.,** On the structure of the cell nucleus, Experimental electron microscopic studies on the isolated nucleus, *Proc. Acad. Sci. U.S.S.R.*, 132, 199, 1960.

81. **Georgiev, G. P. and Chentsov, Y. S.,** On the structural organization of nucleolo-chromosomal ribonucleoproteins, *Exp. Cell Res.,* 27, 570, 1962.
82. **Zbarsky, I. B., Dmitrieva, N. P., and Yermolayeva, L. P.,** On the structure of tumor cell nuclei, *Exp. Cell Res.,* 27, 573, 1962.
83. **Smetana, K., Steele, W. J., and Busch, H.,** A nuclear ribonucleoprotein network, *Exp. Cell Res.,* 31, 198, 1963.
84. **Narayan, K. S., Steele, W. J., Smetana, K., and Busch, H.,** Ultrastructural aspects of the ribonucleoprotein network in nuclei of Walker tumor and rat livers, *Exp. Cell Res.,* 46, 65, 1967.
85. **Berezney, R. and Coffey, D. S.,** Identification of a nuclear protein matrix, *Biochem. Biophys. Res. Commun.,* 60, 1410, 1974.
86. **Berezney, R. and Coffey, D. S.,** Nuclear Matrix. Isolation and characterization of a framework structure from rat liver nuclei, *J. Cell Biol.,* 73, 616, 1977.
87. **Hewish, D. R. and Burgoyne, L. A.,** Chromatin sub-structure. The digestion of chromatin DNA at regularly spaced sites by a nuclear deoxyribonuclease, *Biochem, Biophys. Res. Commun.,* 52, 504, 1973.
88. **Burgoyne, L. A. and Hewish, D. R.,** The regular substructure of mammalian nuclei and nuclear Ca-Mg endonuclease, in *The Cell Nucleus,* Vol. 4, Busch, H., Ed., Academic Press, New York, 1978, 48.
89. **Smith, H. C. and Berezney, R.,** unpublished data, 1983.
90. **Bouvier, D., Dupuy-Coin, A.-M., Bouteille, M., and Moens, P.,** Three-dimensional electron microscopy of the nuclear matrix components of HeLa cells, *Biol. Cellulaire,* 39, 121, 1980.
91. **Comings, D. E. and Okada, T. A.,** Nuclear proteins. III. The fibrillar nature of the nuclear matrix, *Exp. Cell Res.,* 103, 341, 1976.
92. **Kuzima, S., Buddyaeva, T., Troitskaya, L., and Zbarsky, I.,** Characterization and fractionation of rat liver nuclear matrix, *Eur. J. Cell Biol.,* 25, 225, 1981.
93. **Matsura, T., Ueyama, H., Nakayasu, H., and Ueda, K.,** Isolation and characterization of the nuclear matrix from rat liver nuclei, *Cell Struct. Function,* 6, 79, 1981.
94. **Herlan, G. and Wunderlich, F.,** Isolation of a nuclear protein matrix from *Tetrahymena macronuclei, Cytobiologie,* 13, 291, 1976.
95. **Mitchelson, K. R., Bekers, A. G. M., and Wanka, F.,** Isolation of a residual protein structure from nuclei of the myxomycete *Physarum polycephalum, J. Cell Sci.,* 39, 247, 1979.
96. **Snead, H. W., McDonald, T. F., Baker, M. D., and Lanclos, K. D.,** Isolation and characterization of the nuclear matrix from the male *Xenopus laevis* following estrogen administration: kinetics of ^{3}H-uridine incorporation, *J. Supramol. Struct.,* 12, 471, 1979.
97. **Berezney, R., Basler, J., Hughes, B. B., and Kaplan, S. C.,** Isolation and characterization of the nuclear matrix from Zajdela ascites hepatoma cells, *Cancer Res.,* 39, 3031, 1979.
98. **Wolfe, J.,** A possible skeletal substructure of the macronucleus of *Tetrahymena, J. Cell Biol.,* 84, 160, 1980.
99. **Zbarsky, I. B.,** Nuclear skeleton structures in some normal and tumor cells, *Mol. Biol. Rep.,* 7, 139, 1981.
100. **Nakayasu, H. and Ueda, K.,** Isolation and characterization of bovine lymphocyte nuclear matrix, *Cell Struct. Function,* 6, 181, 1981.
101. **Fisher, P. A., Berrios, M., and Blobel, G.,** Isolation and characterization of a proteinaceous subnuclear fraction composed of nuclear matrix, peripheral lamina, and nuclear pore complexes from embryos of *Drosophila melanogaster, J. Cell Biol.,* 92, 674, 1982.
102. **Smith, H. C. and Berezney, R.,** unpublished data, 1983.
103. **Vogelstein, B., Nelkin, B., Pardoll, D., and Hunt, B. F.,** The further evidence for physiological association of newly replicated DNA with the nuclear matrix, in *The Nuclear Envelope and the Nuclear Matrix,* Maul, G. G., Ed., Alan R. Liss, New York, 1982, 169.
104. **Cook, P. R. and Brazell, I. A.,** Supercoils in human DNA, *J. Cell Sci.,* 19, 261, 1975.
105. **Vogelstein, B., Pardoll, D. M., and Coffey, D. S.,** Supercoiled loops and eucaryotic DNA replication, *Cell,* 22, 79, 1980.
106. **Newman, D. and Berezney, R.,** unpublished data, 1983.
107. **Riley, D. E., Keller, J. M., and Byers, B.,** The isolation and characterization of nuclear ghosts from cultured HeLa cells, *Biochemistry,* 14, 3005, 1975.
108. **Riley, D. E. and Keller, J. M.,** The polypeptide composition and ultrastructure of nuclear ghosts isolated from mammalian cells, *Biochim. Biophys. Acta,* 444, 899, 1976.
109. **Keller, J. M., and Riley, D. E.,** Nuclear ghosts: a nonmembranous structural component of mammalian cell nuclei, *Science,* 193, 399, 1976.
110. **Riley, D. E. and Keller, J. M.,** Cell cycle-dependent changes in nonmembranous nuclear ghosts from HeLa cells, *J. Cell Sci.,* 29, 129, 1978.
111. **Detke, S. and Keller, J. M.,** Comparison of the proteins present in HeLa interphase nucleoskeletons and metaphase chromosome scaffolds, *J. Biol. Chem.,* 257, 3905, 1982.
112. **Paulson, J. R. and Laemmli, U. K.,** The structure of histone-depleted metaphase chromosomes, *Cell,* 12, 817, 1977.

113. **Adolph, K. W., Cheng, S. M., Laemmli, U. K.,** Role of nonhistone proteins in metaphase chromosome structure, *Cell,* 12, 805, 1977.
114. **Adolph, K. W., Cheng, S. M., Paulson, J. R., and Laemmli, U. K.,** Isolation of protein scaffold from mitotic HeLa cell chromosomes, *Proc. Natl. Acad. Sci. U.S.A.,* 74, 4937, 1977.
115. **Matsui, S., Antoniades, G., Basler, J., Berezney, R., and Sandberg, A. A.,** Nuclear matrix versus chromosome scaffold: structural basis of eucaryotic chromatin, *J. Cell Biol.,* 91, 60a, 1981.
116. **Herlan, G., Quevedo, R., and Wunderlich, F.,** Structural transformation of the nuclear matrix *in situ, Exp. Cell. Res.,* 115, 103, 1978.
117. **Bekers, A. G. M., Gijzev, H. J., Taalman, R. D. F. M., and Wanka, F.,** Ultrastructure of the nuclear matrix from *Physarum polycephalum* during the mitotic cycle, *J. Ultrastruct. Res.,* 75, 352, 1981.
118. **Capco, D. G., Wan, K. M., and Penman, S.,** The nuclear matrix: three-dimensional architecture and protein composition, *Cell,* 29, 847, 1982.
119. **Adolph, K. W.,** Organization of chromosomes in HeLa cells: isolation of histone-depleted nuclei and nuclear scaffolds, *J. Cell Sci.,* 42, 291, 1980.
120. **Busch, H. and Daskel, Y.,** Methods of isolation of nuclei and nucleoli, *Methods Cell Biol.,* 16, 1, 1977.
121. **Franke, W. W.,** Isolated nuclear membrances, *J. Cell Biol.,* 31, 619, 1966.
122. **Zbarsky, I. B., Perevoshchikova, K. A., Delktorskaya, L. N., and Delektorsky, V. V.,** Isolation and biochemical characteristics of the nuclear envelope, *Nature (London),* 221, 257, 1969.
123. **Kashnig, D. M. and Kasper, C. B.,** Isolation, morphology and composition of the nuclear membrane from rat liver, *J. Biol. Chem.,* 244, 3786, 1969.
124. **Franke, W. W., Deumling, B., Ermen, B., Jarasch, E.-D., and Kleinig, H.,** Nuclear membranes from mammalian liver, I. Isolation procedure and general characterization, *J. Cell Biol.,* 46, 379, 1970.
125. **Agutter, P. S.,** The isolation of the envelope of rat liver nuclei, *Biochim. Biophys. Acta,* 255, 397, 1972.
126. **Stavy, R., Ben-Shaal, Y., and Galun, E.,** Nuclear envelope isolation in peas, *Biochim, Biophys. Acta,* 323, 167, 1973.
127. **Mentre, P.,** Etude des pores nucleaires sur des noyaux et des membranes nucleaires isoles, in *6th Int. Cong. Electron Microscopy,* Vol. 2, Uyeda, R., Ed., Maruzen Co., Tokyo, 1966, 347.
128. **Bornens, M.,** Methode d'isolement de la membrane nucleaire de foie de rat, *C. R. Acad. Sci. Paris,* 266, 596, 1968.
129. **Berezney, R., Funk, L. K., and Crane, F. L.,** The isolation of nuclear membrane from a large scale preparation of bovine liver nuclei, *Biochim. Biophys. Acta,* 203, 531, 1970.
130. **Berezney, R., Macaulay, L. K., and Crane, F. L.,** The purification and biochemical characterization of bovine liver nuclear membranes, *J. Biol. Chem.,* 247, 5549, 1972.
131. **Berezney, R.,** The large scale isolation of nuclear membranes from bovine liver, *Methods Cell Biol.,* 8, 205, 1974.
132. **Price, M. R., Harris, J. R., and Baldwin, R. W.,** A method for the isolation and purification of normal rat liver and hepatoma nuclear "ghosts" by zonal centrifugation, *J. Ultrastruct. Res.,* 40, 178, 1972.
133. **Monneron, A., Blobel, G., and Palade, G. E.,** Fractionation of the nucleus by divalent cations. Isolation of nuclear membrances, *J. Cell Biol.,* 55, 104, 1972.
134. **Kay, R. R., Fraser, D., and Johnston, I. R.,** A method for the rapid isolation of nuclear membranes from rat liver, *Eur. J. Biochem.,* 30, 145, 1972.
135. **Aaronson, R. P. and Blobel, G.,** Isolation of nuclear pore complexes in association with a lamina, *Proc. Natl. Acad. Sci. U.S.A.,* 72, 1007, 1975.
136. **Berezney, R.,** Effect of protease inhibitors on matrix proteins and the association of replicating DNA, *Exp. Cell Res.,* 123, 411, 1979.
137. **van Venrooij, W. J., van Eekelan, C. A. G., Mariman, E. C. M., and Reinders, R. J.,** On the binding of host and viral RNA to a nuclear matrix, in *The Nuclear Envelope and the Nuclear Matrix,* Maul, G. G., Ed., Alan R. Liss, New York, 1982, 235.
138. **Berezney, R.,** unpublished data, 1983.
139. **Kaufmann, S. H., Coffey, D. S., and Shaper, J. H.,** Considerations in the isolation of rat liver nuclear matrix, nuclear envelope, and pore complex lamina, *Exp. Cell Res.,* 132, 105, 1981.
140. **Buchholtz, L. A. and Berezney, R.,** unpublished data, 1983.
141. **Fisher, P. A., Berrios, M., and Blobel, G.,** Characterization of an isolated nuclear matrix fraction obtained from embryos of *Drosophila melanogaster,* in *The Nuclear Envelope and the Nuclear Matrix,* Maul, G. G., Ed., Alan R. Liss, New York, 1982, 145.
142. **Penman, S., Fulton, A., Capco, D., Ben Ze'ev, A., Wittelsberger, S., and Tse, C. F.,** Cytoplasmic and nuclear architecture in cells and tissue: form, functions and mode of assembly, *Cold Spring Harbor Symp. Quant. Biol.,* 46, 1013, 1982.
143. **Zbarsky, I. B. and Georgiev, G. P.,** Cytological characteristics of protein and nucleoprotein fractions of cell nuclei, *Biochim, Biophys. Acta,* 32, 301, 1959.
144. **Steele, W. J. and Busch, H.,** Studies on the nuclear proteins of the Walker tumor and the liver, *Cancer Res.,* 23, 1153, 1963.

145. **Holtzman, E., Smith, I., and Penman, S.,** Electron microscopic studies of detergent-treated HeLa cell nuclei, *J. Mol. Biol.*, 17, 131, 1966.
146. **Colman, A. and Cook, P. R.,** Transcription of superhelical DNA from cell nuclei, *Eur. J. Biochem.*, 76, 63, 1977.
147. **Shelton, K. R., Guthrie, V. H., and Cochran, D. L.,** On the variation of the major nuclear envelope (lamina) polypeptides, *Biochem. Biophys. Res. Commun.*, 93, 867, 1980.
148. **Buchholtz, L. A. and Berezney, R.,** unpublished data, 1983.
149. **Cobbs, C. S., Jr. and Shelton, K. R.,** Major oligomeric structural proteins of the HeLa nucleus, *Arch. Biochem. Biophys.*, 189, 323, 1978.
150. **Gerace, L., Blum, A., and Blobel, G.,** Immunocytochemical localization of the major polypeptides of the nuclear pore complex lamina fraction: interphase and mitotic distribution, *J. Cell Biol.*, 79, 546, 1978.
151. **Krohne, G., Franke, W. W., Ely, S., D'Arcy, A., and Jost, E.,** Localization of a nuclear envelope-associated protein by indirect immunofluorescence microscopy using antibodies against a major polypeptide from rat liver fractions enriched in nuclear envelope-associated material, *Cytobiologie*, 18, 22, 1978.
152. **Ely, S., D'Arcy, A., and Jost, E.,** Interaction of antibodies against nuclear envelope-associated proteins from rat liver nuclei with rodent and human cells, *Exp. Cell Res.*, 116, 325, 1978.
153. **Stick, R. and Hausen, P.,** Immunological analysis of nuclear lamina proteins, *Chromosoma*, 80, 219, 1980.
154. **Berezney, R.,** Fractionation of the nuclear matrix. I. Partial separation into matrix protein fibrils and a residual ribonucleoprotein fraction, *J. Cell Biol.*, 85, 641, 1980.
155. **Gerace, L. and Blobel, G.,** The nuclear envelope lamina is reversibly depolymerized during mitosis, *Cell*, 19, 277, 1980.
156. **Peters, K. E. and Comings, D. E.,** Two-dimensional gel electrophoresis of rat liver nuclear washes, nuclear matrix and hnRNA proteins, *J. Cell Biol.*, 86, 135, 1980.
157. **Newmeyer, D. D., Ohlsson-Wilhelm, B. M., Buchholtz, L. A., Brewster, J. M., and Berezney, R.,** unpublished data, 1983.
158. **Werner, D., Zimmermann, H.-P., Rauterberg, E., and Spalinger, J.,** Antibodies to the most tightly bound proteins in eucaryotic DNA, *Exp. Cell Res.*, 133, 149, 1981.
159. **Dwyer, N. and Blobel, G.,** A modified procedure for the isolation of a pore complex-lamina fraction from rat liver nuclei, *J. Cell Biol.*, 70, 581, 1976.
160. **Comings, D. E. and Wallack, A.,** DNA-binding properties of nuclear matrix proteins, *J. Cell Sci.*, 34, 233, 1978.
161. **Bowen, B., Steinberg, J., Laemmli, U. K., and Weintraub, H.,** The detection of DNA-binding proteins by protein blotting, *Nucleic Acids Res.*, 8, 1, 1980.
162. **Grebanier, A. E. and Pogo, A. O.,** Cross-linking of proteins in nuclei and DNA-depleted nuclei from Friend erythroleukemia cells, *Cell*, 18, 1091, 1979.
163. **Buchholtz, L. A. and Berezney, R.,** unpublished data, 1983.
164. **Gates, D. M. and Bekhor, I.,** DNA binding activity of tightly-bound nonhistone chromosomal protein in chicken liver chromatin, *Nucleic Acids Res.*, 6, 3411, 1979.
165. **Wakabayashi, K. and Hnilica, L. S.,** The immunospecificity of nonhistone protein complexes with DNA, *Nature New Biol.*, 242, 153, 1973.
166. **Fujitami, H., Chiu, J.-C., and Hnilica, L. S.,** Purification of nuclear antigens in Novikoff hepatoma, *Proc. Natl. Acad. Sci. U.S.A.*, 75, 1943, 1978.
167. **Krajewska, W. M., Briggs, R. C., and Hnilica, L. S.,** Cell-specific antigens in chicken erythroid nuclei: species specificity, *Biochemistry*, 18, 5720, 1979.
168. **Krajewska, W. M., Briggs, R. C., Chiu, J.-F., and Hnilica, L. S.,** Immunologically specific complexes of chromosomal nonhistone proteins with deoxyribonucleic acid in chicken erythroid nuclei, *Biochemistry*, 19, 4667, 1980.
169. **Cambell, A. M., Briggs, R. C., Bird, R. E., and Hnilica, L. S.,** Cell specific antiserum to chromosome scaffold proteins, *Nucleic Acids Res.*, 6, 205, 1979.
170. **Berezney, R., Matsui, S., and Eckert, B.,** unpublished data, 1983.
171. **Stein, G. S., Spelsberg, T. C., and Kleinsmith, L. J.,** Nonhistone chromosomal proteins and gene regulation, *Science*, 183, 817, 1974.
172. **Stein, G. and Baserga, R.,** Nuclear proteins and the cell cycle, *Adv. Cancer Res.*, 15, 287, 1972.
173. **Stein, G. S. and Borun, T. W.,** The synthesis of acidic chromosomal proteins during the cell cycle of HeLa 53 cells. I. The accelerated accumulation of acidic residual nuclear protein before the initiation of DNA replication, *J. Cell Biol.*, 52, 292, 1972.
174. **Borun, T. W. and Stein, G. S.,** The synthesis of acidic chromosomal proteins during the cell cycle of HeLa s3 cells, *J. Cell Biol.*, 52, 308, 1972.
175. **Gerner, E. W. and Humphrey, R. M.,** The cell-cycle phase synthesis of nonhistone proteins in mammalian cells, *Biochim. Biophys. Acta*, 331, 117, 1973.

176. **Hemminki, K.,** Labelling of histones and nonhistones in lung nuclear matrix and chromatin fractions, *Hoppe-Seyler's Z. Physiol. Chem.*, 358, 1125, 1977.
177. **Ševaljević, L., Petrović, M., Konstantinović, M., and Krtolica, K.,** Comparative studies of rat liver and sea urchin embryo nuclear matrices: partial fractionation and protein kinase activity distribution, *J. Cell Sci.*, in press, 1982.
178. **Jeanteur, P.,** Enzymatic activities associated with hnRNP, in *The Cell Nucleus*, Vol. 9, Busch, H., Ed., Academic Press, New York, 1981, 145.
179. **Lam, K. S. and Kasper, C. B.,** Selective phosphorylation of a nuclear envelope polypeptide by an endogenous protein kinase, *J. Biol. Chem.*, 18, 307, 1979.
180. **McDonald, J. R. and Agutter, P. S.,** The relationship between polyribonucleotide binding and the phosphorylation and dephosphorylation of nuclear envelope protein, *FEBS Lett.*, 116, 145, 1980.
181. **Steer, R. C., Wilson, M. J., and Ahmed, K.,** Protein phosphokinase activity of rat liver nuclear membrane, *Exp. Cell Res.*, 119, 403, 1979.
182. **Steer, R. C., Goueli, S. A., Wilson, M. J., and Ahmed, K.,** Cobalt-stimulated protein phosphokinase activity of the pore complex-lamina fraction from rat liver nuclear envelope, *Biochem. Biophys. Res. Commun.*, 92, 919, 1980.
183. **Igó-Kemenes, T. and Zachau, H. G.,** Domains in chromatin structure, *Cold Spring Harbor Symp. Quant. Biol.*, 43, 109, 1978.
184. **Benyajati, C. and Worcel, A.,** Isolation, characterization and structure of the folded chromosome of *Drosophila melanogaster, Cell*, 9, 393, 1976.
185. **Warren, A. C. and Cook, P. R.,** Supercoiling of DNA and nuclear conformation during the cell-cycle, *J. Cell Sci.*, 30, 211, 1978.
186. **Ide, T., Nakane, M., Anzai, K., and Andoh, T.,** Supercoiled DNA folded by nonhistone proteins in cultured mammalian cells, *Nature (London)*, 258, 445, 1975.
187. **McCready, S. J., Akrigg, A., and Cook, P. R.,** Electron microscopy of intact nuclear DNA from human cells, *J. Cell Sci.*, 39, 53, 1979.
188. **Razin, S. V ., Mantieva, A. L., and Georgiev, G. P.,** The similarity of DNA sequences remaining bound to scaffold upon nuclease treatment of interphase nuclei and metaphase chromosomes, *Nucleic Acids Res.*, 7, 1713, 1979.
189. **Britten, R. J. and Kohn, D. E.,** Repeated sequences in DNA, *Science*, 161, 529, 1968.
190. **Britten, R. J. and Davidson, E. H.,** Repetitive and non-repetitive DNA sequences and a speculation on the origins of evolutionary novelty, *Quart. Rev. Biol.*, 46, 111, 1971.
191. **Holmes, D. S. and Bonner, J.,** Sequence composition of rat nuclear deoxyribonucleic acid and high molecular weight nuclear ribonucleic acid, *Biochemistry*, 13, 841, 1974.
192. **Basler, J., Hastie, N. D., Pietras, D., Matsui S.-I., Sandberg, A. A., and Berezney, R.,** Hybridization of nuclear matrix attached deoxyribonucleic acid fragments, *Biochemistry*, 20, 6921, 1981.
193. **Berezney, R., Basler, J., Buchholtz, L. A., Smith, H. C., and Siegel, A. J.,** Nuclear matrix organization and DNA replication, in *The Nuclear Envelope and the Nuclear Matrix*, Maul, G. G., Ed., Alan R. Liss, New York, 1982, 183.
194. **Razin, S. V., Mantieva, V. L., and Georgiev, G. P.,** DNA adjacent to attachment points of deoxyribonucleoprotein fibrils to chromosomal axial structure is enriched in reiterated base sequences, *Nucleic Acids Res.*, 5, 4737, 1978.
195. **Jeppesen, P. G. N. and Bankier, A. T.,** A partial characterization of DNA fragments protected from nuclease degradation in histone depleted metaphase chromosomes of the Chinese hamster, *Nucleic Acids Res.*, 7, 49, 1979.
196. **Bowen B. C.,** DNA fragments associated with chromosome scaffold, *Nucleic Acids Res.*, 9, 5093, 1981.
197. **Kuo, M. T.,** Analysis of DNA attached to the chromosome scaffold, *J. Cell Biol.*, 93, 278, 1982.
198. **Matsumoto, L. H.,** Enrichment of satellite DNA on the nuclear matrix of bovine cells, *Nature (London)*, 294, 481, 1981.
199. **Cook, P. R. and Brazell, I. A.,** Mapping sequences in loops of nuclear DNA by their progressive detachment from the nuclear cage, *Nucleic Acids Res.*, 8, 2895, 1980.
200. **Fakan, S. and Hancock, R.,** Localization of newly-synthesized DNA in a mammalian cell as visualized by high resolution autoradiography, *Exp. Cell Res.*, 83, 95, 1974.
201. **Buchholtz, L. A. and Berezney, R.,** unpublished data, 1983.
202. **Hunt, B. F. and Vogelstein, B.,** Association of newly replicated DNA with the nuclear matrix of *Physarum polycephalum, Nucleic Acids Res.*, 9, 349, 1981.
203. **Smith, H. C. and Berezney, R.,** Dynamic domains of or in regenerating rat liver DNA polymerase, *Biochemistry*, 22, 3042, 1983.
204. **Reddy, C. P. V. and Pardee, A. B.,** Multienzyme complex for metabolic channeling in mammalian DNA replication, *Proc. Natl. Acad. Sci. U.S.A.*, 77, 3312, 1980.
205. **Smith, H. C., Buchholtz, L. A., Puvion, E., and Berezney, R.,** Nuclear matrix versus lamina: localization of DNA attachment and replicational sites, *J. Cell Biol.*, 95, 84a, 1982.

206. **Bensch, K. G., Tanaka, S., Hu, S.-Z., Wang, T.-S., and Korn, D.,** Intracellular localization of human DNA polymerase α with monoclonal antibodies, *J. Biol. Chem.*, 257, 8391, 1982.
207. **Steele, W. J. and Busch, H.,** Studies on the ribonucleic acid components of the nuclear ribonucleoprotein network, *Biochim. Biophys. Acta,* 129, 54, 1966.
208. **Steele, W. J. and Busch, H.,** Increased content of high molecular weight RNA fractions in nuclei and nucleoli of livers of thioacetamide-treated rats, *Biochim. Biophys. Acta,* 119, 501, 1966.
209. **Faiferman, I. and Pogo, A. O.,** Isolation of a nuclear ribonucleoprotein network that contains heterogeneous RNA and is bound to the nuclear envelope, *Biochemistry,* 14, 3808, 1975.
210. **Beyer, A. L., Christensen, M. E., Walker, B. W., and LeStourgeon, W. M.,** Identification and characterization of the packaging protein of core 40S hnRNP particles, *Cell,* 11, 127, 1977.
211. **Basler, J. and Berezney, R.,** unpublished data, 1983.
212. **Kuo, M. T.,** Distribution of tightly bound proteins in the chicken ovalbumin gene region, *Biochemistry,* 21, 321, 1982.
213. **Younghusband, H. B. and Maundrell, K.,** Adenoviral DNA is associated with the nuclear matrix of infected cells, *J. Virol.*, 43, 705, 1982.
214. **Nomi, S., Matsuura, T., Ueyama, H., and Ueda, K.,** Effect of vitamin A compounds on the covalent binding of benzo(a)pyrene to nuclear macromolecules, *J. Nutr. Sci. Vitaminol.*, 27, 33, 1981.
215. **Bresnick, E., Vaught, J. B., Chuang, A. H. L., Stoming, T. A., Bockman, D., and Mukhtar, H.,** Nuclear aryl hydrocarbon hydroxylase and interaction of polycyclic hydrocarbons with nuclear components, *Arch. Biochem. Biophys.*, 181, 257, 1977.
216. **Bresnick, E., Boraker, D., and Hassuk, B.,** Intranuclear localization of hepatic cytochrome P-448 by an immunochemical method, *Mol. Pharmacol.*, 16, 324, 1979.
217. **Bresnick, E. Hassuk, B., Liberator, P., Levin, W., and Thomas, P. E.,** Nucleolar cytochrome P-450, *Mol. Pharmacol.*, 18, 550, 1980.
218. **Berezney, R.,** unpublished data.

Chapter 4

PROTEINS OF THE METAPHASE CHROMOSOME

Virginia P. Wray and Wayne Wray

TABLE OF CONTENTS

I. CHROMOSOMAL PROTEINS AND CHROMOSOME STRUCTURE

A. Histones

Both histones and nonhistone chromosomal proteins (NHCP) contribute significantly to the packing of DNA into compact, organized units within the cell (for reviews see References 1 and 2). When metaphase chromosomes are deproteinized, the DNA is released in the form of masses of coiled strands.[3-8]

Histones interact with DNA to provide the first level of compaction of the DNA fiber. DNA-histone complexes in both interphase chromatin and metaphase chromosomes are organized into repeating subunits termed nucleosomes.[9-11] Each nucleosome contains about 200 base pairs of DNA wrapped around an octamer of histones (2 each of histones H2A, H2B, H3, and H4), and 144 base pairs of DNA are closely associated with the histone octamer to form a structure termed the nucleosome core particle. The additional 60 or so base pairs of DNA are referred to as linker DNA and connect adjacent nucleosomes.

Nucleosomes are easily visualized by electron microscopy. but their appearance depends on the method of preparation. When chromatin is prepared under conditions of low ionic strength, the core particles can be seen connected by short strands of linker DNA producing a "beads-on-a-string" appearance. Recent biochemical and electron microscopic studies demonstrated that nucleosome subunit structure of interphase chromatin is conserved in mitotic chromosomes.[12,13]

An understanding of the relationship of nucleosomes to larger fibers, and the larger fibers to metaphase chromosomes, has remained elusive. When chromatin is prepared by detergent spreading, the nucleosomes are closely opposed, forming a 100-Å chromatin fiber.[9-11] Chromatin isolated under physiological conditions appears as a knobby fiber with a diameter of 200 to 300 Å, often referred to as the 250-Å fiber. The principles involved in the coiling and folding of the primary nucleosomal fiber into 100-Å and 250-Å chromatin fibers is unclear. However, the model presently favored suggests that the 100-Å fiber is coiled into a solenoid to form the 250-Å fiber with about 6 nucleosomes per turn.[14-17] The integrity of the 250-Å fiber appears to be stabilized by the lysine-rich histone H1.[17-19] Histone H1 binds to the linker region between nucleosomes and may act by cross-linking nonadjacent nuclesomes through the formation of bridges between superhelical turns.[17,19] Formaldehyde treatment induces trypsin-sensitive DNA-histone cross-linking between contiguous components within chromosomes.[20] After trypsin digestion, the chromosomes lose morphologic integrity, being reduced to 250-Å fibers. These results suggest the exposed, lysine-rich tails of histones may be involved in interaction of adjacent 250-Å coils so that chromatin fiber assembly in metaphase chromosomes is controlled through these DNA-histone interactions. Biochemical and electron microscopic studies have demonstrated that the nucleosomal subunit structure and the 250-Å chromatin fiber of interphase chromatin are conserved during metaphase, but their organization in metaphase chromosomes has yet to be determined.[20,21] Phosphorylation of histones H1 and H3 has been suggested to be involved in the condensation of chromatin into chromosomes during mitosis.[22-26] Some evidence also suggests that protein A24, a conjugate of histone H2A and ubiquitin, is hydrolyzed to free ubiquitin and histone H2A during chromatin fiber condensation thereby allowing ionic interaction between phosphorylated histones and the positively charged H2A.[27]

B. Nonhistone Chromosomal Proteins

Nuclear nonhistone proteins are highly heterogeneous.[28-33] A few prominent species, known to have structural or enzymatic functions, compose about 50% of the total proteins.[9,30] Thus, there are a large number of nonhistone proteins, present in relatively low concentrations, with unknown functions. Some of these nuclear proteins may play a regulatory role in chromosome function while others may be related to specific nonchromosomal nuclear structures.

Most of the NHCP are associated with levels of nucleoprotein fiber organization higher than the nucleosome. Their presence in the metaphase chromosome is in itself an indication of their importance in the maintenance of nucleoprotein supercoiling. Residual metaphase chromosome structure is retained after the release of a large portion of the DNA and histone.[3,34-38] These structures are insensitive to RNase digestion, but they are dissociated by mild protease treatment or urea extraction.[3,34-37,39,40] By other extraction methods, which remove nonhistones as well as histones, the DNA fibers become more uncoiled and unfolded.[5-8,41,42] Thus, these nonhistone proteins could be necessary for the packing of the DNA into 250-Å fibers and the maintenance of the normally observed chromosome structure and packing at metaphase. Early evidence suggests a gene-regulatory function of the residual NHCP (for a general review of the subject see Elgin et al.[43]). The relative amount of NHCP in a given cell type appears to be correlated with transcriptional activity. Certain of the NHCP, those of the high mobility group (HMG), are associated with nucleosomes and spacer regions in interphase chromatin.[44] Subpopulations of nucleosomes have been identified which differ in HMG and histone H1 content.[45] Proteins HMG-1 and HMG-2 binding sites are less certain but do involve single-strand DNA.[39,40,44-48] These latter proteins have a destabilizing influence on the double helix.[45,48] Proteins HMG-14 and HMG-17 bind to intact nucleosomes associated with active gene sequences and appear to be at least partially responsible for their increase DNase I sensitivity.[40,45-48]

There are a number of nonhistone proteins which show species-, tissue-, tumor-, and even DNA sequence-specificity. A number of nonhistone proteins are present in concentrations within the range estimated for regulatory proteins. Certain NHCP fractions contain appreciable amounts of phosphate esterified to serine and threonine; increased temporal phosphorylation has been associated with periods of increased gene activity.

II. PROTEINS OF ISOLATED METAPHASE CHROMOSOMES

A. Mass Composition of Isolated Metaphase Chromosomes

Hearst and Botchan[49] reviewed the gross chemical composition of chromosomes isolated by the reported methods prior to 1970. RNA was reported on almost an equal-weight basis with the DNA of the metaphase chromosome, and the protein to DNA ratio was about 4.0:1.0. It has been shown that the RNA present in these isolated chromosome preparations is primarily the 18S and 2S rRNA, although it has not been demonstrated that these RNAs are on the chromosome as ribosomes.[50,51]

Mendelsohn et al.[52] showed that contaminant RNA and protein remained near the top of a sucrose velocity gradient with the small chromosomes. The fractions enriched for larger chromosomes (A + B fraction) sedimented farther down the gradient than the smaller chromosomes and contained much less RNA and protein. Protein to DNA and RNA to DNA ratios for the fractions containing large-sized chromosomes were 2.4:1.9 and 0.18:1.0, respectively. The reduced amount of RNA in the A + B (large chromosome) fraction supports their previous finding that RNA is not an intrinsic component of metaphase chromosomes.[53]

DNA, RNA, and protein analyses of total heterogeneous chromosomes isolated at neutral pH in hexylene glycol buffer yields a protein to DNA ratio equal to 2.2:1.0 and an RNA to DNA ratio of less than 0.1:1.0[54] Thus, there appeared to be little, if any, adventitiously adsorbed RNA or protein in these chromosomes. This was substantiated by mixing experiments using radiolabeled RNA and cytosol protein.[55]

Metaphase chromosomes exhibit approximately the same mass ratios of the macromolecular constituents (1:0.9-1.2:0.3-1.2:0.1 DNA to histones to nonhistones to RNA) as does isolated interphase chromatin.[56]

Lipids are a constant component of isolated chromosomes.[57] Triglycerides and cholesterol esters prevail while diglycerides, which are a characteristic component of cellular debris are absent from isolated chromosomes except at pH 2.

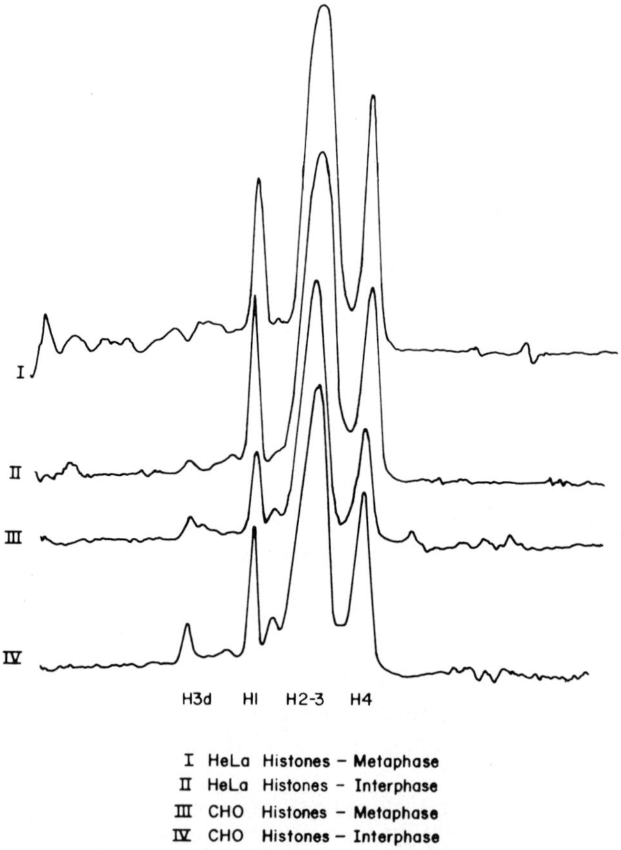

FIGURE 1. Histone complement of metaphase chromosomes and interphase chromatin. Histones from isolated metaphase chromosomes[54] and from chromatin prepared from isolated nuclei[58] were extracted and analyzed according to the methods of Bonner et al.[58] (From Wray, V. P., Elgin, S. C. R., and Wray, W., *Nucleic Acids Res.*, 8, 4155, 1980. With permission.)

B. General Protein Population of Isolated Metaphase Chromosomes

1. Comparison of Metaphase Chromosomal Proteins to Nuclear and Chromatin Proteins

Essentially no difference is noted in the histone patterns of CHO and HeLa interphase chromatin or metaphase chromosomes (Figure 1). A small amount of histone-3-dimer is observed; however, there is no indication that either disulfide-linked H3 multimers or H3-NHCP complexes are formed in the metaphase state, as has been previously suggested.[59] Similar results have been reported by other workers.[60]

The molecular complexity of the metaphase NHCP is as great as that of interphase nuclear NHCP. Figure 2 shows a comparison of NHCP from Chinese hamster (CHO) (Figure 2A) and human (HeLa) (Figure 2B) metaphase chromosomes and interphase nuclei isolated in neutral pH hexylene glycol buffers.[54] The separation profiles obtained by one-dimensional polyacrylamide gel electrophoresis (1D-PAGE) demonstrate species-specific characteristics (see also subsequent figures). For each species there is a strong similarity of patterns between chromosomes and nuclei with respect to the most abundant proteins. However, chromosomes of both species have an increased content of a large protein of about 200,000 mol wt and a smaller protein of about 50,000 mol wt relative to nuclei. There are differences in the separation profiles of proteins about 90,000 to 100,000 mol wt and all proteins of this group are generally increased in chromosomes compared to nuclei. In CHO cells, a protein of

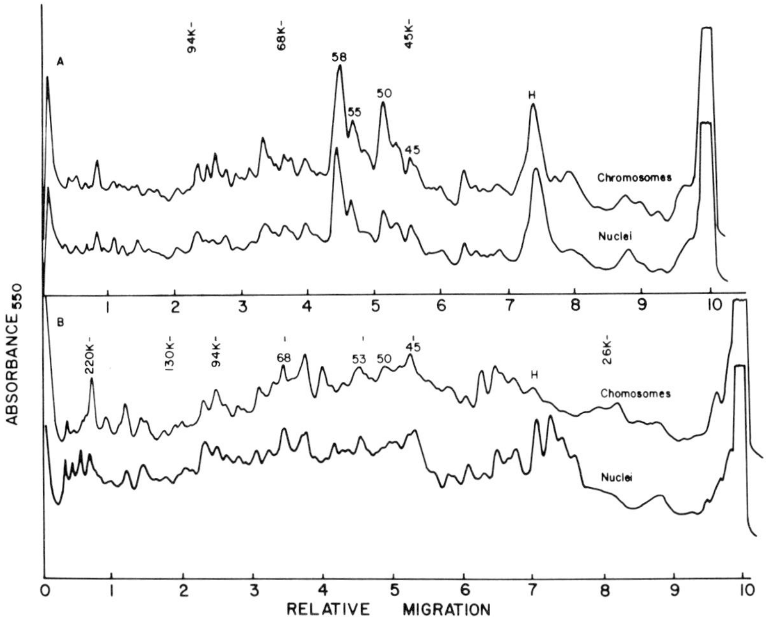

FIGURE 2. Polypeptides of metaphase chromosomes and interphase nuclei. (A) CHO cells. (B) HeLa cells. Metaphase chromosomes and interphase nuclei were isolated by the methods of Wray and Stubblefield in neutral pH hexylene glycol-PIPES-$CaCl_2$ buffers.[54] Chromosomal and nuclear proteins were separated according to molecular weight (40 μg protein per slot) on 0.75 × 10 cm slabs of 9.5% polyacrylamide gels 1D-PAGE[61] and stained with Coomassie® brilliant blue R-250. For these and subsequent gels, combinations of some or all of the following proteins were chromatographed as molecular weight standards. Myosin, 220,000 mol wt; β-galactosidase, 130,000 mol wt; phosphorylase A or B, 94,000 mol wt; bovine serum albumin, 68,000 mol wt; bovine brain tubulin, 56,000 mol wt; glutaraldehyde-3-phosphate dehydrogenase, 53,000 mol wt; hen ovalbumin, 43,000 mol wt; rabbit muscle actin, 45,000 mol wt; liver alcohol dehydrogenase, 42,000 mol wt; carbonic anhydrase, 29,000 mol wt; and α-chymotrypsinogen, 25,600 mol wt. Migration of the molecular weight standards with respect to the NHCP bands are indicated above the scan profiles. Migration is relative to the bromphenol blue tracking dye which is given as an arbitrary value of 10 for purposes of presentation.

about 72,000 mol wt (relative migration 3.3) and proteins of about 45,000 to 60,000 mol wt constitute a larger proportion of the total protein content of chromosomes than of nuclei. HeLa chromosomes have bands at relative migration 0.9 and 6.2 which do not appear in the interphase nuclei fraction.

The differences observed between chromosomes and nuclei were also present when chromosomes and interphase chromatin were compared (Figure 3).[56] Proteins of about 65,000 and 68,000 mol wt are enriched in interphase chromatin while proteins of about 200,000, 90,000 to 100,000, and 50,000 mol wt are more obviously increased in chromosomes compared to chromatin. These observations[56] are in contrast to those of Matsui et al.[27] who found the loss of protein A24 to be the only observable difference between chromosomes and nuclei.

2. *Comparison of Proteins of Chromosomes from Different Species*

Species specificity has been observed in the NHCP obtained from metaphase chromosomes of different species. Regions of species specificity as well as an overall homology have been demonstrated using electrophoretic separations of the protein complements of human (HeLa), Chinese hamster (CHO), and chicken (Marek lymphoma) chromosomes (Figure 4).[62] The protein complements of chromosomes from Chinese hamster and human cell lines other than

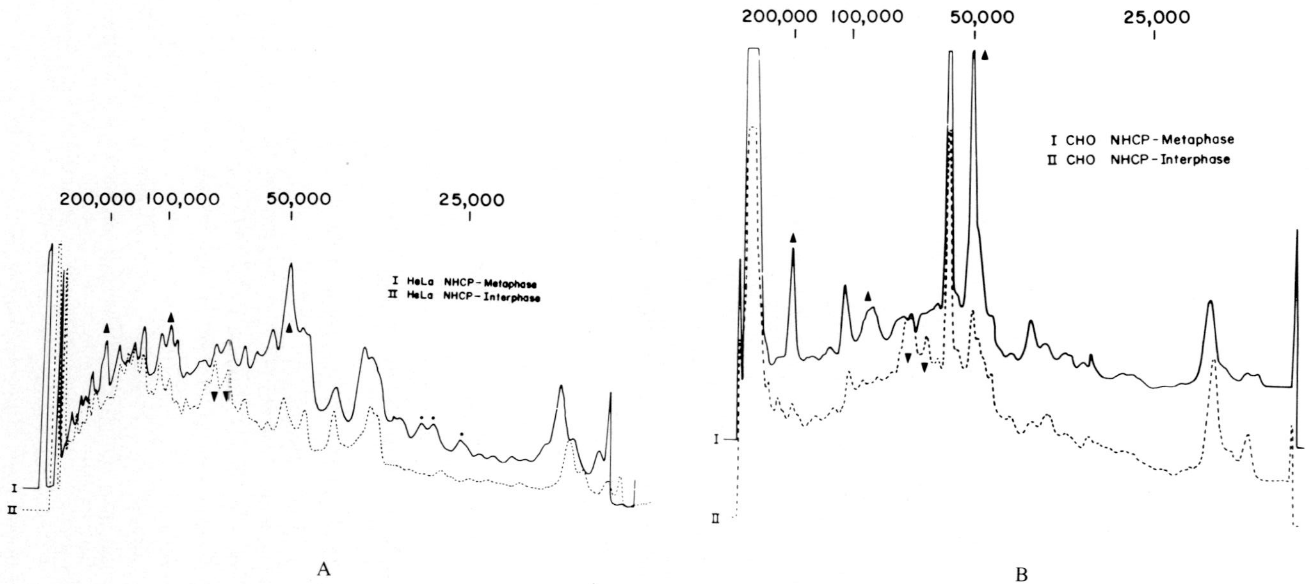

FIGURE 3. Nonhistone chromosomal proteins of isolated metaphase chromosomes and interphase chromatin. (A) HeLa cells. (B) CHO cells. NHCP from isolated metaphase chromosomes[54] and from chromatin prepared from isolated nuclei[54] were extracted and analyzed according to the methods of Bonner et al.[58] Proteins in samples were resolved according to molecular weight in 7.5%, 0.6 × 10 cm cylindrical polyacrylamide gels.[6] Gel origin is to the left. Protein bands of interest are indicated by ▼ and ▲ (From Wray, V. P., Elgin, S. C. R., and Wray, W., *Nucleic Acids Res.*, 8, 4155, 1980. With permission.)

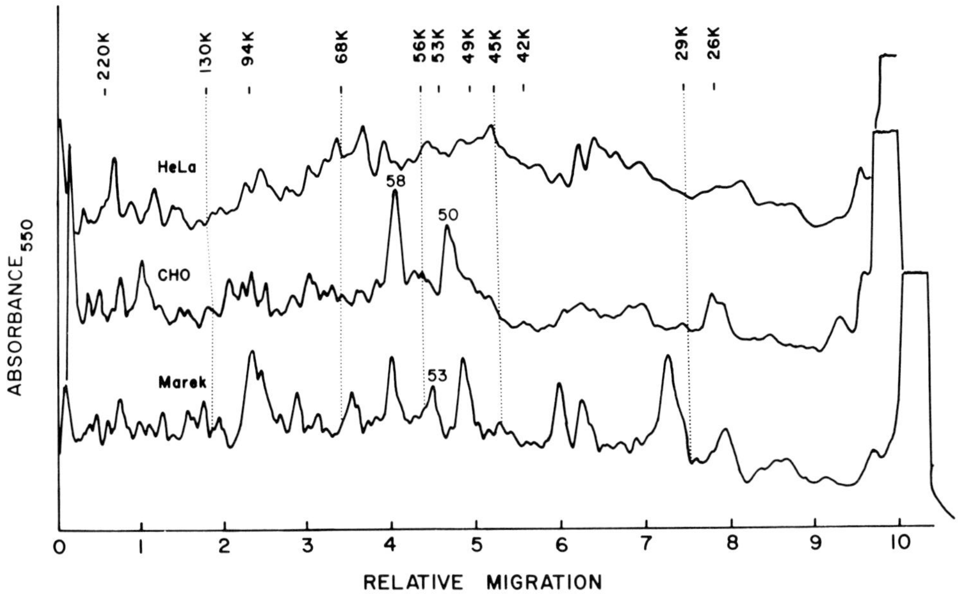

FIGURE 4. Nonhistone chromosomal proteins of human, rodent, and avian species. Metaphase chromosomes of HeLa, CHO, and Marek lymphoma cells are isolated and purified by sucrose density gradient sedimentation.[70] Proteins of the individual fractions were separated by gel electrophoresis as described in Figure 2. Dotted lines indicate migration of protein standards on the different gels.

CHO and HeLa have also been examined (data not shown). The data indicate that different cell lines from the same species have similar protein compositions.

One protein of about 50,000 mol wt is common to chromosomes from all three species. This observation agrees with studies of a chromosome "scaffold" structure by Laemmli et al.[3,36,37] A second protein which is present in all species is of about 200,000 mol wt. This protein is especially enhanced in the CHO line of Chinese hamster cells but also is increased in the other species. A cluster of proteins of 90,000 to 100,000 mol wt show species modulation. Human and hamster cell chromosomes also have a fraction of 45,000 to 50,000-mol wt proteins which are not in chicken cells. Protein p-58 (about 58,000 mol wt), a major protein in CHO cell chromosomes, was decreased or absent in chromosomes from other species. Several variations in the separation patterns of proteins of less than 40,000 mol wt are also present.

Species differences were also apparent when CHO and HeLa NHCP were examined by two-dimensional polyacrylamide gel electrophoresis (2D-PAGE) (Figure 5). Some proteins were common to both species while other proteins seem to be species specific. The separation pattern of CHO chromosome NHCP on a 2D-PAGE gel is shown. The positions of HeLa NHCP from a companion gel are shown by circles drawn on the figure. HeLa proteins not corresponding to CHO proteins are indicated by arrows. The 58,000 mol wt protein in HeLa cell chromosomes, as seen by 1D-PAGE (Figure 2), does not comigrate with the CHO 58,000-mol wt protein by 2D-PAGE.

A human-specific NHCP of 65,000 mol wt is expressed in human/hamster hybrids containing only human chromosome 16.[64] This protein is one of the approximately 70 human-specific NHCP found in isolated human chromatin but not in hamster chromatin. Its presence in the metaphase chromosome has not yet been demonstrated. Another human-specific interphase NHCP of about 300,000 mol wt does not maintain association with the metaphase chromosome.[65] These observations illustrate not only species specificity of NHCP but also add impetus to the need for ultimate *in situ* proof of NHCP association with the metaphase structure.

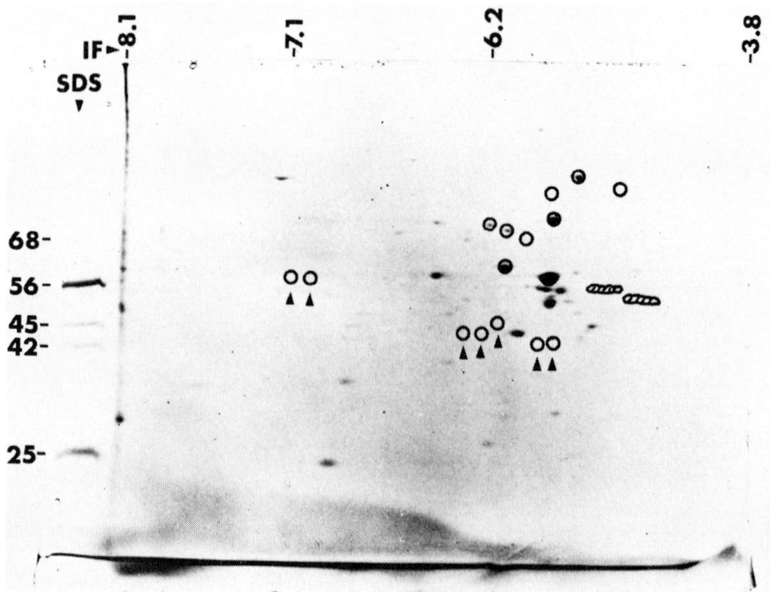

FIGURE 5. Two-dimensional separation of CHO metaphase chromosomal proteins. Proteins from isolated metaphase chromosomes were separated by 2D-PAGE[63] using a 1.5 mm × 8.5 cm, 9.5% polyacrylamide gel in the second dimension. Protein standards were applied in a separate well adjacent to the first-dimension cylindrical gel and run in the second dimension. The positions of the migration of the tubulin monomers on a companion gel are indicated by hatched lines. Migration of HeLa metaphase chromosome NHCP on a third companion gel are indicated by circles drawn on the figure. Proteins unique to HeLa fractions are indicated by arrowheads. (From Wray, V. P., Elgin, S. C. R., and Wray, W., *Nucleic Acids Res.*, 8, 4155, 1980. With permission.)

C. Effects of Isolation Procedures on the NHCP of Metaphase Chromosomes

We used computerized image processing to compare published photographs of 1D-PAGE separations of (CHO) metaphase chromosome proteins from laboratories engaged in studies of metaphase chromosomes. The resulting stain density profiles are presented in Figure 6. We scanned their gels from origin to the center of the slowest migrating H1 band so that each gel would have a similar expansion scale, thereby minimizing differences in total gel length and gel concentration. The gel profiles were normalized to the protein bands identified as 58,000 mol wt. Band matching was done based on regional cluster and group patterns, as indicated by bracketed underline bars (Groups a, b, and c), because most data were not accompanied by molecular weight markers adequate for more detailed comparison. Allowing for variances in band resolution and polyacrylamide concentration, a general similarity in pattern is observed but there is a significant degree of quantitative variation in the electrophoretic profiles. The least variation occurs in data obtained from chromosomes prepared by the same procedure (Figure 6; profiles A, B, C, and D).

Several methods are available for isolations of metaphase chromosomes.[20,54,66-69] Variables in these methods include pH, ion concentrations, buffering agents, and presence or absence of hexylene glycol. Detergents and polyamines have also been included in some of the buffers.[20,67,68] NHCPs from metaphase chromosomes prepared by these different methods were examined and compared in our laboratories. Results from this study are shown in Figures 7 and 8. In Figure 7 the proteins of metaphase chromosomes and nuclei of Chinese hamster DON cells isolated in hexylene glycol-PIPES, pH 6.8 buffer[54,70] or in 50% acetic acid[66] are compared. Chromosomes and nuclei isolated by the two different methods show significantly different protein complements.[62] Histone H1 is removed by 50% acetic acid

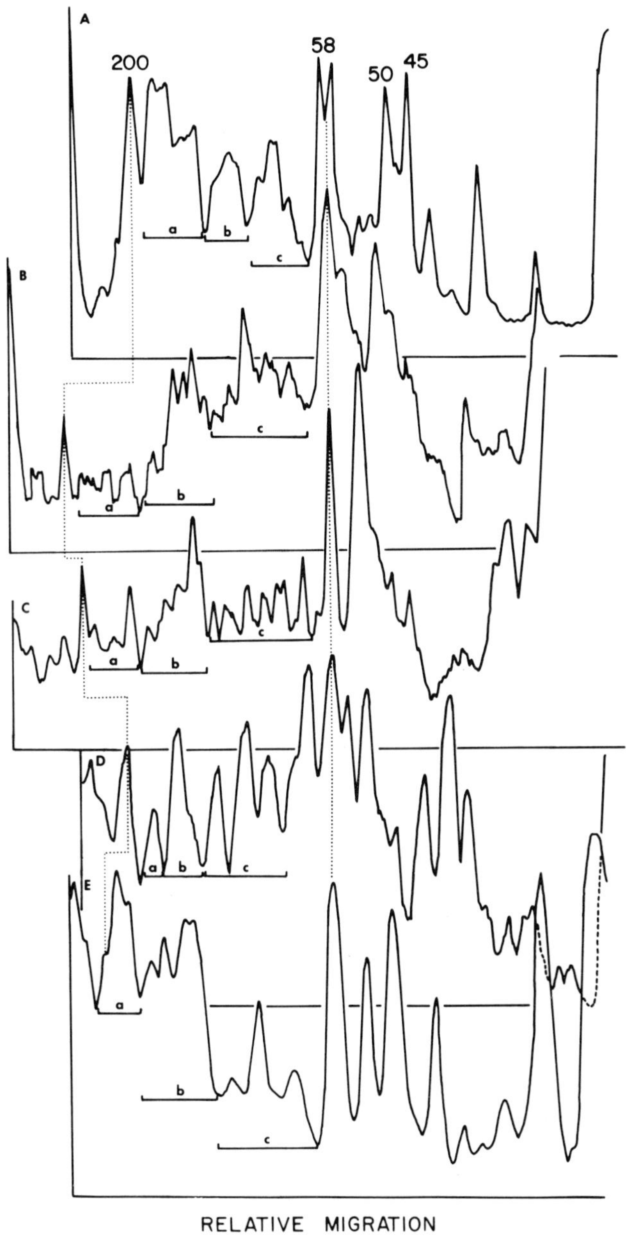

FIGURE 6. Chinese hamster (CHO) metaphase chromosomal proteins: an image-processed composite of published data. Photographs of gels from the listed publications were scanned using a computer-linked image processor. The graphic output scales were adjusted to give maximum peak height to the most densely stained protein band. Maximum gel length was from origin to the lowest mobility histone H1 component which represented a common internal standard. Band matching was based on molecular weight standards (if included), on the major protein constituents, and on general protein separation profile. Bands identified as p-58 and p-200 are connected by dotted lines between gels. (A) Jeppesen, P. G. N., Bankier, A. T., and Sanders, L., *Exp. Cell Res.*, 115, 293, 1978. (B) Wray, V. P. and Wray, W., unpublished data. (C) Okada, T. A. and Comings, D. E., *Chromosoma (Berlin)*, 72, 1, 1979. (D) Hadlaczky, G., Sumner, A. T., and Ross, A., *Chromosoma (Berlin)*, 81, 537, 1981. (E) Goyanes, V. J., Matsui, S.-I., and Sandberg, A. A., *Chromosoma (Berlin)*, 78, 123, 1980.

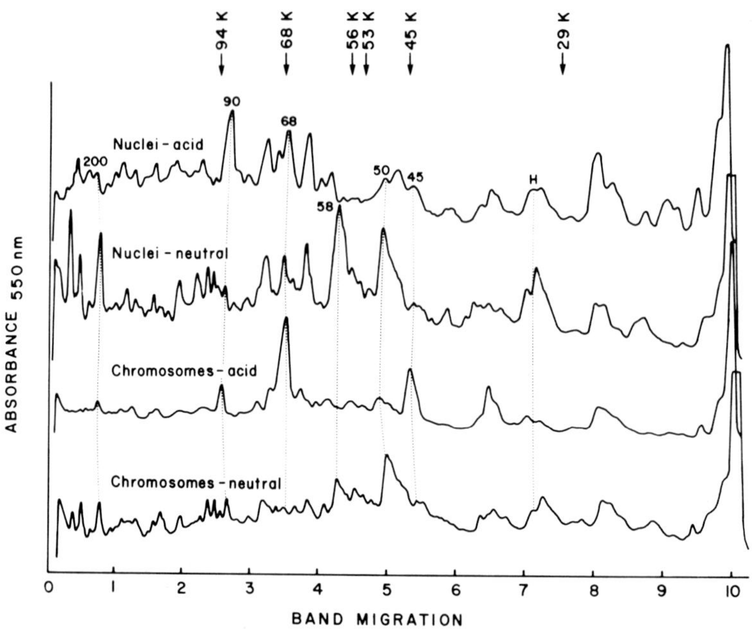

FIGURE 7. Protein complements of nuclei and chromosomes isolated in acid and neutral media. Nuclei and metaphase chromosomes of Chinese hamster DON cells were isolated in hexylene glycol-PIPES-$CaCl_2$, pH 6.8 buffer[54,70] or in 50% acetic acid.[66] Nuclei and chromosomes were purified by sedimentation through dense sucrose-hexylene glycol-PIPES buffer.[54] The sucrose was removed by further washing of pellets with neutral buffer solutions. Each final suspension was adjusted to pH 7.2. Samples were solubilized and analyzed by slab 1D-PAGE as described in Figure 2. Dotted lines indicate migration of protein standards on the different gels.

while the other histones are diminished, but not totally removed (results not shown). Among the NHCP population, modulation including either increase or loss of proteins having about 90,000, 68,000, 58,000, 50,000, and 45,000 mol wt is of significant proportions. Band p-68 (about 68,000 mol wt) is increased substantially in both nuclei and chromosomes isolated in the acidic media. The increase of p-68 in acid-isolated chromosomes has been observed in chromosome preparations from several types of cells. Band p-90, a major component of nuclei isolated in 50% acetic acid is greatly diminished in chromosomes isolated in 50% acetic acid. Protein p-58 is greatly diminished in these preparations also. Reciprocal washing of isolated chromosomes with the different isolation solutions does not produce the same results, except for removal of p-58 by 50% acetic acid. Treatment of acid-isolated chromosomes with neutral pH buffer does not cause release of the p-68 material.

It is not clear at this time whether the acid-induced alterations of NHCP populations are the result of preservation of some unstable or protease-sensitive protein(s) native to the metaphase structure or are due to adsorption of specific extra-chromosomal proteins during the isolation procedure. Observations of Mendelsohn et al.[52] and Wray et al.[56] that contaminant RNA and proteins are removed by sucrose velocity gradient purification of chromosomes suggests the latter is not the case. By phase contrast and electron microscopy the morphology of the chromosomes prepared by either method is quite similar except that chromosomes isolated in acidic media are more compact and cannot be banded by the standard chromosome banding procedures.[66]

In Figure 8A the protein complements of chromosomes isolated in various neutral buffers are compared to those of chromosomes isolated in 50% acetic acid and in the neutral pH hexylene glycol buffers of Wray and Stubblefield.[54] There are several landmark proteins

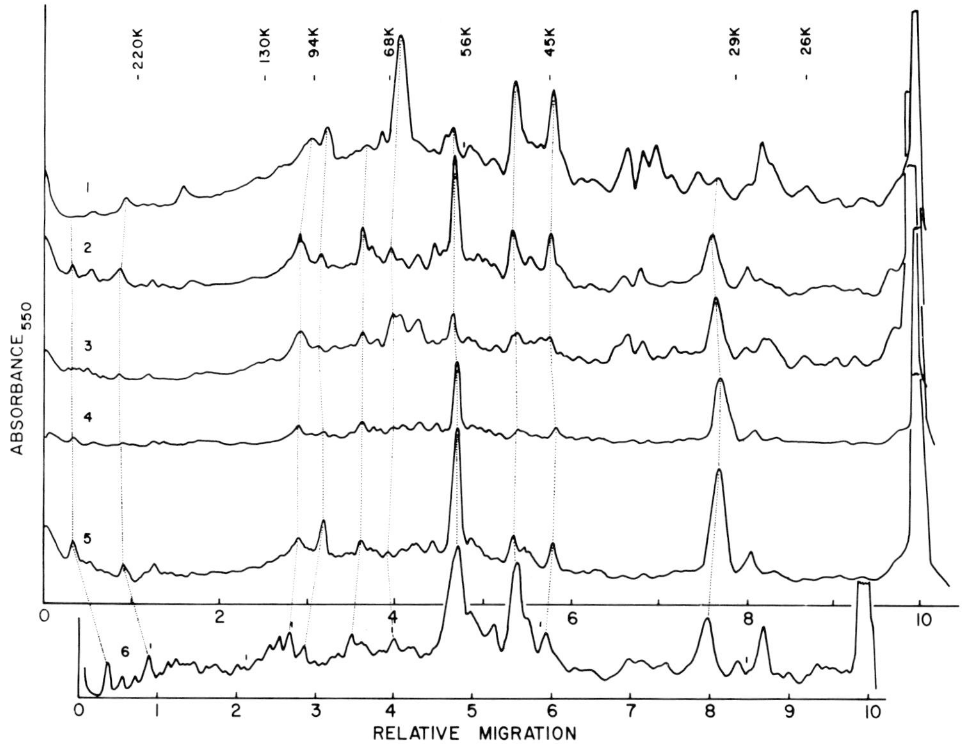

A

FIGURE 8. Protein complements of metaphase chromosomes isolated by neutral and alkaline buffers. Metaphase chromosomes were isolated according to several published methods, purified through dense sucrose solutions and analyzed for protein content by 1D-PAGE as described in Figure 2. Dotted lines indicate migration of protein standards on the different gels. (A1) Acetic acid media as described in Figure 6. (A2) 25 m*M* Tris, 0.75 *M* hexylene glycol, 0.5 m*M* $CaCl_2$, 1.0 m*M* $MgCl_2$, pH 7.5;[69] (A3) 10 m*M* Tris, 2 m*M* $CaCl_2$, 2.5 m*M* $MgCl_2$, 0.5 m*M* PMSF, pH 7.5 with added 5% Triton®-X-100;[20] (A4) 15 m*M* Tris, 0.2 m*M* spermine, 0.5 m*M* spermidine, 2 m*M* EDTA, 0.5 m*M* EGTA, 80 m*M* KCl, 20 m*M* NaCl, 14 m*M* β-mercaptoethanol + 0.34 *M* sucrose, 0.1% digitonin, and 0.5 *M* hexylene glycol, final pH 7.0;[67] (A5) as above + only digitonin, pH 7.2;[67] (A6) 1.0 *M* hexylene glycol 0.1 m*M* PIPES, m*M* $CaCl_2$, pH 6.8.[54] (B slot b) 50 m*M* HEPES(*N*-2-hydroxyethylpiperazine-*N'*-2-ethanesulfonic acid), pH 7.4, 5 m*M* $MgCl_2$, 0.5 m*M* $CaCl_2$, 0.1 m*M* PMSF(phenylmethyl sulfonyl fluoride) + 0.5% NP40 detergent;[68] (B slot c) 80 m*M* NaCl, 5 m*M* HEPES pH 7.4, 0.5 m*M* spermine tetrahydrochloride, 0.15 m*M* spermidine trihydrochloride, 0.2 m*M* EGTA, 0.1 m*M* or 1.0 m*M* PMSF;[68] (B slot d) 0.5 m*M* NaCl, 5 m*M* CAPS (cyclohexylaminopropane sulfonic acid) pH 10.0, 5 m*M* $MgCl_2$, 0.5 m*M* $CaCl_2$, 0.1 m*M* or 1.0 m*M* PMSF;[68] (B slot 3) 80 m*M* NaCl, 5 m*M* CAPS pH 10.0, 0.5 m*M* spermine tetrahydrocloride, 0.15 m*M* spermidine trihydrochloride, 0.2 m*M* EGTA, 0.1 m*M* or 1.0 m*M* PMSF;[68] (B slots a and f) — molecular weight standards as listed in Figure 1.

(connected by dashed lines) which appear to be constant in all the preparations. The approximately 90,000 to 100,000-mol wt grouping of proteins which are enriched in metaphase compared to interphase are also identifiable.

In Figure 8B, the protein complements of chromosomes isolated in neutral and alkaline buffers with and without spermine/spermidine are shown. The profiles are quite similar except for a significant loss of band p-58 material in chromosomes prepared in CAPS, pH 10 buffer containing Mg/Ca. A protein migrating between 130,000 and 220,000 mol wt is present in greater quantity in chromosomes prepared in CAPS pH 10 buffer containing spermine/spermidine. All methods tested have gel profiles which suggest that protein degradation occurs regardless of the addition of protease inhibitor.

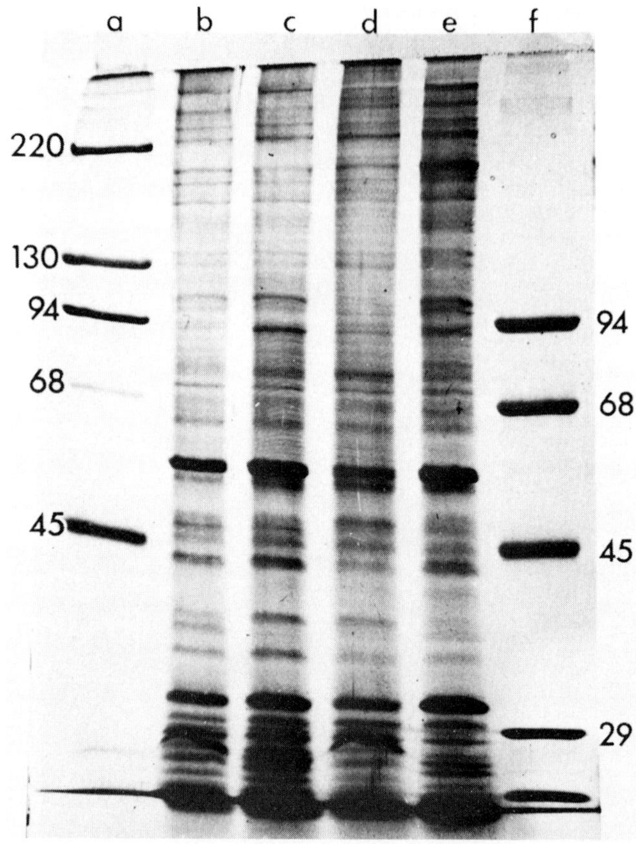

FIGURE 8B

D. Cytoskeletal and Filament Proteins

Contractile proteins and filament proteins such as myosin, actin, and tubulin have not been clearly identified in the NHCP fraction of metaphase chromosomes. Chinese hamster metaphase chromosomes have two prominent protein bands between 45,000 and 60,000 mol wt. Their mobilities are similar to those of tubulin and actin, leading some investigators to consider these proteins to be present.[9] However, our studies[56] show that the mobilities of these proteins are different than those of isolated tubulin isomers and actin (Figure 9). Moreover, by 2D-PAGE, the mobility of p-58 is different than that of the tubulin monomers (Figure 5).

Other biochemical characteristics of p-58 are solubility in acetic acid, insolubility in high NaCl concentrations but removal by urea treatment of metaphase chromosomes, and 2D-PAGE mobility are characteristics similar to those of the intermediate filament protein, decamin, suggesting a molecular relatedness between chromosomal p-58 and intermediate filament material.[71-74] The p-50 protein could not be as thoroughly characterized but was not in a location near that expected for monomer actin.[74]

Duality of function for nuclear proteins (or conversely cytoplasmic proteins) is illustrated in a recent report of a predominantly nuclear protein of about 300,000 mol wt (NuMA) which, during mitosis, becomes a part of the mitotic apparatus.[65] NuMA is human specific and is synthesized in a human/Chinese hamster hybrid cell containing a reduced number of human chromosomes. In the interphase cell, NuMA is localized in the nuclear compartment. As the cell leaves G_2, NuMA reassociates with the mitotic apparatus. It is, therefore, reasonable to expect other NHCPs to exhibit dual functionality similar to this example.

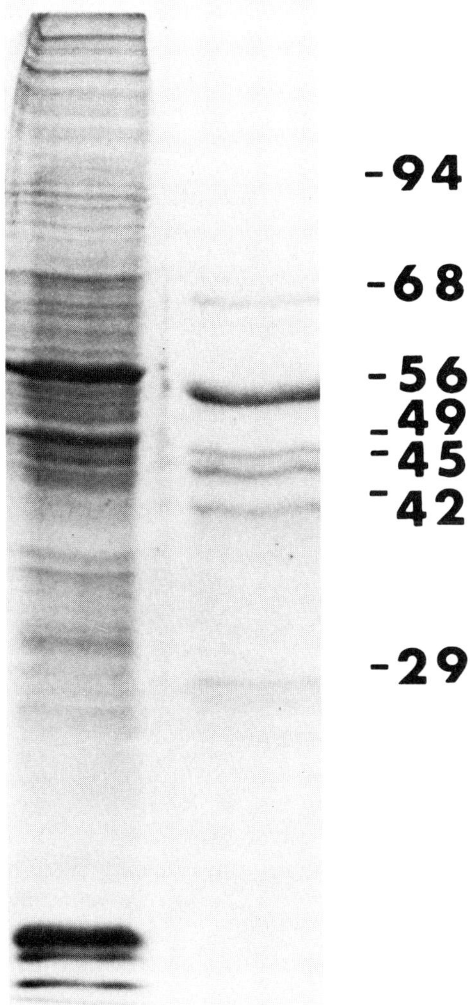

FIGURE 9. Comparison of CHO metaphase chromosomal proteins to standard tubulin and actin markers. Metaphase chromosomes were prepared and analyzed as described in Figure 2. Numbers represent molecular weight × 10^{-3}. Standard protein markers are phosphorylase a, 94,000 mol wt, bovine serum albumin, 68,000 mol wt; bovine brain tubulin, 56,000 mol wt; fumarase, 49,000 mol wt; rabbit muscle actin, 45,000 mol wt; liver alcohol dehydrogenase, 42,000 mol wt; and carbonic anhydrase, 29,000 mol wt. (From Wray, V. P., Elgin, S. C. R., and Wray, W., *Nucleic Acids Res.*, 8, 4155, 1980. With permission.)

III. LOCALIZATION OF PROTEINS ON THE METAPHASE CHROMOSOME

Localization of specific proteins within the overall supramolecular framework of the metaphase chromosome is an important aspect of chromosome structure studies. Depending on their biological role in chromosome organization and gene activity, different proteins could be localized either broadly or specifically with respect to the overall chromosome framework. For example, histones, because of their involvement in the formation of the primary nucleoprotein fiber, should be broadly distributed. Antibodies against histones H2B and H1 localize or bind throughout most chromosomal regions, in agreement with their known general occurrence in nucleosomes and spacer regions.[75,76] Similarly, some nonhistone HMG proteins are also uniformly distributed throughout the chromosome.[76,77]

Our studies correlating specific NHCP and particular size classes of chromosomes within the karyotype suggest that some proteins are ubiquitous and are on all chromosomes while others are differentially distributed among the different-sized chromosomes.[78] Comparison of the densitometry scans of 1D-PAGE gels showed that there are quantitative differences in some proteins while other proteins are quantitatively the same throughout the gradient spectrum. The same results were obtained when the total gradient was more finely subdivided (Figure 10A). To simplify discussion of the data, the protein profiles have been divided into molecular weight groups. Within each group, the proteins have been given identification numbers. The groups are comprised of these estimated molecular weights: α — greater than 130,000; β — 130,000 to 60,000; γ — 60,000 to 40,000; δ — 40,000 to the dye front. The content of each peak was measured with a peak-area integrating densitometer using both liberal and conservative estimates of the start and end of a given peak. At the level of resolution obtained by 1D-PAGE there were no bands unique to a given chromosome fraction. However, quantititive changes are evident. The protein(s) in α-2 and δ-3 show obvious changes in stain density between the different chromosome size classes. Changes in quantities of other peaks are more subtle, but reproducible.

The percentage of individual chromosomes present in each gradient fraction as determined by direct microscopic analysis, was plotted against the percentage of total protein for each of the measured bands. Figure 10B illustrates those proteins which seem to correlate with the chromosomes in the fraction. The relationship is not completely clear because of the complexities involved. Given the relative insensitivity of the analysis system, plus the factors of mixed chromosome populations and the possibility that each stained band represents multiple protein species suggests that for these differences to be observed, there are dramatic underlying variations in the protein complements of individual chromosomes.

Several NHCP are known to have specific sites or regional locations on metaphase chromosomes. Condensed metaphase chromosomes contain several architecturally defined regions whose ultrastructure, staining properties, or biochemistry are different from the remainder of the chromosomes. Nucleolar organizer regions (NOR), which are architecturally unique regions related to ribosomal RNA genes, stain selectively with silver by certain cytochemical methods indicating the presence of a molecular moiety more reactive to silver compared to other chromosomal proteins.[79,80] Recently Brenner and Brinkley have found a NOR-specific antibody in normal human sera.[81] Naturally occurring antibodies against centromeric regions of the metaphase chromosome have also been found.[82,83] An autoantibody described by Tan[84] is reactive with centromeres of cells derived from humans, mice, and Chinese hamsters. The antigen reactivity is sensitive to both protease and DNase. The native antigen for the centromer reactive autoantibody reported by Brenner and Brinkley has not been characterized.[81] Other autoantibodies against nuclear proteins have been found, but their relationships with metaphase chromosome structures have not been characterized.[84]

In metaphase chromosomes glycoproteins are localized in chromosomal bands in a distinctive pattern similar to that observed for quinacrine banding.[85] Wheat germ agglutinin, a plant protein which binds glucosamine, reacts with chromosomes in chromosome spreads. Concanavalin A, a plant protein binding mannose and glucose residues, binds to the interphase nuclear structure but not metaphase chromosomes, indicating these nuclear macromolecules are not incorporated into the metaphase structure.

Further indications of difference in protein distribution along the length of the metaphase

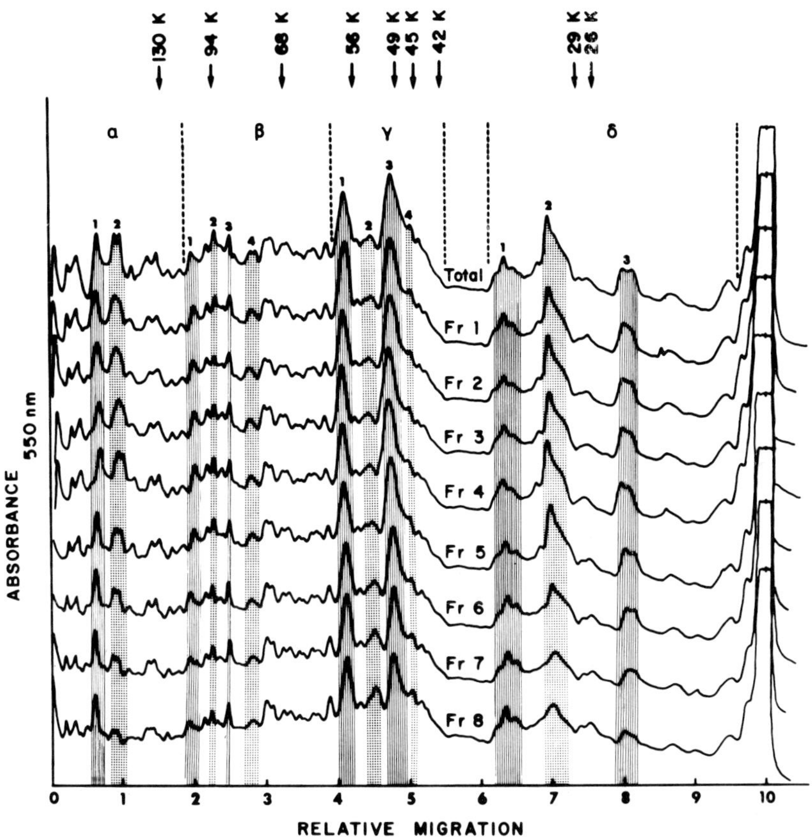

A

FIGURE 10. Nonhistone protein complements of fractions of sized metaphase chromosomes. Metaphase chromosomes were prepared by the method of Wray and Stubblefield[54] in the presence of phenylmethylsulfonylfluoride (PMSF) and separated according to size[70] using centrifugation into a sucrose gradient in a zonal centrifuge rotor. The gradient effluent was collected into 8 fractions with fraction 1 being that which contained the largest chromosomes. Chromosomes contained in the fractions were scored as being in group A (large metacentric), group B (middle-sized metacentric), group C (acrocentric), or group D (small metacentric). (A) Chromosomal proteins were analyzed as described in Figure 2. Protein bands or clusters which were compared quantitatively are indicated by shaded areas. The area under each peak was determined from the automated integration data obtained using an integrating densitometer. The total density per band was expressed as a percentage of the total stainable protein per slot. (B) Correlation of chromosome size groups within the fraction to the protein content of selected NHCP bands and clusters. The shaded area for each protein measurement represent the maximum and minimum areas for each peak measured.

chromosome is demonstrated directly by chromosome banding using the various cytogenetic banding and staining techniques.[86,87] Although the exact mechanisms of chromosome banding are unknown, the bands are clearly related to the regional biochemical properties within the chromosome. Treatment of isolated nuclei with reagents used in chromosome G-banding procedures extracts specific subsets of the total nuclear nonhistone proteins.[88,89] Of the extracted nonhistone, some are common to all treatments while others are unique to a particular treatment. The C-banding techniques produce substantial changes in the nuclear proteins. The R-banding procedures have the least effect of any of the banding techniques tested on the extraction of proteins from isolated nuclei. Only small amounts of five non-

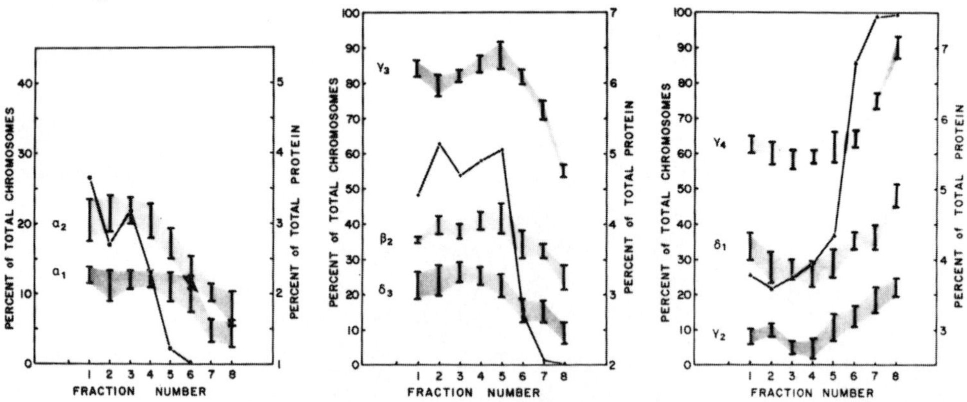

FIGURE 10B

histone proteins are detected in the treatment extract. The simple pattern of released proteins may be reflective of the coding sequences thought to be present in the R-band regions.[90,91] These observations indicate that treatments producing a common type of banding generally have similar effects on the NHCP, while treatments producing different types of banding have substantially different effects on these proteins. These results now need to be confirmed using isolated chromosomes.

Further evidence is derived by inference from immunocytochemical analysis of insect polytene chromosomes. These studies clearly demonstrate differential distribution of NHCP along the length of the polytene chromosome; the specifically localized immunofluorescence yields banding patterns analogous to the banding patterns obtained by other cytochemical techniques.[92]

Other types of investigations have yielded results suggesting certain proteins are associated with DNA sequences sensitive to nuclease digestions.[90,91,93-95] Digestion of metaphase chromosomes with DNase I or micrococcal nuclease is selective, in that the DNA in R-bands is digested extremely rapidly, the DNA in G- or Q-bands is digested more slowly, and the DNA in heterochromatic C-bands is digested most slowly.[93-95] It has been suggested that the differential sensitivity of these chromosome regions to nuclease digestion is due to differences in associated NHCP[93] or chromatin organization.[95] These results are particularly interesting in light of evidence that the DNA sequences which code for cytoplasmic poly (A)-containing mRNA are primarily localized in the nuclease-sensitive R-band regions.[90,91] Thus, it seems likely that although metaphase chromosomes are in a transcriptionally inactive, condensed state they may retain some of the structural characteristics of interphase chromatin such that expressed genes are maintained in a conformation different from that of nonexpressed genes. This may form the basis for chromosome banding and may also be involved in maintaining the differential state of a cell.

One aspect of chromosome structural studies is the identification of specific gene sequences and their associated proteins within the overall architecture of the mammalian metaphase chromosome. Those proteins controlling or modulating a specific gene activity should be specific to the chromosome(s) containing the gene. These gene-related proteins will be relatively scarce compared to the ubiquitously distributed structural proteins. Immunocytochemical studies indicate specific banding for NHCP along the length of insect polytene chromosomes,[92] but there is no immunochemical evidence for chromosome-specific proteins as would be expected if gene-specific proteins occur. However, another type of evidence is available which strongly suggests that such specific sites for proteins do occur. Weideli et al.[96] have provided indirect evidence that certain proteins have sequence-specific interaction with DNA. The isolated NHCP, DB-2, which constitutes less that 0.01% of the total nuclear protein, was used to select specific DNA sequences. The DNA sequences which bound DB-2 were then cloned and subsequently used for *in situ* hybridization on polytene chromosomes. The exogenously added DNA localized to a single band on a single polytene chromosome. By inference, the DB-2 protein is associated with that specific chromosome region. These results support the concept of gene-specific proteins and may be representative of mammalian metaphase chromosome organization.

IV. EVALUATION OF CHROMOSOMAL STRUCTURAL PROTEINS BY BIOCHEMICAL DISSECTION OF METAPHASE CHROMOSOMES

The role of protein in the maintenance of chromosome morphology has been investigated by use of biochemical extractions which alter morphology. Early studies by Maio and Schildkraut,[34] Stubblefield and Wray,[35] and Mace et al.[38] demonstrated that most (up to 91%) of the DNA of chromosomes is removed from isolated chromosomes by solutions which break ionic bonds. Maio and Schildkraut[34] discussed an acid-insoluble structural element of metaphase chromosomes assumed to be a ''residual'' or ''acidic'' protein. Stubblefield and Wray[97] provided morphological evidence using electron microscopy that extraction with 0.2 *N* HCl or chromosome isolation in 50% acetic acid do not disrupt the physical integrity of metaphase chromosomes. Mace et al.[38] also showed that about 60% of the total protein, including histone, is extracted by high salt concentrations.

Scanning electron microscopic (SEM) studies indicate that although a structure retaining the basic shape, size, and morphological characteristics of the metaphase chromosome remains after salt extraction, the fine structure of the chromosome fibers has been altered.[38] The gross morphology of unextracted isolated metaphase chromosomes corresponds well with results obtained from phase contrast and Nomarski optical microscopy, as well as with image reconstructed from serial thin-sectioned metaphase cells and whole-mounted isolated chromosomes examined with scanning electron microscopy (Figure 11A). The centromeric constriction, membranous plates, and chromatid morphology are all evident. Chromatids are comprised of highly coiled nucleoprotein fibers in a 290- to 540-Å microconvular structure over the surface of the chromatid. Metaphase chromosomes extracted with 0.4 *M* NaCl have the morphology shown in Figure 11B. The gross morphology is comparable to that of untreated chromosomes. However, the diameter of the microconvules has been reduced by more than half. As salt concentrations are increased, additional morphological changes occur.[38]

The effect of the salt extractions on the protein content of the chromosomes is shown in Figure 12. There is a preferential extraction of histone H1 by salt concentrations between 0.35 and 0.6 *M* NaCl which correlates with the change in chromosome fiber morphology. Content of other histones could not be determined by the electrophoresis conditions used in the study. Other reports indicate partial to total histone removal by similar extraction methods.[3,4,8,36,37,41,98] A group of proteins clustered at about 25,700 mol wt are also diminished. The residual nonhistones remaining in the chromosome structure are essentially the same for each salt concentration. These experiments suggest that the predominant NHCP in metaphase chromosomes contribute to fiber formation and to maintenance of overall chromosome structure.

More recent studies have shown that the early experiments may have been simplistic in experimental design and interpretation. Laemmli and co-workers,[3,4,36,37] Jeppeson et al.,[98] Okada and Comings,[8] Hadlaczky et al.[41] Wray et al.,[99] and Luck et al.[100] have found that extraction of chromosomes with different concentrations of NaCl, urea, ammonium acetate, sodium deoxycholate, and dextran sulfate-heparin (in combination) removes not only histones but also several nonhistone proteins, producing residual chromosome structures with DNA halos.

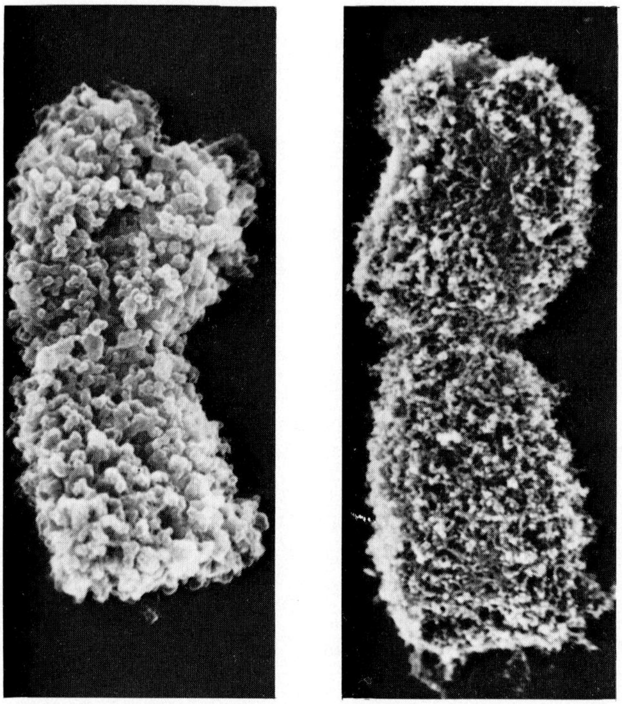

A B

FIGURE 11. Scanning electron microscopy (SEM) studies of intact and salt-extracted metaphase chromosomes. (A) SEM of an isolated metaphase chromosome showing the highly coiled chromatin fibers (microconvules) which comprise the chromatids (× 28,000). (B) SEM of an isolated chromosome which has been extracted with 0.4 *M* NaCl for 30 min showing a reduction in the order of coiling of chromatin fibers, while retaining chromatid configuration and centromeric constriction (× 26,000). (From Mace, M. L., Jr., Daskal, Y., Busch, H., Wray, V. P., and Wray, W., *Cytobios,* 19, 27, 1977. With permission.)

Laemmli et al.[3,4,36,37] have shown that the chromosomal DNA is still highly organized and relatively compact after the histones and most of the NHCP are gently removed from HeLa metaphase chromosomes. The structure of these ''histone-depleted'' chromosomes is thought to be due to the presence of a number of nonhistone proteins which form a central ''scaffold'' that retains the approximate size and shape of intact chromosomes and to which the DNA is attached, predominantly forming loops. The protein ''scaffold'' may be isolated independently of the DNA by treating HeLa chromosomes with micrococcal nuclease before removing the histones.

After dextran sulfate-heparin extraction to remove histones and NHCP, protein-depleted chromosomes from HeLa cells are only partially unfolded.[3,4,36,37] The structures obtained are insensitive to RNase A and are stable in 2 *M* NaCl, but are dissociated by mild trypsin or chymotrypsin treatment or by 4 *M* urea. The histone-depleted chromosomes have a DNA-to-protein ratio of about 3:1. Gel electrophoresis reveals the presence of about 20 nonhistone proteins and the virtual absence of histones. The prominent proteins of the residual chromosome fraction were about 53,000, 51,000, and 35,000 mol wt. The proteins released into the supernantant have an electrophoretic profile considerably different from those of the residual ''scaffold'' structure. The protein constituents of the scaffold are identical when

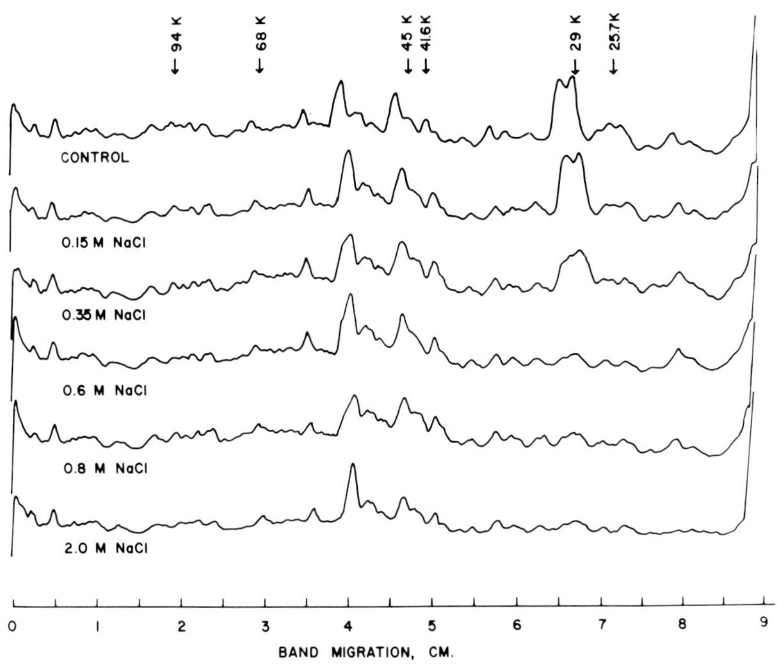

FIGURE 12. Residual proteins of isolated metaphase chromosomes extracted with increasing concentrations of NaCl. Aliquots of isolated metaphase chromosomes[54] were treated with the indicated concentrations of NaCl. Residual structures were collected by high-speed centrifugation. Pellets, rinsed gently to remove salts, were solubilized and analyzed for protein content. Aliquots (50 μg protein per slot) were chromatographed as described in Figure 2 on a 9.5% gel. (From Mace, M. L., Jr., Daskal, Y., Busch, H., Wray, V. P., and Wray, W., *Cytobios*, 19, 27, 1977. With permission.)

the histones are removed with either dextran sulfate or 2 *M* NaCl. Few, if any, of the proteins are associated with the DNA surrounding the dehistonized metaphase structure because nuclease digestion does not alter the protein pattern of the residual ''scaffold'' structure. Jeppesen et al.[98] also found that there is no observable change in the pattern of NHCP bands from acid-extracted chromosomes after incubation with DNase I.

Hadlaczky et al.[42] found that the removal of proteins by these extractions is a very rapid process. Prolonged extraction with either dextran sulfate-heparin or 2 *M* NaCl caused no further release of proteins. Histones are not completely extracted from their preparations, but otherwise their data support those of Paulson and Laemmli. However, they did not see as clear a distinction between the electrophoretic patterns of the nonhistone proteins remaining in scaffolds compared with those extracted, as did Adolph et al.[36,37] and Jeppesen et al.[98]

Extraction of isolated chromosomes by sodium deoxycholate removed several NHCP, mainly in the 45,000 to 70,000 mol wt range, but apparently not the same set of proteins as that extracted by salt treatments.[41] Histones H1, H2A, and H2B are almost all extracted. The nucleoprotein fibers do not spread as far as in salt-extracted preparations.

Luck et al.[100] found that treatment of isolated metaphase chromosomes with 3 or 4 *M* NH_4Ac caused differential release of proteins from the chromosome structure (Figure 13). In particular, a protein of about 50,000 mol wt and other lower-molecular-weight proteins are solubilized while a protein of about 58,000 mol wt and certain higher-molecular-weight proteins remain with the chromosome structure. These findings are in contrast to Okada and Comings[8] who report no difference in NHCP of whole chromosomes compared to the pellet fraction obtained after extraction with 4 *M* NH_4Ac. The patterns of residual protein obtained for Chinese hamster chromosomes are different from those observed for human chromosomes.[3,4,36,37,100]

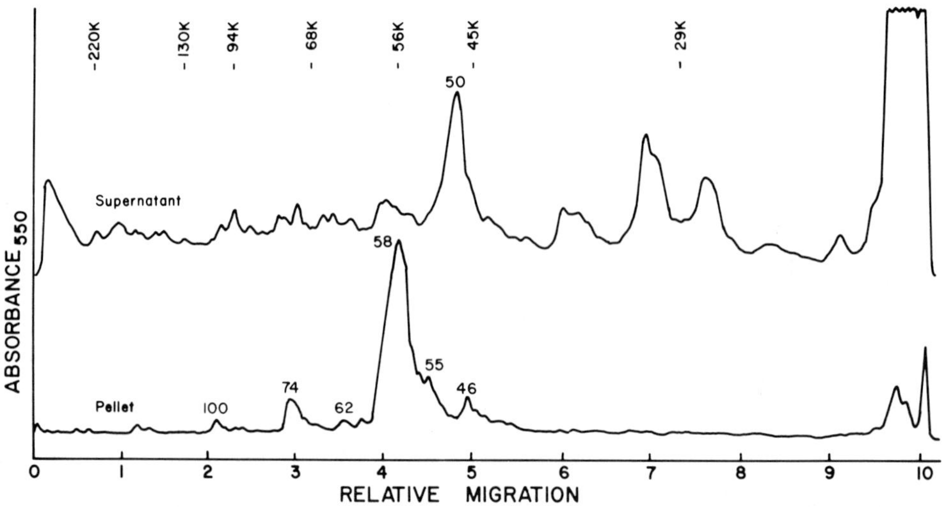

FIGURE 13. Extraction of isolated metaphase chromosomes with 3*M* ammonium acetate. Isolated Chinese hamster (CHO) chromosomes[54] in buffer HPC were extracted by exposure to 3 *M* ammonium acetate for 2 min at 0°C. The residual structures were collected by centrifugation. Protein in each fraction was collected by precipitation with ice-cold 25% trichloroacetic acid for 1 hr. The precipitate was pelleted by centrifugation at 10,000 × g for 10 min and washed three times in 0.5 *M* Tris-HCl buffer, pH 6.8. Aliquots of supernatant or pellet proteins (40 μg/lot) were resolved by 1D-PAGE on a 9.5% gel as described in Figure 2.

Our studies have shown that application of the same extraction method to both chromosomes and nuclei yields similar results using 1D-PAGE, so that extraction of isolated nuclei appears to be a reliable expression of what occurs to chromosomes treated under similar conditions. Figure 14 shows the results obtained from extraction of isolated nuclei with other agents known to cause morphological alterations of metaphase chromosomes. These treatments caused a variety of changes in the protein populations of the supernatant and pellet fractions remaining separated by the extraction.[100] The most obvious of these changes is the partial or differential extraction of the protein(s) of about 58,000 mol wt. Dextran sulfate-heparin extraction results in a split of 58,000-mol wt material into both supernatant and pellet, while extraction with 2 *M* NaCl removes little or none of the proteins of similar size. Conversely, urea seems to cause selective extraction of proteins of about 40,000 to 60,000 mol wt including the approximately 58,000-mol wt material. Examination of the profiles of several of the major peaks suggests that multiple species of similar molecular weight may be separating differentially into pellet and supernatant. Thus 1D-PAGE is probably inadequate for resolution of the individual species involved.

Although it appears that most or all of the histones and many of the NHCP are removed by these extraction procedures, there is no assurance that the order of removal of individual proteins is not of primary importance to the unfolding of the 250-Å fiber and the maintenance of the structure remaining after extractions have been completed. This ordered removal may be obscured by abrupt changes in extract solution concentration. Therefore, it appears that the nonhistones which are extracted may be the proteins primarily responsible for the coiling and folding of the DNA fibers at levels above the nucleosome. These nonhistone proteins may be necessary for the packing of the DNA into 250-Å fibers while the maintenance and packing of the normally observed metaphase structure may be associated with a different group of proteins as originally suggested by Laemmli et al.[3,4,36,37,101]

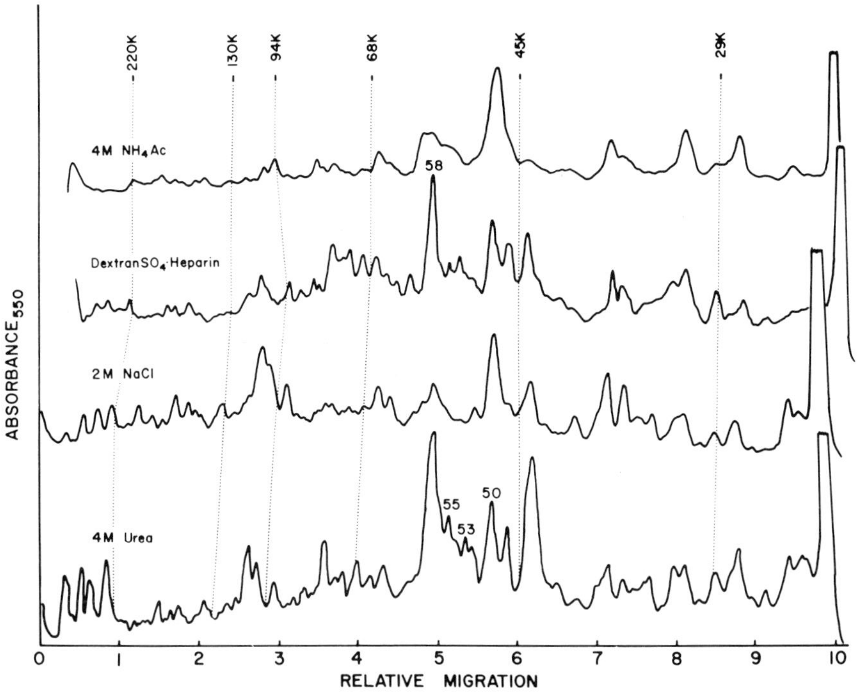

A

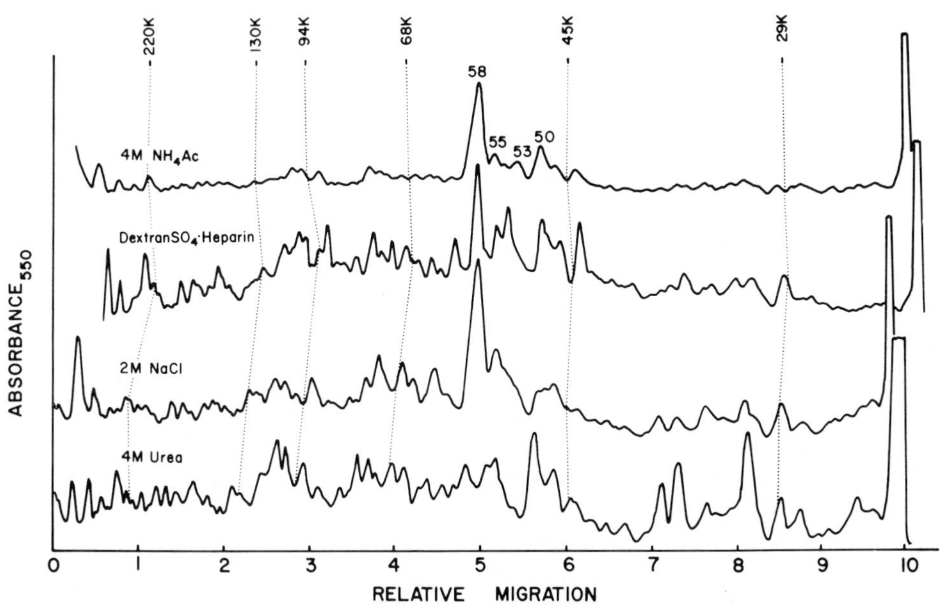

FIGURE 14B

FIGURE 14. Extraction of isolated nuclei with agents causing morphological alterations of metaphase chromosomes. (A) Supernatant fractions. (B) Pellet fractions. Isolated nuclei[54] were extracted with the indicated agents by the methods described in Figure 13 and analyzed as described in Figure 2. Dotted lines indicate migration of protein standards on the different gels.

V. RELATIONSHIP BETWEEN CHROMOSOMES AND THE NUCLEAR MATRIX

In the eukaryotic interphase nucleus, the chromosome fibers must be spatially ordered during replication such that the daughter strands remain untangled yet coupled in a precise topographical fashion for later entry into mitosis. This spatial ordering of interphase chromosome fibers is thought to be derived from their close interaction with a nuclear skeletal structure, the nuclear matrix.[102] The nuclear matrix is composed of three parts: the nuclear pore-lamina complex, the nucleolar matrix, and the intranuclear matrix. The nuclear DNA is attached to all three parts apparently by DNA-protein interactions. Newly replicated DNA is preferentially associated with the nuclear matrix structure, particularly the intranuclear matrix portion. Both the intimate association of the nuclear matrix with sites for DNA synthesis and the biochemical stability of the DNA-matrix protein supramolecular complex suggest that the nuclear proteins which constitute the nuclear skeletal structure should be a part of the metaphase chromosome and may indeed be a portion of a fundamental metaphase chromosome structure.[103]

Protein components of the nuclear matrix are divided into two general groups. One group constitutes 30 to 50% of the total protein and is comprised of a relatively few polypeptide species of about 60,000 to 75,000 mol wt.[102] The three prominent one-dimensional polyacrylamide gel bands have been termed lamins A, B, and C.[103] When examined by 2D-PAGE, each 1D-PAGE lamin band either has isomers or is a composite of more than one polypeptide species.[104,105] Lamins A and C are apparently single species having several isomers of apparent pI 7.5, while lamin B is composed of two polypeptides, one which migrates at about 68,000 mol wt, apparent pI 7.2 to 7.5, and another which migrates at about 68,000 mol wt, apparent pI 6.2 to 6.4. The lamins are reported to be immunologically related.[104] The remaining proteins, which constitue 50 to 70% of the nuclear matrix proteins, are heterogeneous with respect to molecular weight and isoelectric point.

The subnuclear location of these lamins is relevant to studies relating the nuclear skeleton to interphase and metaphase chromosome structure. Currently there are two opposing theories concerning the subnuclear location of the predominant lamin species. Blobel and coworkers[103,107,108] visualize these proteins as belonging preferentially to the membrane-nuclear pore-lamina structure while Berezney and Coffee[102] and others[103] feel that these proteins are components of a larger nuclear fiber structure which extends throughout the nuclear compartment. In the latter instance, the lamins might be expected to be incorporated into the metaphase chromosome. To search for the presence of matrix lamin proteins in metaphase chromosomes we have compared the protein composition of interphase nuclei, metaphase chromosomes, nuclear matrix, and interphase chromatin from nuclei and chromosomes isolated in the neutral buffers by Wray and Stubblefield (Figure 15).

Our findings indicate that nuclear matrix fractions of CHO cells have a protein profile different from, and more complex than, that reported for rat liver nuclei. In particular, the approximately 60,000 to 75,00 lamin proteins (indicated by a bar in Figure 15 A and B) are not of the same prominence relative to the remaining proteins. Other research groups, working with erythrocytes and cultured cells, have also observed this.[109,110] Moreover, in side by side comparisons, the CHO lamin proteins, except for the intermediate-sized component, do not comigrate with rat liver lamin proteins. The decreased content of the lamin proteins in the CHO matrix fraction is apparently not due to protease activity during the course of nuclei and chromosome isolation because inhibition of protease activity by PMSF and sodium tetrathionate does not affect the protein profile. Several recent publications present data which suggest that environmental conditions during preparation of the different fractions have significant effects on the final results.[111] Although most of these conditions have been met in the procedures utilized for this study, their effects on the results must be further evaluated.

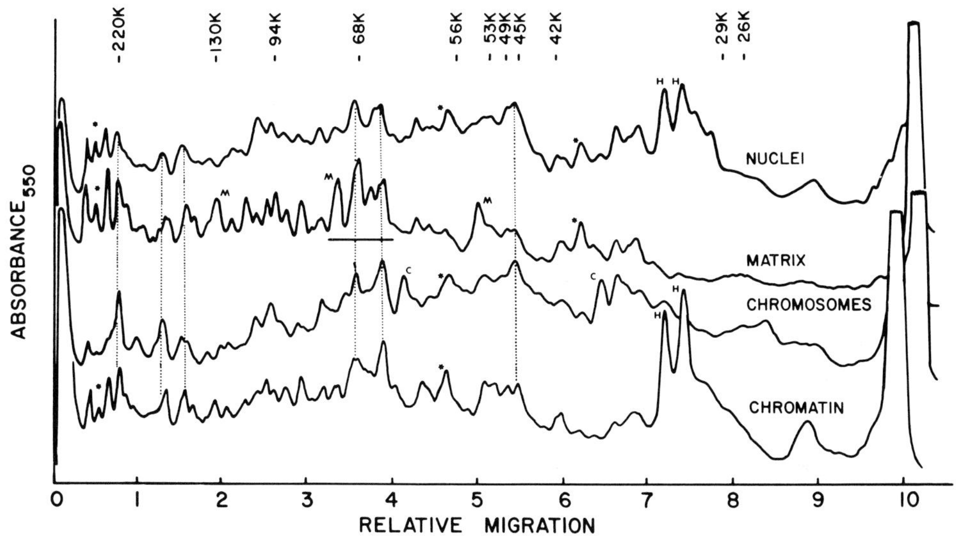

A

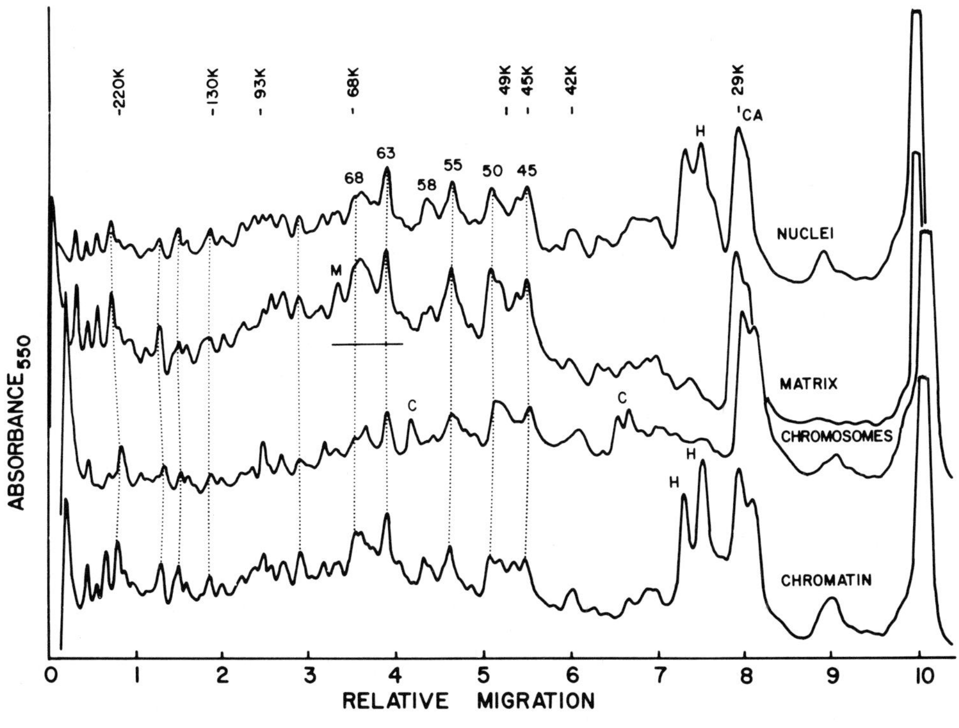

B

FIGURE 15. Comparison of the protein complements of nuclei, matrix, chromosomes, and chromatin. (A) CHO cells. (B) HeLa cells. Nuclei and chromosomes were isolated and purified as described in Figure 2. Chromatin[58] was prepared from isolated nuclei.[54] Nuclear matrix (also known as pore complex-lamina fraction) material was obtained from isolated nuclei by the method of Dwyer and Blobel[108] with RNase included in the nuclease digestion steps. Aliquots (40 μg protein per slot) were electrophoresed as described in Figure 2. The lamin proteins are indicated by an underline bar. Dotted lines connect protein bands common to all fractions: (*) proteins or groupings of proteins present in some, but not all fractions; (M) proteins enhanced in matrix fractions; (C) proteins enhanced in chromosome fractions. Dotted lines indicate migration of protein standards on the different gels.

Six of the more abundant protein species are similar in chromosomes and all the nuclear fractions (dashed lines) (Figure 15A). Other proteins or groupings of proteins are present in some but not all the fractions (asterisk). Some proteins (indicated by M) are enhanced in matrix fractions while other (indicated by C) are enhanced in chromosome fractions. Metaphase chromosomes are deficient in proteins of greater than 220,000 mol wt (indicated by *) compared to all the other fractions.

These results suggest that some, but not all, of the matrix proteins are incorporated into the metaphase structure. In particular, the p-68 protein which is enhanced in CHO nuclei and chromosomes isolated in acidic media appears to be the same as the apparent pI 6.2 to 6.4 component of the intermediate-sized lamin B protein of rat liver nuclear matrix (See Figure 16) This protein is known to be localized in the peripheral chromatin of nuclei.[112] Further experimentation is needed to gain a full understanding of these results in relation to interphase and metaphase chromosome organization.

VI. SUMMARY AND DISCUSSION

Biochemical analysis shows the NHCP of metaphase chromosomes are as numerous as those of isolated chromatin or interphase nuclei. Species-related differences in the general pattern of proteins obtained from both isolated metaphase chromosomes and interphase nuclei are apparent. The method of chromosome isolation has an effect on the content of those abundant proteins detected by 1D-PAGE analysis. There is a general similarity in the band patterns observed, but there is quantitative variation in band profiles. The significant differences in protein content of chromosomes isolated in acid media compared to chromosomes isolated in neutral or alkaline media are not understood. Fractions obtained by any neutral or alkaline pH method are subject to proteolysis unless processed quickly.

The abundant NHCP of chromosomes, chromatin, nuclear matrix, and interphase nuclei are strikingly similar; but differences are also evident. Each of these biologically or biochemically defined fractions have components or properties in common with one or more of the other fractions. Therefore, a commonality of NHCP between the fractions is appropriate. Differences between the NHCP content of the fractions are related to cell cycle events, requirements of gene replication and expression, supramolecular structural elements, and biochemical fractionation methods. The differences in NHCP content demonstrated by relatively insensitive analytical methods should become more apparent and better characterized as more sophisticated analytical technology is used.

At the cytological level, several experimentally diverse investigations demonstrate that there are proteins which are localized on the metaphase chromosome at specific morphologically or spatially defined sites. The evidence is indirect because the proteins involved are characterized only by their immunospecificity or their differential staining properties. Further research is necessary to identify the specific proteins involved.

Information about the biochemistry of the proteins from metaphase chromosomes is becoming more detailed, but further analyses which employ separation by multiple molecular parameters and ultrasensitive detection methods are needed. A limiting corollary which accompanies such a detailed study is the ultimate definition of what constitutes a metaphase chromosomal protein. Not only must various isolation conditions as well as the presence of chromosome-bound fragments of membrane, ribonucleoprotein, residual cytoskeletal elements, and nonspecifically bound soluble cytoplasmic proteins by contended with but dual function proteins present in both nucleus and cytoplasm must also be considered. The proteins common to both nuclei and cytoplasm may have some special role in cell structure and physiology. Nevertheless, because of this commonality, there is no current stringent definition of a nonhistone chromosomal protein. With these factors in mind, our working definition of a metaphase chromosomal protein has been derived from criteria taken from

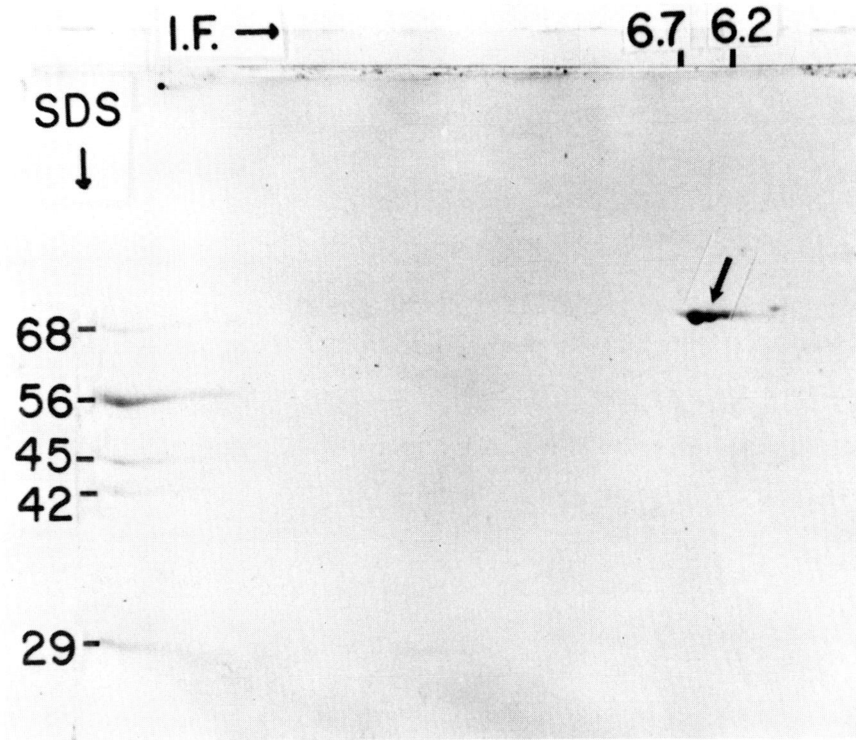

A

FIGURE 16. 2D-PAGE separations of proteins from metaphase chromosomes isolated in 50% acetic acid compared to proteins of nuclear matrix (pore complex-lamina) fractions. (A) Chinese hamster ovary (CHO) metaphase chromosomes. (B) Rat liver nuclear matrix (also known as nuclear pore complex-lamina). Chromosome and matrix fractions were prepared as described in Figure 15. Proteins were separated by 2D-PAGE[63] using a 1.5 mm × 8.5 cm, 9.5% polyacrylamide gel in the second dimension. Protein p-68 is indicated by arrows.

both cell biology and biochemistry disciplines. Thus, a metaphase chromosomal protein is one which remains associated with an isolated, purified metaphase chromosome judged to be morphologically excellent by SEM and localizes to the metaphase chromosome *in situ* by cytochemical methods which are specific and definitive.

Most of the questions relevant to the understanding of the supramolecular organization of the NHCP in metaphase chromosomes and the role of these proteins in maintenance of the metaphase structure and regulation of gene activity have yet to be answered. There is recent evidence demonstrating specific protein associations with unique centromeric and nucleolus organizer regions. The nature of these proteins has not been defined. Still at issue are the complexities and specificities of proteins associated with the telomeric regions. Biochemically or metabolically defined portions of the DNA duplex, such as highly repetitive DNA, middle repetitive DNA, and unique DNA, as well as early and late replicating regions are unexplored with regard to their protein complements. Equally unclear is the possible presence of subsets of the chromosomal protein complement which contribute to the process of condensation at mitosis or to fundamental chromosome structure during the mitotic period of the cell cycle.

The data and concepts derived from studies using interphase nuclei, chromatin, and insect polytene chromosomes must be considered in future experiments. However, the unique nature of metaphase chromosomes prohibits chromosome supramolecular structure to be

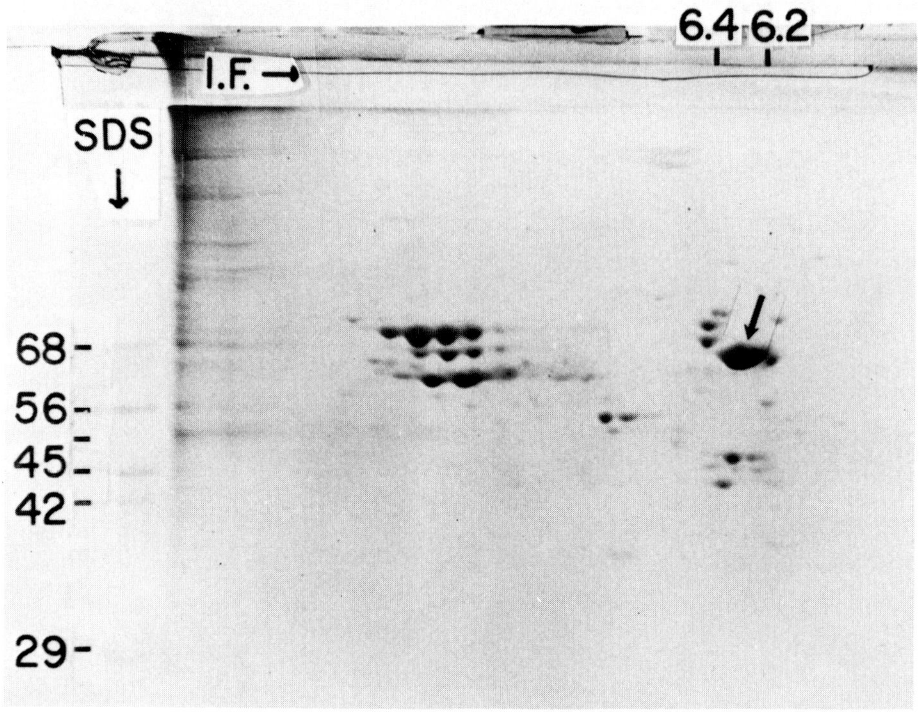

FIGURE 16B

directly inferred from these sources. A metaphase chromosome is highly organized, condensed, and contains one copy of a portion of the gene complement per chromatid; thus it is a unique cellular organelle which the cell biologically fractionates into discrete, identifiable, and consistent units of genetic information.

Current experimental approaches to the study of the biochemistry and supramolecular organization of the NHCP of metaphase chromosomes are multifaceted. Due to the experimental complexities, the process by which all these parameters will coincide represent long-term research goals. With the current rapid acquisition of new techniques such as ultrasensitive multiparameter protein analysis, chromosome isolation and sorting, and of new cell models having unique genetic or chromosome structural features, great progress should be made in the near future.

VII. ADDENDUM

Since the original manuscript for this chapter was written, a particularly important result has been obtained in our laboratory concerning the localization of unique NHCP on individual chromosomes. Also, there have been several significant papers published from other laboratories which cannot be integrated into the text, but are listed below for the reader's convenience.

The chicken lymphoblastoid cell (line MSB-1), established from a spleen tumor in a bird having Marek's disease has a normal karyotype with the exception of a single translocation. Chicken cells contain 28 macrochromosomes and more than 40 microchromosomes which can not be clearly identified under light microscopy. The macrochromosomes vary in size from less than 1 μm in length to about 8 μm and are sufficiently different in size to separate, and morphologically distinct enough to easily recognize. The chromosomes were isolated in 50% acetic acid, and fractionated according to size on a 20 to 40% sucrose gradient. The fractions were monitored with flow cytometry and combined in such a way as to permit efficient flow cytometry sorting of 10 million each of the numbers 1,2,3, and 4 + Z chromosomes. Proteins from chromosomes solubilized in SDS were subjected to single dimension discontinuous polyacrylamide gel electrophoresis on a 9.5% gel. Protein separation patterns obtained using a highly sensitive silver staining technique developed by Wray et al., *Anal. Biochem.*, 118, 197, 1981, demonstrated conclusively that along with a population of ubiquituously distributed proteins, there were certain proteins which were unique to the individual chromosome species. While the staining intensity of many proteins in a molecular weight region remained constant, others were either diminished or enriched in comparison. In some cases, the differences were dramatic. For example, chromosome 3 contained a protein of 58,000 mw not apparent in chromosomes 1,2, or 4 + Z; and chromosomes 4 + Z contained a protein of 80,000 not observed in chromosomes 1,2, or 3.

Adolph, K. W. and Phelps, J. P., Role of non-histones in chromosome structure. Cell cycle variations in protein synthesis, *J. Biol. Chem.*, 257, 9086, 1982.

Bowen, B. C., DNA fragments associated with chromosome scaffolds, *Nuc. Acids Res.*, 9, 5093, 1981.

Detke, S. and Keller, J. M., Comparison of the proteins present in HeLa cell interphase nucleoskeletons and metaphase chromosome scaffolds, *J. Biol., Chem.*, 257, 3905, 1982.

Earnshaw, W. C. and Laemmli, U. K., Architecture of metaphase chromosomes and chromosome scaffolds, *J. Cell Biol.*, 96, 84, 1983.

Engelhardt, P., Plagens, U., Zbarsky, I. B., and Filatova, L. S., Granules 25—30 nm diameter: basic constituent of the nuclear matrix, chromosome scaffold, and nuclear envelope, *Proc. Natl. Acad. Sci. U.S.A.*, 79, 6937, 1982.

Hancock, J. M. and Sumner, A. T., Role of proteins in the production of different types of chromosome bands, *Cytobios.*, 35, 37, 1982.

Jost, E. and Johnson, R. T., Nuclear lamina assembly, synthesis and disaggregation during the cell cycle in synchronized HeLa cells, *J. Cell Sci.*, 47, 25, 1981.

Kaiserman, M. Z. and Burkholder, G. D., Silver-stained core-like structures in Chinese hamster metaphase chromosomes, *Can. J. Genet. Cytol.*, 22, 627, 1980.

Kuo, M. T., Comparison of chromosomal structures isolated under different conditions, *Exp. Cell Res.*, 138, 221, 1982.

Labhart, P., Koller, T., and Wunderli, H., Involvement of higher order chromatin structures in metaphase chromosome organization, *Cell*, 30, 115, 1982.

Lebkowski, J. S. and Laemmli, U. K., Nonhistone proteins and long-range organization of HeLa interphase DNA, *J. Mol. Biol.*, 156, 325, 1982.

Lewis, C. D. and Laemmli, U. K., Higher order metaphase chromosome structure: evidence for metalloprotein interactions, *Cell*, 29, 171, 1982.

Paulson, J. R., Isolation of chromosome clusters from metaphase-arrested HeLa cells, *Chromosoma*, 85, 571, 1983.

Paulson, J. R. and Langmore, J. P., Low angle-X-ray diffraction studies of HeLa metaphase chromosomes. Effects of histone phosphorylation and chromosome isolation procedure, *J. Cell Biol. V.*, 96, 1132, 1983.

Paulson, J. R., Chromatin and chromosomal proteins, in *Electron Microscopy of Proteins*, Vol. 3, Harris, R., Ed., Academic Press, London, 1982, 78.

Paulson, J. R. and Taylor, S. S., Phosphorylation of histones 1 and 3 and nonhistone high mobility group 14 by an endogenous kinase in HeLa metaphase chromosomes, *J. Biol. Chem.*, 257, 6064, 1982.

Paulson, J. R., Sulfhydryl reagents prevent dephosphorylation and proteolysis of histones in isolated HeLa metaphase chromosomes, *Eur. J. Biochem.*, 111, 189, 1980.

Peters, K. E., Okada, T. A., and Comings, D. E., Chinese hamster nuclear proteins. An electrophoretic analysis of interphase, metaphase, and nuclear matrix preparations, *Eur. J. Biochem.*, 129, 221, 1982.

Satya-Prakash, K. L., Hsu, T. C., and Pathak, S., Behavior of the chromosome core in mitosis and meiosis, *Chromosoma*, 81, 1, 1980.

Song, M.-K. and Adolph, K. W., Phosphorylation of nonhistone proteins during HeLa cell cycle. Relationship to DNA synthesis and metotic chromosome condensation, *J. Biol. Chem.*, 258, 3309, 1983.

Wray, W. and Wray, V. P., Individual chromosomes contain unique proteins, *J. Cell Biol.*, in press, 1983.

Zheng, H.-Z. and Burkholder, G. D., Differential silver staining of chromatin in metaphase chromosomes. *Exp. Cell Res.*, 141, 117, 1982.

ACKNOWLEDGMENTS

We wish to express our sincere appreciation to Dr. Dennis Johnston and the Department of Biomathematics of M. D. Anderson Hospital for assisting with computer simulations. Excellent technical assistance was provided by Ms. Cynthia Furlong, Mr. Steve Ladd, and Ms. Elizabeth Berglund. We appreciate the critical review of the manuscript and encouragement given by Drs. Elton Stubblefield, Eric Eastman, and Susanne Gollin. Grant support from N.I.H. CA-18455, and N.S.F. PCM 79-05428 is gratefully acknowledged. W. W. is the recipient of Research Career Development Award CA-00532 from the National Cancer Institute.

REFERENCES

1. **Kornberg, R. D.,** Chromatin structure: A repeating unit of histones and DNA, *Science,* 184, 868, 1974.
2. **Noll, M.,** Subunit structure of chromatin, *Nature (London),* 251, 249, 1974.
3. **Laemmli, U. K., Cheng, S. M., Adolph, K. W., Paulson, J. R., Brown, J. A., and Baumbach, W. R.,** Metaphase chromosome structure: the role of nonhistone proteins, *Cold Spring Harbor Symp. Quant. Biol.,* 42, 351, 1977.
4. **Paulson, J. R. and Laemmli, U. K.,** The structure of histone-depleted metaphase chromosomes, *Cell,* 12, 817 1977.
5. **Mullinger, A. M. and Johnson, R. T.,** Organization of supercoiled DNA from human chromosomes, *J. Cell Sci.,* 38, 369, 1979.
6. **Mullinger, A. M. and Johnson, R. T.,** Packing DNA into chromosomes, *J. Cell Sci.,* 46, 61, 1980.
7. **Okada, T. A. and Comings, D. E.,** Higher order structure of chromosomes, *Chromosoma,* 72, 1, 1979.
8. **Okada, T. A. and Comings, D. E.,** A search for protein cores in chromosomes: Is the scaffold an artifact?, *Am. J. Hum. Genet.,* 32, 814, 1980.
9. **Kornberg, R. D.,** Structure of chromatin, *Annu. Rev. Biochem.,* 46, 931, 1977.
10. **Felsenfeld, G.,** Chromatin, *Nature (London),* 271, 115, 1978.
11. **McGhee, J. D. and Felsenfeld, G.,** Nucleosome structure, *Annu. Rev. Biochem.,* 49, 1115, 1980.
12. **Compton, J. L. Hancock, R., Oudet, P., and Chambon, P.,** Biochemical and electronmicroscopic evidence that the subunit structure of Chinese-hamster-ovary interphase chromatin is conserved in mitotic chromosomes, *Eur. J. Biochem.,* 70, 555, 1976.
13. **Wigler, M. H. and Axel, R.,** Nucleosomes in metaphase chromosomes, *Nucleic Acids Res.,* 3, 1463, 1976.
14. **Finch, J. T. and Klug, A.,** Solenoidal model for superstructure in chromatin, *Proc. Natl. Acad. Sci., U.S.A.,* 73, 1897, 1976.
15. **Carpenter, B. G., Baldwin, J. P., Bradbury, E. M., and Ibel, K.,** Organization of subunits in chromatin, *Nucleic Acids Res.,* 3, 1739, 1976.
16. **Campbell, A. M., Cotler, R. I., and Pardon, J. F.,** Light scattering measurements supporting helical structures for chromatin in solution, *Nucleic Acids Res.,* 5, 1571, 1978.
17. **Schmitz, K. S. and Ramanathan, B.,** Generation of third-order folded structure for chromatin, *J. Theor. Biol.,* 83, 297, 1980.
18. **Manning, G.,** Theory of H1-mediated control of higher orders of structure in chromatin, *Biopolymers,* 18, 2929, 1979.
19. **Thoma, F., Koller, Th., and Klug, A.,** Involvement of histone H1 in the organization of the nucleosome and of the salt-dependent superstructures of chromatin, *J. Cell Biol.,* 83, 403, 1979.
20. **Goyanes, V. J., Matsui, S.-I., and Sandberg, A. A.,** The basis of chromatin fiber assembly within chromosomes studied by histone-DNA crosslinking followed by trypsin digestion, *Chromosoma,* 78, 123, 1980.
21. **Rattner, J. B. and Hamkalo, B. A.,** Higher order structure in metaphase chromosomes. II. The relationship between the 250 Å fiber, superbeads and beads-in-a-string, *Chromosoma,* 69, 373, 1978.
22. **Bradbury, E. M., Inglis, R. J., and Matthews, H. R.,** Control of cell division by very lysine rich histone (F1) phosphorylation, *Nature (London),* 247, 257, 1974.

23. **Bradbury, E. M., Inglis, R. J., Matthews, H. R., and Langan, T. A.,** Molecular basis of control of mitotic cell division in eukaryotes, *Nature (London),* 249, 553, 1974.
24. **Gurley, L. R., Walters, R. A., and Tobey, R. A.,** Cell cycle-specific changes in histone phosphorylation associated with cell proliferation and chromosome condensation, *J. Cell Biol.,* 60, 356, 1974.
25. **Gurley, L. R., D'Anna, J. A., Barham, S. S., Deaven, L. L., and Tobey, R. A.,** Histone phosphorylation and chromatin structure during mitosis in Chinese hamster cells, *Eur. J. Biochem.,* 84, 1, 1978.
26. **Matsumoto, V.-I. Yasuda, H., Mita, S., Marunouchi, T., and Yamada, M.-A.,** Evidence for the envolvement of H1 histone prosphorylation in chromosome condensation, *Nature (London),* 284, 181, 1980.
27. **Matsui, S.-I., Seon, B. K., and Sandberg, A. A.,** Disappearance of a structural chromatin protein A24 in mitosis: Implications for molecular basis of chromatin condensation, *Proc. Natl. Acad. Sci. U.S.A.,* 76, 6386, 1979.
28. **Elgin, S. C. R., and Bonner, J.,** Limited heterogeneity of the major nonhistone chromosomal proteins, *Biochemistry,* 9, 4440, 1970.
29. **MacGillivray, A. J. and Rickwood, D.,** Heterogeneity of mouse-chromatin nonhistone proteins as evidenced by two-dimensional polyacrylamide-gel electrophoresis and ion-exchange chromatography, *Eur. J. Biochem.,* 41, 181, 1974.
30. **Elgin, S. C. R. and Weintraub, H.,** Chromosomal proteins and chromatin structure, *Annu. Rev. Biochem.,* 44, 725, 1975.
31. **Peterson, J. L. and McConkey, E. H.,** Non-histone chromosomal proteins from HeLa cells. A survey by high resolution, two dimensional electrophoresis, *J. Biol. Chem.,* 251, 548, 1976.
32. **Peterson, J. L. and McConkey, E. H.,** Proteins of friend leukemia cells. Comparison of hemoglobin-synthesizing and noninduced populations, *J. Biol. Chem.,* 251, 55, 1976.
33. **Takami, H. and Busch, H.,** Two-dimensional gel electrophoretic comparison of proteins of nuclear fractions of normal liver and Novikoff hepatoma, *Cancer Res.,* 39, 507, 1979.
34. **Maio, J. J. and Schildkraut, C. L.,** Isolated mammalian metaphase chromosomes. II. Fractionated chromosomes of mouse and Chinese hamster cells, *J. Mol. Biol.,* 40, 203, 1969.
35. **Stubblefield, E. and Wray, W.,** Architecture of the Chinese hamster metaphase chromosome, *Chromosoma,* 32, 262, 1971.
36. **Adolph, K. W., Cheng, S. M., and Laemmli, U. K.,** Role of nonhistone proteins in metaphase chromosome structure, *Cell,* 12, 805, 1977.
37. **Adolph, K. W., Cheng, S. M., Paulson, J. R., and Laemmli, U. K.,** Isolation of a protein scaffold from mitotic HeLa cell chromsomes, *Proc. Natl. Acad. Sci. U.S.A.,* 74, 4937, 1977.
38. **Mace, H. L., Jr., Daskal, Y., Busch, H., Wray, V. P., and Wray, W.,** Isolated metaphase chromosomes: scanning electron microscopic appearance of salt-extracted chromosomes, *Cytobios,* 19, 27, 1977.
39. **Isackson, P. J., Fishback, J. L., Bidney, D. L., and Reeck, G. R.,** Preferential affinity of high molecular weight high mobility group non-histone chromatin proteins for single-stranded DNA, *J. Biol. Chem.,* 254, 5596, 1979.
40. **Sandeen, G., Wood, W. I., and Felsenfeld, G.,** Interaction of high mobility protein HMG14 and 17 with nucleosomes, *Nucleic Acids Res.,* 8, 3757, 1980.
41. **Hadlaczky, G., Sumner, A. T., and Ross, A.,** Protein-depleted chromosomes. I. Structure of isolated protein-depleted chromosomes, *Chromosoma,* 81, 537, 1981.
42. **Hadlaczky, G., Sumner, A. T., and Ross, A.,** Protein-depleted chromosomes. II. Experiments concerning the reality of chromosome scaffolds, *Chromosoma,* 81, 557, 1981.
43. **Elgin, S. C. R., Boyd, J. B., Hood, L. E., Wray, W., and Wu, F. C.,** A prologue to the study of the nonhistone chromosomal proteins, *Cold Spring Harbor Symp. Quant. Biol.,* 38, 821, 1973.
44. **Goodwin, G. H., Walker, J. M., and Johns, E. W.,** High mobility group (HMG) nonhistone chromosomal proteins, in *Cell Nucleus,* Vol. 6, Busch, H., Ed., Academic Press, New York, 1978, 181.
45. **Albright, S. C., Wiseman, J. M., Lange, R. A ., and Garrard, W. T.,** Subunit structures of different electrophoretic forms of nucleosomes, *J. Biol. Chem.,* 255, 3673, 1980.
46. **Mathew, C. G. P., Goodwin, G. H., and Johns, E. W.,** Studies on the association of the high mobility group non-histone chromatin proteins with isolated nucleosomes, *Nucleic Acids Res.,* 6, 167, 1979.
47. **Albanese, I. and Weintraub, H.,** Electrophoretic separation of a class of nucleosomes enriched in HMG 14 and 17 and actively transcribed globin genes, *Nucleic Acids Res.,* 8, 2787, 1980.
48. **Mardian, J. K. W., Paton, A. E., Bunick, G. J., and Olins, D. E.,** Nucleosome cores have two specific binding sites for nonhistone chromosomal proteins HMG 14 and HMG 17, *Science,* 209, 1534, 1980.
49. **Hearst, J. E. and Botchan, M.,** The eukaryotic chromosome, *Annu. Rev. Biochem,* 39, 151, 1970.
50. **Huberman, J. A. and Attardi, G. J.,** Isolation of metaphase chromosomes from HeLa cells, *J. Cell Biol.,* 31, 95, 1966.
51. **Maio, J. J. and Schildkraut, C. L.,** Isolated mammalian metaphase chromosomes. I. General characteristics of nucleic acids and proteins, *J. Mol. Biol.,* 24, 29, 1967.

52. **Mendelsohn, J., Moore, D. E., and Salzman N. P.,** Separation of isolated Chinese hamster metaphase chromosomes into three size-groups, *J. Mol. Biol.*, 32, 101, 1968.
53. **Salzman, N. P., Moore, D. E., and Mendelsohn, J.,** Isolation and characterization of human metaphase chromosomes, *Proc. Natl. Acad. Sci. U.S.A.*, 56, 1449, 1966.
54. **Wray, W. and Stubblefield, E.,** A new method for the rapid isolation of chromosomes, mitotis apparatus or nuclei from mammalian fibroblasts at near neutral pH, *Exp. Cell Res.*, 59, 469, 1970.
55. **Wray, W.,** Isopycnic centrifugation of mammalian metaphase chromosomes in metrizamide, *FEBS Lett.*, 62, 202, 1976.
56. **Wray, V. P., Elgin, S. C. R., and Wray, W.,** Proteins of metaphase chromosomes and interphase chromatin, *Nucleic Acids Res.*, 8, 4155, 1980.
57. **Nikolayenko, N. S. and Pinaev, G. P.,** Lipid composition of metaphase chromosomes dependent on the method of their isolation, *Tsitologiya*, 23, 185, 1981.
58. **Bonner, J., Chalkley, G. R., Dahmus, M., Fujimura, F., Huang, R. C., Huberman, J., Jensen, R., Marushige, K., Ohlenbusch, H., Olivera, B., and Widholm, J.,** Isolation and characterization of chromosomal nucleoproteins, *Methods Enzymol.*, 12B, 3, 1968.
59. **Sadgopal, A. and Bonner, J.,** Proteins of interphase and metaphase chromosomes compared, *Biochim. Biophys. Acta*, 207, 227, 1970.
60. **Garrard, W. T., Nobis, P., and Hancock, R.,** Intermolecular disulfide bond formation of F3 histone in chromatin, *Fed. Proc.*, 34, 611, 1975.
61. **Laemmli, U. K.,** Cleavage of structural protein during the assembly of the head of bacteriophage T4, *Nature (London)*, 227, 680, 1970.
62. **Wray, V. P., Wray, W., and Stubblefield, E.,** Protein complement of chromosomes and nuclei isolated in acidic and neutral media, *J. Cell Biol.*, 83, 166a, 1979.
63. **O'Farrell, P. H.,** High resolution two-dimensional electrophoresis of proteins, *J. Biol Chem.*, 250, 4007, 1975.
64. **Bode, U., Deisseroth, A., and Hendrick, D.,** Expression of human and mouse nonhistone chromosomal proteins in hybrid mouse erythroleukemia cells containing a single human chromosoma, *Proc. Natl. Acad. Sci. U.S.A.*, 78, 2815, 1981.
65. **Lydersen, B. K. and Pettijohn, D. E.,** Human-specific nuclear protein that associated with the polar region of the mitotic apparatus: Distribution in a human/hamster hybrid cell, *Cell*, 22, 489, 1980.
66. **Stubblefield, E., Linde, S., Franolich, F. K., and Lee, L. Y.,** Analytical techniques for isolated metaphase chromosome fractions, *Methods Cell Biol.*, 17, 101, 1978.
67. **Blumenthal, A. B., Dieden, J. D., Kapp, L. N., and Sedat, J. W.,** Rapid isolation of metaphase chromosomes containing high molecular weight DNA, *J. Cell Biol.*, 81, 255, 1979.
68. **Adolph, K. W.,** Isolation and structural organization of human mitotic chromosomes, *Chromosoma*, 76, 23, 1980.
69. **Langlois, R. G., Carrano, A. V., Gray, J. W., and VanDilla, M. A.,** Cytochemical studies of metaphase chromosomes by flow cytometry, *Crhomosoma*, 77, 229, 1980.
70. **Wray, W.,** Isolation of metaphase chromosomes, mitotic apparatus, and nuclei, *Methods Cell Biol.*, 6, 283, 1973.
71. **Cooke, P.,** A filamentous cytoskeleton in vertebrate smooth muscle fibers, *J. Cell Biol.*, 68, 539, 1976.
72. **Small, V. J. and Sobieszek, A.,** Studies on the function and composition of the 10-nm (100 Å) filaments of vertebrate smooth muscle, *J. Cell Sci.*, 23, 243, 1977.
73. **Starger, J. M., Brown, W. E., Goldman, A. E., and Goldman, R. D.,** Biochemical and immunological analysis of rapidly purified 10-nm filaments from baby hamster kidney (BHK-21) cells, *J. Cell Biol.*, 78, 93, 1978.
74. **Granger, B. L. and Lazarides, E.,** Desmin and vimentin coexist at the periphery of the myofibril Z disc, *Cell*, 18, 1053, 1979.
75. **Bustin, M., Yamasaki, H., Goldblatt, D., Shani, M., Huberman, E., and Sachs, L.,** Histone distribution in chromosomes revealed by antihistone sera, *Exp. Cell Res.*, 97, 440, 1976.
76. **Bustin, M.,** Immunological approaches to chromatin and chromosome structure and function, *Curr. Top. Microbiol. Immunol.*, 88, 105, 1979.
77. **Smith, B. J., Robertson, D., Birbeck, M. S. C., Goodwin, G. H., and Johns, E. W.,** Immunochemical studies of high mobility group non-histone chromatin proteins HMG1 and HMG2, *Exp. Cell Res.*, 115, 420, 1978.
78. **Wray, W. and Wray, V. P.,** Isolated metaphase chromosomes. II. Proteins of Chinese hamster chromosomes, *J. Histochem. Cytochem.*, 27, 454, 1979.
79. **Goodpasture, C. and Bloom, S. E.,** Visualization of nucleolar organizer regions in mammalian chromosomes using silver staining, *Chromosoma*, 53, 37, 1975.
80. **Hubble, H. R., Rothbloom, L. I., and Hsu, T. C.,** Identification of a silver binding protein associated with the cytological silver staining of actively transcribing nucleolar regions, *Cell Biol. Int. Rep.*, 3, 615, 1979.

81. **Brenner, S. and Brinkley, B. R.**, personal communication.
82. **Brenner, S. Pepper, D. A., Turner, D. S., Boyd, A. E., and Brinkley, B. R.**, Autoantibodies in human serum selectively bind to the centriole region in cultured cells, *J. Cell Biol.*, 87, 240a, 1980.
83. **Moroi, Y., Peebles, C., Fritzler, M. J., Steigerwald, J., and Tan, E. M.**, *Proc. Natl. Acad. Sci. U.S.A.*, 7, 1627, 1980.
84. **Tan, E. M.**, Autoimmunity to nuclear antigens, in *Cell Nucleus*, Vol. 7, Busch, H., Ed., Academic Press, New York, 1979, 457.
85. **Hozier, J. and Furcht, L. T.**, Binding of lectins to mitotic chromosomes and interphase nuclear substructures, *Cell Biol. Int. Rep.*, 4, 1091, 1980.
86. **Comings, D. E.**, Structure and function of chromatin, *Adv. Hum. Genet.*, 3, 237, 1972.
87. **Brown, R. L., Pathak, S., and Hsu, T. C.**, Possible role of histones in mechanisms of chromosomal G-banding, *Science*, 189, 1090, 1975.
88. **Burkholder, G. D. and Duczek, L. L.**, Proteins in chromosome banding. I. Effect of G-banding treatments on the proteins of isolated nuclei, *Chromosoma*, 79, 29, 1980.
89. **Burkholder, G. D. and Duczek, L. L.**, Protein in chromosome banding. II. Effect of R- and C-banding treatments on the proteins of isolated nuclei, *Chromosoma*, 79, 43, 1980.
90. **Yunis, J. J., Kuo, M. T., and Saunders, G. T.**, Localization of sequences specifying messenger RNA to light-staining G-bands of human chromosomes, *Chromosoma*, 61, 335, 1977.
91. **Yunis, J. J. and Tsai, M.-Y.**, Mapping of polysomal messenger RNA and heterogeneous nuclear RNA to the lightly staining G-bands of human chromosomes, *Cytogenet. Cell Genet.*, 22, 364, 1978.
92. **Silver, L. M. and Elgin, S. C. R.**, Immunological analysis of protein distributions in *Drosophila* polytene chromosomes, *Cell Nucleus*, Vol. 5, Busch, H., Ed., Academic Press, New York, 1978, 216.
93. **Burkholder, G. D. and Weaver, M. G.**, DNA-protein interactions and chromosome banding, *Exp. Cell Res.*, 110, 251, 1977.
94. **Rattner, J. B., Krystal, G., and Hamkalo, B. A.**, Selective digestion of mouse metaphase chromosomes, *Chromosoma*, 66, 259, 1978.
95. **Sahasrabuddhe, C. G., Pathak, S., and Hsu, T. C.**, Responses of mammalian metaphase chromosomes to endonuclease digestion, *Chromosoma*, 69, 331, 1978.
96. **Weideli, H., Brack, Ch., and Gehring, W. J.**, Characterization of *Drosophila* DNA-binding protein DB-2: Demonstration of its sequence-specific interaction with DNA, *Proc. Natl. Acad. Sci. U.S.A.*, 77, 3773, 1980.
97. **Stubblefield, E. and Wray, W.**, Biochemical and morphological studies of partially purified Chinese hamster chromosomes, *Cold Spring Harbor Symp. Quant. Biol.*, 38, 835, 1971.
98. **Jeppesen, P. G. N., Bankier, A. T., and Sander, L.**, Non-histone proteins and the structure of metaphase chromosomes, *Exp. Cell Res.*, 115, 293, 1978.
99. **Wray, W., Mace, M. L., Jr., Daskal, Y., and Stubblefield, E.**, Metaphase chromosome architecture, *Cold Spring Harbor Symp. Quant. Biol.*, 42, 361, 1977.
100. **Luck, B. T., Wray, V. P., and Wray, W.**, Ultrastructural and biochemical correlations in CHO metaphase chromsomes, *J. Cell Biol.*, 87, 48a, 1980.
101. **Marsden, M. P. F. and Laemmli, U. K.**, Metaphase chromosome structure: evidence for a radial loop model, *Cell*, 17, 849, 1979.
102. **Berezney, R. and Coffee, D. S.**, Nuclear protein matrix: Isolation, structure, and functions, *Adv. Enzyme Regul.*, 14, 63, 1976.
103. **Gerace, L. and Blobel, G.**, Nuclear envelope lamin is reversibly depolymerized during mitosis, *Cell*, 19, 277, 1980.
104. **Peter, K. L. and Comings, D. E.**, Two-dimensional gel electrophoresis of rat liver nuclear washes, nuclear matrix, and hnRNA proteins, *J. Cell Biol.*, 86, 135, 1980.
105. **Lam, K. S. and Kasper, C. B.**, Electrophoretic analysis of three major nuclear envelope polypeptides, *J. Biol. Chem.*, 254, 11713, 1979.
106. **Gerace, L., Blum, A., and Blobel, G.**, Immunocytochemical localization of the major polypeptides of the nuclear pore complex-lamina fraction, *J. Cell Biol.*, 79, 546, 1978.
107. **Aaronson, R. P. and Blobel, G.**, Isolation of nuclear pore complexes in association with a lamina, *Proc. Natl. Acad. Sci. U.S.A.*, 72, 1007, 1975.
108. **Dwyer, N. and Blobel, G.**, A modified procedure for the isolation of a pore complex-lamina fraction from rat liver nuclei, *J. Cell Biol.*, 70, 581, 1976.
109. **Shelton, K. R., Cobbs, C. S., Povilishock, J. T., and Burkat, R. K.**, Nuclear envelope fraction proteins: Isolation and comparison with the nuclear protein of the avian erythrocyte, *Arch. Biochem. Biophys.*, 174, 177, 1976.
110. **Hodge, L. D., Mancini, P., Davis, F. M., and Heywood, P.**, Nuclear matrix of HeLa S_3 cells. Polypeptide composition during adenovirus infection and in phases of the cell cycle, *J. Cell Biol.*, 72, 194, 1977.

111. **Kaufmann, S. H., Coffey, D. S., and Shaper, J. H.,** Considerations in the isolation of rat liver nuclear matrix, nuclear envelope, and pore complex lamina, *Exp. Cell Res.*, 132, 105, 1981.
112. **Krohne, G., Franke, W. W., and Scheer, U.,** Major polypeptides of the nuclear pore complex, *Exp. Cell Res.*, 116, 85, 1978.

Chapter 5

NONHISTONE PROTEINS AND THE STRUCTURAL ORGANIZATION OF CHROMOSOMES

Kenneth W. Adolph

TABLE OF CONTENTS

I. INTRODUCTION

Three major classes of macromolecules provide the basic components for the structural organization of metaphase chromosomes: DNA, histones, and nonhistone proteins. Other minor species such as RNA and membrane lipids may also have important structural roles, but the fundamental chromosomal substructural must be determined by the predominant macromolecular species. In recent years, most attention has focused on the structures formed by the interaction of histones with DNA.[1] It is clear that the shortest range of structure in mitotic chromosomes results from the assembly of histones with DNA to form a chain of nucleosomes 100 Å wide. This filament is folded, at the next higher order of structure, into a fiber having a thickness of 200 to 300 Å. A major unsettled question is to describe the path of folding of the 200 to 300-Å fiber into the characteristic morphology of metaphase chromosomes. Associated with this question is the possibility that the other major macromolecular components, the nonhistone proteins, help to determine and maintain the metaphase chromosome morphology.

Possible structural roles for nonhistone proteins in maintaining chromosome structure are schematically depicted in Figure 1. One minimal function (Figures 1a and b) might be to stabilize adjacent DNA-histone fibers, especially those at the bases of loops. As will be described in greater detail below, the predominant mode of fiber organization appears to be a radial distribution of loops. This conclusion was derived from electron microscopic studies of intact chromosomes which were slightly expanded to separate the fibers so that their orientation could be deduced,[2-4] and from studies of histone-depleted chromosomes.[5,6] The existence of nonhistone proteins which cross-link DNA-histone loops (Figures 1a and b) could provide a mechanism for stabilizing the structure of condensed chromosomes.

Nonhistones may also be components of specialized regions of mitotic chromosomes. The centromere, for example, must have a substructure which differentiates it from other parts of the chromosome. The centromere region, which is indicated by crosshatching in Figure 1a, is uniquely linked to microtubule proteins as the mitotic spindle develops, and nonhistone proteins could be structural components of the region.

Two possibilities for more extensive localizations of nonhistone proteins are illustrated in Figures 1c and 1d. Figure 1c considers that nonhistone proteins form a loose, fibrous network extending through the volume of the chromatid arms. The protein network could be directly associated with the chromatin fibers through DNA-binding nonhistones, or the network could represent a separate level of structure.

In Figure 1d, the structural nonhistones are shown to comprise a central "core" or "backbone" along the center of each chromatid. The nonhistone core would interact with chromatin fibers to constrain the fibers into the metaphase chromosome shape. Of the various possible structural roles for nonhistone proteins, core models of chromosome structure have been most frequently mentioned in the literature. Stubblefield and Wray[7] have fully developed such a core model, which proposes the existence of "epichromatin" loops attached to central core elements.

In the following section, evidence is presented which suggests a general structural role for nonhistones closest to that shown in Figure 1c. The evidence is based on electron microscopic and biochemical studies of histone-depleted metaphase chromosomes.[5,6] The results of these experiments demonstrate that even though virtually all of the histones can be extracted from isolated chromosomes, the residual material is highly organized and consists of loops of DNA which extend from adjacent points along a "scaffold" of nonhistones. The scaffold consists of a fibrous network which extends through the chromatids and is constructed of a subset of nonhistones.

Another fundamental question discussed in this review concerns the role of nonhistones through other phases of the cell cycle in determining the structural organization of chro-

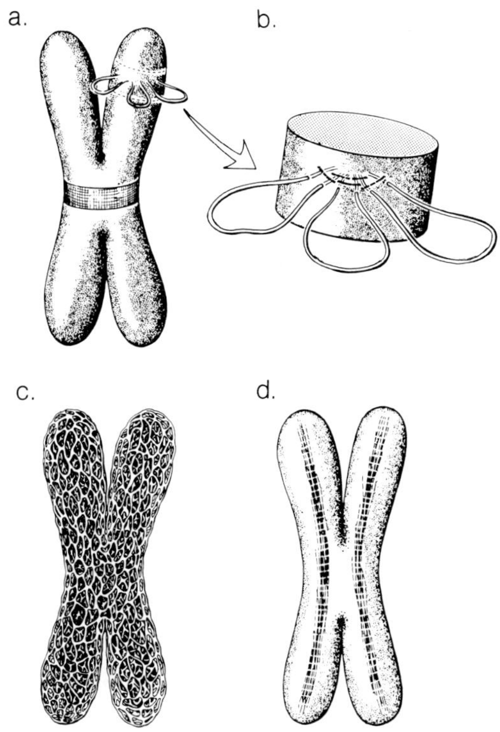

FIGURE 1. Diagram illustrating the possible structural roles of nonhistone proteins in organizing metaphase chromosomes: (a) and (b) show nonhistones stabilizing a chromosome by linking adjacent chromatin fibers, which are probably arranged as loops. Nonhistones could also be structural components of specialized regions of chromosomes such as the centromeres, shown cross-hatched in (a). In (c), the structural nonhistones constitute a loose, fibrous network which extends through the volume of the chromosome; (d) represents the proteins as cores which extend along the center of each chromatid.

mosomal fibers. Although chromatin is relatively diffuse in interphase compared to its state in mitosis, it remains densely packaged in interphase and is likely to be highly organized. One approach to understand the long-range organization of chromatin in the interphase nucleus is to extract histones and other nuclear proteins from isolated nuclei and to characterize the protein composition and structures of the residual particles.[8] This is the same procedure for isolated metaphase chromosomes. The results of such an approach are that the residual particles resemble extracted nuclei which are surrounded by DNA. A small number of nonhistone proteins are associated with the structures. The same proteins are found if the nuclei are treated with nucleases in addition to removing the histones, and the particles still resemble extracted nuclei. The residual structure is therefore termed the ''nuclear scaffold''.[8]

Extracting nuclei with high salt concentrations and nucleases has been used in a number of recent investigations to understand nuclear substructure. The material that remains has been called the nuclear matrix,[9] nuclear pore complex-lamina fraction,[10] nuclear skeleton,[11] and nonmembranous nuclear ghost.[12]

II. CHARACTERIZATION OF HISTONE-DEPLETED METAPHASE CHROMOSOMES

The isolation of histone-depleted metaphase chromosomes allows two basic aspects of chromosome structure to be investigated. The path of folding of the DNA can be determined by electron microscopy, and the residual nonhistones which organize the histone-depleted chromosomes can be analyzed by polyacrylamide gel electrophoresis (PAGE). The important conclusion that emerged from an investigation of histone-depleted metaphase chromosomes from HeLa cells is that the DNA is primarily arranged as loops.[5,6] The loops have lengths of at least 10 to 30 μm (30 to 90 kbases).

To prepare chromosomes for these studies, the hexylene glycol method of Wray and Stubblefield[13] was used. Histones, and many nonhistone proteins were removed with polyanions or high salt concentrations. Polyanions, such as dextran sulfate and heparin sulfate, extract histones by charge competition with DNA. The advantages of the use of polyanions are the efficiency in removing histones and the mild solution conditions which avoid extremes of pH. Virtually all histones can be removed with, in the best experiments, less than 0.1% remaining. High salt concentrations were also used to prepare histone-depleted chromosomes. Typically, histones were extracted with 2 *M* NaCl, 10 m*M* tris-HCl pH 7.4, and 10 m*M* EDTA. The results with polyanions and high salt concentrations were similar.

The bulk properties of histone-depleted chromosomes were characterized by sucrose density gradient sedimentation and PAGE.[5] Histone-depleted HeLa metaphase chromosomes sediment in sucrose density gradients (5 to 20% w/v in a buffer containing 0.1 *M* NaCl) as well-defined peaks with sedimentation coefficients of 4000 to 7000 S. The broad sedimentation coefficients may reflect the heterogeneity in sizes and shapes of human mitotic chromosomes. The material was unfolded by treatment with proteases, high concentrations of urea, and 0.1% SDS (sodium dodecyl sulfate). The results demonstrate that the structures are held together by protein molecules and that the protein interactions are not irreversibly stabilized. The residual nonhistones were termed ''scaffolding'' proteins. The sedimentation rate of histone-depleted chromosomes was not as greatly affected by treatment with ribonuclease. This suggests that RNA molecules are not intimately involved in maintaining the structure of the particles. The amount of protein that is associated with the DNA can be estimated from the density of histone-depleted chromosomes in metrizamide density gradients and, in sucrose gradients, from the fraction of a protein-specific radioactive label bound to the DNA. These methods provide an estimate that 15 to 20% of the mass is protein. The molecular weights of the residual protein species were determined by electrophoresis in polyacrylamide gels containing SDS (SDS-PAGE). A number of major and minor species were detected on one-dimensional gels, and species with molecular weights above 50,000 daltons were especially prominent.

The structures of the particles were characterized by light microscopy and electron microscopy. The light microscope reveals the morphology of particles in solution,[5] while the electron microscope provides the higher resolution needed to view the arrangement of DNA fibers.[6] For light microscopy, the histone-depleted chromosomes were stained with ethidium bromide, a DNA-specific fluorescence stain. The extracted particles were seen to have the characteristic shape of the original chromosomes when observed by epifluorescence. However, much of the DNA was now seen to extend from the chromatid arms in a halo. Individual DNA fibers could not be resolved in the light microscope, and so a surface spreading technique was used with electron microscopy. Samples were mixed with cytochrome c and spread on a hypophase using an ammonium acetate or carbonate method, and then rotary-shadowed with platinum to enhance contrast.[6] The results of these experiments were compatible with the fluorescence studies in showing two basic features of histone-depleted metaphase chromosomes: a central region which retains the shape of the original chromatids and a halo of DNA which surrounds this region. A similar appearance was found for particles produced with dextran sulfate/heparin and with high salt concentrations (2 *M* NaCl). Although individual DNA strands could be resolved, they were difficult to trace for any distance. This is because of the great amount of DNA in each human chromosome, the flexibility of the DNA fibers, and the limited extent of chromosome unfolding. However, in a few favorable electron micrographs, the DNA strands could be followed and were seen to be organized as loops which extended from adjacent points along the central region. The lengths of the loops ranged from less than 5 to almost 50 μm, but most of the DNA was in loops of 10 to 30 μm. The important conclusion from this investigation is that, in native chromosomes containing histones, the DNA is predominantly arranged as a radial array of loops. This general conclusion is supported by further electron microscopic studies of nonextracted chromosomes,[2-4] as is discussed below. Both thin-sectioning techniques and scanning electron microscopy demonstrate the radial orientation of the chromatin fibers.

The central structure of scaffolding proteins does not require the presence of DNA to maintain its integrity. Chromosomes can be pretreated with DNase I or micrococcal nuclease before removing the histones, and a loose, fibrous structure which retains the chromosome morphology can be seen with the light or electron microscope.[14]

III. ADDITIONAL STUDIES OF HISTONE-DEPLETED CHROMOSOMES AND SCAFFOLDING PROTEINS

The results of a number of other investigations are compatible with the observations on histone-depleted chromosomes from HeLa cells. Maio and Schildkraut[15] mentioned that faint structures could be seen following treatment of isolated chromosomes with DNase and RNase and extraction of histones with 0.2 *N* HCl. The particles, although expanded, resembled the original chromosomes. More detailed studies of extracted chromosomes have been presented by Stubblefield and Wray.[7,16] Isolated chromosomes were treated with 2 *M* NaCl, 6 *M* urea, and 0.2 *N* HCl, and a residual "core" structure was noted. Jeppesen et al.[17] studied the particles obtained by removing histones from Chinese hamster chromosomes with 0.2 *N* HCl and by treating acid-extracted chromosomes with DNase I. Acid-extracted chromosomes retained the ability to give characteristic fluorescent banding patterns. DNase I treatment solubilized over 99% of the DNA, while leaving a stable core consisting mainly of non-histones. Incubation with pronase destroyed the visible structure of the cores, while ribonuclease had no effect.

Warren and Cook[18] lysed HeLa cells, synchronized at different stages during the cell cycle, with 1.95 *M* NaCl, 10 m*M* Tris pH 8.0, and 0.5% NP40. In mitotic "nucleoids" the individual chromosomes remained identifiable. The mitotic nucleoids swelled and dispersed when treated with agents such as sodium metrizoate, urea, and sarkosyl, but the individual chromatids were still identifiable and sometimes displayed a banded appearance. Analysis of the protein content of nucleoids on polyacrylamide gels showed the absence of histones and the presence of nonhistone proteins with molecular weights greater than 38,000 daltons. The protein contents of nucleoids prepared at other stages of the cell cycle were similar to the protein content of mitotic nucleoids.

Razin et al.[19] studied the nature of the DNA that remains attached to the residual proteins after removing histones from mitotic chromosomes and treating the material with restriction endonucleases.[19] Metaphase plates from mouse L cells were extracted with 2 *M* NaCl and digested with Bam HI or EcoRI restriction endonucleases. The DNA fragments that remained attached were found by renaturation experiments to be enriched in reiterated base sequences of the mouse satellite fraction and of the most abundant classes of intermediately repeated sequences. The same investigators also applied this procedure to interphase nuclei to determine whether the bound DNA sequences change during the cell cycle.[20] The renaturation properties were found to be essentially the same for interphase and metaphase chromosomes, suggesting that the attachment sites are unchanged during the cell cycle.

Campbell et al.[21] have considered the role of the metaphase chromosomal scaffold in gene expression. Antiserum was prepared that was specific for HeLa cell chromatin, and not to chromatin from other human tissues and cell lines. The antigen was shown to be a DNA-binding protein that is located in the "inactive" fraction of the chromatin. The antigen is not present on nucleosomes or internucleosomal DNA, but is greatly enriched in the non-histone scaffold. The antigen was therefore viewed as a regulatory protein which interacts with the structural elements of the scaffold to control the events at the base of each DNA loop. In this way the scaffold would be involved in the initiation of transcription.

IV. IMPLICATIONS FOR THE STRUCTURAL ORGANIZATION OF METAPHASE CHROMOSOMES

The most significant conclusion from the biochemical and structural studies of histone-depleted HeLa metaphase chromosomes[5,6] was that a radial distribution of loops is a predominant mode of fiber organization in intact chromosomes. With histones present, the DNA loops revealed by EM studies of histone-depleted chromosomes would be condensed into short, thick chromatin fibers. Since the total condensation is a factor of about 40 when the DNA interacts with histones to form fibers of 200 to 300 Å,[22,23] the lengths of loops of nucleohistone in intact chromosomes would be around 0.5 to 2.0 μm. The nucleohistone loops would be tightly bundled in vivo so that it would be difficult to distinguish the general organization by electron microscopy. However, by treating mitotic HeLa cells with a hypotonic buffer containing divalent cations or polyamines, the chromosomes could be slightly swollen to separate the chromatin fibers. Thin-sectioning studies of fiber arrangement showed the primary orientation to be a radial distribution.[3,24] Similar results were also found for isolated chromosomes.[2,4] These investigations will be discussed in greater detail below.

Models of chromosome structure are difficult to compare since all models that have been presented to date are incomplete. This is because of the great complexity of chromosomes in their overall sizes and masses, and because of the intricate organization of chromosomal substructure. Different experimental observations of chromosome structure are therefore not necessarily incompatible, but the speculative nature of chromosome models and the paucity of concrete data raise difficulties in reconciling the various models.

Models of chromosome structure have been of two general types. The structural organization of the fibers has been viewed as resulting either from (a) the direct folding of the fibers into the chromosome morphology, or (b) the coiling of the fibers into higher and higher orders of helixes. A related question concerns the interactions that are responsible for maintaining chromosome structure. The approach of extracting histones from isolated HeLa cell chromosomes[5,6,25] suggested a model of type (a), namely that the chromatin fibers are directly folded into loops, which is the preponderant mode of packaging DNA in metaphase. A very different type of model proposes that the 200 to 300-Å fibers are coiled into a "supersolenoid".[26,27] The supersolenoid is further coiled to give the characteristic chromosome structure.

A number of recent views of chromosome organization have involved models of type (a). The "folded fiber" model considers the mammalian chromosome to result from the folding of chromatin loops into chromomeres by the local accumulation of looping fibers.[28-30] The model also takes into account the presence through the chromatid of longitudinal fibers, which are visible by whole-mount electron microscopy. The looping arrangement may have originated from lampbrush chromosomes of vertebrate oocytes. An early model which included the concept of a protein backbone was given by Taylor.[31] Chromosomes were considered to consist of DNA fibers attached to a two-layered central column of protein. Stubblefield and Wray[7] have presented a detailed model of chromosome architecture based on whole-mount electron microscopy of Chinese hamster metaphase chromosomes. They proposed that the chromatid is constructed of two half-chromatids, each of which contains two DNA-protein ribbons. The ribbons are wound into a "core" consisting of parallel DNA-protein fibers to which "epichromatin" loops are attached.

Another investigation that considered the role of core elements as a fundamental constituent of mitotic chromosomes was presented by Sorsa.[32] The study concluded that the observations support the hypothesis that core fibers are differentiated parts of chromosomal DNA. The fibers persist through the cell cycle and have important roles in the dynamic changes which chromosomes undergo. Abuelo and Moore[33] have published observations on chromatin fiber organization in human lymphocyte chromosomes. Electron microscopic observations of whole-mount specimens suggested that human chromosomes consist of various classes of fibers longitudinal to the chromatid axis. Some fibers, for example, continue uninterrupted along the chromosome, while other fibers cross from one sister chromatid to another and reverse direction. Mullinger and Johnson[34] studied fiber organization in "nucleoids" derived from mitotic HeLa cells. The nucleoids were prepared by extracting metaphase-arrested HeLa cells with 2 *M* NaCl in the presence of the nonionic detergent, Triton® X-100. This procedure was developed by Cook and Brazell[35] to investigate the degree of superhelicity in interphase HeLa cells. Mullinger and Jonhson[34] found, with two-dimensional spreads of mitotic nucleoids, that the individual chromosomes consist of core areas surrounded by a fiber network. Spirals or disks, groups of spirals, radial fibers, and connecting bundles of parallel fibers could all be distinguished in the network. The presence of spirals and the intertwining of pairs of fibers were taken as evidence of DNA supercoiling, and the grouping of disks was seen as reflecting the folding of supercoiled DNA into domains.

A similar approach of extracting chromosomal proteins with high salt concentrations to release the DNA was applied by Okada and Comings[36] to characterize the higher-order structure of chromosomes. The DNA of Chinese hamster chromosomes treated with 4 *M* ammonium acetate was seen to form a series of rosettes connected by interrosette DNA. Nuclear matrix proteins were found to be present in the rosettes.

The higher-order structure of metaphase chromosomes has also been proposed to result from the coiling of the nucleosome-containing fiber into a hierarchy of helixes. Bak et al.[26] observed hollow, cylindrical structures with a diameter of 4000 Å in preparations of partially disintegrated metaphase chromosomes. The structures were considered to represent the untwisting of the final order of coiling of the DNA-histone fibers. Based on this observation, Bak et al. presented a model in which the 300-Å DNA-histone fiber coils into a unit fiber or "supersolenoid" of 4000 Å. A final level of folding of the unit fiber is required to give a chromatid with a diameter of 1.0 μm. The "supersolenoid" could be bent back upon itself and then folded into a double helix at the final level of structure, or a single helix may be more appropriate. Sedat and Manuelidis[27] have published a similar view which interprets light and electron microscopic evidence as suggesting the higher orders of organization of the eukaryotic genome result from higher orders of coiling of chromatin fibers. The proposed diameters of the helixes differed somewhat from those of Bak et al.[26] The nucleosome structure of 100 Å was considered to be coiled to first form a 300- to 500-Å coil, and then a tube of 2000 Å. The 2000-Å tubes would be important in interphase nuclei and also in metaphase chromosomes. In the latter case, the tubes must be wound into a sister chromatid with a diameter of 6000 Å. This model predicts the existence of a hollow center in chromatids, with the central hole having a width of 2000 Å.

Bak et al.[37] and Zeuthen et al.[38] have further characterized the unit fibers ("supersolenoids") obtained from human fetal fibroblasts and *Drosophila* cells. For both systems, long, regular fibers of 4000 Å diameter could be found in statistically significant numbers in preparations of disintegrated, isolated chromosomes.

This discussion of the direct-folding and higher-order coiling models makes clear that there is not yet a consensus view of the general organization of DNA in metaphase chromosomes. And while the different models are not necessarily incompatible, the greatest source of discrepancy is the inadequate data that have been obtained. Even at the level of the 200- to 300-Å chromatin fibers, there remains some question about the arrangement of nucleosomes. Numerous studies have established that the basic chromatin filament consists of a string of nucleosomes 100 Å in diameter.[1] Nucleosomes contain an interior core of two each of the histones H2A, H2B, H3, and H4. About 200 base pairs of DNA are wrapped around the core, and a fifth histone, H1, is attached to the complex. A beads-on-a-string appearance of the intact DNA-histone filament is observed in the electron microscope. The question then arises as to how the 100-Å beads-on-a-string filament is arranged to form the 200- to 300-Å fibers, which are a distinct level of chromosome organization.[39,40]

Finch and Klug[22] have presented a supercoil or "solenoidal" model for superstructure in chromatin. In 0.2 m*M* EDTA, chromatin that was prepared by brief nuclease digestion was seen as a loosely coiled filament with a diameter of 100 Å. The presence of 0.2 m*M* Mg^{2+} caused this basic filament to condense into a supercoiled structure having a thickness of 300 Å and a pitch of 110 Å. The mean number of nucleosomes per turn was about six or seven. Histone H1 is likely to be needed for the formation or stabilization of the condensed structure, since H1-depleted chromatin did not form supercoils when Mg^{2+} was added. Neutron diffraction studies of chromatin in solution also suggested a superhelix model for the higher-order structure of the nucleosome filament.[23] Thoma et al.[41] have systematically studied the structures formed by the nucleosome filament upon increasing the ionic strength with monovalent salt. The filament of nucleosomes was seen to progressively fold up as the monovalent salt concentration was raised from 1 to 100 m*M*. Intermediate higher-order helical structures could be distinguished with 3, 4, and finally 6 to 8 nucleosomes per turn. Above 60 m*M* NaCl (or 0.3 m*M* Mg^{2+}), a 250-Å thick solenoid was formed with a pitch of 110 Å that remained constant as the ionic strength was increased.

An alternate model considers the 200 to 300-Å fibers to result from the formation of globular clusters of nucleosomes. Renz et al.[42] suggested that chromosome fibers are composed of discrete globular clusters and are not found as continuous superhelixes. Each cluster was estimated to contain six to ten nucleosomes. A more precise estimate of eight nucleosomes per cluster was determined by Hozier et al.,[43] and the electron microscope showed the particles to have an average diameter of 220 Å. Sedimentation profiles of mildly digested rat liver chromatin revealed particles with S values of 33, 60, and 90, with 33 S particles representing the "superbead" monomer.[44]

Several distinct patterns of closely apposed arrays of nucleosomes were noted by Rattner and Hamkalo[45] on mouse chromosome fibers. The metaphase chromosomes were found to be composed of a series of loops 1 to 5 μm in length.

The arrangement of nucleosomes in the chromatin of interphase nuclei has also been shown to involve a fiber of 250 to 300 Å. Davies et al.[46] have examined thin sections of chicken erythrocyte nuclei with the electron microscope. The experiments suggested a "superunit thread hypothesis" for chromatin organization. Olins and Olins[47] have also examined thin sections of chicken erythrocyte nuclei, and have observed the parallel alignment of 250-Å chromatin fibers constructed of close-packed arrays of nucleosomes.

V. ELECTRON MICROSCOPIC STUDIES OF FIBER ORGANIZATION IN MITOTIC CHROMOSOMES

Recent electron microscopic studies have strengthened the evidence that a primary mode of organization of the 250 Å fibers in mitotic chromosomes is a radial arrangement of loops. One investigation was with hypotonically swollen HeLa cells arrested in metaphase,[3] while another used preparations of isolated chromosomes.[2]

The procedure with mitotic HeLa cells removed background cellular material with hypotonic swelling, while keeping the chromosomes condensed. The condensed state was maintained by including divalent cations (Mg^{2+}, Ca^{2+}) or polyamines (spermine, spermidine) in the hypotonic buffer. By adjusting the divalent cation or polyamine concentration, the degree of expansion of the chromosomes could be carefully controlled. Slightly expanding the chromosomes separated the fibers so that their general orientation could be observed. In a typical experiment, mitotic cells were resuspended in 10 m*M* NaCl, 5 m*M* PIPES pH 6.8, 1.5 m*M* $MgCl_2$, 0.1 m*M* PMSF, and 0.1% NP40. A $MgCl_2$ concentration of 1.5 m*M* provided the most informative results, since individual fibers were separated but the overall dimensions of the chromosomes were close to their values in growth medium. Reducing the $MgCl_2$ concentration unfolded the chromosomes so that significant conclusions regarding fiber organization in native chromosomes were difficult to obtain.

The most important conclusion from electron micrographs of thin sections of chromosomes in 1.5 m*M* $MgCl_2$ is that the fibers have a primarily radial orientation.[3] This is especially clear toward the peripheries of the chromosomes where the fibers are distinctly separated. Near the centers of the chromatid arms, individual fibers could not be readily distinguished. The radial fibers probably loop back at the chromosome boundaries. The nature of the connections between loops and the possible existence of longitudinal fibers are problems that need to be investigated further.

The arrangement of chromatin fibers in isolated chromosomes from mitotic HeLa cells was also characterized by thin-sectioning and electron microscopy.[2] New procedures were used to isolate metaphase chromosomes. The previous work on histone-depleted chromosomes[5,6] had used samples prepared by the hexylene glycol method of Wray and Stubblefield.[13] The method gives stable particles with good morphology, but the organic solvent used in the isolation may affect the properties of the particles. Because of this possibility, procedures were developed which include more physiological solution conditions and avoid organic solvents.[2] One procedure uses divalent cations to keep the chromosomes in a condensed state during the isolation, while a second procedure uses polyamines. The pH is adjusted to 7.4 with HEPES buffer; alternatively, the pH can be raised to 10.0 with CAPS buffer to inhibit nuclease activity. To isolate chromosomes by the pH 7.4 HEPES-Mg/Ca procedure, colchicine-arrested HeLa cells are resuspended in 50 m*M* NaCl, 5 m*M* HEPES pH 7.4, 5 m*M* $MgCl_2$, 0.5 m*M* $CaCl_2$, and 1.0 m*M* PMSF. The chromosomes are released with a Dounce tissue homogenizer after adding 0.5% NP40 and 0.1% sodium deoxycholate. The sodium deoxycholate can be omitted, although the chromosome preparations will contain more debris. The chromosomes can be partially purified by differential centrifugation. A convenient method for obtaining highly purified samples is to pellet the nonmitotic nuclei and other large debris, and then sediment the chromosomes from the lysate into a gradient of 0.3-0.8*M* metrizamide in isolation buffer containing 0.1% NP40. The chromosomes form a sharp band in the middle of the gradient at their particle density.

Thin sections of isolated chromosomes have a similar appearance in the electron microscope to chromosomes in hypotonically swollen mitotic cells.[2] This demonstrates that chromosome structure is not severely distorted during the isolation procedure. A radial distribution of fibers is the most prominent feature of fiber organization in the micrographs. Longitudinal fibers may be present through the center of the chromatid, since individual fibers are not well resolved in this region. Such longitudinal fibers could function to link adjacent or distant loops.

The results with random sections through mitotic HeLa cell chromosomes have been extended by obtaining serial sections[24] (Figures 2 and 3). With serial sections, the fiber arrangement can be observed through a representative thickness of the chromosome. In addition, grazing sections which cut across the peripheries of the chromosomes can be related to the internal structure. Consecutive sections through the bodies of the chromatids strengthen the conclusion that fiber organization is fundamentally a radial distribution (Figure 2). Grazing, longitudinal sections show a dot pattern as the sections enter the chromatids, which

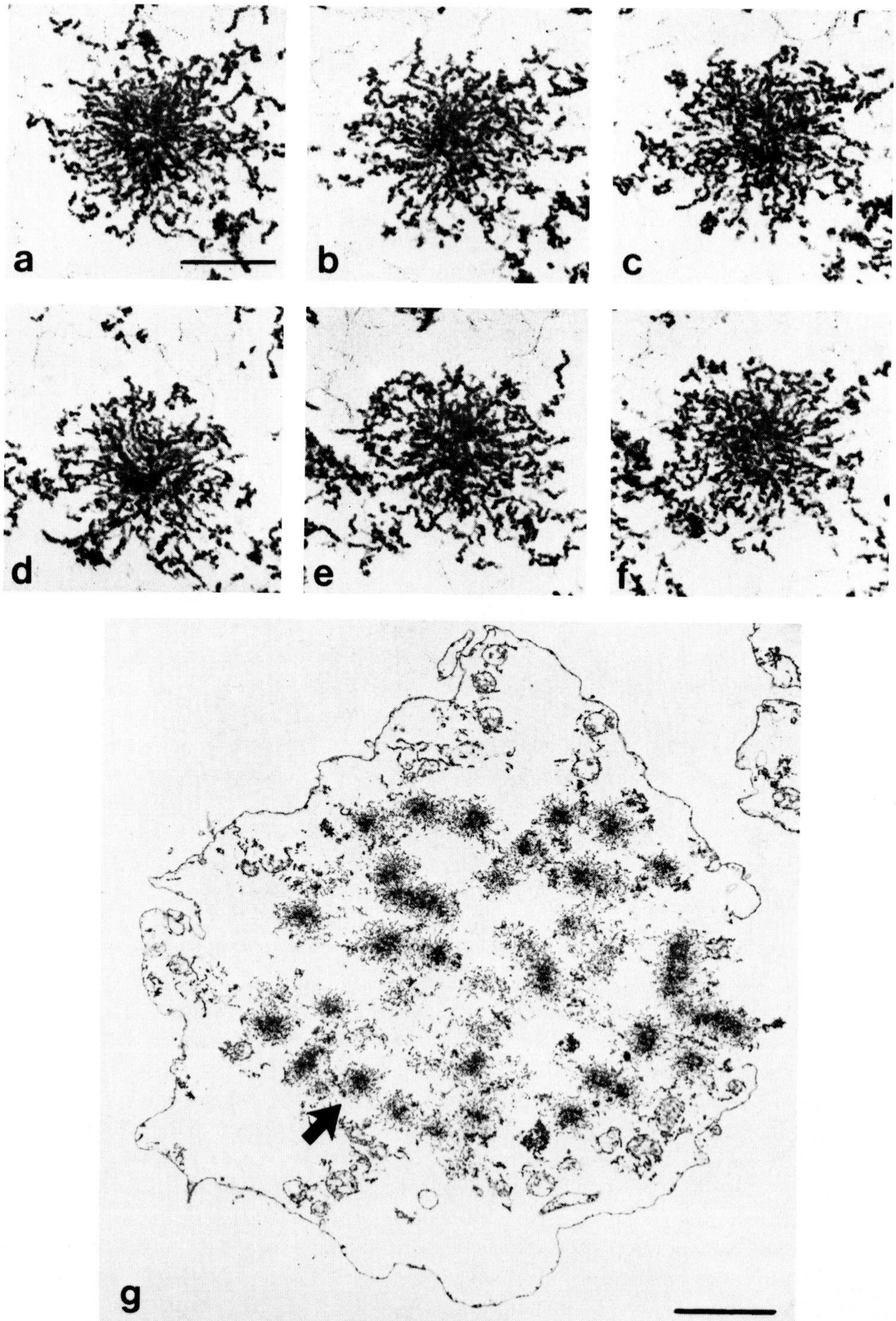

FIGURE 2. Consecutive sections through a mitotic HeLa cell chromosome.[24] The sections intersect the chromatid transversely. An overall view of the mitotic cell is shown in (g) and an arrow points to the chromosome included in (a) to (f). The bar is 0.5 μm in (a) to (f), and 5.0 μm in (g).

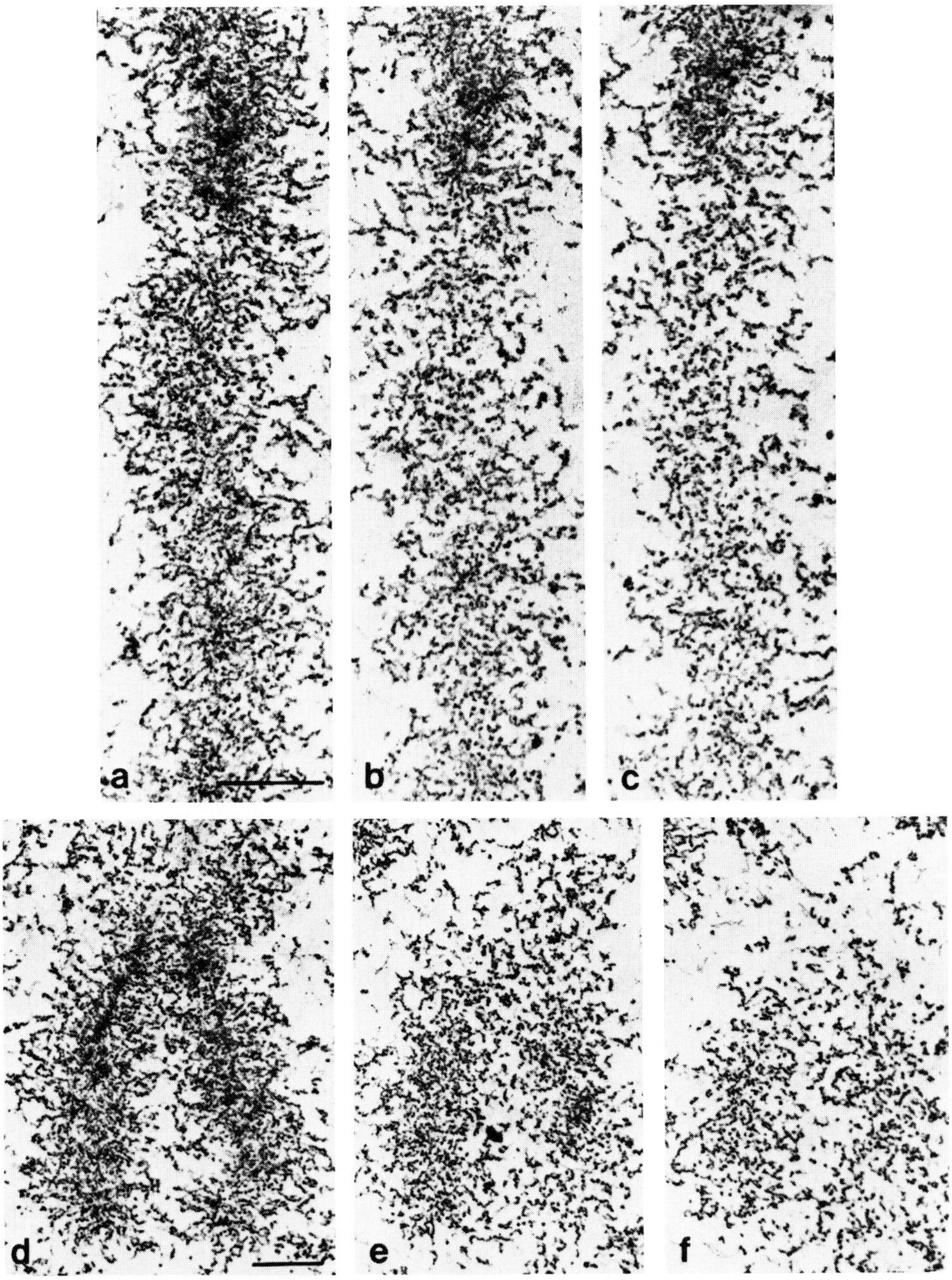

FIGURE 3. Serial sections that graze mitotic HeLa cell chromosomes longitudinally.[24] Two different chromosomes are included in (a) to (c) and (d) to (f). The bar is 0.5 μm.

is transformed near the central axes of the chromatids into patterns of radial fibers in the planes of the sections (Figure 3). These patterns also reflect a radial fiber arrangement. The sections that enter the chromatids cut across the radial fibers but, near the central axes, longitudinal sections must contain fibers that are radially oriented. Serial sections also point out that the predominantly radial arrangement may be more complex than a simple array of loops which originate and end at the precise center of each chromatid. Sections through the broader chromatids, especially, give the impression of a possibly complex distribution of loops.

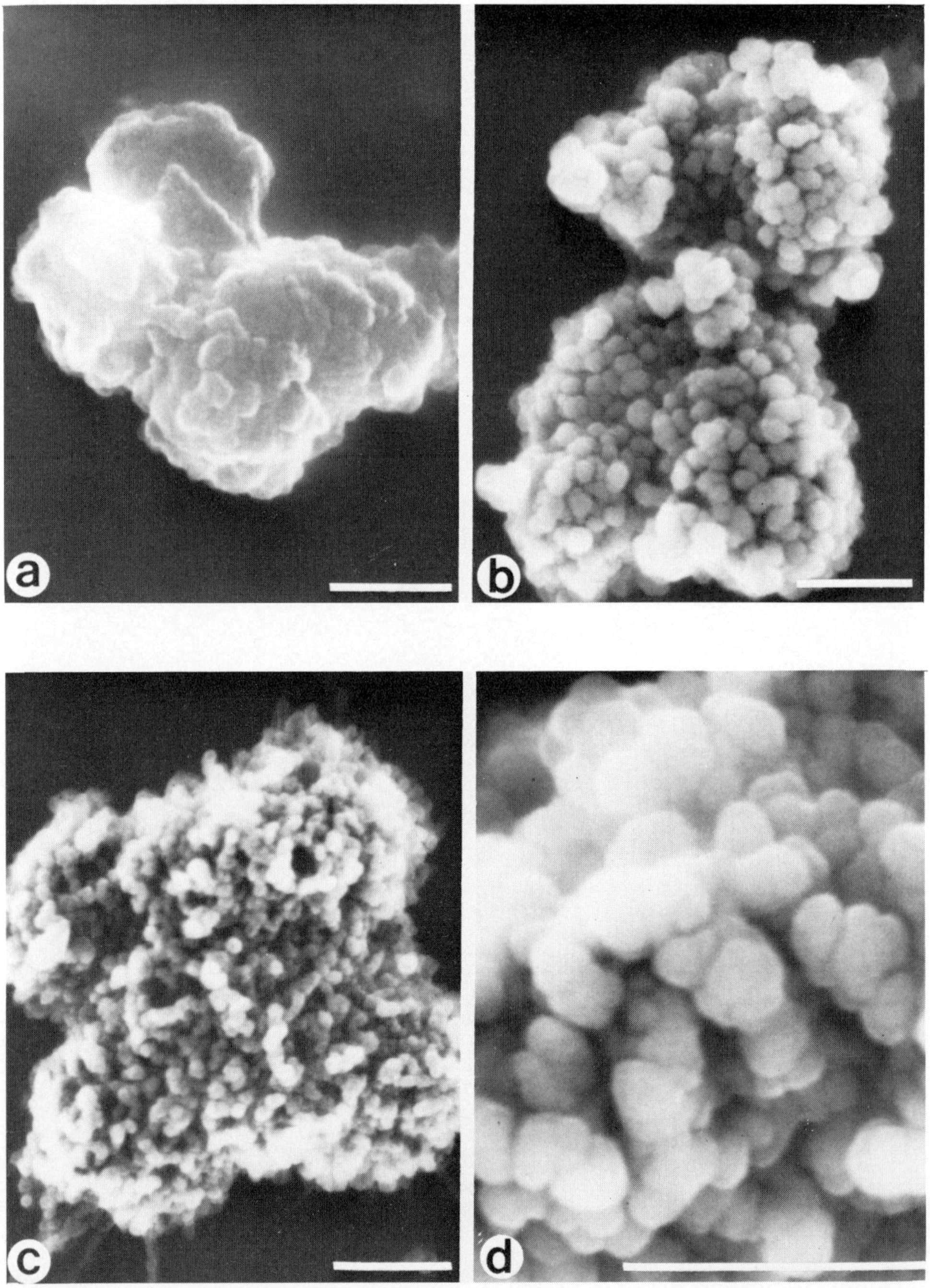

FIGURE 4. Scanning electron micrographs of HeLa metaphase chromosomes isolated by the pH 7.4 HEPES-Mg/Ca method.[2] After isolation, the chromosomes were resuspended in buffer containing: (a) 5.0 m*M* Mg^{2+}, (b) 1.5 m*M* Mg^{2+}, (c) and (d) 0.5 m*M* Mg^{2+}. The bar is 0.5 μm.

Scanning electron microscopy of isolated chromosomes further supports these general conclusions about chromosome organization (Figure 4).[60] The scanning electron microscope reveals information about the surface structure of chromosomes, thereby complementing the sectioning results which show the distribution of material through the chromatids. HeLa metaphase chromosomes were prepared by the pH 7.4 HEPES-Mg/Ca method[2] and resuspended in 5 m*M* HEPES pH 7.4 containing reduced concentrations of Mg^{2+} (1.5, 0.5, 0.15, and 0.05 m*M* Mg^{2+}). Drops of the chromosome suspensions were placed on glass cover slides and, after allowing particles to stick to the surface, the chromosomes were fixed with glutaraldehyde and osmium tetroxide. The fixed specimens were dehydrated through ethanol, critical-point dried, and shadowed with gold/palladium. Micrographs were taken in the scanning mode of a JEOL JEM 100CX electron microscope. The surface appearance of the chromosomes changes greatly as the Mg^{2+} concentration is reduced. In 5.0 m*M* Mg^{2+}, the particles have a dense, smooth, featureless morphology, but in 1.5 m*M* Mg^{2+} a knobby substructure is apparent. The surface protuberances have diameters of close to 700 Å, but this value decreases to 300 Å as the Mg^{2+} concentration is lowered to 0.05 m*M*. The chromosomes progressively expand with decreasing Mg^{2+}, and the underlying 300-Å fibers are revealed. The knobs probably represent the peripheral tips of radial loops, which may be supertwisted to varying extents depending upon the Mg^{2+} concentration.

The newly developed techniques of chromosome isolation[2] should be useful in further biochemical and structural investigations. A number of earlier procedures used acid pH to stabilize the chromosomes during isolation,[48] while more recently, the hexylene glycol procedure of Wray and Stubblefield[13] has been widely applied.

VI. NONHISTONE PROTEINS AND THE RELATIONSHIP OF METAPHASE AND INTERPHASE CHROMOSOME STRUCTURE

Nonhistone proteins may be involved in compartmentalizing interphase chromatin into structural domains. Approximately 2 m of DNA is packaged in each human cell, and nonhistone proteins may help to maintain the higher-order organization of chromosomal fibers in the interphase nucleus. This possibility seems especially likely considering the specific conformational changes interphase chromosomes must undergo in mitotic condensation.

To gain some information about the role of nonhistones in interphase, the approach that was followed in characterizing histone-depleted metaphase chromosomes[5,6] and the metaphase chromosomal scaffold[14] was applied to interphase nuclei.[8] Histones and many other nuclear proteins were extracted from isolated HeLa cell nuclei with high salt concentrations or polyanions, and the particles that resulted were characterized by sucrose density gradient sedimentation, electron microscopy, and PAGE. Nuclei were prepared from randomly growing cultures by resuspending the cells in 10 m*M* NaCl, 10 m*M* tris pH 7.4, 5 m*M* $MgCl_2$, and 1 m*M* PMSF and opening the cells with a Dounce homogenizer with the aid of detergents.

Sucrose gradient sedimentation of histone-depleted nuclei labeled with [^{3}H]-thymidine showed rapidly sedimenting peaks which contained greater than 80% of the label. The sedimentation coefficient was about 12,000 S when using 2 *M* NaCl to extract the nuclear proteins and 6000 S when using dextran sulfate (2 mg/mℓ). Treating the histone-depleted nuclei with protease (trypsin at 10 μg/mℓ) unfolded the structures. A lesser effect upon the sedimentation rate resulted from treating samples with ribonuclease (RNase A at 100 μg/mℓ). The structure of the residual particles was characterized by light microscopy and electron microscopy. The distribution of DNA for particles in solution was determined by epifluorescence after staining with ethidium bromide. Histone-depleted nuclei, prepared with 2 *M* NaCl, were seen to consist of a halo of DNA surrounding a central region which retained features of the original nucleus. The nuclear periphery was visible and internal structures, such as the nucleoli, remained. Polyanions had a more potent effect in dissociating nuclear substructure. In addition to removing histones and other nuclear proteins, dextran sulfate at pH 9.0 disrupted the nuclear periphery. Although the periphery of histone-depleted nuclei could no longer be distinguished by phase contrast microscopy or epifluorescence, the DNA

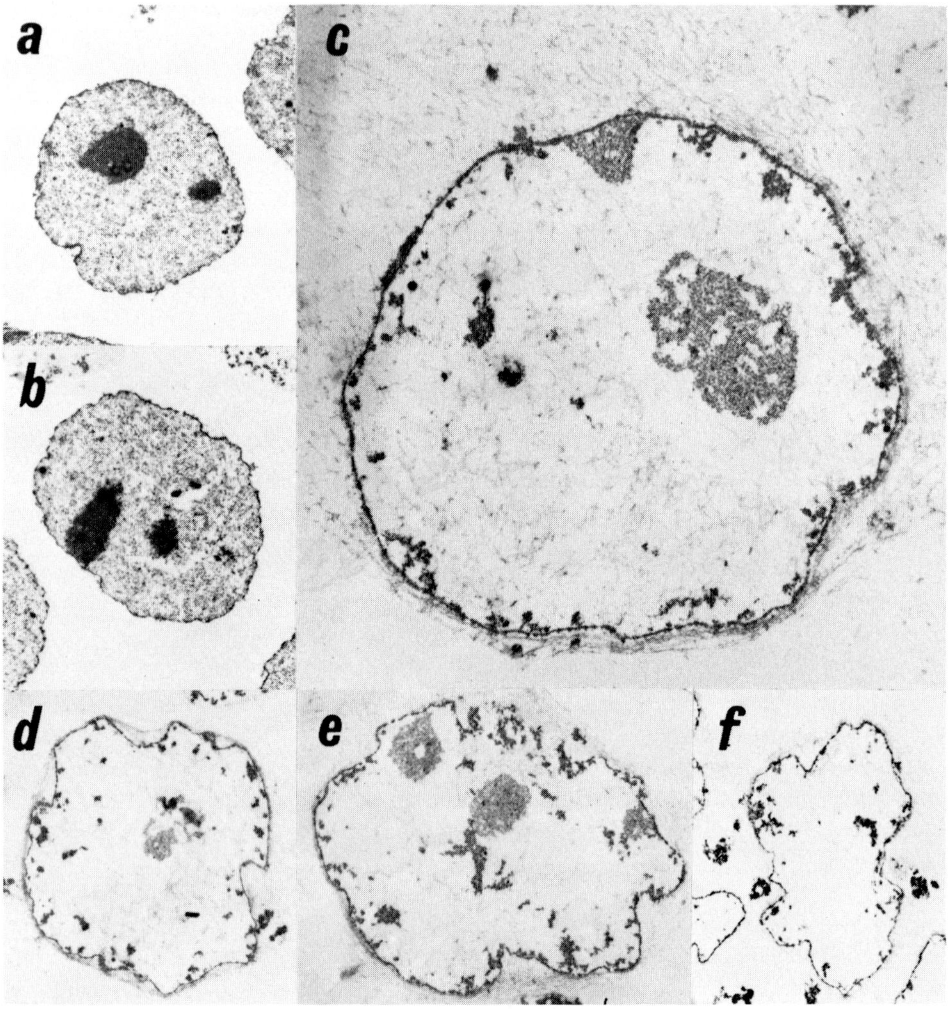

FIGURE 5. Thin sections of histone-depleted HeLa cell nuclei extracted with 2 *M* NaCl at pH 7.4 [(c) to (f)].[8] Nonextracted nuclei are included in (a) and (b).

remained constrained into a circular shape. The diameters of the particles were twice that of histone-depleted nuclei prepared with 2 *M* NaCl. Electron microscopy provided a higher resolution description of these structures. Examination of thin sections of 2 *M* NaCl particles in the electron microscope showed DNA strands to extend through the nuclear volume and also to form a shell around the nuclear periphery (Figure 5).

''Nuclear scaffolds'' were prepared by treating nuclei with DNase I, DNase I plus RNase A, or micrococcal nuclease, each at a concentration of 100 μg/mℓ, in addition to extracting proteins. The particles that were seen after a 2 *M* NaCl extraction still showed structural features of the original nuclei. In particular, the nuclear periphery was apparent in the phase contrast and electron microscope. The nature of the internal structure depended on the type of nuclease that was used. With DNase I, remnants of the nucleoli and other internal material were present, while micrococcal nuclease treatment resulted in the loss of some internal material. With RNase A and DNase I, most if not all of the internal structure could be removed.

The proteins that are associated with histone-depleted nuclei and nuclear scaffolds were characterized by SDS-PAGE. The proteins separate on the basis of molecular weight and major groups of proteins could be distinguished with histone-depleted nuclei prepared with 2 *M* NaCl. One group has molecular weights of 65,000 to 70,000 daltons and appears to be the same proteins reported to be components of the nuclear matrix[9] and nuclear pore complex-lamina fraction.[10] Nuclear scaffolds made with DNase I and 2 *M* NaCl show the same pattern of major proteins as histone-depleted nuclei. Fewer major proteins are found when polyanions are used.

The proteins associated with histone-depleted nuclei and nuclear scaffolds were further characterized by the two-dimensional PAGE technique of O'Farrell[49] (Figure 6). Two-dimensional gels provide a higher-resolution separation since the proteins separate by two independent parameters, isoelectric point and molecular weight. The pattern of protein species found with histone-depleted nuclei and nuclear scaffolds is seen to be more complex than was apparent with one-dimensional polyacrylamide gels. About 12 major protein species are detected, though another 10 minor proteins may also be present. Most proteins have molecular weights of 45,000 to 70,000 daltons, and isoelectric points of between 5 and 6. Proteins that have isoelectric points above 7 would not be detected by this procedure.

Comparing the gel patterns of the proteins of nuclear scaffolds with those of metaphase chromosomal scaffolds suggests that the same protein species may be present. More precise studies will be required to prove this possibility, but some of the major scaffolding proteins may be conserved through the cell cycle. They could serve a similar function around the cell cycle in organizing the highest order of chromatin structure. Although the same major protein species are found, they may differ in their posttranslational modifications. Such modifications could determine the differences in chromatin packaging in metaphase chromosomes and interphase nuclei. Preliminary experiments have shown that some of the major nuclear scaffold proteins are phosphorylated.[60] Phosphorylation may thus be important in determining the structural interactions of the proteins in metaphase and interphase.

VII. INVESTIGATIONS OF NUCLEAR SUBSTRUCTURE

Many recent studies have aimed at understanding chromatin organization in the interphase nucleus and the nature of protein substructures in the nucleus. The lengths of structural domains of DNA have been estimated by Igó-Kemenes and Zachau[50] from micrococcal nuclease and restriction nuclease digestion studies. The average domain length was estimated to be up to 70,000 nucleotide pairs, and the chromatin strands were proposed to be attached to a supporting structure of nuclear matrix proteins. Benyajati and Worcel[51] determined the average size of supercoiled DNA loops in the interphase genome of *Drosophila melanogaster* to be about 85,000 base pairs. A model was suggested in which supercoiled loops containing an average of 400 nucleosomes are constrained by protein-DNA and RNA-DNA interactions.

The composition and structure of interphase "nucleoids" have been characterized by Cook and co-workers.[18,35,52] Cells lysed in a solution containing 1.95 *M* NaCl produced "nucleoids" which contained intact, supercoiled, and circular DNA. The particles lack histones and most other chromatin proteins, but retain some proteins with mobilities on SDS-polyacrylamide gels between actin and tubulin. Brasch and Setterfield[53] reported that following extraction of all histones from chicken erythrocyte nuclei, remnants of condensed chromatin structure were observable for nuclei in saline. The condensed chromatin structure dispersed in water, but reformed on being returned to saline.

Other investigations of nuclear substructure have combined nuclease digestion and protein extraction steps. Aaronson and Blobel[10] isolated a fraction which includes nuclear pore complexes and the protein lamina which lies under the inner nuclear membrane. The three major protein components of the fraction have molecular weights of 60,000, 67,000, and 70,000 daltons, and have been localized by immunocytochemical techniques as components of the peripheral nuclear lamina.[54,55] The lamina was shown to be reversibly depolymerized

A

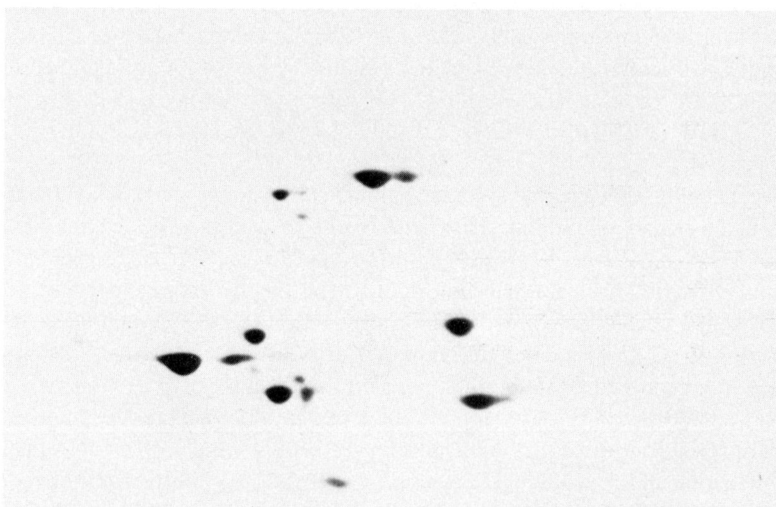

B

FIGURE 6. Two-dimensional polyacrylamide gel of: (A) total proteins of isolated HeLa cell nuclei, and (B) nuclear scaffold proteins. The proteins were separated by isoelectric focusing in the horizontal dimension, and by electrophoresis in a SDS polyacrylamide slab gel (8%) in the vertical dimension. The proteins were labeled with [^{35}S]-methionine, and the gels were developed by fluorography.

during cell division.[56] The three predominant polypeptide components are phosphoproteins, and have a substantially higher level of phosphorylation during mitosis than during interphase. Engelhardt and Pusa[57] proposed that nuclear pore complexes are involved in the attachment of chromosomes to the nuclear envelope by a "press-stud" mechanism.

A nuclear protein matrix was identified by Berezney and Coffey[9] as the structural framework of the rat liver nucleus. Three acidic polypeptides with molecular weights of 60,000 to 70,000 daltons were found to be the primary components of the nuclear matrix. Hodge et al.[58] isolated a nuclear matrix from HeLa S_3 cells which contains 30 to 35 distinguishable protein bands on SDS-polyacrylamide gels, with major peptides having molecular weights of 14,000 to 18,000 and 45,000 to 75,000. Miller et al.[11] showed the presence of a nuclear skeleton containing HnRNA by extracting chromatin with high-salt buffer and DNase. A network of matrix fibrils, which are complex heteropolymers of nonhistone proteins, has been seen to pass through the whole of DNA-depleted nuclei prepared by a mild procedure from Friend erythroleukemia cells.[59]

VIII. CONCLUSIONS

This review has outlined the evidence that nonhistone proteins have structural roles in organizing chromatin in metaphase chromosomes and interphase nuclei. Roles for nonhistones in metaphase seem likely because of the specific and complex structures of mitotic chromosomes. And even though chromatin is relatively diffuse in interphase, it is densely packaged and must be highly organized. The possibility that DNA and histones are entirely responsible for metaphase and interphase chromosome structure seems remote, and so experimental approaches were undertaken to characterize the structural roles of nonhistones.

A fundamental approach was to reduce the complexity of chromosome structure by extracting chromosomal proteins, and then to characterize the residual particles both biochemically and structurally. Applying this procedure to metaphase chromosomes isolated from HeLa cells showed that the chromosomes were only partially unfolded by removing histones and many nonhistones with polyanions or high-salt concentrations.[5,6] A subset of nonhistone ''scaffolding'' proteins was associated with the residual particles and organized the DNA into loops of 10 to 30 μm. With intact chromosomes, the DNA loops would be compacted with histones to form the dense bundles of chromatin that are observed by electron microscopy. The model of chromosome organization that is suggested by these results is supported by thin-sectioning[24] and scanning electron microscopy studies.

Roles for nonhistones in compartmentalizing chromatin in the interphase nucleus were also suggested by removing histones and other nuclear proteins with polyanions or high-salt concentrations. In some conditions, the histone-depleted particles resembled extracted nuclei with the DNA constrained into a halo by the residual nuclear scaffold proteins. Some of the residual proteins had mobilities on SDS-polyacrylamide gels that were similar to the predominant proteins of histone-depleted metaphase chromosomes. Nuclei can be treated with nucleases in addition to agents which remove proteins, and the nuclear scaffolds that remain can possess the nuclear periphery and internal material. A similar approach to nuclear substructure has been used by a number of investigators to produce material usually called the nuclear matrix or nuclear pore complex-lamina fraction.

REFERENCES

1. **McGhee, J. D. and Felsenfeld, G.,** Nucleosome structure, *Annu. Rev. Biochem.*, 49, 1115, 1980.
2. **Adolph, K. W.,** Isolation and structural organization of human mitotic chromosomes, *Chromosoma (Berlin)*, 76, 23, 1980.

3. **Adolph, K. W.,** Organization of chromosomes in mitotic HeLa cells, *Exp. Cell Res.*, 125, 95, 1980.
4. **Marsden, M. P. F. and Laemmli, U. K.,** Metaphase chromosome structure: evidence for a radial loop model, *Cell,* 17, 849, 1979.
5. **Adolph, K. W., Cheng, S. M., and Laemmli, U. K.,** Role of nonhistone proteins in metaphase chromosome structure, *Cell,* 12, 805, 1977.
6. **Paulson, J. R. and Laemmli, U. K.,** The structure of histone-depleted metaphase chromosomes, *Cell,* 12, 817, 1977.
7. **Stubblefield, E. and Wray, W.,** Architecture of the Chinese hamster metaphase chromosome, *Chromosoma (Berlin),* 32, 262, 1971.
8. **Adolph, K. W.,** Organization of chromosomes in HeLa cells: isolation of histone-depleted nuclei and nuclear scaffolds, *J. Cell Sci.*, 42, 291, 1980.
9. **Berezney, L. and Coffey, D. S.,** Identification of a nuclear protein matrix, *Biochem. Biophys. Res. Commun.*, 60, 1410, 1974.
10. **Aaronson, R. P. and Blobel, G.,** Isolation of nuclear pore complexes in association with a lamina, *Proc. Natl. Acad. Sci. U.S.A.*, 72, 1007, 1975.
11. **Miller, T. E., Huang, C.-Y., and Pogo, A. O.,** Rat liver nuclear skeleton and ribonucleoprotein complexes containing HnRNA, *J. Cell Biol.*, 76, 675, 1978.
12. **Keller, J. M. and Riley, D. E.,** Nuclear ghosts: a nonmembranous structural component of mammalian cell nuclei, *Science,* 193, 399, 1976.
13. **Wray, W. and Stubblefield, E.,** A new method for the rapid isolation of chromosomes, mitotic apparatus, or nuclei from mammalian fibroblasts at near neutral pH, *Exp. Cell Res.*, 59, 469, 1970.
14. **Adolph, K. W., Cheng, S. M., Paulson, J. R., and Laemmli, U. K.,** Isolation of a protein scaffold from mitotic HeLa cell chromosomes *Proc. Natl. Acad. Sci. U.S.A.*, 74, 4937, 1977.
15. **Maio, J. J. and Schildkraut, C. L.,** Isolated mammalian metaphase chromosomes. I. General characteristics of nucleic acids and proteins, *J. Mol. Biol.*, 24, 29, 1967.
16. **Stubblefield, E. and Wray, W.,** Biochemical and morphological studies of partially purified Chinese hamster chromosomes, *Cold Spring Harbor Symp. Quant. Biol.*, 38, 835, 1974.
17. **Jeppesen, P. G. N., Bankier, A. T., and Sanders, L.,** Non-histone proteins and the structure of metaphase chromosomes, *Exp. Cell Res.*, 115, 293, 1978.
18. **Warren, A. C. and Cook, P. R.,** Supercoiling of DNA and nuclear conformation during the cell cycle, *J. Cell Sci.*, 30, 211, 1978.
19. **Razin, S. V., Mantieva, V. L., and Georgiev, G. P.,** DNA adjacent to attachment points of deoxyribonucleoprotein fibril to chromosomal axial structure is enriched in reiterated base sequences, *Nucleic Acids Res.*, 5, 4737, 1978.
20. **Razin, S. V., Mantieva, V. L., and Georgiev, G. P.,** The similarity of DNA sequences remaining bound to scaffold upon nuclease treatment of interphase nuclei and metaphase chromosomes, *Nucleic Acids Res.*, 7, 1713, 1979.
21. **Campbell, A. M., Briggs, R. C., Bird, R. E., and Hnilica, L. S.,** Cell specific antiserum to chromosome scaffold proteins, *Nucleic Acids Res.*, 6, 205, 1979.
22. **Finch, J. T. and Klug, A.,** Solenoidal model for superstructure in chromatin, *Proc. Natl. Acad. Sci. U.S.A.*, 73, 1897, 1976.
23. **Carpenter, B. G., Baldwin, J. P., Bradbury, M., and Ibel, K.,** Organization of subunits in chromatin, *Nucleic Acids Res.*, 3, 1739, 1976.
24. **Adolph, K. W.,** A serial sectioning study of the structure of human mitotic chromosomes, *Eur. J. Cell Biol.*, 24, 146, 1981.
25. **Laemmli, U. K., Cheng, S. M., Adolph, K. W., Paulson, J. R., Brown, J. A., and Baumbach, W. R.,** Metaphase chromosome structure: the role of non-histone proteins, *Cold Spring Harbor Symp. Quant. Biol.*, 42, 351, 1978.
26. **Bak, A. L., Zeuthen, J., and Crick, F. H. C.,** Higher order structure of human mitotic chromosomes, *Proc. Natl. Acad. Sci. U.S.A.*, 74, 1595, 1977.
27. **Sedat, J. and Manuelidis, L.,** A direct approach to the structure of eukaryotic chromosomes, *Cold Spring Harbor Symp. Quant. Biol.*, 42, 331, 1978.
28. **DuPraw, E. J.,** Evidence for a "folded-fiber" organization in human chromosomes, *Nature (London),* 209, 577, 1966.
29. **DuPraw, E. J. and Bahr, G. F.,** The arrangement of DNA in human chromosomes, as investigated by quantitative electron microscopy, *Acta Cytol.*, 13, 188, 1969.
30. **Bahr, G. F.,** Chromosomes and chromatin structure, in *Molecular Structure of Human Chromosomes,* Yunis, J. J., Ed., Academic Press, New York, 1977, chap. 5.
31. **Taylor, J. H.,** Duplication of chromosomes, *Sci. Am.*, 198, no. 6, 36, 1958.
32. **Sorsa, V.,** The role of the chromosome axis, *Hereditas,* 79, 109, 1975.
33. **Abuelo, J. G. and Moore, D. E.,** The human chromosome. Electron microscopic observations on chromatin fiber organization, *J. Cell Biol.*, 41, 73, 1969.

34. **Mullinger, A. M. and Johnson, R. T.,** The organization of supercoiled DNA from human chromosomes, *J. Cell Sci.*, 38, 369, 1979.
35. **Cook, P. R. and Brazell, I. A.,** Supercoils in human DNA, *J. Cell Sci.*, 19, 261, 1975.
36. **Okada, T. A. and Comings, D. E.,** Higher order structure of chromosomes, *Chromosoma (Berlin)*, 72, 1, 1979.
37. **Bak, P., Bak, A. L., and Zeuthen, J.,** Characterization of human chromosomal unit fibers, *Chromosoma (Berlin)*, 73, 301, 1979.
38. **Zeuthen, J., Bak, P., and Bak, A. L.,** Chromosomal unit fibers in *Drosophila*, *Chromosoma (Berlin)*, 73, 317, 1979.
39. **Ris, H.,** Chromosomal structure as seen by electron microscopy, *Ciba Found. Symp.*, 28, 7, 1973.
40. **Ris, H.,** Higher order structures in chromosomes, *Proc. 9th Int. Congr. Electron Microsc.*, 3, 545, 1978.
41. **Thoma, F., Koller, T., and Klug, A.,** Involvement of histone H1 in the organization of the nucleosome and of the salt dependent superstructure of chromatin, *J. Cell Biol.*, 83, 403, 1979.
42. **Renz, M., Nehls, P., and Hozier, J.,** Involvement of histone H1 in the organization of the chromosome fiber, *Proc Natl. Acad. Sci. U.S.A.*, 74, 1879, 1977.
43. **Hozier, J., Renz, M., and Nehls, P.,** The chromosome fiber: evidence for an ordered superstructure of nucleosomes, *Chromosoma (Berlin)*, 62, 301, 1977.
44. **Strätling, W. H., Müller, U., and Zentgraf, H.,** The higher order repeat structure of chromatin is built up of globular particles containing eight nucleosomes, *Exp. Cell Res.*, 117, 301, 1978.
45. **Rattner, J. B. and Hamkalo, B. A.,** Higher order structure in metaphase chromosomes. I. The 250 A fiber, *Chromosoma (Berlin)*, 69, 363, 1978.
46. **Davies, H. G., Murray, A. B., and Walmsley, M. E.,** Electron-microscope observations on the organization of the nucleus in chicken erythrocytes and a superunit thread hypothesis for chromosome structure, *J. Cell Sci.*, 16, 261, 1974.
47. **Olins, A. L. and Olins, D. E.,** Stereo electron microscopy of the 25-nm chromatin fibers in isolated nuclei, *J. Cell Biol.*, 81, 260, 1979.
48. **Hanson, C. V.,** Techniques in the isolation and fractionation of eukaryotic chromosomes, in *New Techniques in Biophysics and Cell Biology*, Vol. 3, John Wiley & Sons, New York, 1977, chap. 2.
49. **O'Farrell, P. H.,** High resolution two-dimensional electrophoresis of proteins, *J. Biol. Chem.*, 250, 4007, 1975.
50. **Igó-Kemenes, T. and Zachau, H. G.,** Domains in chromatin structure, *Cold Spring Harbor Symp. Quant. Biol.*, 42, 109, 1978.
51. **Benyajati, C. and Worcel, A.,** Isolation, characterization, and structure of the folded interphase genome of *Drosophila melanogaster*, *Cell*, 9, 393, 1976.
52. **Cook, P. R., Brazell I. A., and Jost, E.,** Characterization of nuclear structures containing superhelical DNA, *J. Cell Sci.*, 22, 303, 1976.
53. **Brasch, K. and Setterfield, G.,** Structural organization of chromosomes in interphase nuclei, *Exp. Cell Res.*, 83, 175, 1974.
54. **Gerace, L., Blum, A., and Blobel, G.,** Immunocytochemical localization of the major polypeptides of the nuclear pore complex-lamina fraction. Interphase and mitotic distribution, *J. Cell Biol.*, 79, 546, 1978.
55. **Ely, S., D'Arcy, A., and Jost, E.,** Interaction of antibodies against nuclear envelope-associated proteins from rat liver nuclei with rodent and human cells, *Exp. Cell. Res.*, 116, 325, 1978.
56. **Gerace, L. and Blobel, G.,** The nuclear envelope lamina is reversibly depolymerized during mitosis, *Cell*, 19, 277, 1980.
57. **Engelhardt, P. and Pusa, K.,** Nuclear pore complexes: "press-stud" elements of chromosomes in pairing and control, *Nature New Biol.*, 240, 163, 1972.
58. **Hodge, L D., Mancini, P., Davis, F. M., and Heywood, P.,** Nuclear matrix of HeLa S_3 cells. Polypeptide composition during adenovirus infection and in phases of the cell cycle, *J. Cell Biol.*, 72, 194, 1977.
59. **Long, B. H., Huang, C.-Y., and Pogo, A. O.,** Isolation and characterization of the nuclear matrix in Friend erythroleukemia cells: chromatin and hnRNA interactions with the nuclear matrix, *Cell*, 18, 1079, 1979.
60. **Adolph, K. W.,** unpublished data.

Chapter 6

NUCLEOLAR PROTEINS: PURIFICATION, ISOLATION, AND FUNCTIONAL ANALYSES

Harris Busch

TABLE OF CONTENTS

I. INTRODUCTION

The development of methods for isolation of nucleoli[1-3] provided the basis for many investigations on nucleolar structures, their components, and their functions.

Although, initially, studies were made on nucleolar histones, with improving technology the opportunity existed to define the numbers and types of nucleolar proteins and to evaluate their changes in the course of altered nucleolar functions. There are many more nucleolar proteins than were originally suspected. These act in a harmonious relationship to provide adequate numbers of ribosomes for cytoplasmic functions.

In any consideration of nucleolar function, it must be born in mind that the major activity of the nucleolus is the production of ribosomes[3] (Figure 1). To this end, a variety of nucleolar functions are necessarily highly coordinated including synthesis of preribosomal RNA, modification reactions, cleavage reactions, three-dimensional spatial coupling reactions, and organization of products into an appropriate three-dimensional array. In addition, it is clearly essential that the nucleolus transports these fundamental elements in the nucleolar ribonucleoprotein network (matrix) for further modification and transport to the cytoplasm.

So much remains to be learned about the synthesis, modification, packaging, transport, and other nucleolar functions that the future is far more important than the past in these studies. Nonetheless, the work of the last 15 years has offered a base for further understanding of nucleolar roles for a number of types of molecules.

II. NUMBERS OF NUCLEOLAR PROTEINS

The earliest estimates of the numbers of nucleolar proteins noted that nucleoli contained histones, synthetic enzymes such as the RNA polymerases, and proteins that were in preribosomal or ribosomal particles. It was clearly uncertain, however, as to the number and types of other proteins that might exist. As is the case for nuclear proteins, it was not clear which, if any, were specific nucleolar proteins and which constituted proteins that were simply equilibrated between the nucleus and the nucleolus. The key problem in investigation of the nonhistone proteins of the nucleolus is their limited solubility, particularly in media that were commonly used for extraction including saline solutions of various types and ionic strengths.

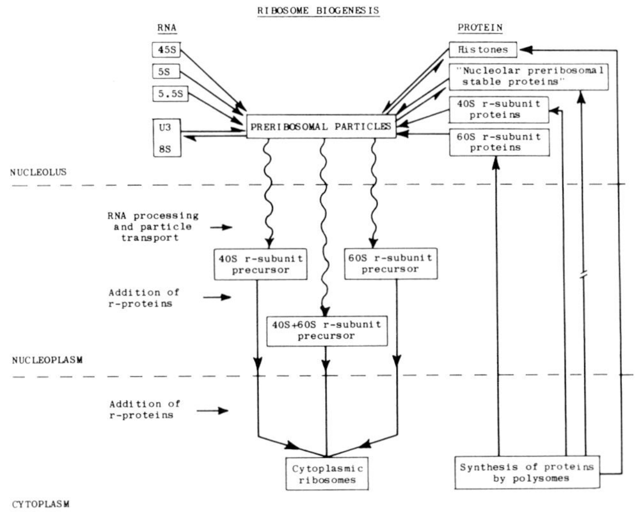

FIGURE 1. Schematic representation of factors involved in ribosome biosynthesis and transport in eukaryotic cells.

With the development of the urea-salt extraction methods and the 6 to 8 *M* urea solutions extraction,[4] it became possible to evolve more satisfactory solubilization of nuclear and nucleolar proteins. It was also possible to make use of acid-urea solutions for extraction and gel studies to define more carefully the various nucleolar components.

In 1973 we developed a two-dimensional gel method for separation of nucleolar elements.[5] This method involved (Figure 2) initial fractionation of nucleolar proteins on gels containing 10% polyacrylamide and 6 *M* urea. The fractionation of the nucleolar proteins in this dimension indicated that there were approximately 30 to 40 bands. A second-dimensional gel containing 12% polyacrylamide-0.1% SDS provided a further separation which permitted identification of many more proteins. With the aid of this method, approximately 100 distinct protein spots were identified. A number were histones, but many were proteins which had never been seen previously. These included elements such as the remarkable proteins A24, B23, C23, and others which have subsequently been investigated further.

Inasmuch as there were a large number of proteins, a numbering system was devised in which the proteins of the group which migrated fastest had the prefix A and the slowest group had the prefix C. The intermediate group had the prefix B. Within each group, the proteins were numbered in order of increasing number for the slower migration rate, so that protein A1 would be the fastest migrating protein in the group. Other proteins which migrated more slowly had higher numbers. It was of considerable interest that most of the proteins in the normal liver were also present in the Novikoff hepatoma and vice versa. However, a few proteins were distinctive in these tissues as indicated in Table 1.

With the evolution of two-dimensional gel technique, it became clear that many more nucleolar proteins existed. The latest technology we employed is based in part on the improved two-dimensional (Figure 3) method developed by Hirsch et al.[6] and Takami and Busch.[7] The improved salt extraction techniques[8] were helpful in these studies. The essence of these studies was that the nucleolar proteins probably number more than 200, and possibly as many as 300 when a variety of enzymes are included in the total number.

III. FRACTIONATION OF NUCLEOLAR PROTEINS

A number of procedures have been devised to separate nucleolar elements. Among these are methods which provide for treatment with DTT (dithiothreitol) or PBS (polyvinyl sulfate)

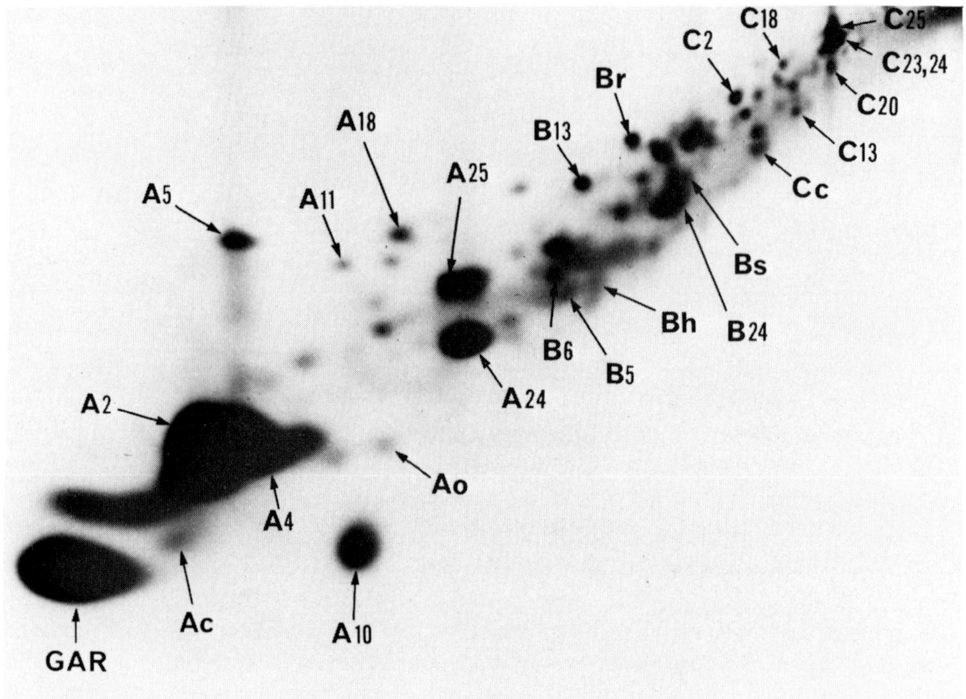

A

FIGURE 2. (A) Two-dimensional gel electrophoresis of nuclear proteins by the method of Orrick et al.[5] In the first dimension, the 10% polyacylamide gel contains 6 *M* urea and in the second dimension the 12% polyacrylamide gel contain 0.1% SDS. The proteins were stained with Coomassie® blue. Note the dense spot of protein C23, C24. Protein B23 is in the dark spot containing B24. (B) Autoradiograms of nucleolar phosphoproteins. These proteins were labeled with $^{32}P_i$ and then subjected to two-dimensional gel electrophoresis as in Figure 2A. Note the dense C23 spot and B23 spots in the longer exposure and the less-dense spots in the shorter exposure of the inset.

or other agents such as EDTA. Such methods are designed to separate the nucleolar particles from the remainder of nucleolar structures and constitute "release" procedures for nucleolar-soluble substances and particles, such as reductive cleavage of $-S-S-$ bonds.

In our laboratory, a multistep method was devised in which nucleoli were extracted in six steps (Table 2, Figure 4). This procedure provided separation of enzymes, histones, and a variety of particles and proteins.

IV. PURIFICATION OF NUCLEOLAR PARTICLES

One of the important procedures in analyses of nucleolar proteins is the separation of particulate elements from soluble and other nucleolar elements. There are two types of nucleolar particles that are readily separable from the remainder of nucleolar structures, namely, preribosomal RNP particles and the U3 RNP particles. Both of these types of particles are obtainable from the procedure devised by Rothblum et al.,[8] in which the 0.01 *M* Tris-HCl (pH 8.0) extract was sedimented at 105,000 × g for 18 hr over a 1 *M* sucrose cushion. The pellet, which composed approximately 10% of the extracted proteins, contained the nucleolar RNP particles. Their protein patterns are shown in Figure 4; the distribution and number of the proteins corresponded well to that found by Prestayko et al.[9,10] and Daskal et al.[11]

It is interesting and perhaps important that most U3 RNP particles were also in this fraction.

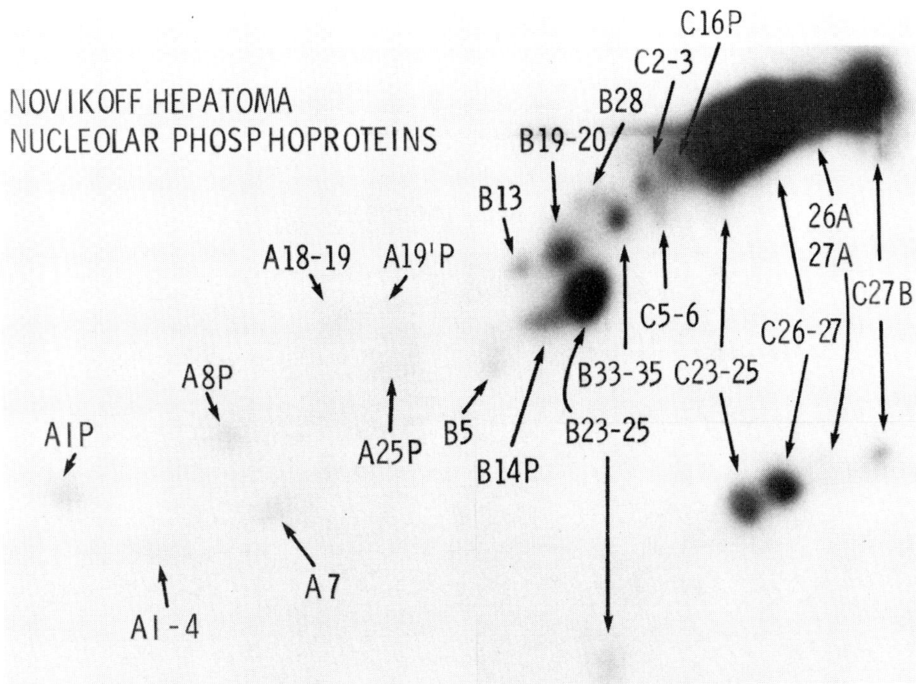

FIGURE 2B

Scanning electron micrographs of these pellets indicated that these particles were essentially homogeneous, with diameters similar to those reported earlier by Koshiba et al.[12] In earlier studies,[9,10] nucleolar preribosomal RNP particles were prepared by extraction of nucleoli with a buffer containing 0.01 *M* NaCl/0.005 *M* $MgCl_2$/0.01 *M* Tris HCl/pH 7.4 followed by addition of DNase to this preparation. Slow stirring for 15 min at 0° provided a mixture which, on sedimentation, resulted in a pellet that was resuspended in 10 m*M* DTT along with the buffer indicated above. This treatment released the nucleolar particles which remained in a supernatant that had been sedimented at 20,000 × g. This supernatant was underlayered with 1 *M* sucrose in the same buffer, and after sedimentation at 90,000 × g for 16 hr the pellet contained the preribosomal nucleolar particles.

A third method for extraction of the preribosomal particles involved treatment of the nucleoli with a buffer containing 0.25 *M* sucrose/0.002 *M* $MgCl_2$/0.05 *M* KCl/0.01 *M* sodium acetate/pH 6.0. This preparation was treated with DNase and centrifuged as noted above. Then 40 μg/mℓ PVS (polyvinylsulfate) was added. The procedure was essentially the same as noted above except that 0.0005 *M* $MgCl_2$ was used in place of the 0.002 *M* $MgCl_2$. The supernatant from this sedimentation was underlayered with 1 *M* sucrose in the same buffer and the mixture was sedimented for 16 hr at 90,000 × g. Under the conditions of these experiments, highly purified nucleolar particles were obtained.[9,10] The preribosomal RNP particles had a very specific pattern on two-dimensional polyacrylamide gels. Some of the preparations appeared to contain histones but others did not. Accordingly, it was uncertain whether the histones represented a defined component of these preparations.

What was most important about these studies was they indicated the presence of a number of proteins of the small 40S ribosomal subunits and the larger 60S ribosomal subunits. Approximately 10 proteins of the 40 subunits were present and 22 proteins of the 60S subunits were identifiable in these particles.

The relationship of the remainder of the proteins to the ribosomal proteins has not been

Table 1
SPOT DENSITY DIFFERENCES FOR THE HEPATOMA AND LIVER PATTERNS

	Density of spot	
Spot number	**Hepatoma**	**Liver**
A10	+ + + +	+
A10L	+ +	+ + +
A11	+	+ + +
B1	+	+ + + +
B4L	Absent	+
B5L	Absent	+
B18	+ +	+ + +
C8	+ + +	+ +
C9	+	Absent
C11	+	+ + +

From Orrick, L. R., Olson, M. O. J., and Busch, H., *Proc. Natl. Acad. Sci. U.S.A.*, 70, 1316, 1973. With permission.

clarified to this point (Figure 5). The possibility exists that the proteins in the preribosomal particles represent different proteins, or that they may represent precursor proteins, that they may are processing or transport proteins.

V. SYNTHETIC REACTIONS IN THE NUCLEOLUS

The nucleolus is the site of synthesis of preribosomal RNA. Among the important elements of the nucleolar machinery involved in these synthetic reactions are

1. The rDNA template
2. The enzyme RNA polymerase I[13-16]
3. A variety of systems which make available nucleoside triphosphates which are necessary as precursors of the RNA
4. Processing enzymes which produce many modifications of the RNA
5. Cleavage enzymes which split the newly synthesized RNA into precursors of the 28 and 18S rRNA
6. Trimming enzymes involved in preparation of the final product
7. Specific proteins involved in structural interaction with the RNA precursors so as to produce the final conformationally correct state of the RNA

Although the nucleolar RNA polymerase was identified almost 20 years ago[17] and subsequently has been purified in a number of laboratories, the exact composition and arrangement of the elements and subunits of this important enzyme are still unknown. The elegant work of Roeder's group[14-16] and Chambon and associates[13] showed that this enzyme contains 11 or more subunits. The enzyme has been readily purified by ammonium sulfate fractionation and by DEAE cellulose fractionation techniques. Although RNA polymerase I can be further purified, unfortunately it becomes unstable with increasing purity. The overall enzyme is quite large, with a molecular weight of approximately 0.5×10^6. Good information is not available on the composition, properties, or interactions of its many subunits (Figure 6).

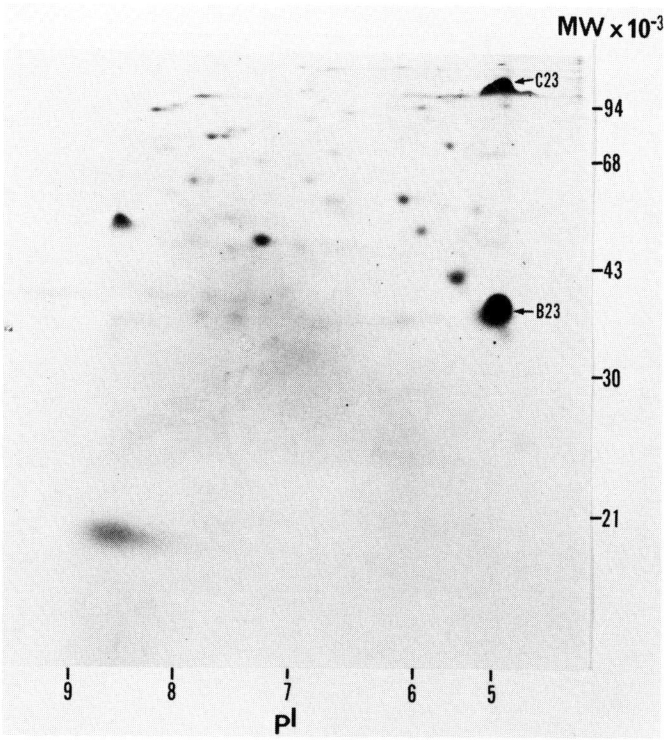

FIGURE 3. Two-dimensional polyacrylamide gel electrophoresis of total Novikoff hepatoma nucleolar proteins. This first dimension was isoelectric focusing and the second dimension was a 10% SDS gel (see Takami and Busch[7]).

The role of RNA polymerase I, as isolated, has clearly been established to be that of an elongation enzyme which permits reading of DNA chains, but intracellularly it only serves the role of a specific reading element for rDNA.

A. Initiation of Pre-rRNA Transcription

A major question in rRNA synthesis has been to define the controls involved in the *initiation* of synthetic events for rRNA synthesis. Many workers have attempted to clarify this problem with increasing success, as reviewed by Grummt.[18] There must exist substances or enzymes not yet identified that may either serve as topoisomerases, nick-closing enzymes, or catenases of specific types that may in fact control nucleolar reactions either directly or indirectly. At this moment it is not possible to specifically state which, if any, function is subserved by such enzymes and which, if any, are specific for rDNA.

B. Termination of rDNA Readouts

As for the initiation reactions, there is little known about the termination signals or enzymes involved in nucleolar readouts. It is certain, however, that usually no mistakes are made and no extramolecular portions are produced. Evidence from our laboratory[19] showed that after treatment with low salt concentrations, nucleoli produced normal products which are indistinguishable from the in vivo products as determined by T_1-RNase digests of the high-molecular RNA (Figure 7). However, when the nucleolus was treated with increasing salt concentrations, additional T_1-RNase products were produced which differed from the normal products. It seemed likely that these new products reflected the production of extra elements in the pre-rRNA chains by reading into flanking sequences of the rDNA which are normally not read because of the termination signals.

Table 2
DISTRIBUTION OF NUCLEOLAR PROTEINS AND SPECIFIC ANTIGENS

Fraction	% Total nucleolar protein	% Total dissociated nucleolar chromatin protein	Protein kinase (cpm X 10^{-5})	RNA polymerase (units X 10^{-3})
0.075 NaCl-0.025 *M* EDTA	19 ± 3		11.2 (40)	10.4 (12.5)
10m*M* Tris-HCl (pH 8.0)	35 ± 5		18.4 (60)	67.9 (87.5)
Chromatin	46 ± 5	100		
0.15 *M* NaCl	5 ± 1	14 ± 1.4		
0.35 *M* NaCl	13 ± 4	27 ± 4.0		
0.6 *M* NaCl	14 ± 4	30 ± 4.8		
3 *M* NaCl-7 *M* urea	14 ± 6	28 ± 7.6		

From Rothblum, L. I., Mamrack, P. M., Olson, M. O. J., and Busch, H., *Biochemistry*, 16, 4716, 1977. With permission.

In the process of synthesis of pre-rRNA there are three types of elements involved, namely: *initiation, elongation* and *termination* elements. Despite the intense efforts that have been made in a number of laboratories to characterize these elements further, progress has been extremely slow.[20] The reason for this may be that the systems employed are too simple or that the products important in the biosynthetic reaction are either short-lived or rapidly destroyed.

C. Electron Microscopy of Nucleolar Synthetic Events

Miller and associates[21,22] demonstrated the "Christmas tree" pattern of nucleolar synthetic events. The nucleolar system in amphibia provided the best type of ultrastructural elements for such studies. Long rDNA chains were the centers of branched structures consisting primarily of preribosomal RNA. The spaces existing between the Christmas tree structures represented "nontranscribed spaces" which varied in length. As shown by Brown and associates,[25-27] they also varied in composition from species to species.

The transcribed elements of the Christmas trees had high orders of homology and reproducibility. The fact that virtually every sequence had the same appearance led to the concept of the master gene. This meant that either a single gene was in control of the entire tandem array of 200 to 400 genes, or that one gene having been read, the remaining several hundred genes would also be read.[28,29]

A most important observation in these studies, which has not been clarified, relates to the existence of small particles at the ends of the branches. These small particles are of special interest in view of the recent studies made in our laboratory on proteins B23 and C23, which are the silver-staining proteins. It seems possible that these particles might be unique nucleolar structures involved in early phases of readouts by interaction with the genes themselves or the "TAATAA" box elements upstream of the main coding reagents. The relationship of the C23 and B23 elements and the particles to "fibrillar portions" of the nucleolus has not yet been fully defined. The elegant electron microscopic pictures from Trendelenburg and McKinnell[30] clearly indicate that the Christmas trees are the central elements of the fibrillar portions of the nucleolus. Figure 8 shows the fibrillar mass of the nucleolus is in fact a highly conglomerated segment of these fibrillar Christmas tree masses which can be unfolded in the presence of proper detergents.

VI. PHYSIOLOGICAL AND PATHOLOGICAL EFFECTS ON NUCLEOLAR PROTEINS

The development of two-dimensional gel analysis of nucleolar proteins led to further studies on the alteration of specific components during physiological events.[31,32] One ap-

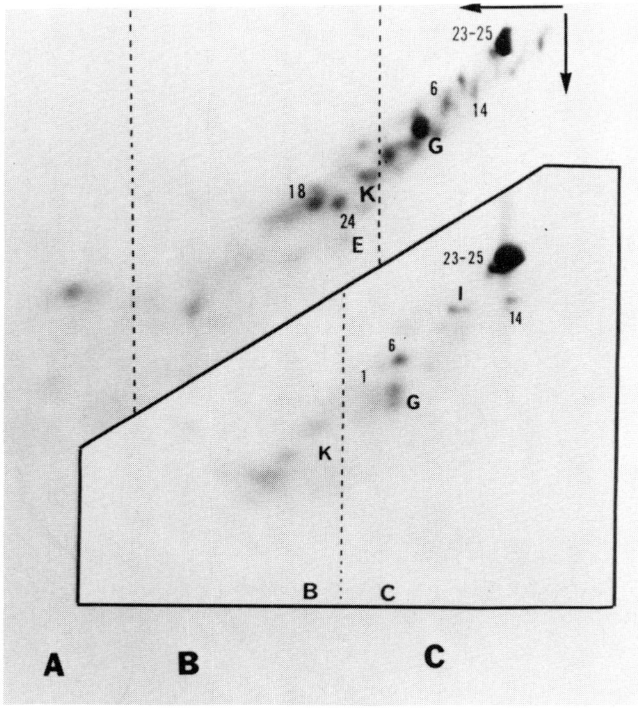

A

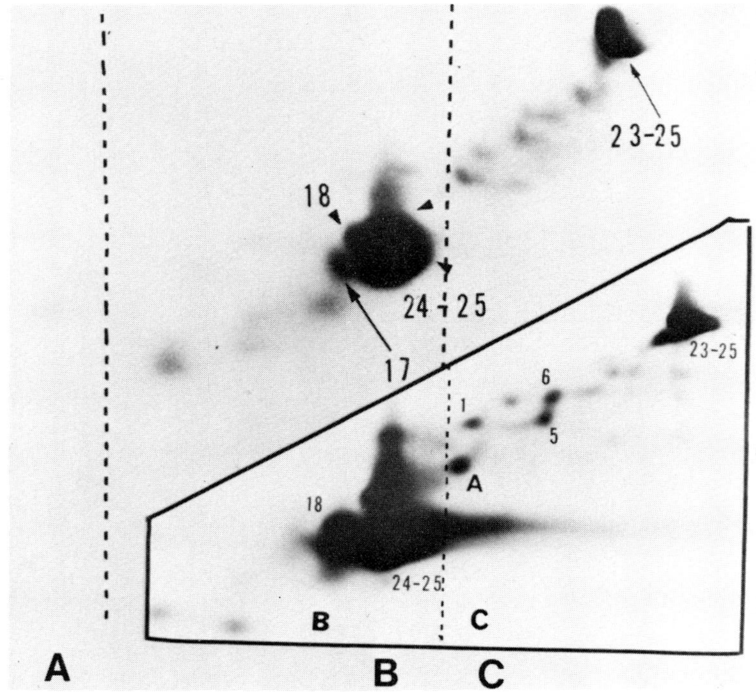

B

FIGURE 4. Two-dimensional gel electrophoresis of 200 mg of Novikoff hepatoma nucleolar proteins. These were done by the method of Orrick et al[5] (A) NaCl-EDTA extract; (B) 0.01 *M* Tris-HCl/pH 8 extract; (C) preribosomal particle proteins sedimented from 0.01 *M* Tris-HC1/pH 8 extract; (D) proteins extracted were 0.15 *M* NaCl from nucleolar chromatin; (E) 0.35 *M* NaCl extract from nucleolar chromatin; (F) 0.6 *M* NaCl from nucleolar chromatin; (G) 3 *M* NaCl/7 *M* urea extract of chromatin residue. These data show the variability of proteins extracted with the different solvents employed. (From Rothblum, L. I., Mamrack, P. M., Olson, M. O. J., and Busch, H., *Biochemistry*, 16, 4716, 1977. With permission.)

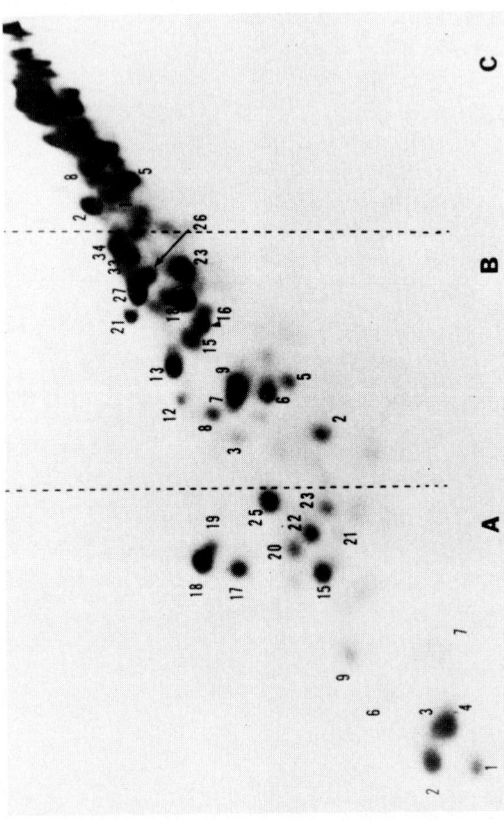

FIGURE 4C

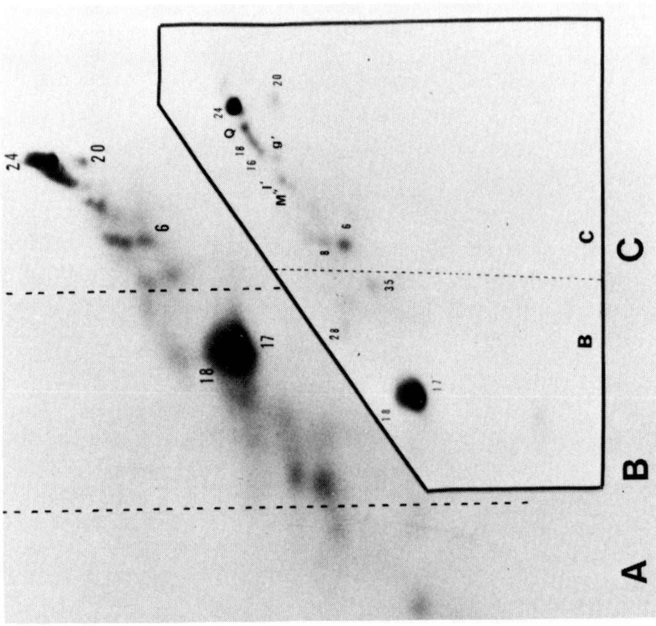

FIGURE 4D

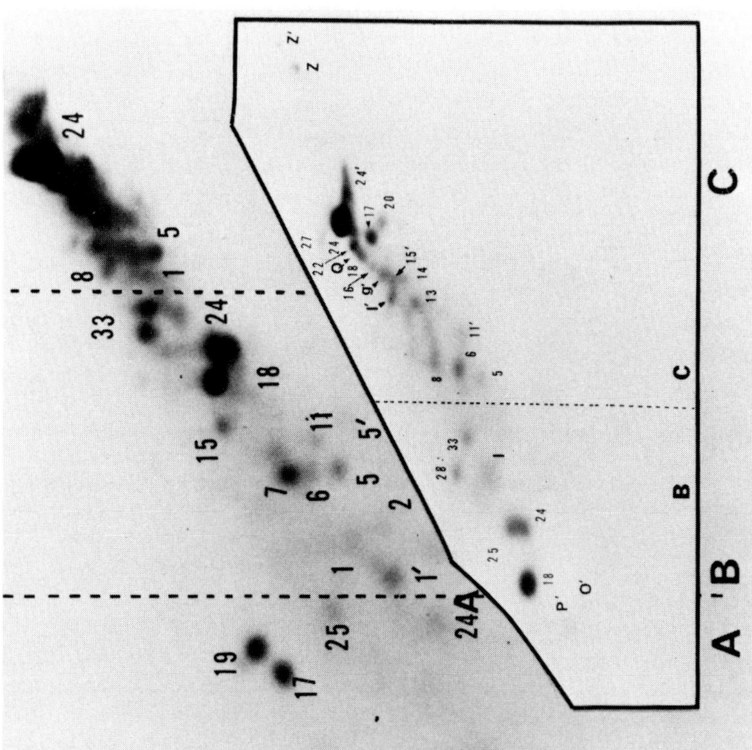

FIGURE 4E

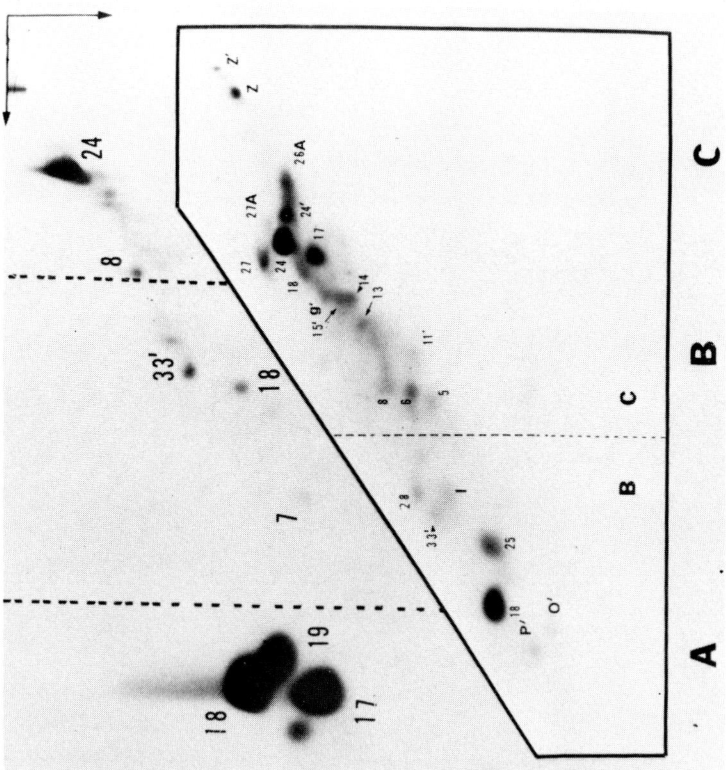

FIGURE 4F

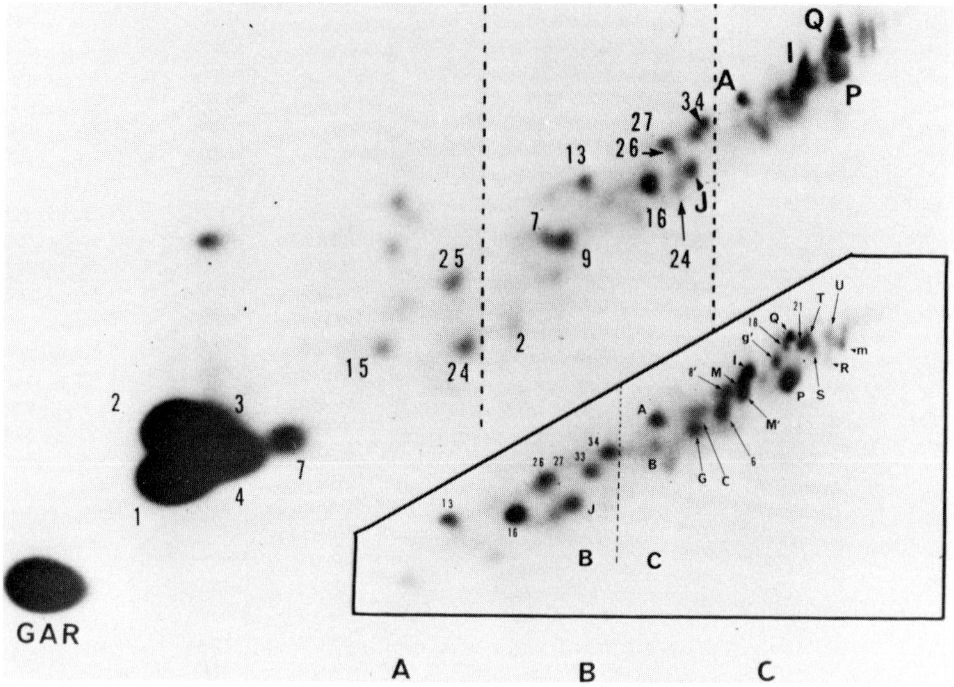

FIGURE 4G

proach to such analysis was determination of increases and decreases in specific proteins during thioacetamide treatment.[31] The most notable alterations were marked decreases in protein A24 and A11. Two other proteins referred to as C13 and C14 decreased to a lesser extent (Figure 9).

On the other hand, a series of proteins referred to as B23-25 and proteins C23-25 remarkably increased following thioacetamide treatment.

A number of these proteins were not found in either ribosomes or nucleolar RNP particles and accordingly it was of interest to determine whether these proteins might reflect alteration in chromatin structure or alterations in the polymerase activity within the nucleolar elements.

The analysis of protein changes in liver regeneration revealed very similar changes in nucleolar proteins to those observed when thioacetamide-treated liver samples were studied. There was a marked decrease in protein A24 and protein A11 during the period of 8 to 24 hr after partial hepatectomy. In regenerating liver, protein A25 also diminished in size.

Both proteins C13 and C14 were markedly decreased in amount. Proteins C23 to C25 were increased in size as were proteins B23 to B25.

In the study on regenerating liver,[32] another phenomenon was observed with ^{32}P-labeling. Under these conditions there was a marked increase in the radioactivity in proteins C23 and C24 and in proteins B23 and B24. These enhanced phosphorylations were important in the development of further information on these proteins (Figure 2B).

A. Protein A24

When protein A24 was purified,[33] it was noted to have a remarkable similarity in composition to histone 2A which has now been found to be the parent protein to which ubiquitin is conjugated. In addition to the similarity in composition, virtually all the peptides of protein A24 were identical to those in protein 2A as shown by tryptic and chrymotryptic maps.[34] The further purification of protein A24 was achieved by a series of methods including one-dimensional preparative gel electrophoresis.

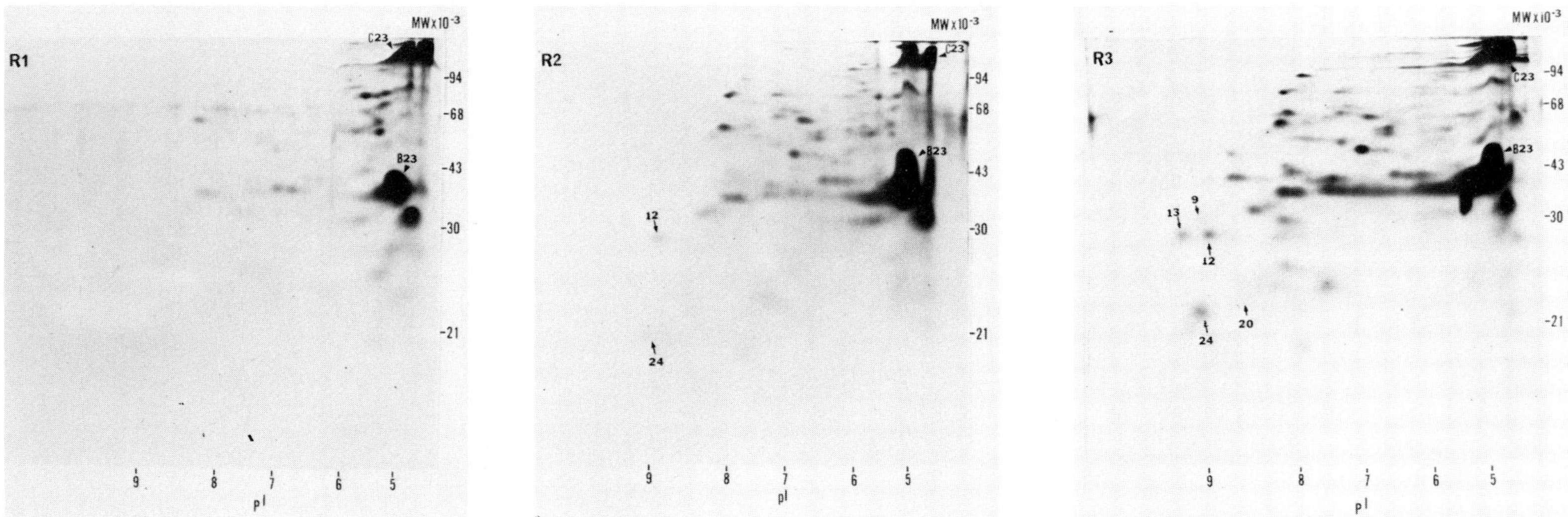

FIGURE 5. Proteins were extracted from fractions of sucrose density gradients of solubilized nucleoli (see Figure 28). The two-dimensional gel system of Hirsch et al.[6] was used. The isoelectric focusing gel was run for 1 hr at 50V, 1 hr at 100V, and 40 min at 200V. Approximately 150 μg of protein was applied to the gels. The approximate molecular weight and pI are shown. The gels were stained with Coomassie® brilliant blue R. Proteins C23 and B23 are labled. The numbered spots are ribosomal proteins. These results show the progressive addition of special proteins to the initial complexes formed in the nucleolus with increased density from R1 to R6.

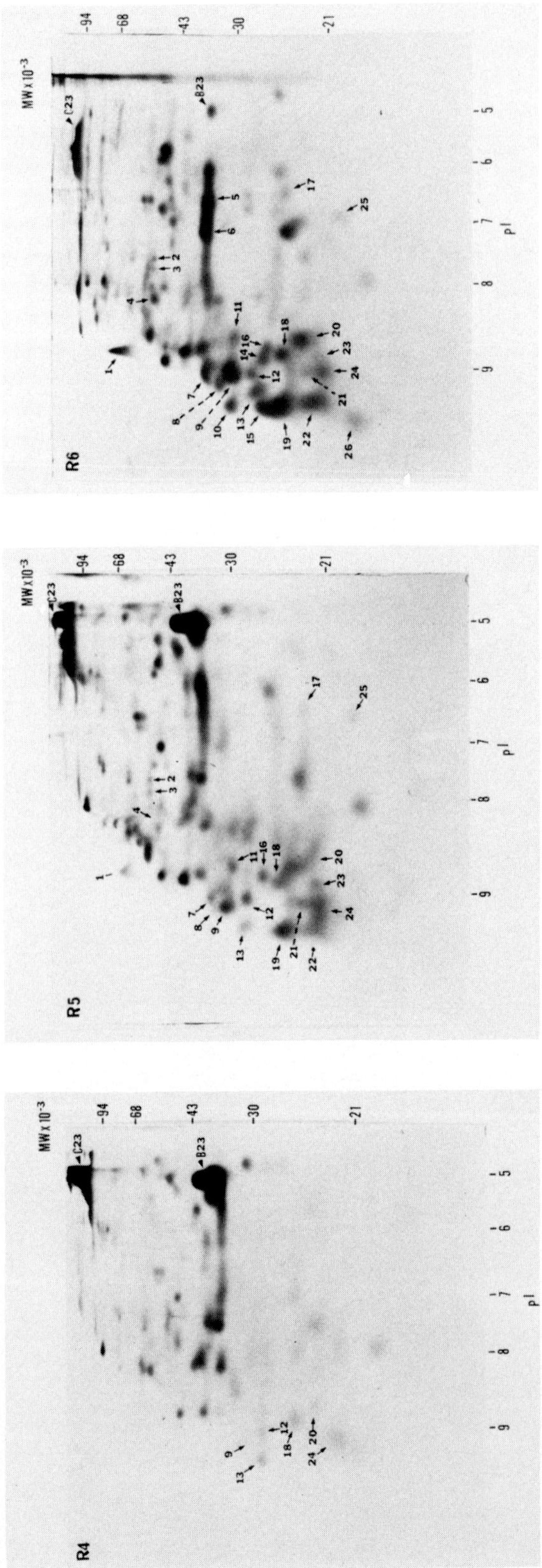

FIGURE 5(continued)

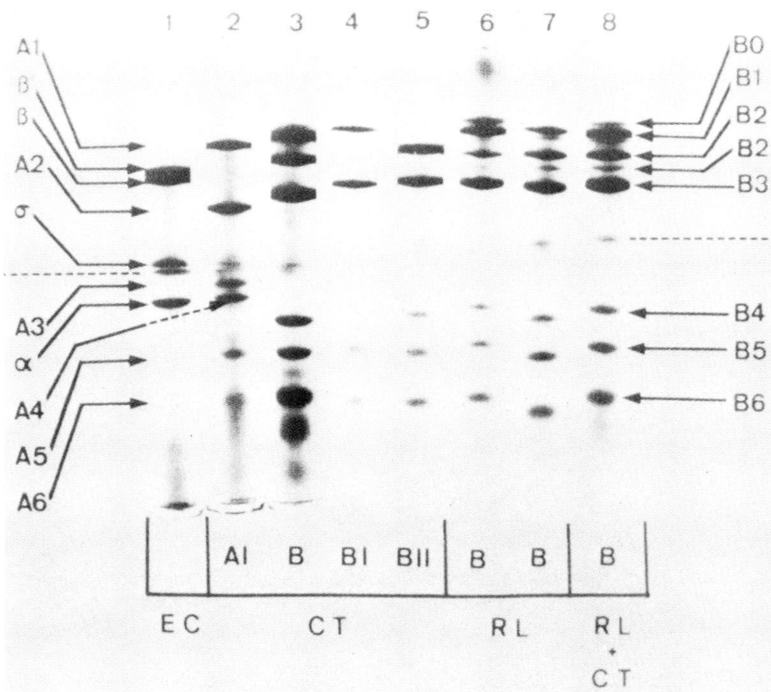

FIGURE 6. Subunit pattern of calf thymus RNA polymerases AI, BI, BII, and a mixture of purified rat liver B enzymes. Comparison with *Escherichia coli* holoenzyme. About 1.5 activity unit of each enzyme was denatured and SDS-polyacrylamide gel electrophoresis was carried out using mixed gels (5 and 10% in the upper and lower parts of the gel, respectively), in order to resolve both the high- and low-molecular weight subunits. The gels belong to two series run independently (series A: gels 1, 2, and 3; series B: gels 4, 5, 6, 7, and 8). Gel 1, *E. coli* holoenzyme; gel 2, CT RNA polymerase AI; gel 3, mixture of purified CT B enzymes; gel 4, CT RNA polymerase BI; gel 5, CT RNA polymerase BII; gel 6, mixture of purified rat liver B enzymes; gel 7, mixture of purified rat liver B enzymes; gel 8, coelectrophoresis of B enzymes from calf thymus and rat liver. EC, *E. coli* holoenzyme; CT, calf hymus; RL, rat liver. (From Chambon, P., Gissinger, F., Kedinger, C., Mandel, J. L., and Meilhac, M., in *The Cell Nucleus*, Vol. 3, Busch, H., Ed., Academic Press, New York, 1974, chap. 7. With permission.)

The greatest surprise with respect to this protein emerged following the publication of the initial amino terminal sequence analysis of protein A24.[35] Although the amino terminal of histone 2A was blocked, as found in our studies on the amino acid sequence of this protein,[36] the terminus of protein A24 was not blocked and was methionine. The remainder of the structure was sequenced with little difficulty up to position 37. It was almost immediately found that this sequence was identical to that of ubiquitin.[37]

It was exceptionally interesting that the amino terminal sequence of protein A24 was totally different from that of histone 2A, but the C terminal sequence was identical.[38] Protein A24 had only one C terminal sequence (Figure 10).

The most interesting result obtained was that histone 2A was linked by an isopeptide linkage to ubiquitin through an isopeptide linkage to lysine 119. This result arose from isolation of the specific linking peptide, which indicated that the nonhistone ubiquitin was linked to histone 2A in a remarkable "Y-shape".[39] This structure was of particular interest because the protein A24, like histone 2A, was demonstrated to be present in nucleosomes. It is associated with the fundamental chromatin structure but represents an alteration that is a three-dimensional variant within chromatin. Exactly how and why this protein undergoes this modification is not clear, but the branched histone was the first such structure demonstrated.

Subsequently, a variety of ubiquitin-containing molecules were found. These include a variant of histone 2A, two variants of histone 2B, and a series of proteins linked to ubiquitin in the cytoplasm. The functions of these linkages are under active investigation.[40]

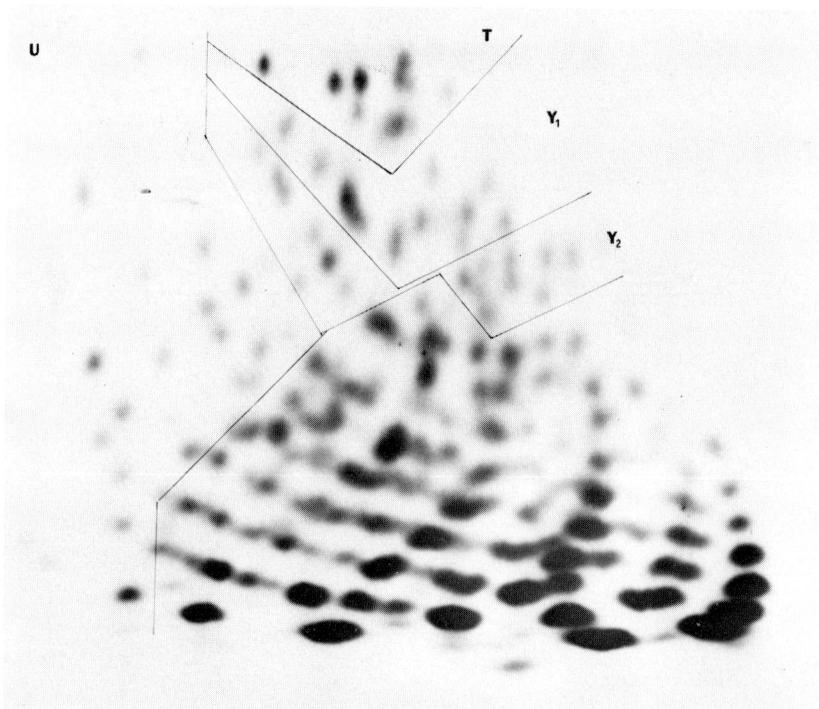

A

FIGURE 7. (A) Homochromatography of T_1 RNase digests of 45S pre-rRNA labeled in vivo. For convenience, the pattern was divided into regions: U (uridylate-rich), T (top triangle), Y_1 (Y-shaped), Y_2 (below Y_1), and D (approximately decanucleotides). (B) Diagrammatic map of the oligonucleotide homochromatogram in (A). Closed circles represent oligonucleotide spots found in 18S rRNA; hatched circles represent fragments found in 28S rRNA; crosshatched circles represent fragments present in both 18 and 28S rRNA; open circles represent oligonucleotide fragments present in the spacer segments of pre-rRNA. Spot T-9 was a variable spot which was occasionally missing from the digests.

As noted above, the variants of the histones are altered in various types of alteration of cell function. In the nucleolus, with increased activity there is a decreased content of protein A24. However, a decreased content of protein A24 has also been noted in chromatin when metaphase chromosomes are being formed or in the transition from the active chromatin of chicken erythrocytes to the inactive form.[41] Accordingly, the reticulocytes contain protein A24 while the mature reticulocytes do not. The ubiquitin, then, has a short half-life in its linkage to protein 2A and is the dynamic constituent of this structure.[42]

It has been suggested that protein A24 might be a structural component of active chromatin (Figure 11). A correlation between activity and structure has not been demonstrated.

B. Protein A11

The lysine-rich histones constitute a minimum of three and possibly more components. These histones are characterized (Table 3) by high contents (27% or more) of lysine residues in each of the histones of the group A17 and A18/19 complex.[5] Each has a high content of alanine. Protein A11 is a different protein in the sense that its alanine content is much lower. The remainder of the amino acids are essentially the same except that the arginine contents are somewhat higher than that of other proteins. Protein A11 was purified in this laboratory essentially to homogeneity (Figure 12); its amino acid composition is similar to protein $H1^0$ that was reduced in actively proliferating tissues but was present in tissues with little cell division.[43]

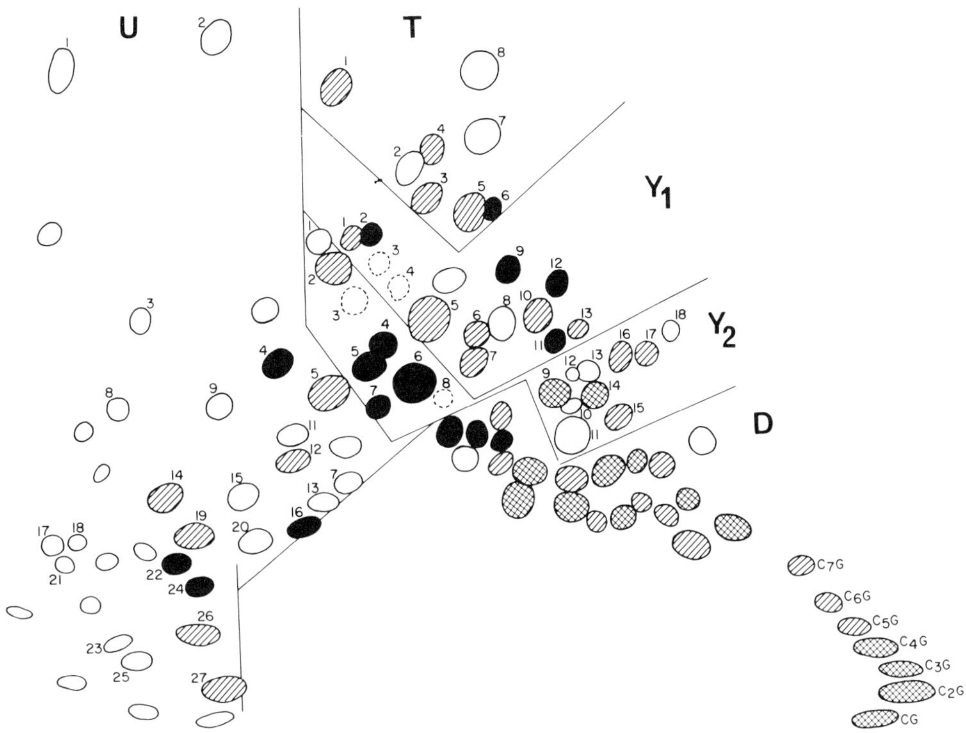

FIGURE 7B

In the studies in this laboratory, regenerating rat liver was found to contain less histone A11 than other proteins.[31,32] Histone A11 also decreased in thioacetamide-treated liver.

One question was whether this protein was phosphorylated. The evidence from our studies was that it was phosphorylated in rapidly growing tissues, and its apparent disappearance resulted from its incorporation into the protein A17 spot because of the negative charge induced by phosphorylation. Further studies on this protein will be of interest because of the interesting relationship between structure and function.[44,45]

C. Protein C14

Protein C14 was one of the proteins that underwent significant changes in concentration during the course of stimulation of nucleolar activity in rat liver.[46] This protein was of sufficient interest for its isolation in our laboratory. Isolation was done by nucleolar purification, 0.35 *M* NaCl extraction of chromatin and ammonium sulfate fractionation (30 to 60%). The 30 to 60% ammonium sulfate fraction was placed on Sephadex® G-75 columns in a buffer consisting of 0.35 *M* NaCl/10 m*M* Tris-HCl/pH 8.0/0.1 m*M* PMSF. This fraction was dialyzed against 0.9 *N* acetic/10 *M* urea/1% 2-mercaptoethanol at room temperature and was then placed on 6% polyacrylamide slab gels. The conditions used for fractionation of the proteins on the gel were described earlier.[33] Protein C14 was eluted from gel slices and was dialyzed against the buffer containing 0.35 *M* NaCl. This protein was shown by two-dimensional gel analyses to be essentially electrophoretically homogeneous (Figure 13).

The amino acid analyses of protein C14 is shown in Table 4. This protein was shown to be moderately rich in glutamic and aspartic acids, and had a molecular weight of approximately 70,000 as determined by gel electrophoresis in 0.1% SDS. The amino terminal amino acid was lysine as demonstrated by the dansyl method. A peptide map was obtained by fluorography.

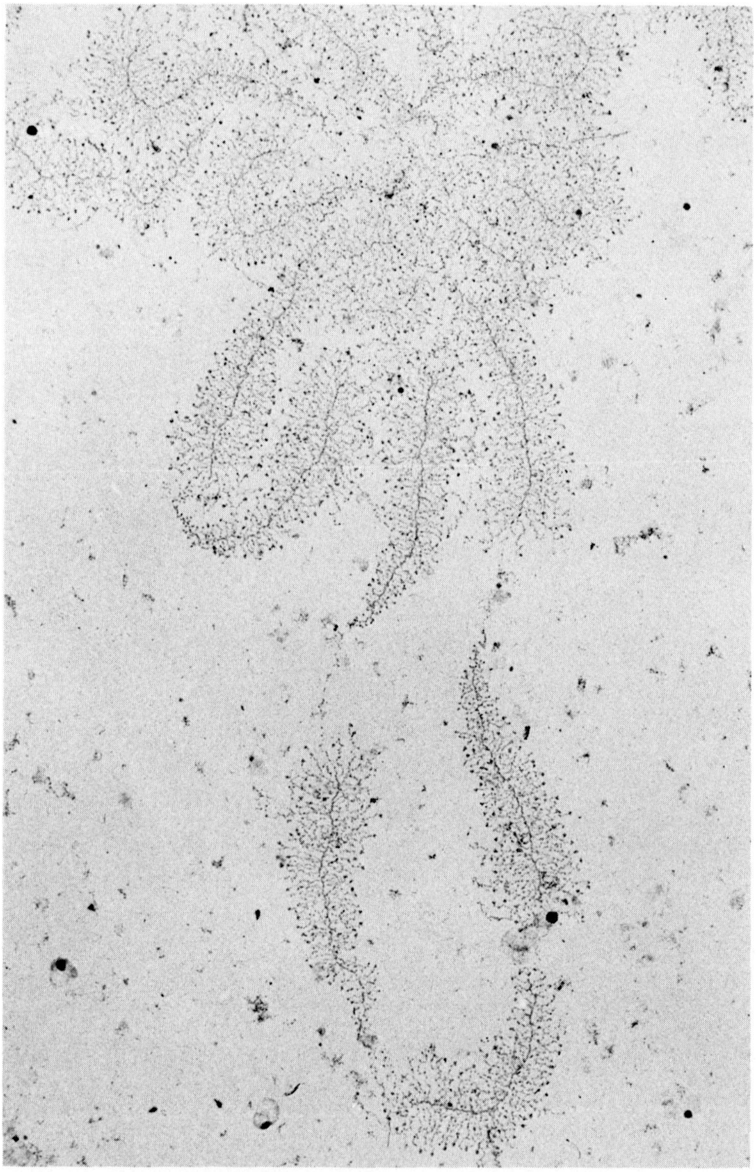

A

FIGURE 8. (A) Portion of isolated core of nucleolus from *Xenopus laevis* oocyte-presumptive rRNA cistrons. Uranyl acetate stain. The DNA fibers are visualized at the centers of elongated periodic masses. Along these masses there are RNA fibers of progressively increasing chain length. The distance between these RNA fibers presumably represents spaces occupied by RNA polymerase or other synthetic elements. (× 20,000.) The DNA in the spaces between the synthetic clusters has been referred to as ''spacer DNA'' but its structure and function are unknown. (B) Survey electron micrograph of spread nucleolar chromatin from a *Rana pipiens* oocyte nucleus after completion of oogenic growth. Under the dispersion conditions used, the nucleolar chromatin is not fully spread, i.e., spheroidal nucleolar material is converted into a two-dimensional structure, which permits the determination of the surface area occupied by individual nucleoli and the number of transcribed pre-rRNA genes per nucleolar unit. In preparations of this type, nucleolar chromatin axes are identified and can be traced only at the periphery of nucleoli, and exhibit the typical tandem arrangement of transcribed (fibril covered) ''matrix units'' and untranscribed (fibril fee) ''spacer'' intercepts (higher magnification shown in the insert; *arrows* denote 'smooth' configuration of spacer segments; nonbeaded chromatin structure of a large fibril free region within an incomplete transcriptional unit is denoted by *arrow-heads).* Transcribed pre-rRNA genes in central nucleolar regions are identified by the enhanced contrast of polymerase-covered matrix units. The two nucleoli shown provide an example of adjacent nucleoli of similar size and an almost fully active state of their pre-rRNA genes.

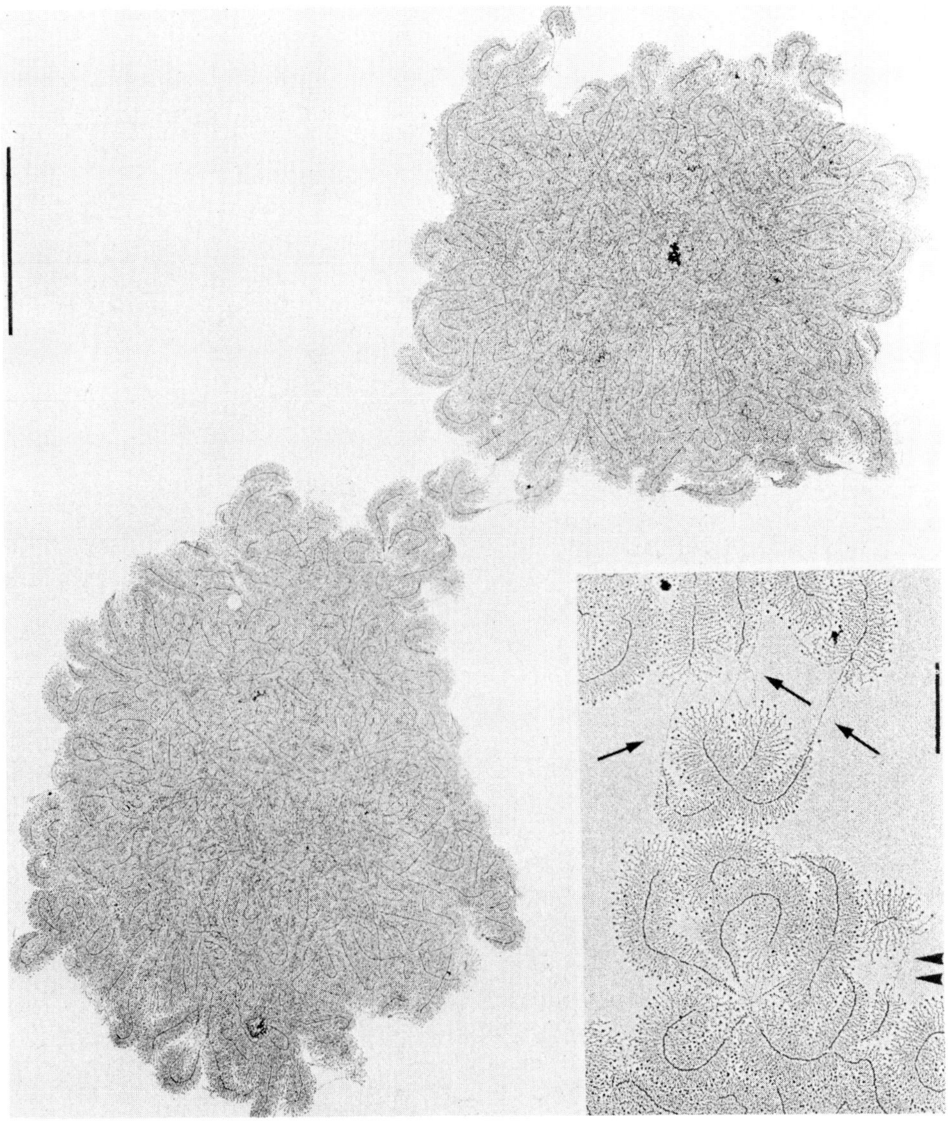

FIGURE 8B

This protein was interesting because it stimulated the uptake of nucleotides into RNA in the RNA polymerase systems. This increased activity was demonstrable with increasing concentrations of protein C14 (Figure 14). The stimulatory activity was due to this particular protein and not to overall protein content. It was notable that stimulation occurred also when an *Escherichia coli* DNA-dependent RNA polymerase was employed.[46]

D. Polyamine-Mediated Nucleolar Protein Phosphorylation

Kuehn et al.[47] reported that they isolated a nucleolar protein from *Physarum* which has virtually the same amino acid composition as protein C14; its molecular weight was also approximately 70,000. This protein was phosphorylated by *Physarum* nucleoli when polyamines were added; the phosphorylation of this phosphoprotein was detected by labeling with [^{32}P]-ATP. In the absence of the polyamines, 0.33 m*M* putrescine, spermidine, or spermine, phosphorylation was reduced.

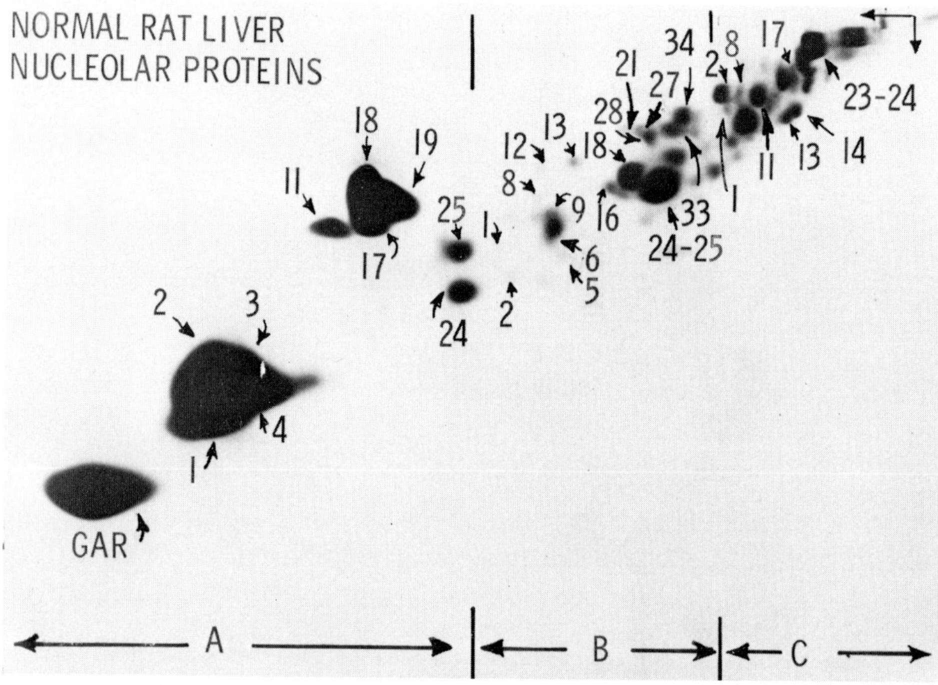

A

FIGURE 9. (A) Two-dimensional polyacrylamide gel electrophoresis of normal rat liver nucleolar acid-soluble proteins. Protein samples (750 μg) were run in the first dimension in gel rods of 10% polyacrylamide/4.5 *M* urea/0.9 *N* acetic acid at 120 V for 6 hr. For the second dimension, a 12% polyacrylamide-0.1% sodium dodecyl sulfate slab gel was run for 15 hr at 50 mA per slab. Gels were stained with Coomassie® brilliant blue R. (B) Two-dimensional polyacrylamide gel electrophoresis of nucleolar acid-soluble proteins of livers of rats after 4 hr of partial hepatectomy. Conditions were the same as those described in A. (C) Two-dimensional polyacrylamide gel electrophoresis of nucleolar acid-soluble proteins of livers of rats after 8 hr of partial hepatectomy. Conditions were the same as those described in A. (D) Two-dimensional polyacrylamide gel electrophoresis of nuclear acid-soluble proteins of livers of rats after 24 hr of partial hepatectomy. Conditions were the same as those described in A.

This nucleolar phosphoprotein was related to stimulation of transcription of the nucleolar rDNP complex (Figure 15). Increasing amounts of the protein increased activity in terms of [^{3}H]-UMP corporation. This result was similar to that found with protein C14.[46] All of these reactions were linear with respect to time. The product consisted mainly of rRNA.

A restriction endonuclease map of rDNA of *Physarum* showed that the DNA had a length of approximately 38 × 10^6. It consists of a palindrome, and the protein apparently localizes to the center of the palindrome (Figure 16). The native phosphoprotein bound to the Hind III restriction fragment but the dephosphorylated protein did not bind, indicating that it is not nonspecific. *E. coli* DNA and calf thymus DNA did not compete for this protein. Accordingly, the DNA of the *Physarum* contained regions for binding that have a high order of specificity. This instance of high-specificity binding for these DNA fragments represents a new finding in this important field. The phosphorylation is of special interest, but it remains to be proven as to whether there is a highly specific activation of this particular gene. Proof that the binding is related to stimulation of readouts of these rDNA sequences would be most interesting.

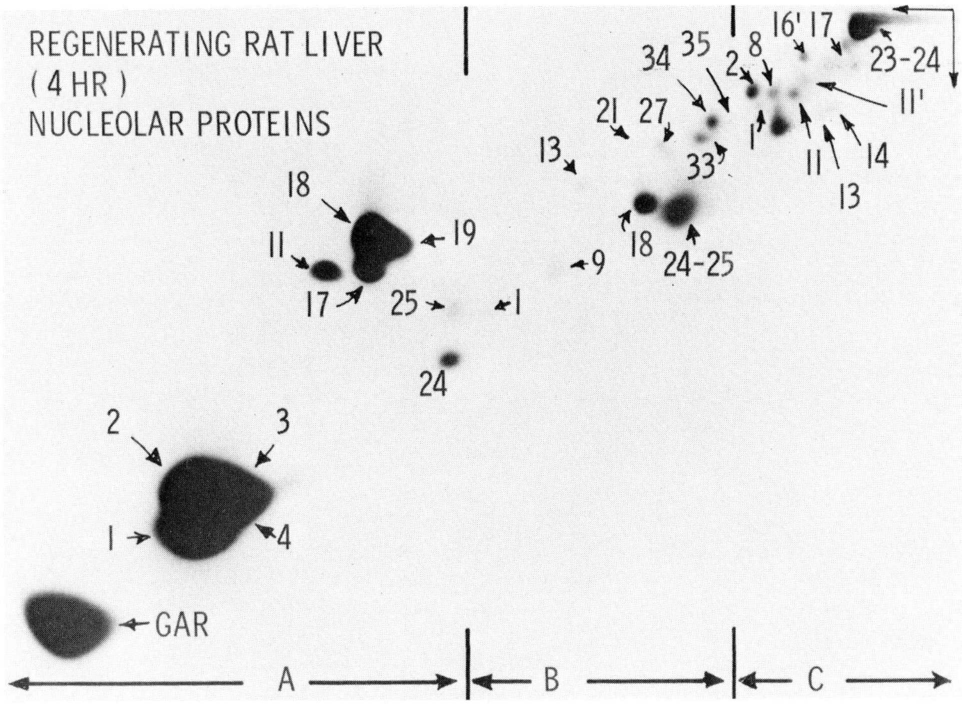

FIGURE 9B

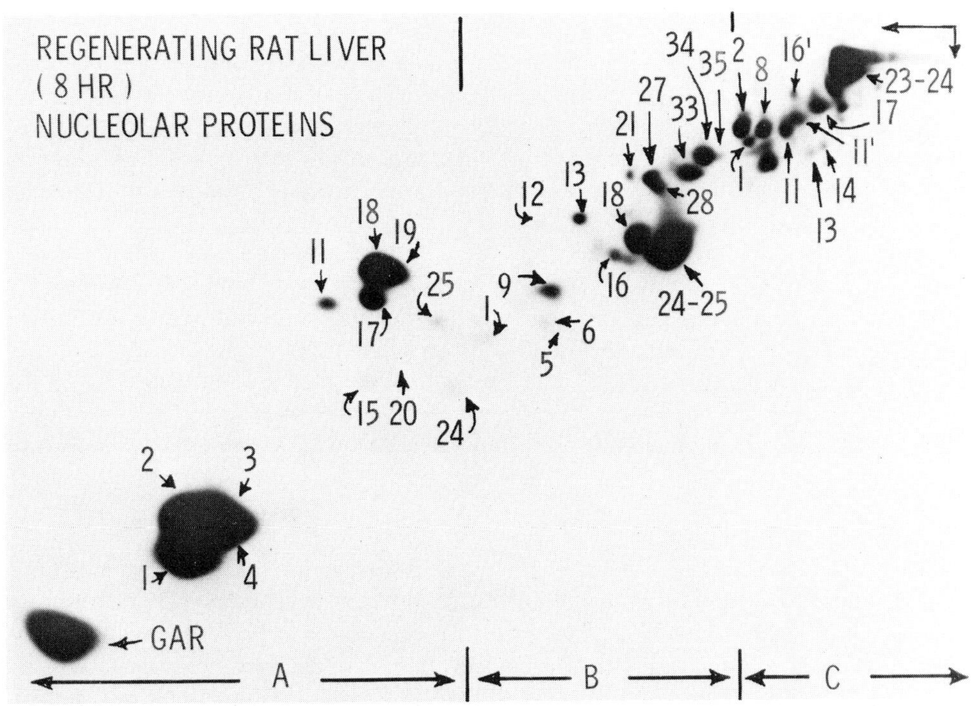

FIGURE 9C

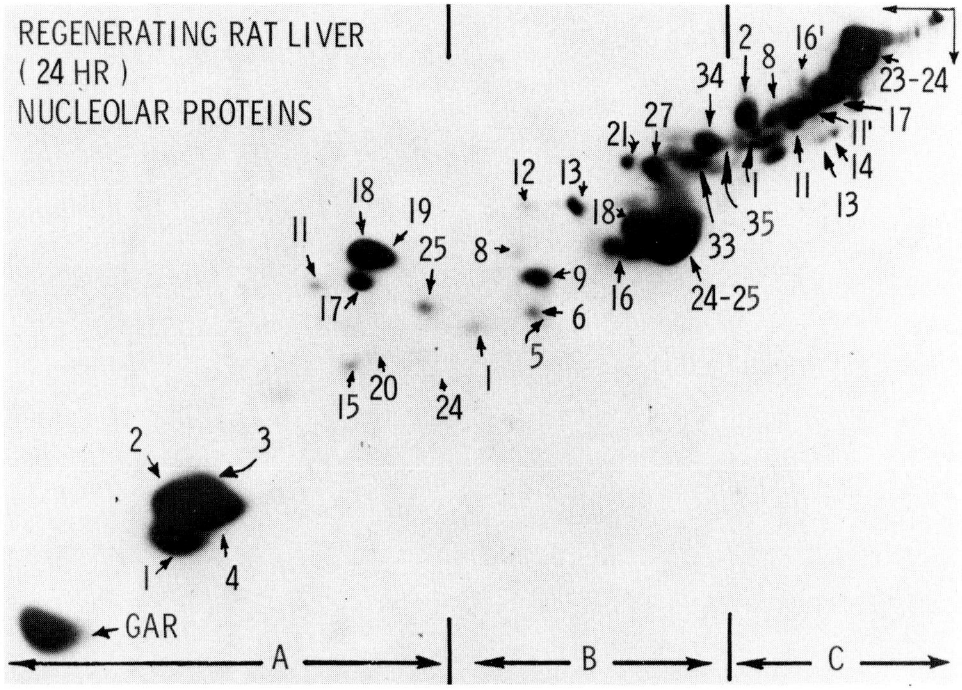

FIGURE 9D

VII. PROTEIN BA

Interest in protein BA was first provided by two-dimensional polyacrylamide gel comparison studies.[48-51] Protein BA was found in the chromatin of nongrowing cells, but was either absent or greatly reduced in amount of the chromatin of active, growing cells. The marked decrease of protein BA in the chromatin of regenerating rat liver and PHA-stimulated human lymphocytes suggested that this protein may be associated with growth repression.

A. Distribution of Protein BA

A two-dimensional gel of the ''chromatin fraction'', containing protein BA, is shown in Figure 17. Protein BA (Figure 17) was found in the saline-EDTA (0.075 *M* NaCl/0.025 *M* EDTA/pH 8.0) extract of normal rat liver nuclei and in the NHP proteins of chromatin.[52] Table 5 shows that approximately 3% of the total nuclear protein is accounted for by BA. For isolation of protein BA, preparative slab gel electrophoresis was performed. The absence of other protein spots in Figure 17 (inset) shows that electrophoretically isolated protein BA was highly purified.

B. Analysis of Protein BA

The molecular weight of protein BA is 31,000, by comparison of its relative mobility to that of standard proteins of known molecular weight. Amino acid analyses (Table 6) showed glutamic acid and lysine were the most abundant. Protein BA had an A/B amino acid ratio of 0.9. Unlike the histones, protein BA contains tryptophan and was not extractable in 0.4 N H_2SO_4. The NH_2 terminal of protein BA was blocked. Hydrazinolysis yielded a C-terminal lysine.

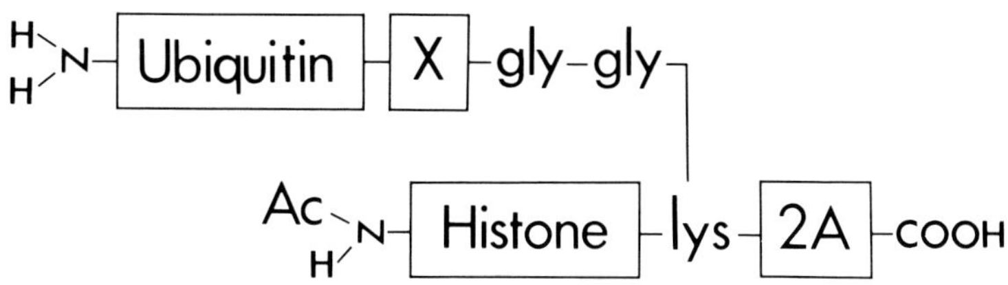

FIGURE 10. The structure of protein A24. Protein A24 has a branched structure through which its two polypeptides, histone 2A and ubiquitin, are joined. The carboxyl terminus of ubiquitin is connected to lysine residue 119 of histone 2A through glycylglycine in an isopeptide linkage to the E-NH_2 group of the lysine. The portion of the molecule designated X indicates the carboxyl is in peptide linkage to the two glycine residues.

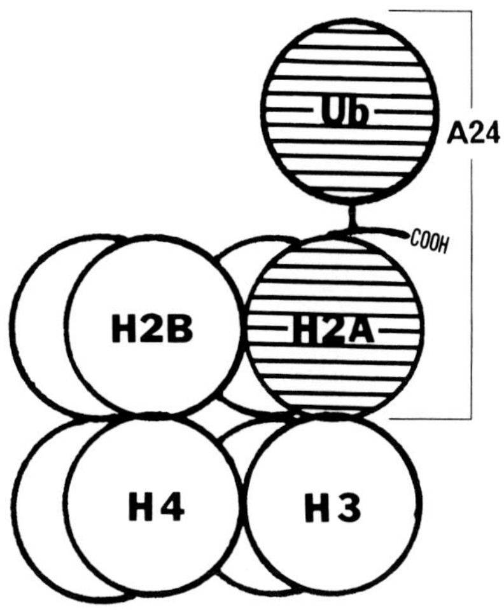

FIGURE 11. Model of protein A24-containing nucleosome. The branched protein A24 (hatched spheres) probably fits into its nucleosome so that the histone 2A portion takes the place of a histone 2A molecule in the nucleosome core while the ubiquitin (Ub) is outside the core.

C. DNA Binding Characteristics

The nitrocellulose assay for binding to double-stranded DNA[53,54] was used. The binding of protein BA to DNA was optimal in the pH range 7 to 8; pH 8.0 was utilized for further investigations. The ionic strength of the incubation buffer was critical. At concentrations greater than 0.2 *M* NaCl, the binding of protein BA to DNA rapidly diminished.

D. Immune Microscopy

The indirect immunofluorescence investigation of rat liver cryosections indicated that the antibodies to protein BA produced both a cytoplasmic cell membrane fluorescence and a nucleoplasmic fluorescence.[55] Within the nucleoplasm, the nucleoli were observed to be slightly more fluorescent than the nucleoplasm. It could not be determined by this technique whether this nucleolar surface protein was generally distributed throughout the nucleolus. With the peroxidase-labeling technique, electron microscopic analysis of the enzymatically labeled sample showed that the nucleolar fluorescence resulted from the localization of protein BA in the perinucleolar heterochromatin. In addition protein BA was shown to be present with the heterochromatic regions continuous to the inner layer of the nucleolar envelope as well as in small chromocenters throughout the nucleoplasm. The general heterochromatic localization was thought to be responsible for the general nuclear fluorescence observed in Figure 18.

Table 3
AMINO ACID COMPOSITIONS OF PROTEINS A11, A17, AND A18/19

	Mol %		
	A11	**A17**	**A18/19**
Asx	3.7	2.8	2.4
Thr	5.3	4.8	6.4
Ser	8.6	7.3	7.6
Glx	5.2	3.4	3.2
Pro	8.9	8.3	8.9
Gly	5.0	6.5	7.1
Ala	17.4	25.6	24.5
Val	6.0	5.3	3.2
Met	0.6	1.0	0
Ile	2.8	1.2	1.3
Leu	2.6	3.9	4.2
Tyr	1.4	0.5	0.5
Phe	0.8	0.8	0.7
Lys	27.0	27.1	28.0
His	1.7	0	0
Arg	3.4	1.6	2.1
Trp	0	n.d.	n.d.
Acid base	0.28	0.22	0.19
NH_2-terminal	Blocked	Blocked	Blocked
COOH-terminal	Lys	Lys	Lys

VIII. NOR PROTEINS

One of the key questions is why do certain proteins seek rDNA and in the same vein, how do specific macromolecules localize in the nucleolar structure? Many cytologists have followed the work of Heitz[56] and McClintock[57] who initially defined the NOR region, which has now been recognized as being the rDNA locus in the genome. This region was assigned the function of formation of the nucleolus although it is clear that DNA, per se, cannot subserve such a function. Accordingly, it seems evident that rDNA must have the property of binding specifically with both structural and functional macromolecules to provide the basic elements of the nucleolus. Following the development of many methods for silver staining,[58,59] Howell[59] noted that the silver staining of the NOR region in the chromosomes was the result of the binding of the silver to the NOR proteins rather than DNA. He also showed that the nonhistone proteins, rather than the histones, were responsible for this binding. Extraction of the histones did not affect the staining with the silver method, but destruction of the nonhistone proteins with proteases eliminated the silver staining.

As part of our studies, one of the approaches was to determine which proteins were phosphorylated in vivo and in vitro. It was remarkable that proteins B23 and C23 (also called C18 at one point) were highly phosphorylated both in vivo and in the in vitro systems used for labeling.[60-62] In the in vitro systems, it was particularly interesting that when Mn^{++} or Co^{++} was added to the system, protein C23 was the main product, but when Zn^{++} was added to the system, protein B23 was the main labeled product. It was also noteworthy that none of the fast-moving proteins of the A region became phosphorylated (Figure 2B).

Because of the intrinsic interest of these rapidly labeled proteins, two approaches were made to their structure and functional roles. In the first, attempts were made to isolate nucleolar RNP particulate elements and determine whether they were present. Proteins B23 and C23 were both found in the nucleolar preribosomal particles, but neither of these proteins

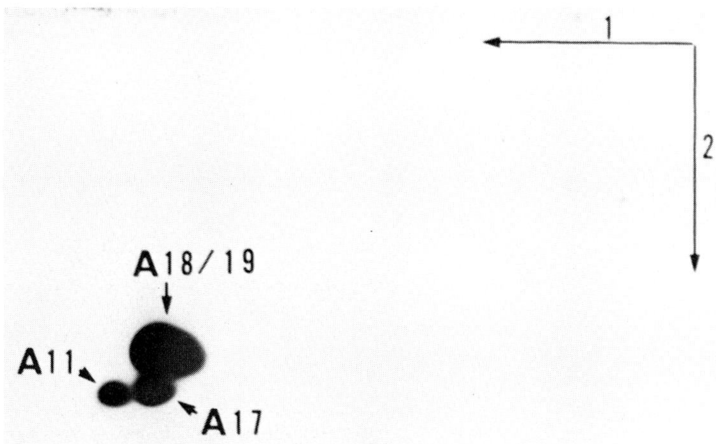

A

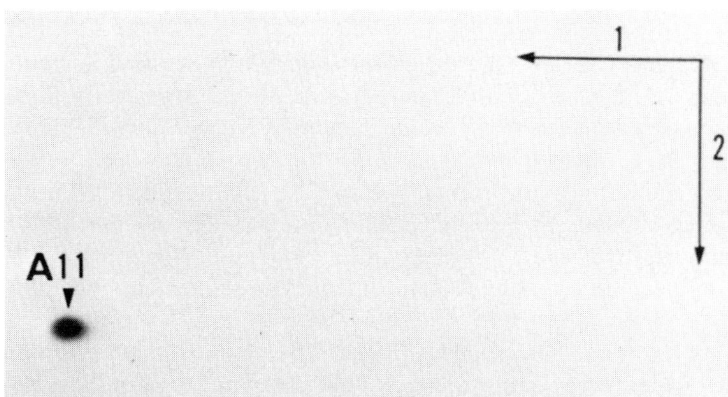

B

FIGURE 12. (A) Purification of protein A11, A17, A18/19. Two-dimensional gel electrophoresis of 50 μg of rat liver nuclear PCA extract. Samples dissolved in 10 *M* urea-0.9 *N* acetic acid, and 1% 2-mercaptoethanol were first loaded on tube gels of 10% acrylamide, 6 *M* urea and run in the first dimension (right to left) for 6.5 hr at 120 V constant voltage. For the second dimension (top to bottom), the tube gel was adapted to 0.1% SDS and polymerized to a 12% acrylamide-0.1% SDS slab gel which was then run for 14.5 hr at 50 mA constant amperage. Protein spots were stained with Coomassie® brilliant blue R. (B) Two-dimensional gel electrophoresis of 20 μg of purified protein A11. Conditions were identical to those in Figure 12A.

was found in the cytoplasmic ribosomes.[9,10] Accordingly, it seemed that these proteins would have a special role in the processing or transport of the nucleolar RNP products but would not have a functional role in the ribosomes. Although these proteins were present in the nucleolus, they were not found in the extranucleolar portion of the nucleus; this finding supports the idea that whatever function they have, it is limited to a nucleolar localization. In this sense (Table 7), these molecules fall into the class of nucleolus-restricted macromolecules. These proteins are responsive to cellular demands as shown by the finding that only a weakly phosphorylated spot was found in normal liver chromatin preparations, but a highly phosphorylated spot was found in regenerating liver chromatin preparations.[63]

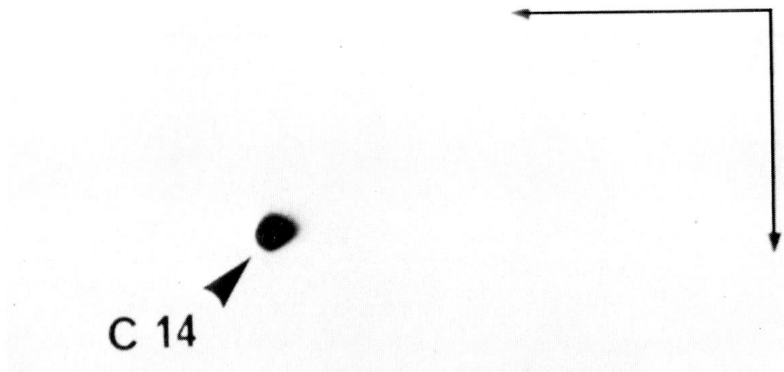

FIGURE 13. Purification of C14. Two-dimensional polyacrylamide gel electrophoresis of protein C14 extracted by the 0.35 *M* NaCl buffer from Novikoff nuclear chromatin. The first dimension (6% polyacrylamide) was run from right to left, and the second (8%) from top to bottom; protein spots were stained with Coomassie® blue. The proteins were initially obtained from the 30 to 60% ammonium sulfate precipitate of the 0.35 *M* NaCl extract. Electrophoretically isolated protein C-14 migrated as a single spot on a two-dimensional polyacrylamide gel (see James et al.[46]).

A. Proteins B23 and C23 are the Major Nucleolar Silver-Staining Proteins

In association with T. C. Hsu and H. Hubbell of the M. D. Anderson Hospital in Houston, we developed an interest in the silver-staining proteins of the nucleolus (Figure 19). Inasmuch as two-dimensional gel techniques were available for separation of the nucleolar proteins, it was relatively simple for Lischwe et al.[64] to stain the gels with the same methods evolved for silver staining of NORs by Howell[59] (Figure 20). Under the conditions of these experiments, only proteins B23 and C23 were stained (Figure 20). It was particularly important that this staining was done under the same conditions as those used for staining on microscopic specimens.

Within a very short time, Hubbel et al.,[65] working jointly in our laboratory and at the M. D. Anderson Hospital, provided conformatory evidence for these results. Although they reported that another silver-staining protein was present besides B23 and C23, that result is probably the result of degradation of protein C23, as was found in earlier studies.[60,61]

B. Nucleolar Localization of Silver-Staining Proteins

To determine the relationships of nucleolar silver-staining proteins to other nucleolar elements, studies were done on interphase and metaphase cells. In the nucleoli of Novikoff hepatoma cells, silver-staining granules were present in large numbers and varied arrays in the nucleoli. They were not found in the cytoplasm or extranucleolar fraction as expected from the previous results with proteins B23 and C23. The number of these granules in the tumor nucleoli, approximately 50 per cell, was notable (Figure 21).

C. Silver-Stained Granules in Tumor Cells

Figure 21 is a composite showing various formation of silver granules in a number of tumor cells.[66] The number of granules varied considerably from nucleolus to nucleolus and cell to cell in these preparations. The mean number was 53 granules per nucleus and 21 granules per nucleolus in the populations of Novikoff hepatoma ascites cells, 13 granules per nucleolus and 20 granules per nucleus in KB cells, or 11 granules per nucleolus and 39 granules per nucleus in HeLa cells (Table 8). Quite frequently, the nucleolar granules appeared to be isolated free of reticulum although many were in a reticulum. Frequently, they were arrayed in one or more rows containing several granules per row (Figures 21 to 23) in each of the types of tumors studied.[66]

Table 4
AMINO ACID ANALYSES OF THE PHOSPHOPROTEIN FROM *PHYSARUM POLYCEPHALUM* AND PROTEIN C-14 FROM NOVIKOFF HEPATOMA ASCITES CELLS

	***Physarum* phosphoprotein**		**Novikoff hepatoma protein C-14**	
Amino acid	**Mol%**	**Residues/ 70,000 Mol Wt**	**Mol %**	**Residues/ 70,000 Mol Wt**
Lys	8.3	55	7.8	50
His	3.2	21	1.2	8
Arg	4.2	28	4.5	29
Trp	0	0	0	0
Asx	11.0	73	11.9	76
Thr	4.5	30	6.9	44
Ser	7.2	48	7.8	50
Glx	16.2	107	13.9	89
Pro	4.1	27	4.2	27
Gly	12.2	81	9.2	59
Ala	7.7	51	7.8	50
1/2-Cys	0.2	1	1.4	9
Val	5.0	33	3.9	25
Met	0.8	5	2.2	14
Ile	3.5	23	4.4	28
Leu	7.4	49	7.5	48
Tyr	1.7	11	2.0	13
Phe	2.9	19	3.3	21
A/B[2]		1.7		1.9

[2] Ratio of acidic to basic amino acids.

From Kuehn, G. C., Affolter, H. -U., Atmar, V. J., Seeback., Gubler, U., and Braun, R., *Proc. Natl. Acad. Sci. U.S.A.*, 76, 2541, 1979. With permission.

When comparable studies were made on relatively inactive cells such as circulating lymphocytes, as few as one silver-stained granule was found. In normal liver (Figure 22A) and in the 6-hr (Figure 22B) and 18-hr regenerating liver (Figure 22C), the nucleolus contained the characteristic reticulum described as a "nucleolonema". The nucleolar granules were thickenings of this network at a number of points; thus the structure resembles a reticulum. Unlike the tumors, the liver nucleoli did not contain individual granules or the types of rows noted above. Both the density of the staining and the size of the nucleoli were markedly increased in the regenerating liver nucleoli (Figure 22A to C). An analysis of the numbers of "granules" present per nucleolus indicated they increased from approximately 4 per nucleolus and 13 per nucleus in the normal liver to approximately 15 per nucleolus and 33 per nucleus in the 18-hr regenerating liver. These values correlate well with the increase reported in 45S RNA synthesis from 3 to 5 to 15 fg/min/nucleolus, a 3- to 5-fold increase, in the 18-hr regenerating liver.[3,66]

D. Nucleolar Granules in Cell Division

Figure 23 shows that during the course of cell division there were marked changes in the number and appearance of the silver-staining nuclear granules. During metaphase (Figure 6A), "doublets" of granules associated with chromosomes in the NOR were present around

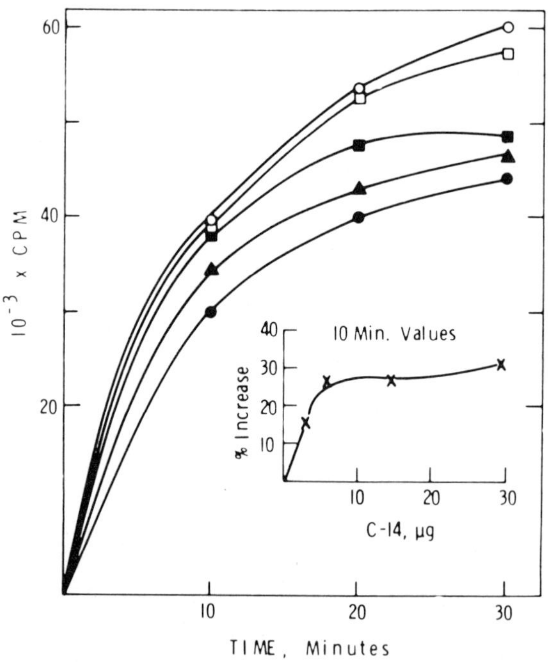

FIGURE 14. Simulation of incorporation of [^{3}H]UMP by protein C-14 in an RNA polymerase I system with Novikoff nucleolar DNA as template. The values are averages of two experiments. Each sample contained 5 μg of DNA which was rate limiting for the reaction: control without protein C-14; (▲—▲) 3 μg; (■—■) 6μg; (□—□) 15 μg; (○—○) 30 μg of protein C-14. The inset shows the stimulatory effect of increasing amounts of protein C-14 (see James et al.[46]).

the metaphase equatorial plates, and their total number was approximately 8 to 11. The size and density of these granules were approximately the same as that of the nucleolar granules, but it is not clear whether they contain the same silver-staining proteins.

During anaphase, the number of granules appeared to have increased but the granules were largely single, i.e., the doublets appear to have been separated (Figure 23). In telophase (Figure 23), when the cytoplasm of the daughter cells had not separated, there was a marked increase in the density and the number of the silver granules in the newly formed nuclei. As noted in Figure 23 (arrow), there was a reappearance of the linear arrays of the separated nucleolar granules. Apparently, even before the two daughter cells had fully formed, the nuclei contained the large number of dense granules found in interphase cells (Figure 23).

E. Electron Microscopic Analysis of Localization of the Silver Granules

Even with relatively low-power electron microscopic analysis, it was noted that the silver granules had discrete localizations and were composed of large numbers of individual silver grains (Figure 24).

The silver grains were distributed throughout the nucleolar mass but were intensely concentrated in the same distribution and overall densities as the silver grains in the light microscopic pictures (Figure 24). It appeared from the associated morphological elements that the silver grains were not in the nucleolar light spaces and not in the preribosomal RNA particles of the granular elements.

To test this idea more definitively, the Novikoff hepatoma cells employed in this study were treated with adriamycin to induce segregation of the granular and fibrillar elements.[67] Silver staining of this preparation (Figure 25) indicated that the granular elements were

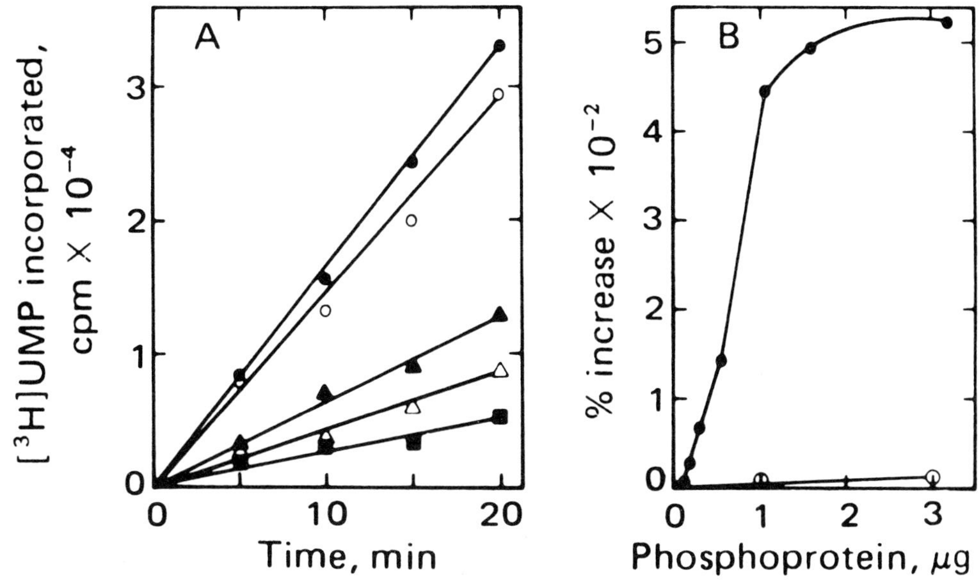

A B

FIGURE 15. Stimulation of in vitro transcription by the rDNP complex in the presence and absence of the phosphoprotein. Purified rDNP complex provided both RNA polymerase I activity and rDNA template for each assay. (A) Assay components included the "high UTP" assay mixture previously described with the omission of calf thymus DNA template $[^3H]$UTP (0.41 Ci/mmol) as the labeled nucleotide, approximately 0.4 μg of rDNA as the rDNP complex, and the following quantities of phosphoprotein: ■, control, no phosphoprotein; △, 0.26 μg; ▲, 0.53 μg; ○, 1.06 μg; ●, 3.18 μg. The reaction volume was 0.2 mℓ. (B) Stimulatory effect of increasing amounts of phosphoprotein (●) and no effect by the dephosphorylated acidic protein (○) after treatment of the phosphoprotein with alkaline phosphatase-agarose, which hydrolyzed 5.9 mol P_i/70,000 mol wt of protein. (C) Acidic phosphoproteins associated with the rDNP complex derived from nucleoli that had been phosphorylated with $[\gamma\text{-}^{32}P]$ATP in the presence (C) and absence (c') of the polyamines. Nucleoli were isolated in Percoll gradients from 50 surface cultures and were divided into two equal portions. (c) One portion of nucleoli was incubated for 60 min in a phosphorylation mixture containing $[\gamma\text{-}^{32}P[$ATP (3.18×10^6 cpm/μmol) and 0.33 m*M* each of putrescine, spermine, and spermidine. The nucleoli were collected by centrifugation at 1500 × g for 10 min and were solubilized in 0.5 mℓ of a buffered EDTA solution containing 1 *M* urea/10 m*M* Na_2MoO_4. The soluble fraction and the insoluble nucleolar debris were separated by centrifugation at 3000 × g for 10 min. Total protein in the insoluble debris was solubilized by $NaDodSO_4$/phenol extraction, dialyzed, and fractionated by $NaDodSO_4$/polyacrylamide rod gel electrophoresis (0). The rDNP complex was isolated by sucrose gradient sedimentation from the nucleolar fraction that was soluble in buffered EDTA solution. The isolated rDNP complex was extracted with $NaDodSO_4$/ phenol, dialyzed, and also fractionated on a polyacrylamide gel (●). (c') The second portion of nucleoli was similarly treated except that no polyamines were supplied to the phosphorylation mixture. Phosphoproteins associated with the rDNP complex (●) and the insoluble nucleolar debris (○) are shown. After electrophoresis, all gels were sectioned into 2-mm slices and solubilized in 30% H_2O_2, and radioactivity was measured. (See Kuehn et al.[47]).

unstained, but the fibrillar elements exhibited dense staining. These results are in good agreement with results of Puvion and colleagues, but other reports have suggested that the silver grains are more specifically localized to the NOR regions, which have been suggested to be the nucleolar fibrillar bodies.[68,69]

IX. CHEMISTRY AND PROTEINS C23 AND B23

When proteins B23 and C23 were shown to be highly phosphorylated, it was a logical extension of the initial studies to isolate and attempt to characterize the proteins and their phosphorylated regions. Initially protein C23 was isolated by a two-step procedure in which the nucleolar proteins soluble in 0.4 *N* H_2SO_4 were fractionated on acid/urea-polyacrylamide

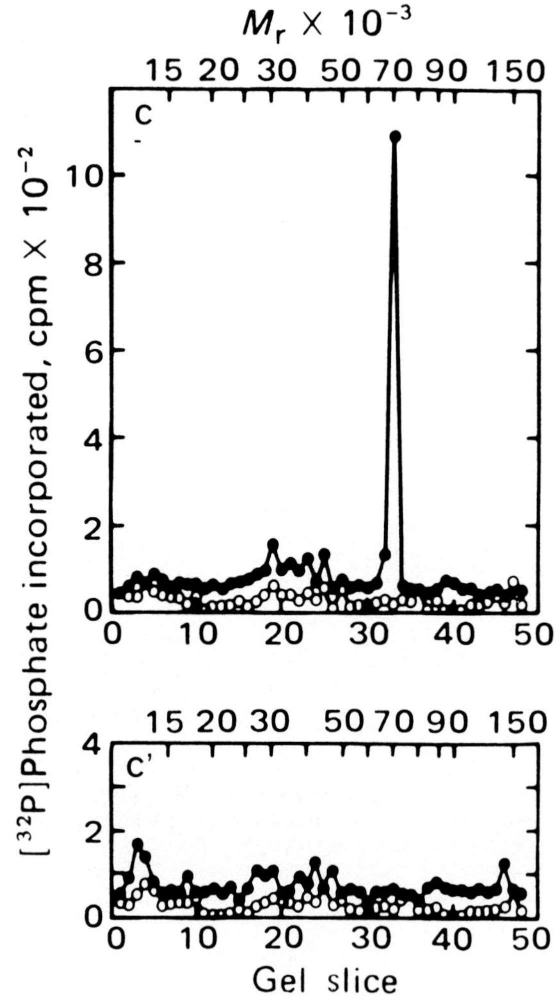

FIGURE 15C

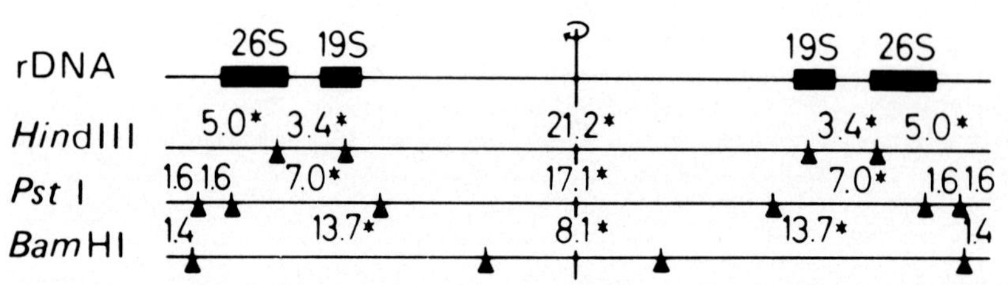

FIGURE 16. Restriction endonuclease map of rDNA from *Physarum polycephalum*. The approximate molecular weight of DNA is 38 × 10^6. The top figure depicts the linear length of rDNA and the locations of 19S and 26S rRNA coding sequences on the palindrome relative to the axis of symmetry (arrow and vertical line). The bottom three lines show the respective cleavage sites (▲) for restriction endonucleases *Hind* III, *Pst* I, and *Bam*HI. The number between cleavage sites above each frangment is the molecular weight × 10^{-6} of the respective fragment. Numbers followed by an asterisk represent the fragments that were tested in the protein binding assay. (From Kuehn, G. C., Affolter, H.-U., Atmar, V. J., Seeback, T., Gubler, V., and Braun, R., *Proc. Natl. Acad. Sci. U.S.A.*, 76, 2541, 1979. With permission.)

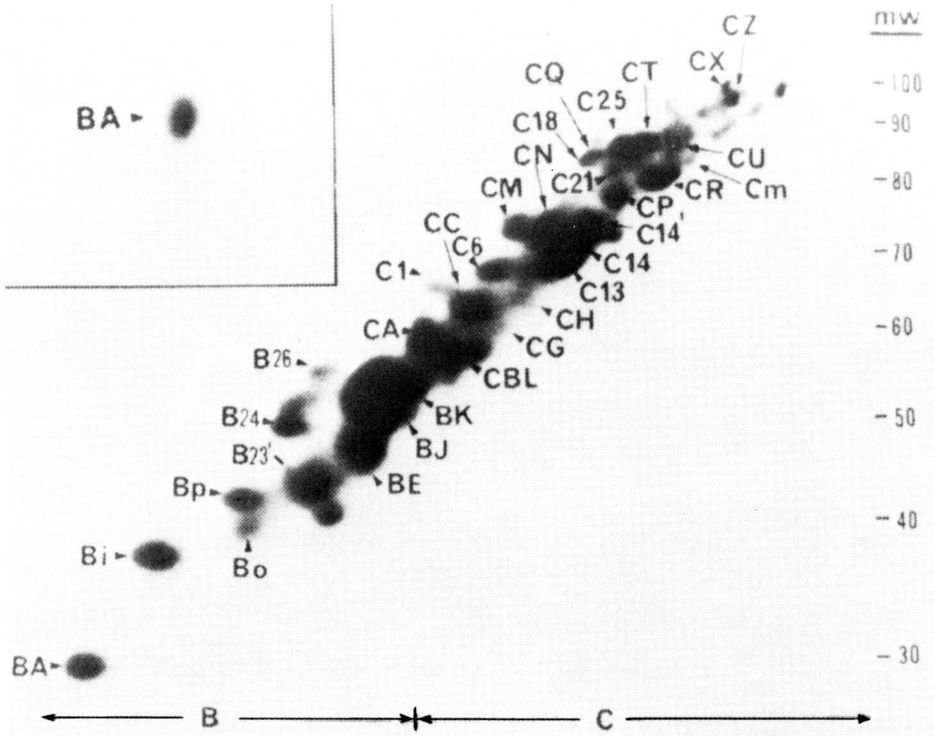

FIGURE 17. Two-dimensional polyacrylamide gel electrophoresis of 500 μg of liver chromatin residual proteins. The first dimension of 6% polyacrylamide was run from right to left and the second dimension of 8% polyacrylamide from top to bottom. Protein spots were stained with Coomassie® brilliant blue R. The values on the right-hand margin indicate approximate molecular weights. Inset shows two-dimensional polyacrylamide gel electrophoresis of 20 μg of highly purified nuclear protein BA. (From Catino, J. J., Yoeman, L. C., Mandel, M., and Busch, H., *Biochemistry*, 17, 983, 1978. With permission.)

gels (6% polyacrylamide), eluted from the gels, and then placed on hydroxylapatite columns which were eluted with increasing concentrations of phosphate. The fraction which was eluted with 0.25 *M* phosphate was electrophoretically homogeneous (Figure 26) and was used for further studies. Table 9 presents the amino acid composition of this protein.[70]

A. Improved Purification of Protein C23

Nucleoli from Novikoff hempatoma ascites cells were homogenized in 8 *M* urea/2 m*M* PMSF/0.2 m*M* leupeptin at a concentration of ≈50 A_{260}/mℓ. One volume of 6 *M* LiC1 was added. The suspension was vigorously homogenized and the proteins were separated from the residue by centrifugation at 27,000 × g for 20 min. The supernatant was dialyzed against Buffer A: 20 m*M* Tris malate/6 *M* urea/1 m*M* EDTA/1 m*M* DTT/1 m*M* PMSF/0.1 m*M* leupeptin/pH 6. The dialysate was centrifuged at 27,000 × g for 15 min and the supernatant was applied to a DEAE cellulose colum (1 × 25 cm) equilibrated with the buffer. The proteins were eluted with an NaCl gradient of 0.0 to 0.5 *M* at a flow rate of 1 m≈/10 min; 3-mℓ fractions were collected. Protein C23 was the last protein eluted from the column. Further purifications was obtained with a Bio Rad® AG3-X4A column (1 × 3 cm) equilibrated in buffer A. The fractions containing protein C23 from the DEAE-cellulose column were pooled, dialyzed against the same buffer, and placed on the AG3-X4A column. The column was eluted stepwise with 20-mℓ aliquots of the buffer, 0.1 *M* NaCl, 0.6 *M* NaCl, 1 *M* NaCl, and 2 *M* NaCl, all in the buffer. The 2 *M* NaCl wash contained the highly purified protein C23.

Table 5
DISTRIBUTION OF PROTEIN BA IN NUCLEAR EXTRACTS

Nuclear fraction	% Nuclear protein	% BA	% BA of total nuclear protein
Saline-EPTA	14.0	10.4	1.5
Tris	11.5		
0.4 *N* H_2SO_4	33.3		
PCA supernatant	0.6		
CF II	40.0	3.7	1.5
CF III	0.4		
Whole nuclei	100.0		3.0

From Catino, J. J., Yeoman, L. C., Mandel, M., and Busch, H., *Biochemistry,* 17, 983,1978. With permission.

Table 6
AMINO ACID COMPOSITION OF PROTEIN BA

Amino acid	Mol %
Lys	9.7
His	4.7
Arg	8.3
Asp	8.3
Thr	3.8
Ser	4.8
Glu	11.0
Pro	6.5
Gly	7.3
Ala	5.2
Cys	0.0
Val	4.1
Met	2.5
Ile	3.9
Leu	9.1
Tyr	4.9
Phe	5.9
Trp	0.5
Acid/base	0.9
N terminal	Blocked
C terminal	Lysine

From Catino, J. J., Yeoman, L. C., Mandel, M., and Busch, H., *Biochemistry,* 17, 983, 1978. With permission.

For studies on the phosphorylated protein C23 with ^{32}P-labeling, purification was done on 5% acid-urea slab gels. Because of the acidity of the peptides produced by tryptic digestion, it was possible to fractionate these peptides on DEAE-Sephadex® columns in 7 *M* urea/0.05 *M* Tris/pH 7.5 with salt gradients from 0.1 to 0.6 *M*. Under these conditions, 3 peaks were eluted which had net negative charges of −9 to −17. It was particularly notable that these peaks had high contents of aspartic and glutamic acids (Table 9) which

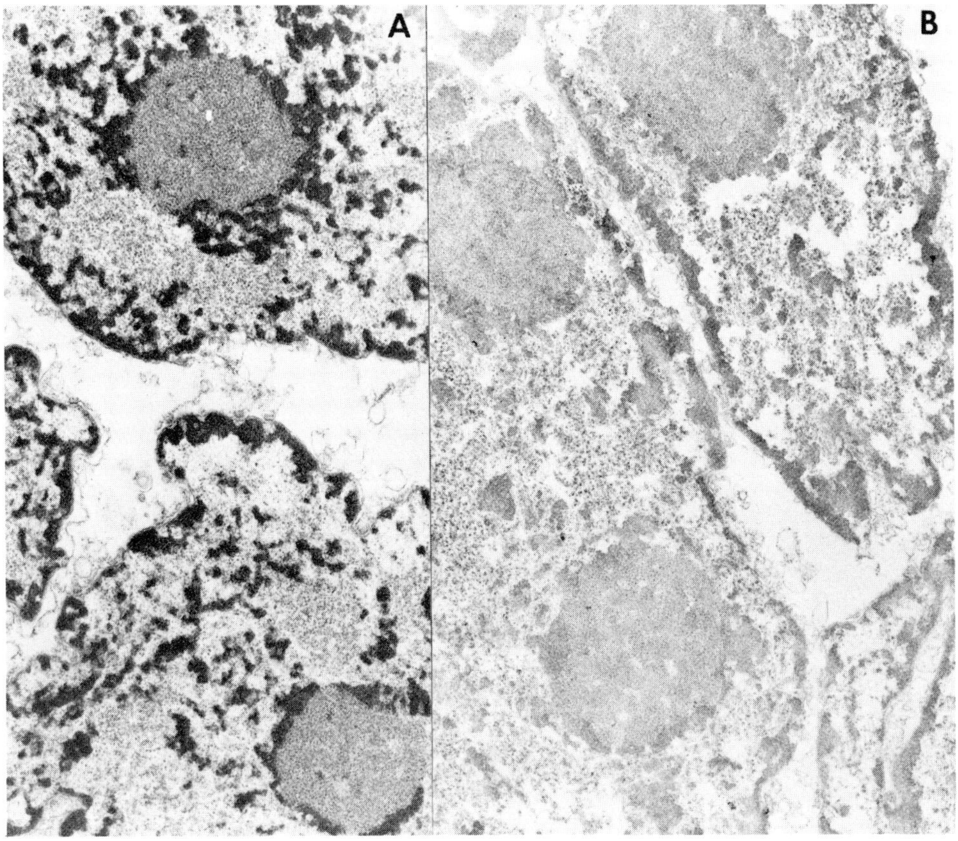

A B

FIGURE 18. Immunoenzymatic localization of protein BA in normal rat liver nuclei. An electromicrograph of normal rat liver nuclei was analyzed by the unlabeled antibody enzyme method utilizing peroxidase enzyme. (A) Shows the immunoenzymatic staining with anti-BA serum, and (B) Shows analysis with preimmune serum (× 15,000). (From Catino, J. J., Busch, H., Daskal, Y., and Yeoman, L. C., *J. Cell Biol.*, 83, 462, 1979. With permission.)

totaled as much as 65% of the amino acid residues in these peaks.[71] Some of the acidic peptides obtained by tryptic digestion had molecular weights approximating 11,000. Their differing amino acid compositions suggested that there might be several such peptides distributed throughout protein C23.

Comparable studies were done on protein B23, which showed two such peaks rather than the three found for protein C23. This protein also had a peptide which contained 65% aspartic and glutamic acids.

B. Acidic Peptide C23-1

To derive further information on the structure of the acidic peptides of proteins B23 and C23, these tryptic peptides were subjected to electrophoresis and separated on the basis of charge density. One such peptide has the sequence shown in Figure 27. It contained 10 glutamic acid residues, 11 aspartic acid residues, and 3 phosphoserines in a total of 40 amino acids. This very highly acidic peptide apparently existed in several states of phosphorylation involving the three phosphoserine residues, each of which may or may not be phosphorylated. Although this particular peptide may not be completely characteristic of the other phosphopeptides in these proteins, it is likely that it is at least representative.

Table 7
NUCLEOLUS-SPECIFIC MACROMOLECULES

Proteins

Silver staining NOR proteins, B23, C23 (37/5.1, 110/5.2)
RNA polymerase I
45S RNA cleavage and modification enzymes
U3 RNA binding proteins
Special proteins of pre-rRNP particles
Specific liver and tumor nucleolar antigens

Nucleic Acids

rDNA
U3 RNA
45S pre-rRNA

C. Relationships of Protein C23 and B23 to Synthesis of rRNP Particles

To analyze the relative positions of proteins in the progression of nucleolar events, two approaches were used. The nucleoli were treated with 0.01 *M* KC1/0.1 m*M* $MgCl_2$/0.01 *M* triethanolamine/0.01 *M* DTT/0.1% Brij® 35/pH 7.4, centrifuged at 20,000 × g for 20 min, and the supernatant was subjected to sucrose density gradient centrifugation.

The sucrose density gradient profile obtained in these studies is shown in Figure 28. Peaks 6 and 7 were previously shown to contain the granular elements which include the 60S and 80S preribosomal particles while peak 3 contains fibrillar structures.[11]

When the incorporation of [^{3}H]-uridine into the gradient fractions was analyzed, after 15 min of incubation, peak 1 had the highest label. After 45 min of incubation the label shifted down the gradient and peaks 3 and 4 were equally labeled. Incorporation of [^{3}H]-uridine in the granular region was evident after 90 min. When cells were labeled from 18 hr with [^{32}P]-orthophosphate, the granular region was mainly labeled. These results are in agreement with those obtained previously with [^{32}P]-orthophosphate.[11]

These proteins in the peaks were extracted with 3 *M* LiCl/4 *M* urea and analyzed on two-dimensional polyacrylamide gels.[5] The protein patterns clearly showed that the rapidly labeled peak 1 as well as peaks 2 to 4 contained high concentrations of proteins C23 and B23. The granular components in peaks 6 and 7 contained less protein C23 and B23. These results showed that proteins C23 and B23 sediment with rapidly labeled nucleolar RNA. A unique characteristic of the protein patterns was the increase in basic proteins with an increase sedimentation value.

The low sedimentation value of peak 1 suggested the protein components could be sedimenting as individual species or as a complex with RNA. Fractions containing [^{3}H]-uridine-labeled peak 1 from a gradient were pooled, concentrated, and placed on Sepharose® 2B columns. Figure 29 shows the RNA and protein eluted in a single peak, suggesting they existed in an RNA-protein complex.

X. STUDIES WITH ANTIBODIES TO PROTEIN B23

For these studies the improved isolation procedure used for protein B23 is shown in Figure 30. More than 90% of the protein of the Novikoff hepatoma nucleoli was extracted with 4 *M* urea/3 *M* LiCl. Much of the DNA was sedimented by centrifugation at 27,000 × g. Among the many proteins in this extract (Figure 30) were protein C23 (molecular weight 103,000) and protein B23 (molecular weight 37,000). In the next step of the purification procedure, dialysis was carried out against 20 m*M* Tris-malate pH 5.5. In this step, most nucleolar proteins precipitated including protein C23. The supernatant containing most of

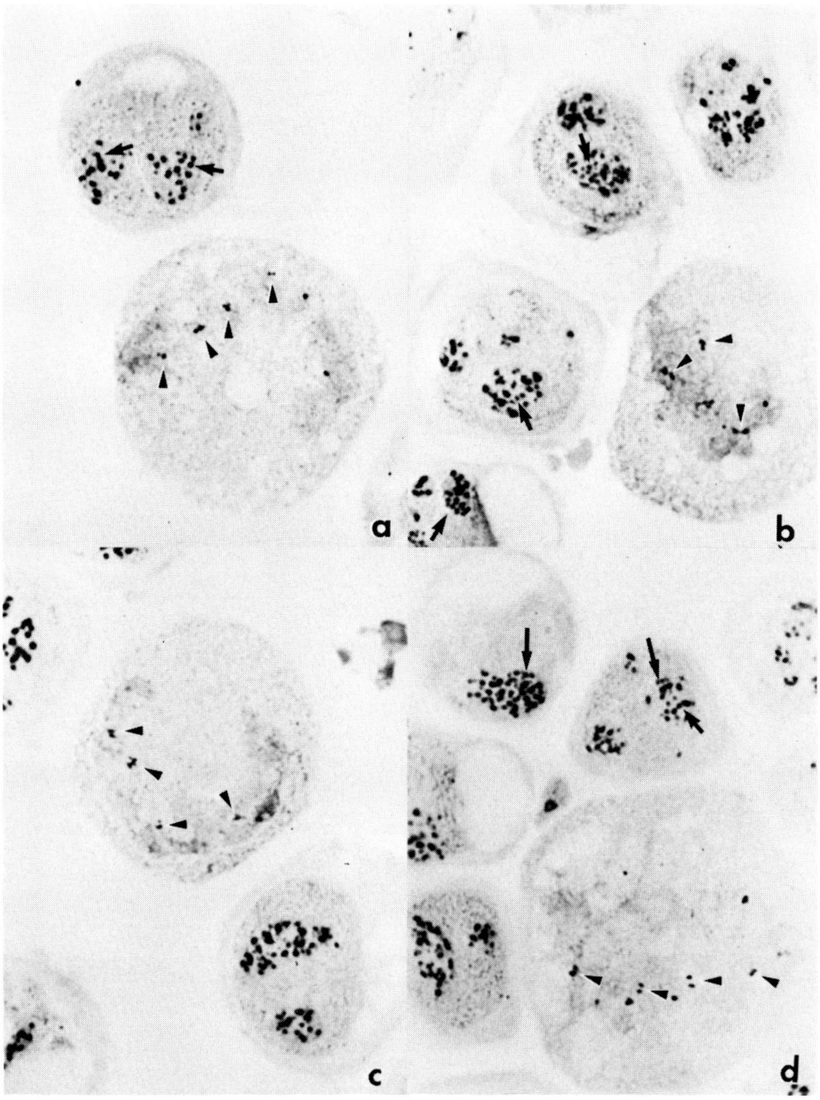

FIGURE 19. (a) Silver stained hepatoma cell in metaphase containing 4 pairs (arrowheads) and 2 single dense granules, presumably at NOR regions of "nucleolar chromosomes". Adjacent to this cell is an interphase cell containing 40 nucleolar granules in 3 nucleoli. The nucleoli contain rows of dense granules (arrows) (b) A similar cell in metaphase showing the presence of 3 pairs of dense granules (arrowheads) and 3 cells with rows of nucleolar dense granules (arrows). (c) Novikoff hepatoma cell in metaphase containing 4 pairs of dense granules (arrowheads). (d) Large Novikoff hepatoma cell in metaphase containing several doublets of dense granules in metaphase cells (× 1400).

protein B23 was loaded on a DEAE-cellulose column which was then eluted with 0 to 0.5 *M* NaCl gradient in 20 m*M* Tris-malate pH 6.5 (Figure 31). The contaminants were eluted at low-salt concentrations and protein B23, the most acidic protein, was eluted with 0.12 *M* NaCl; it was the last major peak (Figure 31). A one-dimensional Laemmli gel (Figure 30B) showed only one band. A two-dimensional, isoelectric focusing SDS gel showed a large dense spot (Figure 32) in the position of protein B23 and a minor spot of the same molecular weight slightly to the left of the major spot.

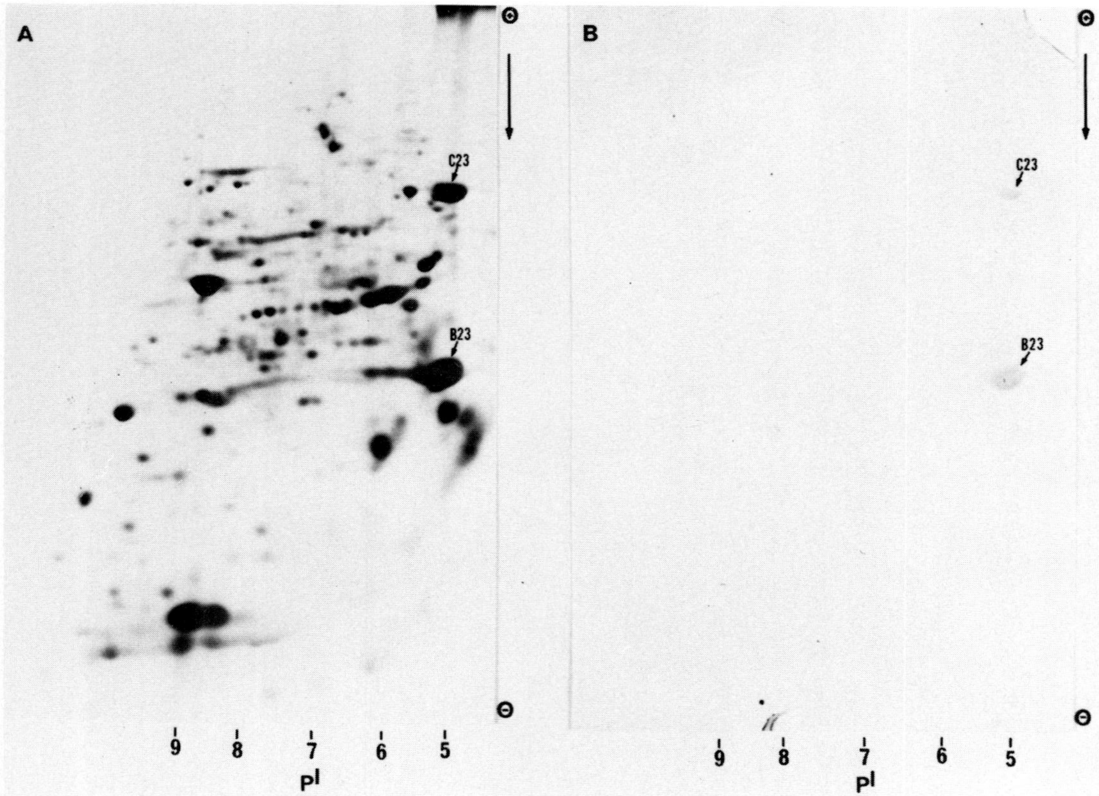

FIGURE 20. Two-dimensional polyacrylamide gel electrophoresis of the 2 *M* potassium acetate-5 *M* urea extract of nucleoli. Two-dimensional gel electrophoresis isoelectric focusing acid-urea polyacrylamide (6%) gel electrophoresis was performed. The proteins (160 μg) were stained with Coomassie® blue (panel A) or with silver (panel B). (From Lischwe, M. A., Smetana, K., Olson, M. O. J., and Busch, H., *Life Sci.*, 25, 701, 1979. With permission.)

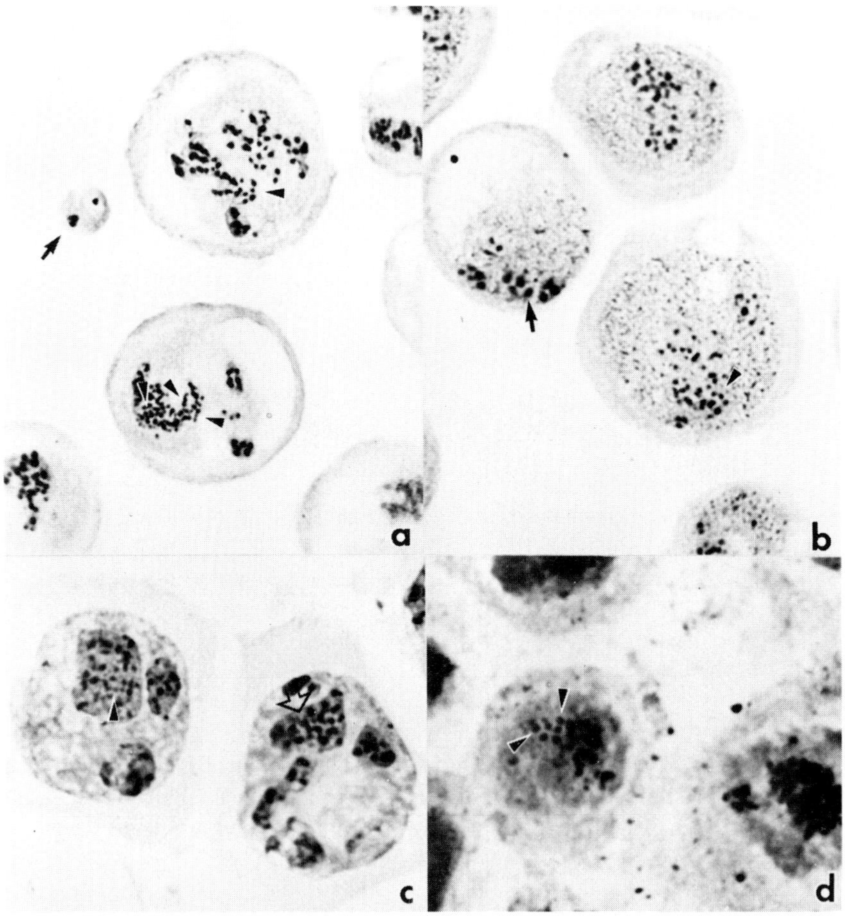

FIGURE 21. Silver-staining granules in nucleoli of tumor cells. The diameter of the silver staining granules range from 0.10 to 0.16 μg. (a) Two Hovikoff hepatoma cells with large numbers of nucleolar granules, a number of which are in rows of 3 or more (arrowheads). Top nucleus, 64 granules; bottom nucleus, 46 granules, all of which are restricted in location to the nucleolus. A small adjacent WBC contains 4 granules (arrow). (b) Several Novikoff hepatoma cells containing silver granules. These granules differ in density, size, and distribution (arrowhead, arrow). (c) HeLa cell nucleus on the left contains 43 granules in the large nucleolus (arrowhead) and 17 granules in the smaller. In the cell on the right, there is a total of 58 granules. In the nucleolus marked with an arrowhead, some granules are in a linear array, others are in a reticular arrangement (arrow). (d) Silver-stained nucleus of a KB cell showing the linear arrangement (arrowheads) of separated silver granules similar to that noted in (a). This cell contains 31 granules (× 1400).

A. Production of Anti-B23 Antiserum in Rabbits

Inasmuch as protein B23 was only weakly antigenic, long immunization periods and high doses of antigen were required for production of antibodies. One of the four rabbits immunized produced an antibody that formed an immunoprecipitin complex with the antigen (Figure 33). Ouchterlony gel analysis of the reactions of anti-B23 IgG with purified protein B23 (wells 1 and 4). Novikoff hepatoma nucleolar 0.01 *M* Tris/pH 8.0 extract (well 3), and 0.35 *M* NaCl/0.01 *M* Tris/ pH 8.0 extract (well 2), is shown in Figure 33. Neither cytosol proteins nor protein C23 formed immunoprecipitates with anti-B23 IgG (wells 5 and 6, respectively).

The specificity of the antibody reaction to protein B23 was further analyzed by immunoelectrophoresis. Figure 34 (lane 2) shows the formation of a single precipitin arc of protein

Table 8
THE NUMBER OF ARGYROPHILIC GRANULES IN NUCLEOLI OF VARIOUS CELLS

Cells	Grains/nucleolus	Grains/nucleus	No. of cells analyzed
Rat hepatocytes	4.4 ± 0.3	13.1 ± 1.9	150
Regenerating liver hepatocytes—18 hr	15.3 ± 0.9	33.4 ± 1.7	150
Rat Novikoff hepatoma cells	21.0 ± 0.1	53.0 ± 7.5	150
HeLa cells	11.2 ± 3.3	38.6 ± 6.3	30
KB cells	13.5 ± 5.2	19.6 ± 4.8	25

From Busch, H., Daskal, Y., Gyorkey, F., and Smetana, K., *Cancer Res.*, 39, 857, 1979. With permission.

B23 with the IgG fraction. The cytosol (Figure 34, lane 4) did not cross-react with the anti-B23 IgG. Figure 34 (lane 3) shows that the nucleolar 0.01 *M* Tris extract formed an arc with a sharp spur. Figure 34 (lane 1) shows that the nucleolar 0.01 *M* Tris extract formed a single immunoprecipitin arc after the 0.01 *M* Tris extract was digested with RNase A (50 μ/mℓ for 30 min at 37°).

B. Cellular Localization

The indirect immunoperoxidase staining technique was used to study the localization of protein B23 in the sections of normal rat liver, thioacetamide-treated rat liver, and Novikoff hepatoma cells (Figure 35). When the sections or cells were incubated with normal rabbit IgG (Figure 35) no positive immunostaining was observed. When the immune IgG was used, nucleoli of normal rat liver cells were positively immunostained and were seen as small, compact structures (Figure 35B). The nucleoli of thioacetamide-treated rat liver (Figure 35C) were also positively stained and much enlarged as noted in earlier studies, after thioacetamide administration.[3] In Novikoff hepatoma cells, in addition to the dense nucleolar staining, some staining of the nuclear envelope was observed in some cells (Figure 35D). This latter effect may have resulted from nonspecific IgG binding, since in the tumor cells treated with preimmune IgG, a similar nuclear envelope staining was observed.

C. Localization of Protein C23 to the Nucleolus, Nucleolar Granules, and Nucleolus Organizer Regions

Protein C23 was isolated from the nucleoli of Novikoff hepatoma ascites cells by DEAE-cellulose and Bio Rad® Ag3X4A ion Exchange column chromatography as described above. New Zealand white rabbits were immunized with purified protein C23 and an Ig fraction was isolated. The presence of an antibody to protein C23 is evident with the precipitating band obtained on Ouchterlony plates (Figure 36A). The Ig fraction did not contain precipitating antibodies to protein B23 as shown in Figure 36A,f.

The immunoperoxidase method was used to localize protein C23 in cells. The nucleoli of Novikoff hepatoma ascites cells were darkly stained when a 1/50 antibody dilution was used (Figure 36B,a). If the concentration of Ig was further diluted (1/75), granular structures were evident within the nucleoli (Figure 36B, b and c). Similar structures are evident when nucleoli are silver stained (Figure 36B, d). It has been demonstrated that these structures are the fibrillar components of the nucleolus or the site of rRNA synthesis.

Metaphase chomosomes from Novikoff hepatoma tissue culture cells were also stained with silver and by the immunoperoxidase methods with the antibody to protein C23. An unusual chromosome was consistently stained by both methods, which contained four rather than the usual two NORs (Figure 37). These results illustrate that protein C23 site of action is nucleolar and that this protein is localized on the NORs or the site of rRNA synthesis.

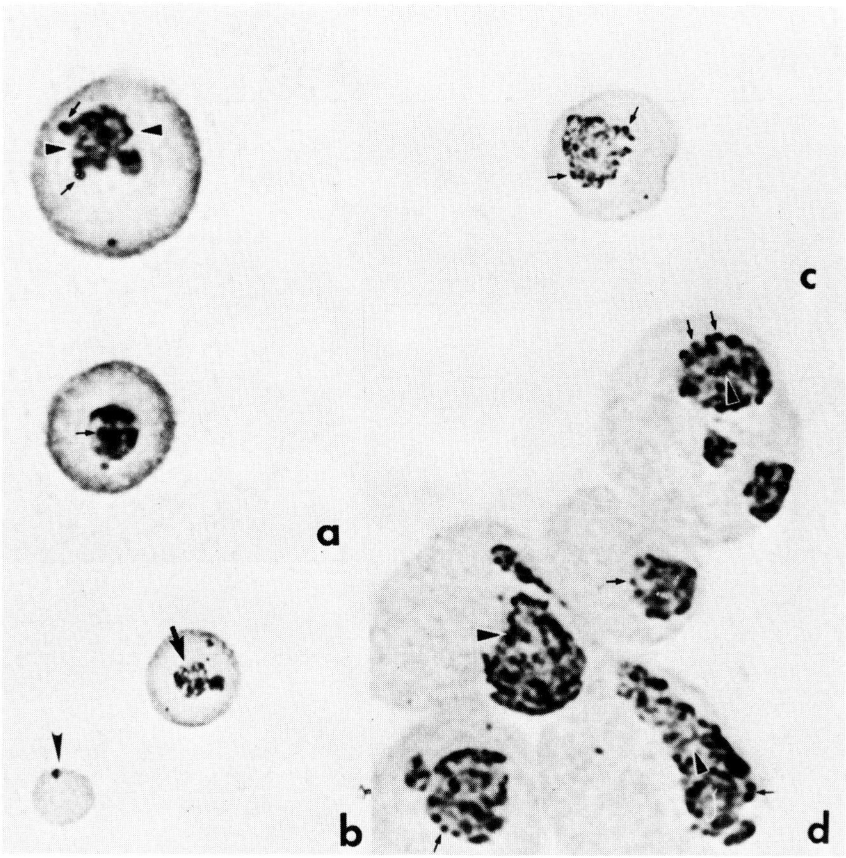

FIGURE 22. (a) Silver stain of spread liver cell nuclei showing a silver-stained reticulum (arrowheads) connecting a few silver-stained granules (arrows) (× 1250); (b) Two liver nuclei, one (arrowhead) with a single granule — in the other, the denser silver-stained granules are associated with a less-dense reticulum (arrow) (× 1000); (c) 6-hr regenerating liver showing increased silver-staining reticulum of the enlarged nucleolus — arrows indicate granules (× 1000); (d) 18-hr regenerating liver: markedly enlarged nucleoli are noted in these cells. These cells contain larger numbers of granules (arrows) associated with the nucleolar silver-stained reticulum (pointers) (× 1250).

XI. DISCUSSION

The current state of our knowledge regarding the proteins of nucleus and nucleolus is quite primitive. There are probably more than 300 enzymes and other proteins as well as RNA elements involved in nuclear function. Accordingly, it is very clear that the number of variables far exceeds the information available today. With respect to nuclear proteins, good information is now available on the histones, HMG proteins, proteins A24 and A11, proteins BA and B23, and proteins C14 and C23. This is a very small number of the total array of such structures.

In the nucleolus, running through the substructure is a fundamental thread-like element; the ''nucleolonemal structure'' or the ''nucleolonema''. This structure contains a number of cluster ''fibrillar'' elements which contain the rDNA reading frame and the ''Christmas trees'' which compose the silver-staining granules. These silver-staining granules should be called NM granules to refer to their nucleolonemal origins. The nucleolus contains the PR granules (preribosomal granules) which were called granular elements to distinguish them from the fibrillar elements.

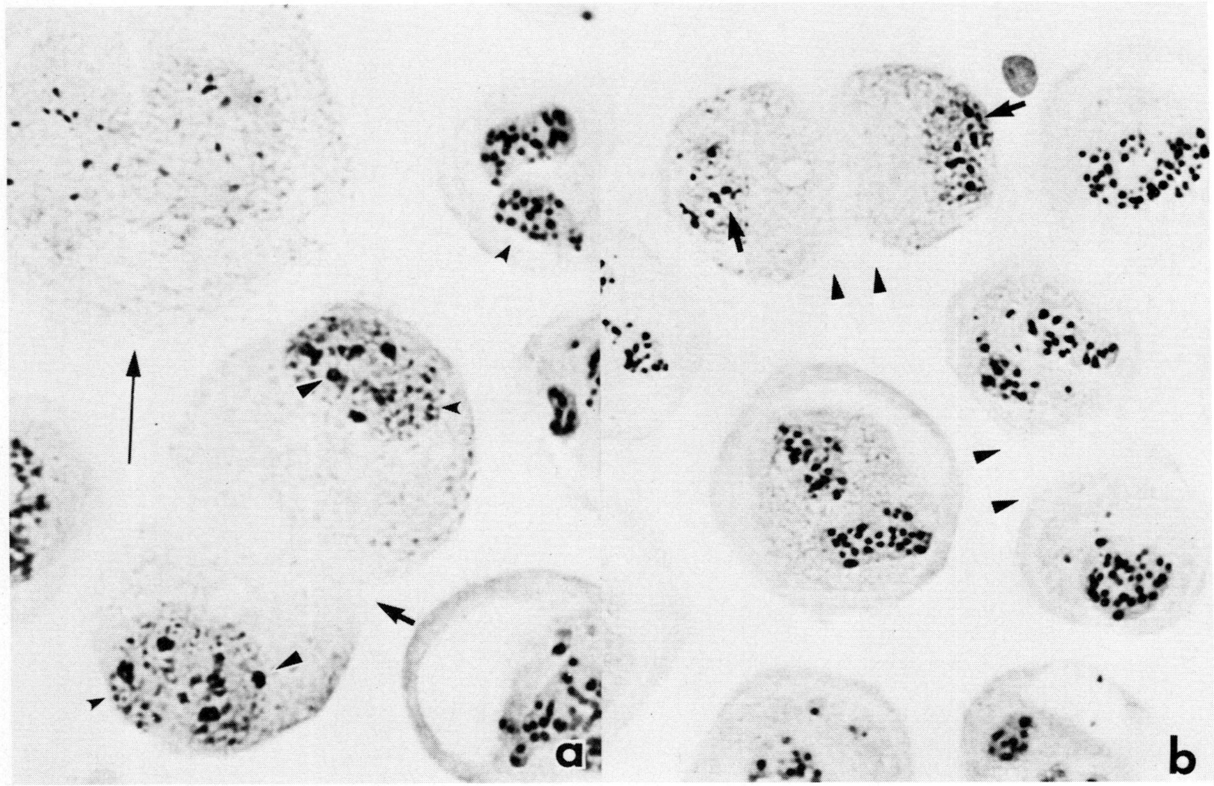

FIGURE 23. (a) Two Novikoff hepatoma cells in mitosis showing anaphase and telophase states. *Thin arrow*, anaphase; *thick arrow*, telophase. In the anaphase cell, there are 11 dense granules in each nucleus. In the telophase cell the nucleus has reformed, and many dense granules are visible in each nucleus. Some are arranged in rows (black arrowheads). Some unusually dense rectangular formations are visible (white-edged arrowheads). Rows of granules are also visible in some interphase cells. (b) Novikoff hepatoma cells in telophase (double arrowheads) showing multiple granules (arrows) in nuclei of the daughter cells. The number of granules in these cell pairs are 28 and 67. In adjacent interphase cells, rows of granules are noted. In a pair of cells apparently in very late telophase, similar rows of granules are visible (× 1400).

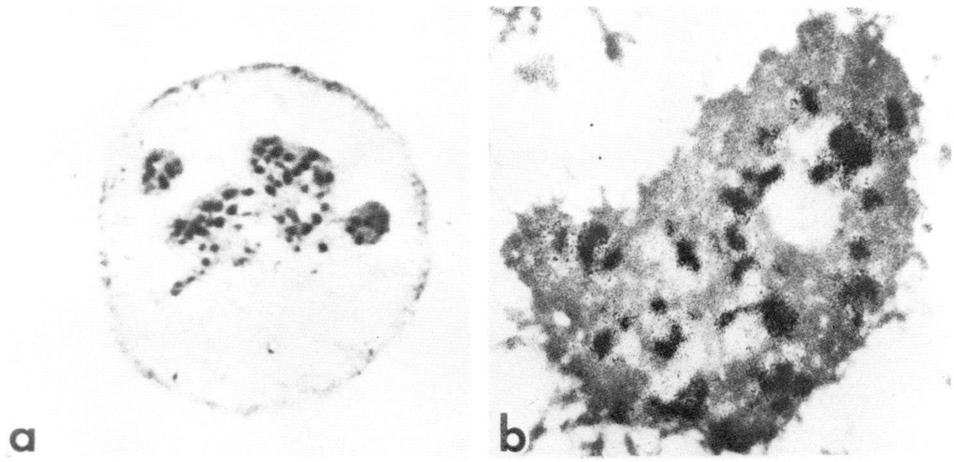

FIGURE 24. (a) Light micrograph of Novikoff hepatoma cell and stained with silver. The silver grains localized mainly to the nucleolus and were distributed in the form of distinct, dense grains. Methanol:acetic acid fixation. (b) Electron micrograph of nucleolus prepared as in Figure 7A. The silver grains, when examined with the electron microscope, are composed of numerous grains about 100 Å in diameter. The grain distribution appears random with respect to nucleolar topography. No electron counter staining.

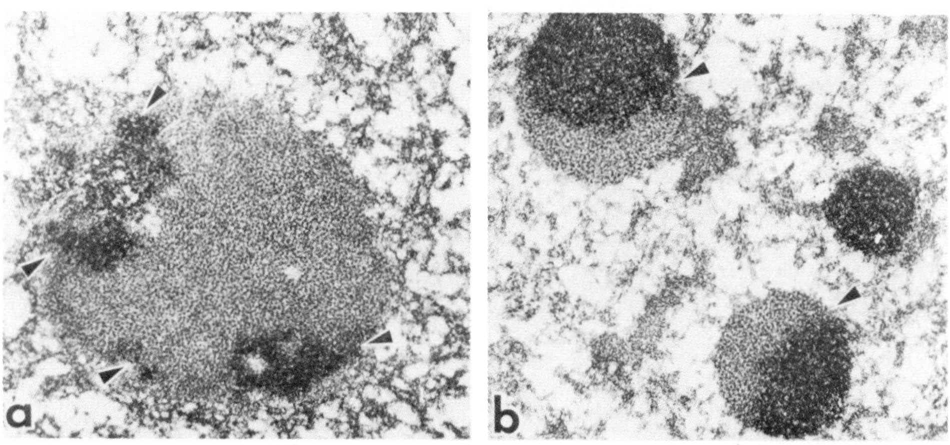

FIGURE 25. (a) A segregated nucleolus prefixed with 1% formaldehyde followed by methanol:acetic acid and silver stained. Note the preservation of nucleolar granular elements. The silver granules are localized to the fibrillar fraction only (pointers). (b) Fragments of segregated nucleoli silver stained after prefixation with 1% formaldehyde followed by methanol:acetic acid. Fragments composed of both granular and fibrillar elements are "capped" by silver localized to the fibrillar components (pointers). Fibrillar fragments were totally covered with silver.

Where do proteins C23, B23, BA, and the nucleolar antigens fit in these structures? The present studies in this laboratory indicate that proteins B23 and C23 are the acidic phosphoproteins, which are characterized by their high content of phosphate groups and amino acids and their association with the NM granules. These proteins are also part of the PR granules but are not released into the nucleoplasm or the cytoplasm.

Protein BA[55] is interesting in that it apparently forms a complex on the nucleolar surface of liver cells. It is also present in other structures in the nucleus. It decreases in rapidly growing tissues such as regenerating liver as well as in a variety of tumors studied. This, and possibly other proteins, may constitute a potential block to penetration of the nucleolus

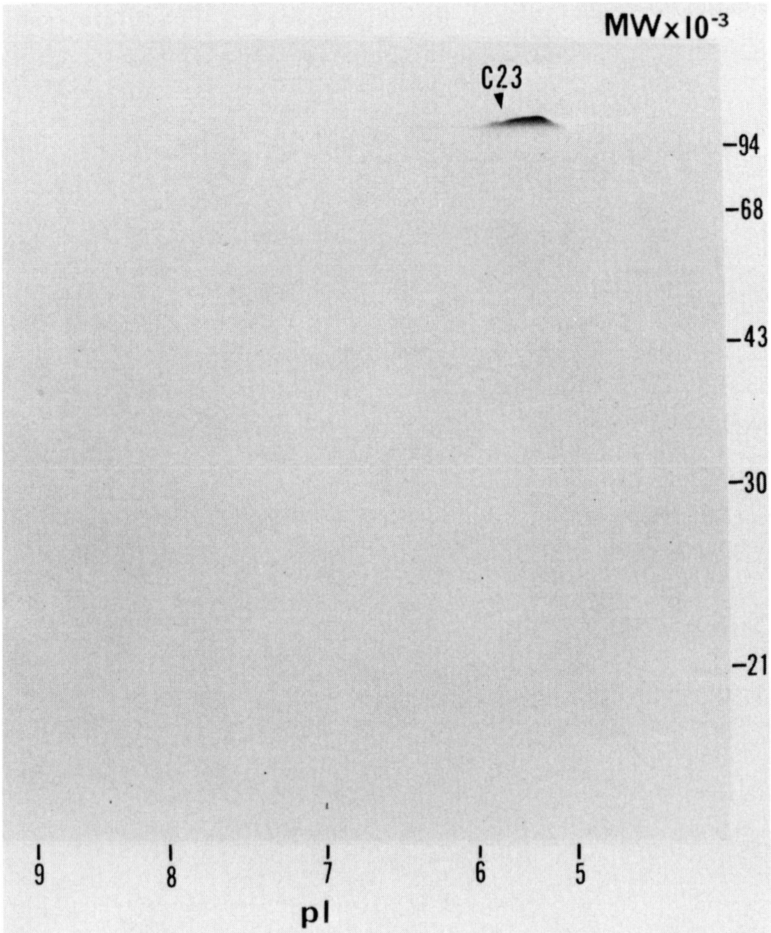

FIGURE 26. Isolated protein C23. Protein C23 was isolated by DEAE-cellulose and AG3-X4A ion-exchange column chromatography.

by antibodies and may account in part for the failure of the nucleolus to be visualized by antinucleolar antibodies. Such proteins may also be related to the round or spherical appearance of nucleoli in normal cells.

A basic finding of molecular oncology has been the pleiomorphism of nucleoli of cancer cells.[3] The biochemical events have not been defined but now the pleiomorphism seems associated with the presence of the nucleolar antigen or, at least, its visualization. Under conditions where cells are going to grow and divide, such as the human neoplasms studied and occasional cells in tissue culture, there has been a ready recognition of the structures by the human tumor nucleolar antibodies. This suggests that the pI 6.3 antigen is associated with nucleolonemas and becomes visible as part of the protein events associated with cell reproduction.[72,73]

An important question is whether there are differences between these nucleolonemal structures and the spherical nucleoli seen in normal liver, in which a weak fluorescence is observed with the antiHeLa nucleolar antibodies. Recently, other antibodies have provided bright staining of liver nucleoli. For example, antibodies to rat protein C23 produced bright fluorescence in the liver nucleoli which may be related either to elements on the surface of those nucleoli like protein BA or other nucleolar surface elements. The weak staining observed may reflect a homogeneous distribution of some elements, possibly the surface

Table 9
AMINO ACID COMPOSITION OF PROTEIN C23

Amino acid	Mol %
Asx	12.0
Thr	5.6
Ser	5.4
Glx	18.5
Pro	5.3
Gly	10.0
Ala	10.3
1/2-Cys	0.0
Val	6.0
Met	1.2
Ile	2.5
Leu	5.6
Tyr	0.8
Phe	3.5
Lys	11.8
His	0.5
Trp	0.0
Arg	2.9
Asx, Glx	2.0
Lys, His, Arg	

From Mamrack, M. D., Olson, M. O. J., and Busch, H., *Biochemistry*, 18, 3381, 1979. Wth permission.

5 10
Ala-Ala-Pro-Ala-Ala-Pro-Ala-Ser-Glu-Asp-Glu-
15 20
Asp-Glu-Glu-Asp-Asp-Asp-Asp-Glu-Asp-Asp-Asp-Asp-
25 30
Asp-Ser-Gln-Glu-Ser-Glu-Glu-Glu-Asp-Glu-Glu-
35 40
Val-Met-Glu-Ile-Thr-Pro-Ala-Lys (1)

FIGURE 27. Amino acid sequence of a highly acidic peptide (C23-1) of protein C23.

structures. In contrast, the malignant tumors contain the lacy, irregular, and tortuous network of the nucleolonemas. These findings suggest that two events are occuring in the tumors, namely, excessive production of the nucleolar pI 6.3 antigen and diminished production or loss of nucleolar surface proteins.

ACKNOWLEDGMENTS

These studies were supported by the Cancer Research Center Grant CA 10893 awarded by the National Cancer Institute, Department of Health, Education, and Welfare, the Pauline Sterne Wolfe Memorial Foundation, the William Stamps Farish Fund, the Bristol-Myers Fund, the M. E. DeBakey Foundation, and the Taub Foundation.

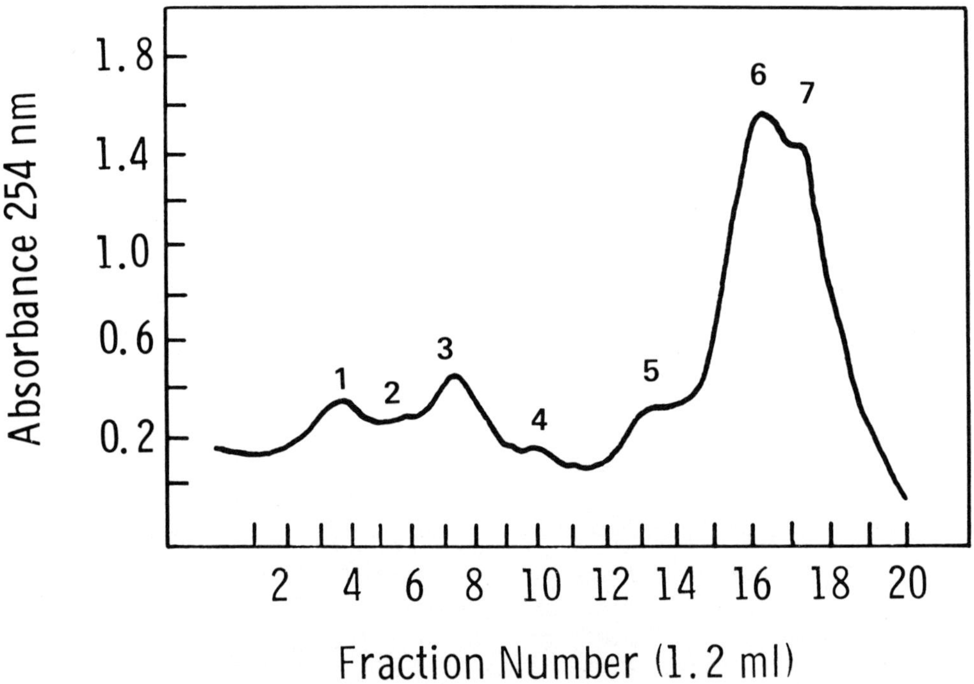

FIGURE 28. Sucrose density gradient profile of nucleolar extracts. Nucleoli were treated with 0.01 *M* KC1/0.1 m*M* $MgC1_2$/0.01 *M* triethanolamine/0.01 *M* DDT/0.1% Brij® 35/pH 7.4, centrifuged at 20,000 × g for 20 min and the supernatant was subjected to sucrose density gradient centrifugation.

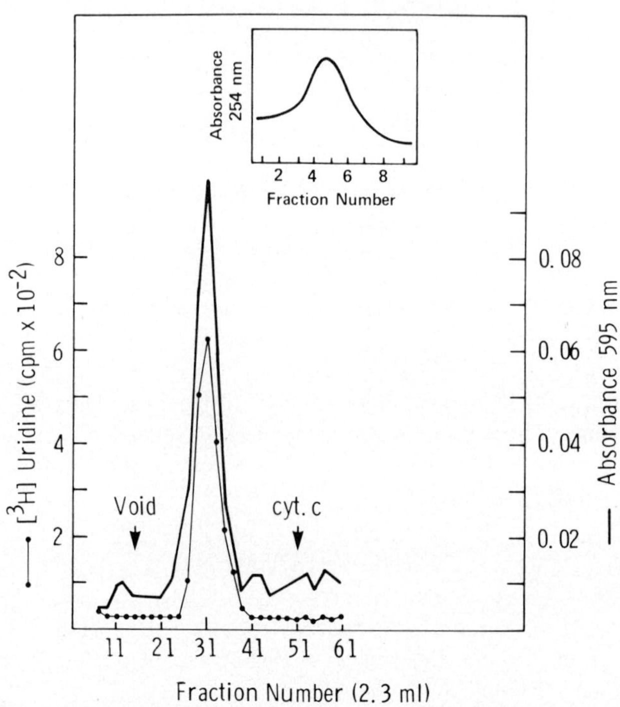

FIGURE 29. Integrity of the complex of fraction 1, (Figure 28) which contained protein C23. The complex was passed through Sepharose® 4B and the complex retained its integrity indicating that protein C23 is linked to other bound nucleolar macromolecules.

NOVIKOFF HEPATOMA ASCITES CELLS

↓

NUCLEOLI

↓

4M UREA/3M LICI EXTRACTION

↓

DIALYSIS IN pH 5.5

↓

DEAE-CELLULOSE

A

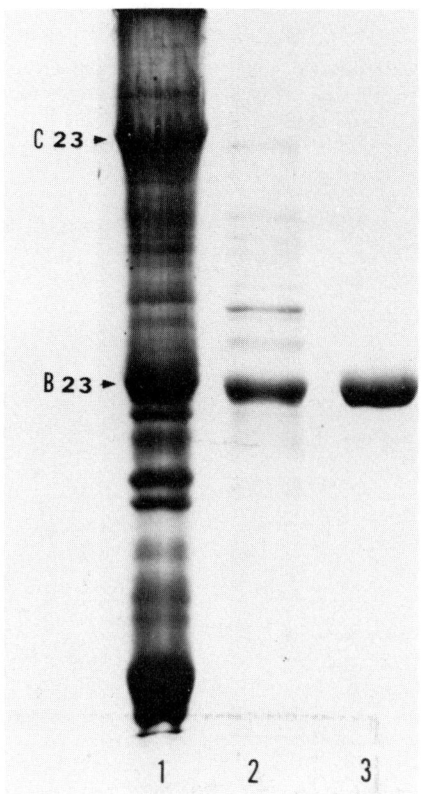

B

FIGURE 30. A. Isolation scheme for nucleolar protein B23: 1 g of the Novikoff hepatoma nucleoli was extracted with 8 to 10 volumes of 4 *M* urea/3 *M* LiC1/0.5 m*M* PMSF. The 27,000 × g supernatant of this extract was dialyzed against 20 m*M* Tris-malate buffer/4 *M* urea/1 m*M* DTT/1 m*M* EDTA/0.5 m*M* PMSF/pH 5.5 for 16 hr with two changes of the buffer. The 27,000 × g supernatant was adjusted to pH 6.5 by dialysis against the same buffer at pH 6.5 and loaded on DEAE-cellulose column. B. Polyacrylamide gel electrophoresis of proteins at various purification steps of protein B23. The gel was a 10% Laemmli gel containing 0.1% SDS. Lane 1— 4 *M* urea/3 *M* LiC1 nucleolar extract; lane 2—supernatant after dialysis at pH 5.5; lane 3—protein B23 eluted from the DEAE-cellulose column.

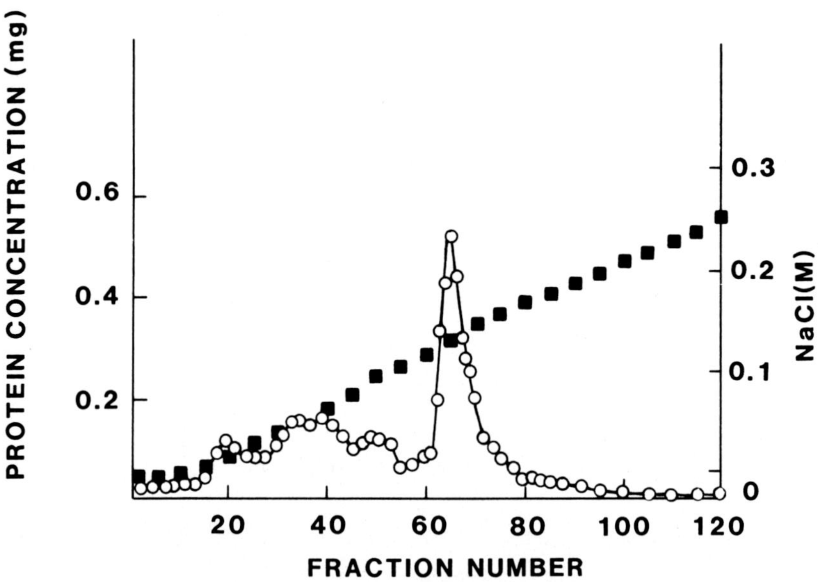

FIGURE 31. DEAE-cellulose elution profile of the supernatant after pH 5.5 dialysis of the 4 *M* urea/3 *M* LiCl nucleolar extract. The protein (18.7 mg) in 20 m*M* Tris-malate buffer/ *M* urea/ 1 m*M* DTT/1 m*M* EDTA/0.5 m*M* PMSF/pH 6.5 was loaded on a 1 × 25 cm DEAE-cellulose column which was eluted with 0 to 0.5 *M* NaCl gradient (total volume—300 mℓ) in the same buffer; 2.0-mℓ fractions were collected.

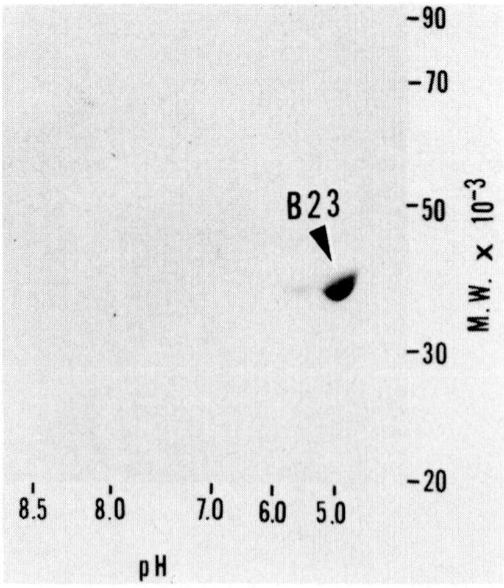

FIGURE 32. Two-dimensional, isoelectro-focusing sodium dodecyl sulfate gel of purified protein B23. A total of 5 μg of protein was loaded on the gel; the only apparent contaminant is a faint spot of the same molecular weight to the left of the main B23 spot, which may reflect minor dephosphorylation of protein B23.

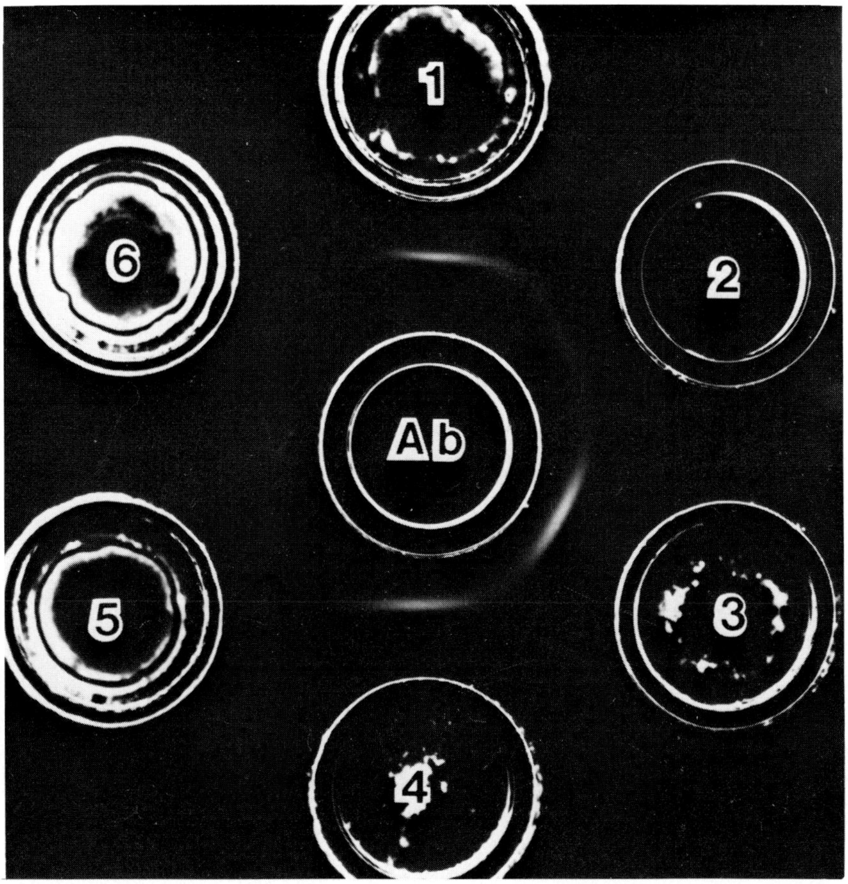

FIGURE 33. Ouchterlony double-diffusion analysis of rabbit IgG to protein B23. The center well contained 35 μℓ of the IgG fraction at a concentration of 26 mg/mℓ. Wells 1 and 4 contained 1.5 μg of protein B23; 2, 18 μg nucleolar 0.35 *M* NaCl extract; 3, 8.5 μg nucleolar Tris extract; 5, 22 μg cytosol; 6, 1.5 μg protein C23. Im-unoprecipitin bands were detected after 18 hr. These results demonstrate the specificity of the antibody for protein B23.

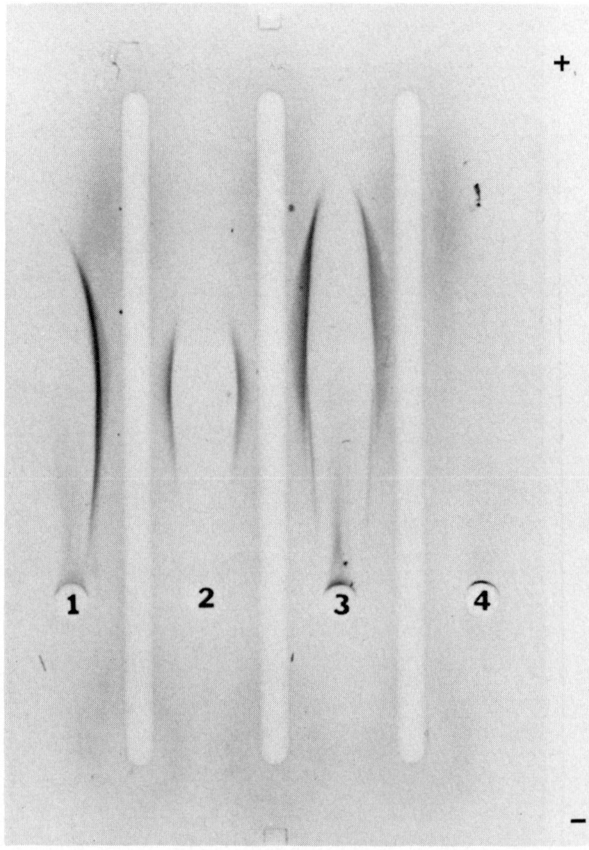

FIGURE 34. Immunoelectrophoretic analysis of the nucleolar and cytosol extracts of Novikoff hepatoma cells. The wells contained; 1, nucleolar Tris extract, digested with RNase A (50 μg/mℓ); 2, protein B23; 3, nucleolar Tris extract; 4, cytosol. The slots contained rabbit anti-protein B23 IgG.

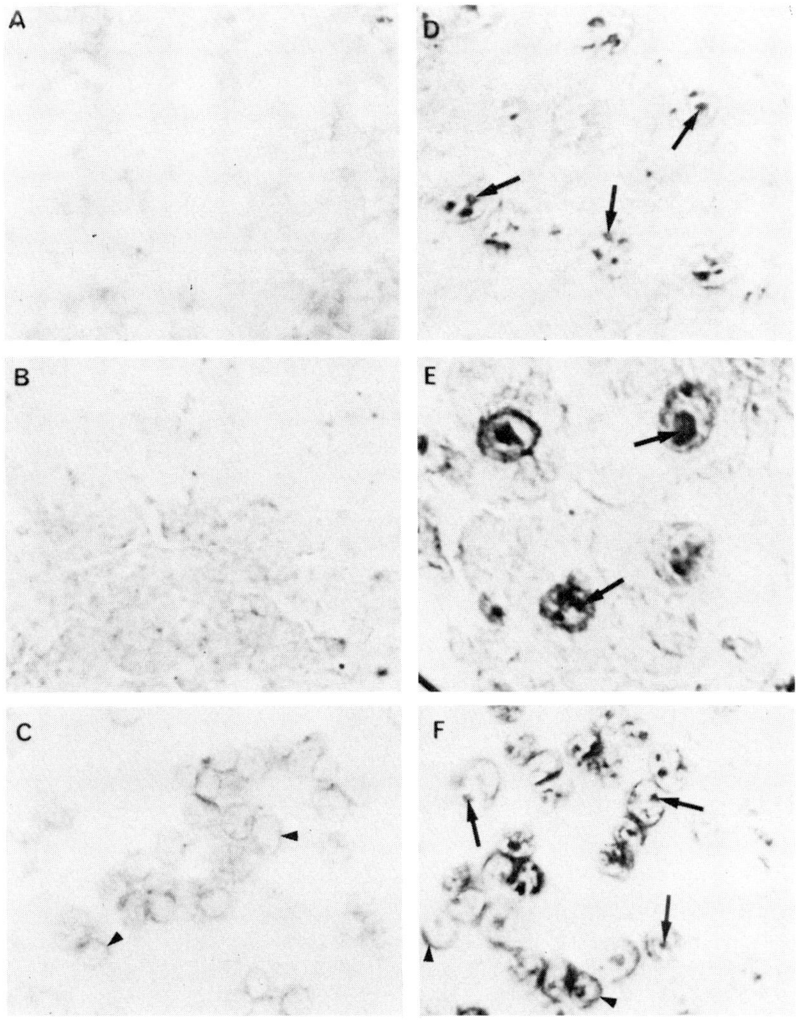

FIGURE 35. Cellular localization of protein B23 in Novikoff hepatoma cells, normal rat liver, and thioacetamide-treated rat liver sections by indirect peroxidase immunostaining. Figures 35 A, D show normal rat liver sections (× 840); B,E — thioacetamide-treated rat liver sections (× 840); C,F — Novikoff hepatoma cells, × 400), incubated with preimmune (A,B,C) and immune (D,E,F) IgG. The arrows show nucleolar positive staining, arrowheads point to weak, nonspecific nuclear envelope staining in Novikoff hepatoma cells. The cells or the sections were fixed on slides with acetone and incubated with anti-B23 or preimmune IgG fractions diluted 1:60 or 1:20 for cells and sections, respectively. After washing with PBS, the cells were treated with peroxidase-conjugated goat anti-rabbit IgG diluted 1:50. The complexed B23 was visualized by the peroxidase reaction, after treatment of the cells with 0.25 mg/mℓ diaminobenzidine/0.01% H_2O_2 for 30 min.

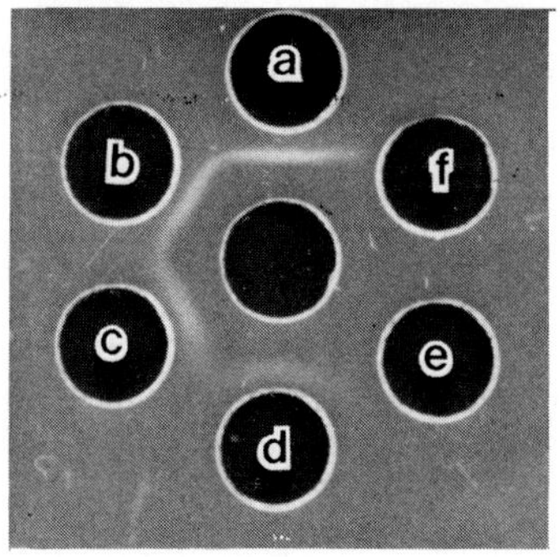

A

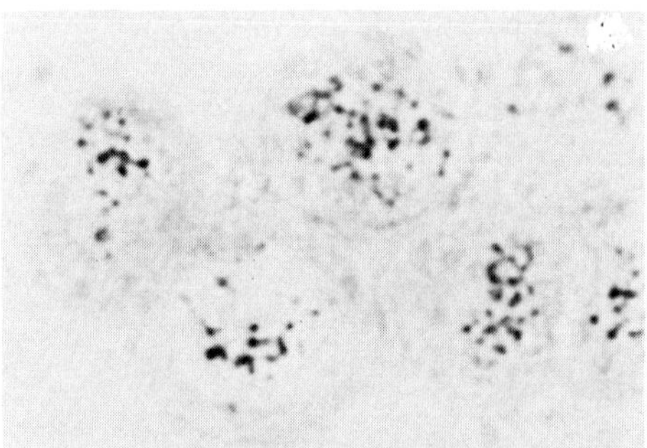

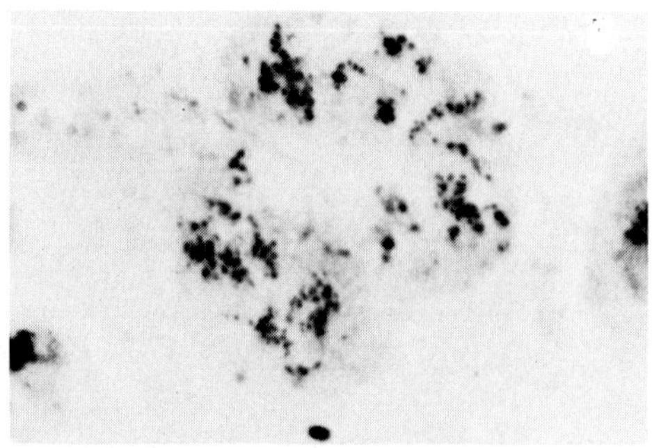

B C

FIGURE 36. (A) Antibodies to protein C23. Approximately 1.5 mg of protein C23 was injected into a rabbit in 6 doses. The rabbit Ig (40 mg/mℓ) was isolated from serum (center well); (a) 5 μg protein C23; (b) 4 μg protein C23; (c) 3 μg protein C23; (d) 2 μg protein C23, (e) 1 μg protein C23; (f) 12.2 μg protein B23. (B) Comparison of silver staining of Novikoff hepatoma nuclei (B) with immunoperoxidase staining (A) of Novikoff hepatoma ascites cells with antibodies to protein C23. The specificity of these antibodies is indicated in Figure 36A.

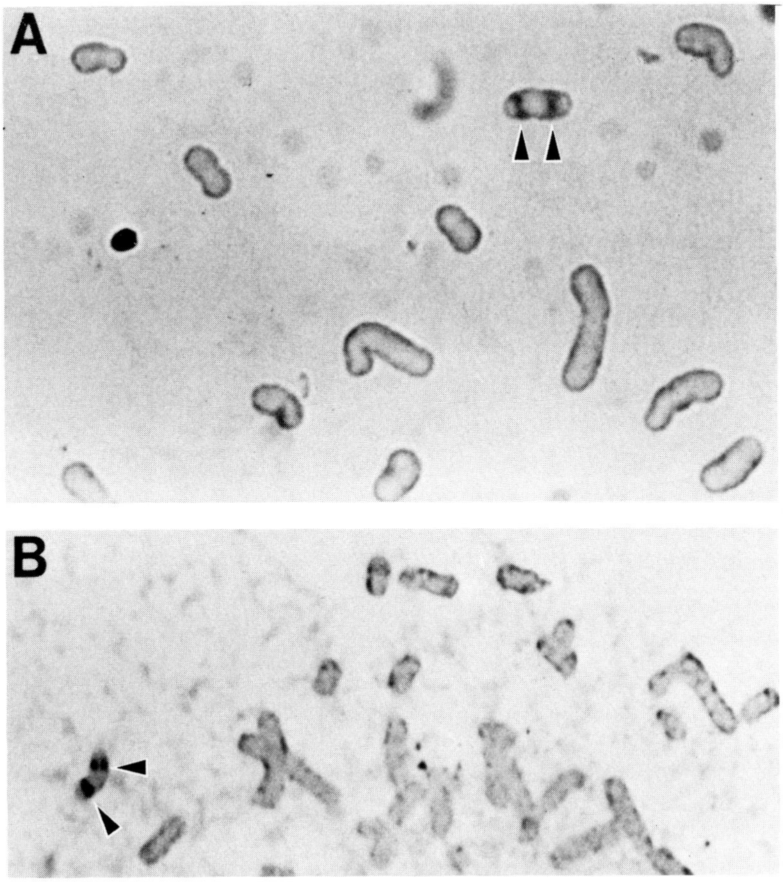

FIGURE 37. Immunocytochemical localization of protein C23 in NORs of Novikoff hepatoma tissue culture cell chromosomes showing the identity between the silver staining of the NORs (B) and the bands visualized by immunoperoxidase staining (A).

REFERENCES

1. **Busch, H., Muramatsu, M., Chien, S. C., and Somers, C.,** Isolation and metabolic activity of nucleoli of the Walker tumor, *Proc. Am. Assoc. Cancer Res.*, 3, 308, 1962.
2. **Muramatsu, M., Smetana, K., and Busch, H.,** Quantitative aspects of isolation of nucleoli of Walker carcinosarcoma and liver of the rat, *Cancer Res.*, 23, 510, 1963.
3. **Busch, H. and Smetana, K.,** *The Nucleolus*, Academic Press, New York, 1970.
4. **Gronow, M. and Griffiths, G.,** Rapid isolation and separation of non-histone proteins of rat liver nuclei, *FEBS Lett.*, 15, 340, 1971.
5. **Orrick, L. R., Olson, M. O. J., and Busch, H.,** Comparison of nucleolar proteins of normal rat liver and Novikoff hempatoma ascites cells by two-dimensional polyacrylamide gel electrophoresis (acid-extracted nucleolar proteins), *Proc. Natl. Acad. Sci. U.S.A.*, 70, 1316, 1973.
6. **Hirsch, F. W., Nall, K. N., Busch, F. N., Morris, H. P., and Busch, H.,** Comparison of abundant cytosol proteins in rat liver, Novikoff and Morris hepatomas by two-dimensional gel electrophoresis, *Cancer Res.*, 38, 1514, 1978.
7. **Takami, H. and Busch, H.,** Two-dimensional gel electrophoretic comparison of proteins of nuclear fractions of normal liver and Novikoff hepatoma, *Cancer Res.*, 39, 507, 1979.

8. **Rothblum, L. I., Mamrack, P. M., Olson, M. O. J., and Busch, H.,** Fractionation of nucleoli. Enzymatic and two-dimensional polyacrylamide gel electrophoretic analysis, *Biochemistry,* 16, 4716, 1977.
9. **Prestayko, A. W., Klomp, G. R., Schmoll, D. J., and Busch, H.,** Comparison of proteins of ribosomal subunits and nucleolar preribosomal particles from Novikoff hepatoma ascites cells by two-dimensional polyacrylamide gel electrophoresis, *Biochemistry,* 13, 1945, 1974.
10. **Prestayko, A. W., Olson, M. O. J., and Busch, H.,** Phosphorylation of proteins of ribosomes and nucleolar preribosomal particles in vivo in Novikoff hepatoma ascites cells, *Fed. Eur. Biochem. Soc. Lett.,* 44, 131, 1974.
11. **Daskal, Y., Prestayko, A. W., and Busch, H.,** Ultrastructural and biochemical studies of the isolated fibrillar component of nucleoli from Novikoff hepatoma ascites cells, *Exp. Cell Res.,* 88, 1, 1974.
12. **Koshiba, K., Chandra, T., Daskal, I., and Busch, H.,** Ultrastructural and biochemical studies of ribonucleic particles of isolated nucleoli in livers of thioacetamide treated rats, *Exp. Cell Res.,* 68, 235, 1971.
13. **Chambon, P., Gissinger, F., Kedinger, C., Mandel, J. L., and Meilhac, M.,** Animal nuclear DNA-dependent RNA polymerases, in *The Cell Nucleus,* Vol. 3, Busch, H., Ed., Academic Press, New York, 1974, chap. 7.
14. **Reeder, R. J. and Roeder, R. G.,** Ribsosomal RNA synthesis in isolated nuclei, *J. Mol. Biol.,* 67, 433, 1972.
15. **Roeder, R. G. and Rutter, W. J.,** Specific nucleolar and nucleoplasmic RNA polymerases, *Proc. Natl. Acad. Sci. U.S.A.,* 65, 675, 1970.
16. **Roeder, R. G., and Rutter, W. J.,** Multiple ribonucleic acid polymerases and ribonucleic acid synthesis during sea urchin development, *Biochemistry,* 9, 2543, 1970.
17. **Ro, T. S. and Busch, H.,** In vitro labeling of RNA in isolated nucleoli of the Walker tumor and liver, *Cancer Res.,* 24, 1630, 1964.
18. **Grummt, I.,** In vitro synthesis of pre-rRNA in isolated nucleoli, in *The Cell Nucleus,* Vol. 5, Busch, H., Ed., Academic Press, New York, 1978, chap. 11.
19. **Ballal, N. R., Choi, Y. C., Mouche, R., and Busch, H.,** Homochromatographic evidence of fidelity of preribosomal RNA synthesis in isolated nucleoli and nucleolar chromatin, *Proc. Natl. Acad. Sci. U.S.A.,* 74, 2446, 1977.
20. **Muramatsu, M., Matui, T., Onishi, T., and Mischima, Y.,** Nucleolar RNA polymerase and transcription of nucleolar chromatin, in *The Cell Nucleus,* Vol. 7, Busch, H., Ed., Academic Press, New York, 1979, chap. 4.
21. **McKnight, S. L., Martin, K. A., Beyer, A. L., and Miller, O. L., Jr.,** Visualization of functionally active chromatin, in *The Cell Nucleus,* Vol. 7, Busch, H., Ed., Academic Press, New York, 1979, chap. 3, 97—122.
22. **Miller, O. L., Jr. and Beatty, B. R.,** Visualization of nucleolar genes, *Science,* 164, 955, 1969.
23. **Scheer, U., Spring, H., and Trendelenburg, M. F.,** Organization of transcriptionally active chromatin in lampbrush chromosome loops, in *The Cell Nucleus,* Vol. 7, Busch, H., Ed., Academic Press, New York, 1979, chap. 1.
24. **Franke, W. W.,** Scheer, U., Spring, H., Trendelenburg, M. F., and Zentgraf, H., *Organization of nucleolar chromatin, in* The Cell Nucleus, Vol. 7, Busch, H., Ed., Academic Press, New York, 1979, chap. 2.
25. **Brown, D. D. and Dawid, I. B.,** Specific gene application in oocytes, *Science,* 160, 272, 1968.
26. **Brown, D. D., Wensink, P. C., and Jordan, E.,** Comparison of the ribosomal DNA's of *Xenopus laevis* and *Xenopus nulleri:* The evolution of tandem genes, *J. Mol. Biol.,* 63, 57, 1972.
27. **Brown, D. D., Caroll, D., and Brown, R. D.,** The isolation and characterization of a second oocyte 5S DNA from *Xenopus laevis, Cell,* 12, 1045, 1977.
28. **Fuke, M. and Busch, H.,** Hind III sensitive sites present once in four repeats of ECO R1 sensitive sites in Novikoff rat hepatoma DNA, *FEBS Let.,* 99, 136, 1979.
29. **Fuke, M., Davis, F. M., and Busch, H.,** Localization of highly repeated Hind III fragments of Novikoff rat DNA to the nucleolus, *FEBS Lett.,* 102, 46, 1979.
30. **Trendelenburg, M. F. and McKinnell, R. G.,** Transcriptionally active and inactive regions of nucleolar chromatin in amplified nucleoli of fully grown oocytes of hibernating frogs, *Rana pipiens* (Amphibia, Anura), *Differentiation,* 15, 73, 1979.
31. **Ballal, N. R., Goldknopf, I. L., Goldberg, D. A., and Busch, H.,** The dynamic state of liver nucleolar proteins as reflected by their changes during administration of thioacetamide, *Life Sci.,* 14, 1835, 1974.
32. **Ballal, N. R., Kang, Y. -J., Olson, M. O. J., and Busch, H.,** Changes in nuclear proteins and their phosphorylation patterns during liver regeneration, *J. Biol. Chem.,* 250, 5921, 1975.
33. **Goldknopf, I. L., Taylor, C. W., Baum, R. M., Yeoman, L. C., Olson, M. O. J., Prestayko, A. W., and Busch, H.,** Isolation and characterization of protein A24, a "histone-like" nonhistone chromosomal protein, *J. Biol. Chem.,* 250, 7182, 1975.
34. **Goldknopf, I. L. and Busch, H.,** Remarkable similarities of peptide fingerprints of histone 2A and nonhistone chromosomal protein A24, *Biochem. Biophys. Res. Commun.,* 65, 951, 1975.

35. **Olson, M. O. J., Goldknopf, I. L., Guetzow, K. A., James, G. T., Hawkins, T. C., Mays-Rothberg, C. J., and Busch, H.,** The NH_2- and COOH-terminal amino acid sequence of nuclear protein A24, *J. Biol. Chem.*, 251, 5901, 1976.
36. **Yeoman, L. C., Olson, M. O. J., Sugano, N., Jordan, J. J., Taylor, C. W., Starbuck, W. C., and Busch, H.,** Amino acid sequence of the center portion of the arginine-lysine-rich histone from calf thymus and the total sequence, *J. Biol. Chem.*, 247, 6018, 1972.
37. **Schlesinger, D. H., Goldstein, G., and Niall, H. D.,** The complete amino acid sequence of ubiquitin, and adenylate cyclase stimulating polypeptide probably universal in living cells, *Biochemistry*, 14, 2214, 1975.
38. **Goldknopf, I. L. and Busch, H.,** Modification of nuclear proteins: The ubiquitin-histone 2A conjugate, in *The Cell Nucleus*, Vol. 6, Busch, H., Ed., Academic Press, New York, 1978, chap. 5.
39. **Goldknopf, I. L. and Busch, H.,** An isopeptide linkage between the nonhistone and histone 2A polypeptides of chromosomal conjugate-protein A24, *Proc. Natl. Acad. Sci. U.S.A.*, 74, 864, 1977.
40. **West, M. H. P. and Bonner, W. M.,** Histone 2A, a heteromorphous family of eight protein species, *Biochemistry*, 19, 3238, 1980.
41. **Goldknopf, I. L., Wilson, G., Ballal, N. R., and Busch, H.,** Chromatin conjugate protein A24 is cleaved and ubiquitin is lost during chicken erythropoiesis, *J. Biol. Chem.*, 255, 10555, 1980.
42. **Goldknopf, I. L., Sudakar, S., Rosenbaum, F., and Busch, H.,** Timing of ubiquitin synthesis and conjugation into protein A24 during the HeLa cell cycle, *Biophys. Biochem. Res. Commun.*, 95, 1253, 1980.
43. **Panyim, S. and Chalkely, R.,** A new histone found only in mammalian tissues with little cell division, *Biochem. Biophys. Res. Commun.*, 37, 1042, 1969.
44. **Marsh, W. H. and Fitzgerald, P. J.,** Pancreas acinar cell regeneration. XIII. Histone synthesis and modification, *Fed. Proc.*, 32, 2119, 1973.
45. **Varricchio, F., Mobogunje, O., Kim, D., and Fitzgerald, P. J.,** Pancreas acinar cell regeneration and histone (H1 and $H1^0$) modifications after partial pancreatectomy or after a protein-free ethionine regimen, *Cancer Res.*, 37, 3964, 1977.
46. **James, G. T., Yeoman, L. C., Matsui, S., Goldberg, A. H., and Busch, H.,** Isolation and characterization of nonhistone chromosomal protein C-14 which stimulates RNA synthesis, *Biochemistry*, 16, 2384, 1977.
47. **Kuehn, G. C., Affolter, H.-U., Atmar, V. J., Seeback, T., Gkler, U., and Braun, R.,** Polyamine-mediated phosphorylation of a nucleolar protein from *Physarum polycephalum* that stimulates rRNA synthesis, *Proc. Natl. Acad. Sci. U.A.*, 76, 2541, 1979.
48. **Yeoman, L. C., Taylor, C. W., Jordan, J. J., and Busch, H.,** Differences in chromatin proteins of growing and nongrowing tissues, *Exp. Cell Res.*, 91, 207, 1975.
49. **Yeoman, L. C., Taylor, C. W., Jordan, J. J., and Busch, H.,** Two-dimensional polyacrylamide gel electrophoresis of chromatin proteins of normal rat liver and Novikoff hepatoma ascites cells, *Biochem. Biophys. Res. Commun.*, 53, 1067, 1973.
50. **Yeoman, L. C., Taylor, C. W., and Busch H.,** Two-dimensional gel electrophoresis of acid extractable nuclear proteins of regenerating and thioacetamide treated rat liver, Morris 9618A hepatoma and Walker 256 carcinosarcoma, *Cancer Res.*, 34, 424, 1974.
51. **Yeoman, L. C., Taylor, C. W., Jordan, J. J., and Busch, H.,** Early and late changes in nonhistone chromatin proteins accompanying rat liver regeneration, *Cancer Res.*, 35, 1249, 1975.
52. **Catino, J. J., Yeoman, L. C., Mandel, M., and Busch, H.,** Characterization of a DNA binding protein from rat liver chromatin in which decreases during growth, *Biochemistry*, 17, 983, 1978.
53. **Riggs, A. D. and Bourgeois, S.,** On the assay, isolation and characterization of the *lac* repressor, *J. Mol. Biol.*, 34, 361, 1968.
54. **Riggs, A. D., Bourgeois, S., Newby, R. F.,and Cohn, M.,** DNA binding of the *lac* repressor, *J.Mol. Biol.*, 34, 365, 1968.
55. **Catino, J. J., Busch, H., Daskal, Y., and Yeoman, L. C.,** Subcellular localization of DNA-binding protein BA by immunofluorescence and immunoelectron microscopy, *J. Cell Biol.*, 83, 462, 1979.
56. **Heitz, E.,** Uber totale und partielle somatische heteropyknose, sowie strukturelle geschlechtsch romosomen bei Drosophila funebris *Z. Zellforsch. Mikroskop. Anat.*, 19, 720, 1933.
57. **McClintock, B.,** The relation of a particular chromosomal element to the development of the nucleoli in zea mays, *Am. Nat.*, 95, 294, 1934.
58. **Goodpasture, C. and Bloom, S. E.,** Visualization of nuclear organizer regions in mammalian chromosomes using silver staining, *Chromosoma*, 53, 37, 1975.
59. **Howell, M. W.,** Visualization of the ribosomal gene activity: Silver stains proteins associated with rRNA transcribed from oocyte chromosomes, *Chromosoma*, 62, 361, 1977.
60. **Olson, M. O. J., Orrick, L. R., Jones, C. E., and Busch, H.,** Phosphorylation of acid soluble nucleolar proteins of Novikoff hepatoma ascites cells in vivo, *J. Biol. Chem.*, 249, 2823, 1974.

61. **Olson, M. O. J., Prestayko, A. W., Jones, C. E., and Busch, H.,** Phosphorylation of proteins of ribosomes and nucleolar preribosomal particles from Novikoff hepatoma ascites cells, *J. Mol. Biol.*, 90, 161, 1974.
62. **Kang, Y.-J., Olson, M. O. J., and Busch, H.,** Phosphorylation of acid-soluble proteins in isolated nucleoli of Novikoff hepatoma ascites cells: Effects of divalent cations, *J. Biol. Chem.*, 249, 5580, 1974.
63. **Ezrailson, E. G., Olson, M. O. J., Guetzo, K. A., and Busch, H.,** Phosphorylation of nonhistone chromatin proteins in normal and regenerating rat liver Novikoff hepatoma and rat heart, *Fed. Eur. Biol. Soc. Lett.*, 62, 69, 1976.
64. **Lischwe, M. A., Smetana, K., Olson, M. O. J., and Busch, H.,** Proteins C23 and B23 are the major nucleolar silver staining proteins, *Life Sci.*, 25, 701, 1979.
65. **Hubbell, H. R., Rothblum, L. I., and Hsu, T. C.,** Identification of a silver binding protein associated with the cytological silver staining of actively transcribing nucleolar regions, *Cell Biol. Int. Rep.*, 3, 615, 1979.
66. **Busch, H., Daskal, Y., Gyorkey, F., and Smetana, K.,** Silver staining of nucleolar granules in tumor cells, *Cancer Res.*, 39, 857, 1979.
67. **Daskal, Y., Smetana, K., and Busch, H.,** Evidence from studies on segregated nucleoli that nucleolar silver staining proteins C23 and B23 are in the fibrillar component, *Exp. Cell Res.*, 127, 285, 1980.
68. **Goessens, G.,** Localization of nucleolus organizing regions and interphase cells, *Cell Tissue Res.*, 200, 159, 1979.
69. **Hernandez-Verdun, D., Hubert, J., Bourgeois, C. A., and Bouteille, M.,** Ultrastructural localization of Ag-NOR stained proteins in the nucleolus during the cell cycle and in other nucleolar structures, *Chromosoma*, 74, 349, 1980.
70. **Mamrack, M. D., Olson, M. O. J., and Busch, H.,** Amino acid squence and sites of phosphorylation in a highly acidic region of nucleolar nonhistone protein C23, *Biochemistry*, 18, 3381, 1979.
71. **Mamrack, M. D., Olson, M. O. J., and Busch, H.,** Negatively charged phosphopeptides of nucleolar nonhistone proteins from Novikoff hepatoma ascites cells, *Biochem. Biophys. Res. Commun.*, 76, 150, 1977.
72. **Busch, H., Gyorkey, F., Busch, R. K., Davis, F. M., and Smetana, K.,** A nucleolar antigen found in a broad range of human malignant tumor specimens, *Cancer Res.*, 39, 3024, 1979.
73. **Chan, P. -K., Feyerabend, A., Busch, R. K., and Busch, H.,** Identification and partial purification of human tumor nucleolar antigen 54/6.3, *Cancer Res.*, 40, 3194, 1980.

INDEX

A

B

C

D

E

F

G

H

I

J

K

L

M

N

O

P

R

S

T

U

V

W

Y